PRINCIPLES OF HEART VALVE ENGINEERING

PRINCIPLES OF HEART VALVE ENGINEERING

Edited by

ARASH KHERADVAR

University of California, Irvine, CA, United States

ACADEMIC PRESS

An imprint of Elsevier

Notices

Knowledge and best practice in this field are constantly changing. As new research and experience broaden our understanding, changes in research methods, professional practices, or medical treatment may become necessary.

Practitioners and researchers must always rely on their own experience and knowledge in evaluating and using any information, methods, compounds, or experiments described herein. In using such information or methods they should be mindful of their own safety and the safety of others, including parties for whom they have a professional responsibility.

To the fullest extent of the law, neither the Publisher nor the authors, contributors, or editors, assume any liability for any injury and/or damage to persons or property as a matter of products liability, negligence or otherwise, or from any use or operation of any methods, products, instructions, or ideas contained in the material herein.

Library of Congress Cataloging-in-Publication Data
A catalog record for this book is available from the Library of Congress

British Library Cataloguing-in-Publication Data
A catalogue record for this book is available from the British Library

ISBN: 978-0-12-814661-3

For information on all Academic Press publications visit our
website at https://www.elsevier.com/books-and-journals

Publisher: Mara Conner
Acquisition Editor: Fiona Geraghty
Editorial Project Manager: Joshua Mearns
Production Project Manager: Mohana Natarajan
Cover Designer: Greg Harris

Typeset by TNQ Technologies

Working together
to grow libraries in
developing countries

www.elsevier.com • www.bookaid.org

I would like to dedicate this book to my wife Ladan for her constant love and support, to my children Aryana and Ario for being the reason I would never give up, and to my beloved parents for their never-ending devotion and care.

Contents

Contributors

Hamza Atcha
Department of Biomedical Engineering, University of California Irvine, Irvine, CA, United States; The Edwards Lifesciences Center for Advanced Cardiovascular Technology, University of California Irvine, CA, United States

Ali N. Azadani
Department of Mechanical & Materials Engineering, Ritchie School of Engineering and Computer Science, University of Denver, Denver, CO, United States

Stefanie V. Biechler
Director of Marketing Collagen Solutions PLC Minneapolis, MN, United States

Carlijn V.C. Bouten
Department of Biomedical Engineering, Eindhoven University of Technology, Eindhoven, the Netherlands; Institute for Complex Molecular Systems (ICMS), Eindhoven University of Technology, Eindhoven, the Netherlands

Lakshmi Prasad Dasi
Department of Biomedical Engineering, The Ohio State University, Columbus, OH, United States

Linda L. Demer
Departments of Medicine, Physiology, & Bioengineering, University of California, Los Angeles, Los Angeles, CA, United States

Craig J. Goergen
Weldon School of Biomedical Engineering, Purdue University, West Lafayette, IN, United States

Richard L. Goodwin
Biomedical Sciences, University of South Carolina School of Medicine, Greenville, SC, United States

K. Jane Grande-Allen
Department of Bioengineering, Rice University, Houston, TX, United States

Boyce E. Griffith
Department of Mathematics, Carolina Center for Interdisciplinary Applied Mathematics, Computational Medicine Program, and McAllister Heart Institute, University of North Carolina, Chapel Hill, NC, United States

Elliott M. Groves
Division of Cardiology, College of Medicine, University of Illinois at Chicago, Chicago, IL, United States

Megan Heitkemper
Department of Biomedical Engineering, The Ohio State University, Columbus, OH, United States

Svenja Hinderer
Natural and Medical Sciences Institute (NMI), University of Tübingen, Reutlingen, Germany

Geoffrey D. Huntley
Department of Cardiovascular Medicine, Mayo Clinic, Rochester, MN, United States

Harkamaljot S. Kandail
Department of Biomedical Engineering, Eindhoven University of Technology, Eindhoven, the Netherlands

Arash Kheradvar
Department of Biomedical Engineering, University of California Irvine, Irvine, CA, United States; Department of Medicine, Division of Cardiology, University of California Irvine, CA, United States; The Edwards Lifesciences Center for Advanced Cardiovascular Technology, University of California Irvine, CA, United States

Wendy F. Liu
Department of Biomedical Engineering, University of California Irvine, Irvine, CA, United States; Department of Chemical Engineering and Materials Science, University of California Irvine, CA, United States; The Edwards Lifesciences Center for Advanced Cardiovascular Technology, University of California Irvine, CA, United States

Wenbin Mao
Tissue Mechanics Laboratory, The Wallace H. Coulter Department of Biomedical Engineering, Georgia Institute of Technology and Emory University, Atlanta, GA, United States

Petra Mela
Department of Biohybrid & Medical Textiles (BioTex), AME — Institute of Applied Medical Engineering, Helmholtz Institute, RWTH Aachen University, Aachen, Germany; Medical Materials and Implants, Department of Mechanical Engineering, Technical University of München, München, Germany

Madeline Monroe
Department of Bioengineering, Rice University, Houston, TX, United States

Daisuke Morisawa
Department of Biomedical Engineering, University of California Irvine, Irvine, CA, United States; The Edwards Lifesciences Center for Advanced Cardiovascular Technology, University of California Irvine, CA, United States

Vuyisile T. Nkomo
Department of Cardiovascular Medicine, Mayo Clinic, Rochester, MN, United States

Niema M. Pahlevan
Department of Aerospace & Mechanical Engineering, University of Southern California, CA, United States

Evan H. Phillips
Weldon School of Biomedical Engineering, Purdue University, West Lafayette, IN, United States

Mohammad Sarraf
Division of Cardiovascular Disease, School of Medicine, University of Alabama at Birmingham, Birmingham, AL, United States

Anthal I.P.M. Smits

Department of Biomedical Engineering, Eindhoven University of Technology, Eindhoven, the Netherlands; Institute for Complex Molecular Systems (ICMS), Eindhoven University of Technology, Eindhoven, the Netherlands

Wei Sun

Tissue Mechanics Laboratory, The Wallace H. Coulter Department of Biomedical Engineering, Georgia Institute of Technology and Emory University, Atlanta, GA, United States

Jeremy J. Thaden

Department of Cardiovascular Medicine, Mayo Clinic, Rochester, MN, United States

Yin Tintut

Departments of Medicine (Cardiology), Physiology & Orthopaedic Surgery, University of California, Los Angeles, CA, United States

Ivan Vesely

Class III Medical Device Consulting, Gaithersburg, MD, United States; CroiValve Limited, Dublin, Ireland

Amadeus Zhu

Department of Bioengineering, Rice University, Houston, TX, United States

Preface

Heart valves are living tissue structures that ensure adequate blood flow passes from one heart chamber to the next without the possibility of backflow. Native heart valves are among the body's most enduring tissues, with the ability to grow during the pediatric years. However, these tissue structures cannot regenerate or repair themselves spontaneously. Although heart valve disease is etiologically diverse—it can be acquired or congenital—phenotypically, it results in either valve stenosis or regurgitation. However, since the valves cannot repair themselves, medical interventions are always required to remedy these diseases. No drug currently exists to cure heart valve disease, and all interventions are based on surgical or transcatheter repair or replacement of the diseased valve. These issues have inspired researchers to seek effective and long-lasting means of mitigating damaged heart valves.

The very first successful heart valve repair is reported to have been performed by Dr. Walton Lillehei in 1957, followed by the first successful artificial heart valve implantation, by Dr. Albert Starr, in 1960s. Since then, heart valve engineering has been behind all major advances in treating patients with heart valve disease. Successful translation of a technology for heart valve replacement or repair from a research lab to a patient's heart takes many steps, from bench testing and acute animal studies to chronic animal studies and major clinical trials. These steps require many years of hard work by a multidisciplinary team, not to mention enormous amounts of funding.

To engineer a heart valve, many technical issues must be considered. Issues such as creating tissue structures that are able to resist deterioration and having designs that avoid thrombus formation, to name just two among many examples, challenge us to develop heart valve technologies that last longer while working seamlessly. Therefore, experts from diverse backgrounds—such as, but not limited to, mechanobiology, cardiology, physiology, tissue culture, mathematical modeling, fluid dynamics, and polymer science—often form core teams to develop new heart valve—related technologies. A new heart valve technology should be tested according to regulatory authorities' safety and efficacy guidelines before it can be used in humans. These verification, validation, and preclinical and clinical feasibility studies depend on close collaborative efforts among a group of expert engineers, regulatory bodies, and physicians from academia and industry, a group whom this book aims to address.

A few years ago, I led the publication of a series of four review articles on emerging trends in heart valve engineering in the *Annals of Biomedical Engineering*. During that effort, I realized that the field lacked a comprehensive textbook addressing this ever-expanding area of research and development. Inspired by that realization, and with the intention of disseminating the science and knowledge of heart valve engineering, I asked experts in different areas of heart valve research and development to assist me in this crucial effort. The present work is the first of its kind to comprehensively bring together a variety of techniques and disciplines from the current state of the art in heart valve engineering within a single book that can be used by students, scholars, engineers, and physicians. An elite group of internationally known experts on different aspects of heart valves contributed to it. Completing all the book chapters entailed over 2 years of efforts working on topics ranging from mechanobiology, engineering, epidemiology, and imaging to heart valve—focused therapies, among others. We hope this volume will generate new conversations among educators and scholars and spur continual improvements in these technologies.

I am indebted to the many outstanding faculty members from a range of disciplines and to leading heart valve experts, who donated their time and effort to produce carefully crafted chapters on topics at the cutting edge of work that is critical to research and development in heart valve engineering. I remain grateful to the whole Elsevier team, especially to *Joshua Means, Sheela B. Josy*, and *Mohanambal Natarajan*, for their support, patience, and guidance during all the stages of book production. Finally, I would like to thank all my past mentors, present and past trainees, and collaborators, whose guidance, hard work, and friendship have inspired me to pursue research in heart valve engineering, a field that continues to astonish and reward me. I hope our efforts in this book will likewise be beneficial for educators and scholars around the world who are interested in heart valve research and development.

Arash Kheradvar
University of California,
Irvine, CA, United States

CHAPTER 1

Clinical anatomy and embryology of heart valves

Richard L. Goodwin[1], Stefanie V. Biechler[2]
[1]Biomedical Sciences, University of South Carolina School of Medicine, Greenville, SC, United States; [2]Director of Marketing Collagen Solutions PLC Minneapolis, MN, United States

Contents

1.1 Atrioventricular valves

1.1.1 Embryology

The heart is first formed as a simple tube from anterior lateral splanchnic mesoderm as the flat, trilaminar embryonic disc rolls into a cylinder. The growing prosencephalon and the closing gut tube endoderm bring the left and right lateral mesoderms together ventrally at the midline of the developing embryo [1]. At this stage or even a bit before, the primitive myocardium begins to spontaneously contract. The formation of the primitive heart tube is critical to further development of the embryo as it relies on effective hemodynamics to support the ontogenesis of other structures.

Though the cardiac valves play a central role in the maintenance of unidirectional blood flow for the entire cardiovascular system, other tissues have valves, including some veins and lymphatic vessels. It is important to note that a valve-like structure is formed, transiently, between the left and right atria known as the foramen ovale. This structure allows placentally derived oxygen- and nutrient-rich blood to pass from the right atrium to the left atrium, allowing it to be distributed systemically during fetal development. Following the first breath and perfusion of the pulmonary vascular, the blood pressure of the right side of the circulation drops below that of the systemic left side blood pressures, physiologically closing the foramen. Over time, the septum primum

Principles of Heart Valve Engineering
ISBN 978-0-12-814661-3, https://doi.org/10.1016/B978-0-12-814661-3.00001-0

1

and the septum secundum fuse, leaving a thumbprint-shaped indentation on the atrial septum known as the fossa ovale. Failure of this foramen to close results in atrial septal defects of varying degree and severity.

The acelluar cushions are largely composed of the glycosaminoglycans hyaluronan and chondroitin sulfate, which yield a soft, jelly-like consistency, giving it the name, *cardiac jelly* (Fig. 1.1). Nonetheless these soft, pliable atrioventricular (AV) cushions do contribute to unidirectional blood flow in the early embryonic circulation. The myocardium of the AV junction produces the initial extracellular matrix (ECM) of the cushions. This provides the substrate that cells will use to migrate into the cushions and produce the tissues of the mature valves and supporting structures.

The majority of the cells populating the AV cushions are derived from endocardial cells of the inferior and superior AV cushions as well as significant contribution of epicardially derived cells that have undergone an epithelial-to-mesenchymal transformation (EMT) (Fig. 1.1C). During this process, cells detach from the simple epithelium that lines the interior and exterior of the heart and migrate into the matrix-filled cushions. The cells of this newly formed mesenchyme become VICs, which remodel and maintain the ECM into the complex, stratified valve leaflets [2]. The endocardial cells that cover the valves have been reported to retain their ability to undergo EMT throughout adult life [3]. Under pathological conditions, these endothelial cells transform and migrate into the mesenchyme of the valve leaflets and adversely contribute to valve disease. The roles that other cell types, such as macrophages and other immune cells, play in development and disease of valve tissues are beginning to gain increased attention by investigators, as they appear to be key regulators of homeostasis and pathology [4].

The inferior and superior AV cushions fuse, forming the septum intermedium, which physically separates the left (systemic) and right (pulmonary) sides of the circulatory system. As development continues, lateral AV cushions emerge on the left and right sides and fuse with the inferior and superior endocardial tissues, providing the cells that will go on to form the AV septum, AV valve leaflets, and supporting tissues. In lineage tracing studies, neural crest cells were detected in the AV septum and shown to have migrated from the top of the neural tube into the heart via pharyngeal arches 3, 4, and 6. The roles that specific cells play in the differentiation and their contributions to eventual adult cardiac structures are not clear despite intensive and ongoing efforts. It is critical that these studies be brought to their fruition, as defects in the AV valvuloseptal tissues are amongst the most lethal.

During normal development of the AV septum, the ostia of the atria anatomically align with the AV valves and the ventricular chambers. Subsequent fibroadipose development of the AV septum provides a foundation for the remodeling of the endocardial cushions into the valve leaflets. The AV septum and its fibrous *cardiac skeleton* also act as an electrical insulator that allows for the atrial, ventricular delay of the cardiac cycle. Housing the AV node of the cardiac conductance pathway, malformations of this region impact cardiac rhythm and function and are thus critically pathological.

Development of the valve leaflets and tension apparatus of AV valves is generally thought to be driven by a remodeling process in which cushion cells differentiate into ECM-producing VICs that create the stratified fibroelastic connective tissue of the valve leaflets and the fibrocartilage-like chordae tendineae. This remodeling occurs in humans during infancy and early childhood. The mechanisms that drive the differentiation of cells into VICs versus cells of the chordae tendineae are not clear [5]. Hemodynamically driven differentiation is an attractive, though, not well-tested mechanism. Malformations of these structures include prolapse, stenosis, and atresia.

The molecular mechanisms that create and maintain the tissues of the cardiac valves have a long history of investigation. Decades of research studies using a variety of model systems have delineated the molecular signaling pathways that are critical for the induction, differentiation, and maturation of cardiac valves [2,5]. These processes can be divided into four stages: endocardial cushion formation, endocardial transformation, growth and remodeling, and stratification (Fig. 1.2).

AV valve formation is initiated when the myocardium of the AV canal produces the cardiac jelly that fills the superior and inferior AV cushions. Along with the ECM proteins, these cells secrete morphogens that activate overlying endocardial cells to disconnect from neighboring endothelial cells and migrate into the ECM of the cushions. Myocardially derived BMP2 signals initiate transformation of the AV canal endocardial cells, while canonical Wnt and TGF-β signaling are critical for sustaining EMT [6]. Endocardially derived Notch and VEGF signaling are also required for EMT, and several other well-characterized signaling pathways that are summarized in Fig. 1.2.

In addition to the molecules above, transcription factors Twist1 and Tbx20 are critical for the proliferation and differentiation of newly formed mesenchymal cells. Interestingly, VEGF becomes a negative regulator of VIC proliferation at the post-EMT stage of valve development [6]. During this stage of valve development, the matricellular protein, periostin, becomes highly expressed in the developing cushions and is necessary for the differentiation of VICs into ECM-producing fibroblasts within valve cushions [2]. As its name denotes, periostin is also involved in bone development. In fact, valve development involves a number of molecules that have been implicated in the development of bone and cartilage. Another similarity between bone and cardiac valve development is the BMP-driven expression of Sox9 [6]. However, there is a tendon-like gene expression pattern in the differentiation of the chordae tendineae of AV valves, involving Fibroblast Growth Factor (FGF), scleraxis, and tenascin.

As the valve leaflets mature, they become more complex with specific combinations of ECM proteins deposited in different locations within the valve [2]. This results in the formation of three distinct layers within valves, which are discussed in detail below. Here, it is important to note that the bone-like expression pattern remains in the collagen-rich fibrosa layer, which is dependent on NFATc1, whereas a cartilage-like expression pattern has been found in the proteoglycan- and glycosaminoglycan-rich spongiosa layer.

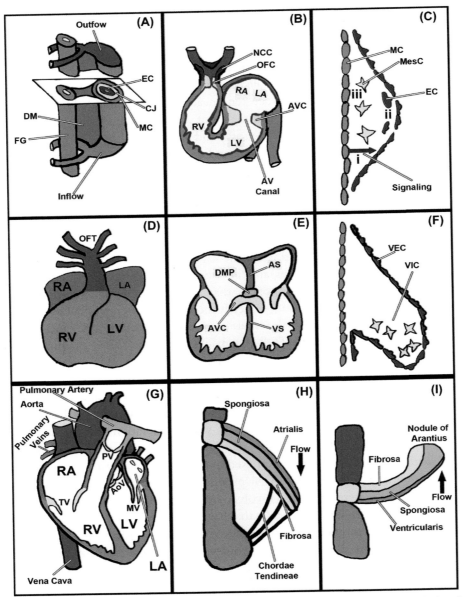

Figure 1.1 *Overview of cardiac development.* (A) The heart initiates as a tube composed of endothelial cells (ECs), cardiac jelly (CJ), and myocyte cells (MCs). The tube is initially linked to the foregut (FG) via the dorsal mesocardium (DM), but this connection later breaks away as looping occurs. (B) As the tube bends and twists, cushions filled with CJ form. Atrioventricular cushions (AVCs) form in the AV canal, and outflow cushions (OFCs) form at the heart outlet where they receive a cellular contribution from neural crest—derived cells (NCCs). The future right and left ventricles, RV and LV, and the future right and left atria, RA and LA, are defined. (C) (i) Signaling from the MCs to the ECs induce EMT.

The third layer, which is on the flow facing side of the valve leaflet, has a smooth muscle-like ECM, which contains elastin and collagen [2]. However, the molecular regulation of this layer has yet to be clearly defined.

1.1.2 Morphology

As discussed above, the AV valves differ from the semilunar valves in that the AV valve leaflets have a tension apparatus that consists of the chordae tendineae and the papillary muscles. The chordae tendineae are string-like extensions off of the valve leaflets that connect the AV leaflets to the papillary muscles of the ventricles (Fig. 1.3). The papillary muscles are invested with conductive tissue that are closely connected to the branch bundles and thus are amongst the earliest regions to contract in the ventricles, tensing the leaflet and preparing the structure to withstand systole. Failure to do so results in regurgitation of blood back into the atrium, resulting in loss of cardiac output. Generally, the orifice of the left ostia is bigger than the right, though this can change as a result of malformation or pathology.

The mitral valve, or bicuspid valve, separates the left atrial and ventricular chambers and has two valve leaflets, the anterior (aortic) and posterior (mural). The chordae tendineae of these leaflets coalesce into two well-defined papillary muscles located near the apex of the left ventricular chamber. During systole, the two leaflets have one zone of apposition that seals off the AV ostia and prevents regurgitation back into the left atria. Clinically, this crescent-shaped zone is divided into the anterolateral commissure and the posteromedial commissure, which enables anatomical description of areas of regurgitation or prolapse [7].

The tricuspid valve separates the right atrial and ventricular chambers and has three valve leaflets: the anterior; posterior; and the mural. The chordae tendineae from these leaflets coalesce into three clusters of papillary muscles in the right ventricle. The papillary muscles of the right ventricle are less organized and more variable than those of the left

---

(ii) Activated ECs lose their cell—cell junctions and invade the CJ. (iii) The activated ECs begin to express mesenchymal cell (MesC) markers. (D) The future heart chambers are in their final location when the outflow tract (OFT), atria, and ventricles begin to septate. (E) Muscular protrusions grow from the heart wall to form the atrial septum (AS) and ventricular septum (VS). The AS protrusion has a cap, the dorsal mesenchymal protrusion (DMP), that connects with the AVCs. (F) After EMT, the AVCs elongate into leaflets that are lined with valvular endothelial cells (VECs) and valvular interstitial cells (VICs). (G) Blood flows through the developed heart in the following order: vena cava, RA, tricuspid valve (TV), RV, pulmonary valve (PV), pulmonary artery, lungs, pulmonary veins, LA, mitral valve (MV), LV, aortic valve (AoV), aorta, body (H) Atrioventricular (AV) valves are composed of three layers: the elastin-rich atrialis, the water-rich spongiosa, and the collagenous fibrosa. The leaflet tip is tethered to the heart wall via the chordae tendineae. (I) Semilunar (SL) valves have the same three layers, but the elastin-rich layer is referred to as the ventricularis. The leaflet cusps end in thick, fibrous tips known as the nodules of Arantius in the AoV or nodules of Morgagni in the PV. The trilaminar leaflets and associated support structures function to withstand flow-induced forces.

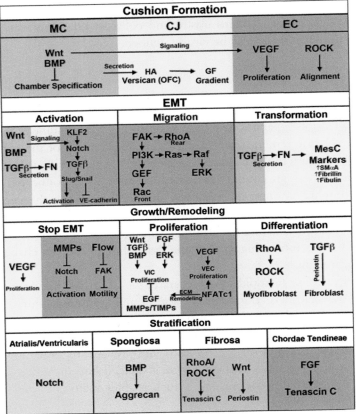

Figure 1.2 *Signaling in cardiac development.* Each phase of cardiac development is associated with biochemical signals that are still being fully elucidated. During cushion formation, myocyte cells (MCs) signal to the endothelial cells (ECs) and induce alignment and proliferation. At the same time, the MCs secrete hyaluronic acid (HA) and, in the outflow cushions (OFCs), versican to fill the cardiac jelly (CJ) space. The CJ maintains a gradient of growth factors (GFs), allowing different stages of development to be triggered at different times. Epithelial-to-mesenchymal transformation (EMT) begins with EC activation and MC secretion of fibronectin (FN). The activated ECs phenotypically change as they lose their cell–cell junctions and migrate into the CJ. Inside the CJ, these cells begin to express mesenchymal cell (MesC) markers. Flow and proteinases stop EMT and the cushions grow and remodel. The VICs in the CJ have regulated proliferation, and EC proliferation is thought to slow as extracellular matrix (ECM) is increasingly deposited in the CJ. MesCs differentiate into ECM-secreting cells characteristic of mature valves. A mature valve exhibits three distinct layers that are regulated with unique signaling mechanisms, and the atrioventricular (AV) valves have a tendon-like support apparatus that undergoes signaling similar to cartilage and tendons.

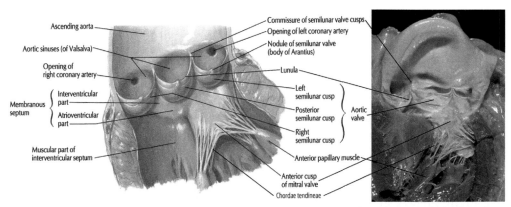

Figure 1.3 *Clinical anatomy of left heart valves.* The semilunar and atrioventricular valves have unique structures that provide support and anchor the valves to the wall. The three leaflets of the semilunar valves have commissures at the wall juncture (depicted for the aortic valve) and the atrioventricular valves are attached to the papillary muscle via a tension apparatus, chordae tendineae (depicted for the mitral valve). *((Left) From Frank H. Netter, Atlas of Human Anatomy — 4th Edition, 2006; (Right) CNRI/Science Photo Library.)*

papillary muscles [7]. The moderator band, an important cardiac conductance tissue, is incorporated within the septomarginal trabecula, which is a myocardial structure that connects the anterior papillary muscle of the right ventricle to the interventricular septum. Being trifoliate, there are three zones of apposition and three commissures in the tricuspid valve: the anteroseptal; the anteroposterior; and the posterior.

1.1.3 Histology

The tissues that make up valve leaflets of both AV and semilunar valves have a similar overall design. A common feature of all cardiac valve leaflets is that they are organized into three layers (Fig. 1.1, panels H and I). The first layer of cells under the endocardial epithelium on the flow side of the leaflet contains densely packed cells that are surrounded by an elastic connective tissue. This layer is named the atrialis in AV valves and the ventricularis in semilunar valves. Elastic fibers are radially oriented from the hinge of the leaflet to the coapting edge [7]. The composition and organization of the matrix allows for extension and recoil of this layer as the valve opens and closes. The middle layer of valve cells is called the spongiosa and contains sparsely distributed cells embedded in ground substance, which is largely composed of proteoglycans. This layer is thought to carry out a cushioning function for the valves. The layer on the back (nonflow) side of the valve leaflets is called the fibrosa. As its name implies, it is a dense connective tissue containing large bundles of insoluble, fibrous Type I collagen, giving it a comparatively stiff quality. These fibers are circumferentially oriented, providing tensile strength to the leaflet. Together, the three layers of the valve leaflets provide a balanced mix of stiffness, pliability, and recoil, giving it the mechanical properties necessary for healthy valves to be competent when closing and compliant when opening. Alterations in the composition and organization of these layers are associated with numerous valve pathologies including valve calcification and myxomatous valves.

Another difference between the mitral and tricuspid AV valves, in addition to the number of leaflets and the size of the annuli, is the thickness of the valve leaflets. The mitral valve leaflets are thicker than those of the tricuspid. However, this difference is not evident until after birth, indicating that the increasing hemodynamic load experienced on the systemic side of the circulation is driving this morphogenesis.

The annuli of the valves are composed of dense connective tissue and provide a firm foundation to anchor the hinges of the leaflets. Type I collagen is the main ECM protein of the valve annuli, forming the major components of the cardiac skeleton. With the exception of the pulmonary valve, cardiac valve annuli are embedded in the AV septum with the aorta being wedged between the tricuspid and mitral valves, making this a highly complex region of the heart.

1.2 Semilunar valves

1.2.1 Embryology

The cushions that go on to contribute to the semilunar valves appear just after the cushions of the AV canal. These conotruncal cushions form as oppositely opposed ridges that spiral down the truncus arteriosus, which is the single outflow vessel, or arterial pole, of the tubular heart. This single outflow tract is divided into the pulmonary and aortic arteries as the conotruncal cushions become populated with cells, grow, and fuse at midline, creating the septum intermedium, which physically separates left and right sides of the arterial pole of the heart and sets the stage for complete aorticopulmonary septation [2].

Septation of the truncus arteriosus occurs from the inside out, with the tunica intima of the two newly formed vessels forming first, followed by the generation of their tunica medias, and, finally, their adventitias. In this way, the single outflow vessel is divided into two completely formed great arteries. The completeness of this separation is particularly evident in the transverse pericardial sinus, which separates the infundibulum of the pulmonary trunk from the aorta. Failure of proper formation and progression of the aorticopulmonary septum can result in life-threatening lesions including persistent truncus arteriosus or subclinical lesions such a small ventricular septal defect. This is because the aorticopulmonary septum fuses with the muscular ventricular septum that arises between the left and right ventricular chambers, becoming the membranous component of the ventricular septum. Neural crest cells have been found to contribute to both the AV valves and the semilunar valves; however, their role in morphogenesis of the outflow tract is particularly critical. Failure of neural cells to populate and migrate with the aorticopulmonary septum results in persistent truncus arteriosus.

In combination with the remodeling of the conotruncal cushions of the embryonic outflow tract, the leaflets of the semilunar valves develop from another set of cushions in the outflow tract, the intercalated cushions, which form adjacent to the conotruncal cushions. Once again in a manner similar to the AV valves, cardiac jelly filled swellings appear between the myocardium and endocardium, become cellularized by endocardial EMT, and undergo ECM remodeling over time into the highly organized valve leaflets. Much remains unknown about mechanisms that regulate the number and positioning of semilunar valve leaflet anlagen that differentiate into the trifoliate adult semilunar valves. Defects in these structures result in bicuspid and stenotic semilunar valves.

1.2.2 Morphology

The two semilunar valves have similar structures with three pocket-like leaflets arranged such that they are competent without the tension apparatus that is found in the AV valves. The three leaflets of the semilunar valves have three commissures, which act as anchoring points to support the juncture to the wall at the base of the leaflets (Fig. 1.3). The semilunar leaflet geometry creates spaces behind the leaflets known as the sinuses of Valsalva. The U-shaped base of the semilunar valve leaflets creates triangular-shaped areas in the walls of the great arteries that are not occupied by either valve tissue or valve sinuses. Thickened nodes are present at the tip of each leaflet, known as the nodules of Arantius in the aortic valve and the nodules of Morgagni in the pulmonary valve. These nodules exhibit an enlarged spongiosa layer making them characteristically elastic structures that can act to support the extreme hemodynamic forces present at the point of valve closure or the valvular orifice (Fig. 1.11I). The pulmonary valve differs from the aorta in that it has a column of myocardium, the infundibulum, to support its root. However, the aortic root is embedded in the connective tissue of AV septum. The aortic valve leaflets are thicker than those of the pulmonary, but again, this difference appears to develop postpartum as a response to the increased load that is required for systemic circulation. Another difference between this class of cardiac valves is the presence of coronary artery ostia in two of the aortic valve sinuses.

The aortic valve is situated in the middle of the AV septum, wedged between the mitral and tricuspid annuli and wrapped by the pulmonary infundibulum. The left and right sinuses of the aorta have the openings of the left and right coronary arteries. Therefore, the sinus of the posterior leaflet is known as the noncoronary sinus. Interestingly, this leaflet is the only aortic leaflet that does not appear to have any contribution of neural crest cells; however, this may be due to the fact this leaflet is deeply embedded in between the mitral and tricuspid annuli and may be inaccessible to neural crest cells migrating down the aorticopulmonary septum during development.

As mentioned above, the aortic valve, unlike the pulmonary valve, is surrounded by fibrous tissues. A portion of the aortic valve is continuous with the fibrous aspects of the mitral valve, including its anterior leaflet. The left fibrous trigone of the aortic root is also continuous with the mitral valve. The right fibrous trigone is continuous with the membranous portion of the ventricular septum, which is derived from the aortico-pulmonary septum. The aortic root is bulbous in appearance and contains the annulus and sinuses of the valve leaflets. At the distal attachments of the valve leaflets, the aorta becomes cylindrical and is called the sinotubular junction. This marks the end of the aortic valve and the beginning of the ascending portion of the aorta.

The pulmonary valve is positioned at a distinctly different angle than the other three cardiac valves. The elongated, funnel-shaped infundibulum of the right ventricular outlet wraps around the aortic root. This sleeve of myocardium acts to support the pulmonary valve root. The three leaflets of the pulmonary valve are the left, right, and anterior. Common malformations of this valve include atresia and stenosis, which can be associated with other cardiac lesions, as in the case of tetralogy of Fallot, and result in cyanosis.

1.2.3 Histology

As described above, the histological organization of the semilunar valve leaflets is similar to that of the AV valves. They share a common tissue architecture, having the three tissue layers of the fibrosa, spongiosa, and elastic layer, known as the ventricularis in the semilunar valves. The semilunar leaflets are significantly thinner than the AV leaflets. Another difference between the AV and semilunar leaflets is that the distal edges of the semilunar leaflets are thick and contain a bulbous structure in the middle of the free edge known as the nodule of Arantius. These modifications and their overall geometry allow for the semilunar valves to seal as they close during diastole, preventing regurgitation.

Recently a new anatomical structure has been described in the root of the aorta. This "prelymphatic" organ appears to be distinct from the adventitia and is thought to provide a "shock absorber" function. This structure contains a series of interconnected "vessels" that are produced by bundles of fibrous collagen and sporadically lined with CD34 + cells that have not been well-characterized [8]. This initial report did not appear to investigate whether this structure is present in the pulmonary trunk.

1.3 Epigenetic factors in heart valve formation

As previously indicated, hemodynamics is thought to play an important and fundamental role in the morphogenesis of the cardiac valves. In particular, the shear stresses and pressures of the developing cardiovascular system appear to play a formative role in the remodeling of the EMC of the cardiac cushions as they morph into the fibrous tissues of the valves [9]. While substantial clinical and experimental literature support the "no flow/no grow" hypothesis, the specific molecular mechanisms that are used to transduce these mechanical signals into distinct cellular activities are not well-characterized.

These studies are frequently confounded by the chicken/egg paradox. For instance, individuals with bicuspid aortic valves have a high risk of developing calcified leaflets that become incompetent. Is this pathological calcification caused by the abnormal geometry that results in an aberrant aortic flow field or by the same process that drove the altered anatomy? New 3D in vitro model systems in which mechanical forces can be carefully controlled are coming on line that will allow for the direct testing of mechanotransduction pathways in specific mechanical environments [10,11]. These studies will be important in designing new therapies for these deadly valve diseases.

Other investigations using surgically created hemodynamic abnormalities have been carried out on avian embryos and have revealed that at its earliest stages, valve development is regulated by blood flow by affecting EMT [12]. Not surprisingly, similar approaches have found that altered hemodynamics drives alterations in the expression and deposition of critical ECM proteins [13].

The mechanisms by which developing valve tissues sense and transduce mechanical signals have been aided by the discovery of primary cilia on VICs within the developing valve cushions [14]. Primary cilia have been implicated in valve development previously, but only in early endocardial cells [15]. Primary cilia have long been known as cellular structures that sense and mediate responses to mechanical forces. Thus, their discovery in cushion mesenchyme indicates that forces other than shear stress, such as deformation-inducing pressures, could be important regulators of valve development.

Gestational diabetes has recently been reported to be an epigenetic regulator of valve development. Fetal hyperglycemia results in the increase of reactive oxygen species, which has been reported to result in cardiac malformations. Specifically, malformations of the outflow tract were associated with hyperglycemia and decreased nitrous oxide signaling [16].

Much progress has been made in delineating the genetic pathways that are critical in the progression of valve development and disease. However, the roles that epigenetic factors play in these processes are in their infancy and require more intense study.

References

[1] Sylva M, van den Hoff MJ, Moorman AF. Development of the human heart. Am J Med Genet A June 2014;164A(6):1347−71. https://doi.org/10.1002/ajmg.a.35896. Epub 2013 Apr 30. Review. PMID: 23633400.

[2] de Vlaming A, Sauls K, Hajdu Z, Visconti RP, Mehesz AN, Levine RA, Slaugenhaupt SA, Hagège A, Chester AH, Markwald RR, Norris RA. Atrioventricular valve development: new perspectives on an old theme. Differentiation July 2012;84(1):103−16. https://doi.org/10.1016/j.diff.2012.04.001. Review. PMID: 22579502.

[3] Bischoff J, Aikawa E. Progenitor cells confer plasticity to cardiac valve endothelium. J Cardiovasc Transl Res December 2011;4(6):710−9. https://doi.org/10.1007/s12265-011-9312-0. Epub 2011 Jul 26. Review. PMID: 21789724.

[4] Hulin A, Anstine LJ, Kim AJ, Potter SJ, DeFalco T, Lincoln J, Yutzey KE. Macrophage transitionsinheart valve development and myxomatous valve disease. Arterioscler Thromb Vasc Biol March 2018;38(3):636—44. PMID: 29348122.

[5] Koenig SN, Lincoln J, Garg V. Genetic basis of aortic valvular disease. Curr Opin Cardiol February 2, 2017. https://doi.org/10.1097/HCO.0000000000000384. PMID: 28157139.

[6] Hinton RB, Yutzey KE. Heart valve structure and function in development and disease. Annu Rev Physiol 2011;73:29—46. https://doi.org/10.1146/annurev-physiol-012110-142145. Review. PMID: 20809794.

[7] Spicer DE, Bridgeman JM, Brown NA, Mohun TJ, Anderson RH. The anatomy and development of the cardiac valves. Cardiol Young December 2014;24(6):1008—22. https://doi.org/10.1017/S1047951114001942 [Review].

[8] Benias PC, Wells RG, Sackey-Aboagye B, Klavan H, Reidy J, Buonocore D, Miranda M, Kornacki S, Wayne M, Carr-Locke DL, Theise ND. Structure and distribution of an unrecognized interstitium in human tissues. Sci Rep March 27, 2018;8(1):4947. https://doi.org/10.1038/s41598-018-23062-6.

[9] Wu B, Wang Y, Xiao F, Butcher JT, Yutzey KE, Zhou B. Developmental mechanisms of aortic valve malformation and disease. Annu Rev Physiol February 10, 2017;79:21—41. Review. PMID: 27959615.

[10] Tan H, Biechler S, Junor L, Yost MJ, Dean D, Li J, Potts JD, Goodwin RL. Fluid flow forces and rhoA regulate fibrous development of the atrioventricular valves. Dev Biol February 15, 2013;374(2):345—56. PMID: 23261934.

[11] Biechler SV, Junor L, Evans AN, Eberth JF, Price RL, Potts JD, Yost MJ, Goodwin RL. The impact of flow-induced forces on the morphogenesis of the outflow tract. Front Physiol June 17, 2014;5:225. 2014.

[12] Menon V, Eberth JF, Goodwin RL, Potts JD. Altered hemodynamics in the embryonic heart affects outflow valve development. J Cardiovasc Dev Dis 2015;2(2):108—24. Epub 2015 May 15.

[13] Rennie MY, Stovall S, Carson JP, Danilchik M, Thornburg KL, Rugonyi S. Hemodynamics modify collagen deposition in the early embryonic chicken heart outflow tract. J Cardiovasc Dev Dis December 20, 2017;4(4). pii: E24. PMID: 29367553.

[14] Toomer KA, Fulmer D, Guo L, Drohan A, Peterson N, Swanson P, Brooks B, Mukherjee R, Body S, Lipschutz JH, Wessels A, Norris RA. A role for primary cilia in aortic valve development and disease. Dev Dynam August 2017;246(8):625—34. https://doi.org/10.1002/dvdy.24524. Epub 2017 Jun 28. PMID: 28556366.

[15] Egorova AD, Khedoe PP, Goumans MJ, Yoder BK, Nauli SM, ten Dijke P, Poelmann RE, Hierck BP. Lack of primary cilia primes shear-induced endothelial-to-mesenchymal transition. Circ Res April 29, 2011;108(9):1093—101. https://doi.org/10.1161/CIRCRESAHA.110.231860. PMID: 21393577.

[16] Basu M, Zhu JY, LaHaye S, Majumdar U, Jiao K, Han Z, Garg V. Epigenetic mechanisms underlying maternal diabetes-associated risk of congenital heart disease. JCI Insight October 19, 2017;2(20). https://doi.org/10.1172/jci.insight.95085. pii: 95085.

CHAPTER 2

Heart valves' mechanobiology

Madeline Monroe, Amadeus Zhu, K. Jane Grande-Allen
Department of Bioengineering, Rice University, Houston, TX, United States

Contents

2.1 Introduction

Heart valves are perhaps the most mechanically active connective tissues within the human body. With every heartbeat, the valve tissues stretch out to cover the valve orifice, press their edges against each other to maintain valve closure, and then rapidly open by retracting back to an unstretched state when the blood pressure load is removed. The opening and closing motions of these essential tissues are so forceful and dynamic that they can be heard from outside the body. This function is made possible by the extracellular matrix (ECM) microstructure within the valve tissues. This same ECM also transmits mechanical strains and stresses to the valve cells. Like many cell types that exist within connective tissues, heart valve cells are mechanosensitive. Understanding their mechanobiology is relevant to healthy valves, diseased valves, and tissue–engineered valves. This chapter

Principles of Heart Valve Engineering
ISBN 978-0-12-814661-3, https://doi.org/10.1016/B978-0-12-814661-3.00002-2

13

will review the valve cell phenotypes and how these are influenced by the cell micro-environment (specific signaling pathways as well as the specific ECM) and will describe how these cell microenvironments become altered in valve diseases. The chapter will conclude with a summary of mechanobiology considerations for tissue-engineered heart valves and 3D culture models of heart valve disease.

2.2 Valvular interstitial cells

Heart valves are responsible for the unidirectional flow of blood via precise opening and closing of associated leaflets. Through this constant cyclic motion and under a range of hemodynamic conditions, it is vital that leaflets maintain their structural and compositional integrity. Cells that comprise the leaflets play a substantial role in the valves' maintenance and functioning [1]. Leaflets contain two primary cells types: valvular endothelial cells (VECs), which line the outer surfaces of the leaflets, and valvular interstitial cells (VICs), residing within the bulk of the leaflet. VICs, the predominant cell population of valve leaflets, are thought to play a fundamental role in the main-tenance and continued proper functioning of the valves. VICs are characterized by a diverse range of phenotypes, each associated with a specific function and contribution toward overall matrix structure and function [2].

The main role of VICs is in maintaining structural and compositional integrity through synthesis and degradation of the ECM. Most VICs in the native, healthy valve are considered quiescent and fibroblastic. However, there is a distinct subpopulation of VICs that have the ability to transition to a number of different phenotypes based on ECM environmental conditions, as well as in response to injury or for maintaining homeostasis. In total, there are five distinct VIC phenotypes: progenitor endothelial/mesenchymal cells (eVICs), quiescent VICs (qVICs), activated VICs (aVICs), post-developmental adult progenitor VICs (pVICs), and osteoblastic VICs (obVICs).

2.2.1 Valvular interstitial cell phenotypes

eVICs are involved in embryogenesis and develop into qVICs in the maturing heart, potentially through an active process. This transition in eVICs is characterized by endothelial-to-mesenchymal transformation (EndMT), initiating valve formation within the embryonic cushion [3].

qVICs are the most common VIC phenotype found in a healthy adult heart valve, and they help maintain normal ECM structure and leaflet functionality. The ability of qVICs to degrade and synthesize matrix plays a vital role in this process. Additionally, qVICs have been shown to communicate among themselves via intercellular junctional complexes, gap junctions, and adhesion junctions [4]. It has been hypothesized that net-works of these junctions are involved in allowing VICs to sense and respond rapidly to external mechanical forces. Additionally, qVICs have been shown to inhibit angiogenesis within the leaflets, which is important in the context of disease progression. Normally, human heart valves are avascular; however, in pathological valve conditions such as calcification progress, neovascularization occurs. This change is linked to the mobilization of VECs and the upregulation of proangiogenic factors, including vascular endothelial growth factor, and downregulation of antiangiogenic factors such as chondromodulin-1, a factor that qVICs have been shown to secrete abundantly in healthy heart valves [5].

In diseased or injured valves, qVICs are signaled to transition toward an aVIC phenotype. aVICs, the precursor phenotype to obVICs, demonstrate an increased production of alpha smooth muscle actin (αSMA), an isoform of actin that is not found in qVICs [6]. Additionally, this phenotype exhibits markedly increased cellular proliferation, ECM remodeling, and migration capabilities as compared with the quiescent phenotype [7]. Specifically, in regard to matrix remodeling capabilities, aVICs demonstrate increased degradation via upregulation of matrix metalloprotease 1 (MMP-1) and synthesis via increased secretion of tissue inhibitor of MMP-1. Dysfunction with regard to matrix regulation and apoptosis contributes significantly to valve disease progression. Although the reason for increased proportion of aVICs over the course of disease development within valve leaflets is largely unknown, factors including valve injury and abnormal mechanical forces are thought to influence this phenomenon [8].

The increase in the number of VICs that is observed as a response to injury or to abnormal mechanical stimuli has also been attributed to a variety of sources. One source contributing to these elevated numbers is pVICs, which are thought to invade the valve and help aVICs repair their environment. These cells have a variety of origins; some of them arise from the bone marrow or from circulating blood, whereas others originate within the valves themselves. pVICs are thought to have the potential to infiltrate the valve leaflet during injury and thereby provide support to aVICs in repairing the valve environment via transformation into qVICs or aVICs. The phenotype of this specific subpopulation is characterized by cellular expression of CD34, CD133, NG2, and CD338. The relationships between pVICs, qVICs, and aVICs, as well as the propensity of valves toward disease in relation to the relative quantities of these types of VICs comprising the valve, are topics that remain to be studied in depth [9,10].

Finally, obVICs—the most pathological phenotype—promote calcification in the heart valve by secreting factors such as alkaline phosphatase (ALPL), osteocalcin, osteopontin, and bone sialoprotein. These factors drive the formation of calcific (calcium- and phosphate-containing) nodules throughout the valve leaflet. The transition from aVICs to obVICs is characterized by an upregulation in runt transcription factor 2 (RUNX2), matrix Gla protein (MGP), ALPL, and sex-determining region Y-box 9 (Sox9), expression of which positively correlate with degree of calcification [11]. In addition, calcified heart valves have been shown to express numerous osteogenic growth factors and cytokines, including transforming growth factor beta 1 (TGFβ1), bone morphogenic protein (BMP), and tumor necrosis factor alpha (TNFα). Recent studies have implicated

TNF-α as a critical driver of osteogenic VIC transformation, specifically by influencing the upregulation of RUNX2, and, moreover, the valve calcification process [12,13].

2.2.2 Influence of environmental mechanics

The specific phenotype demonstrated by VICs is largely dependent on the mechanical environment and stimulation to which the valvular cells are subjected. In vivo, heart valves are exposed to large pressure gradients and stresses—specific mechanical conditions that play a significant role in dictating VIC phenotype. For instance, it has been shown that porcine VICs isolated from the left side of the heart (mitral and aortic valves) have higher inherent stiffness than the pulmonary and tricuspid VICs, as measured by micropipette aspiration and atomic force microscopy [14]. Additionally, mechanical stiffness and anisotropy are important substrate features that have been shown to drastically alter VIC morphology, as demonstrated in Fig. 2.1 [15]. Furthermore, VICs behave differentially based on the valve that they are derived from and correspondingly based

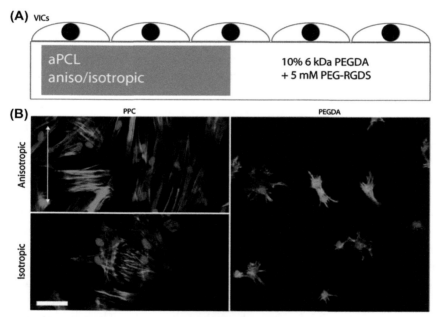

Figure 2.1 Aortic VICs seeded on materials of varying mechanics exhibit differential morphology and expression. (A) Polyethylene glycol (PEG) gels were embedded with sheets of either isotropic or aniso-tropic electrospun acrylated-polycaprolactone (aPCL), resulting in a composite PEG—aPCL material (PPC). (B) VICs seeded on anisotropic PPC, isotropic PPC, and PEG without aPCL exhibit differential morphology and alignment based on mechanical characteristics of the respective substrate. Here, VICs are stained with DAPI (blue) and phalloidin (green). Scale bar = 50 μm. *(Credit: Adapted with permission from Tseng, Hubert, et al. Anisotropic poly(ethylene glycol)/polycaprolactone hydrogel—fiber composites for heart valve tissue engineering. Tissue Engineering Part A 2014;20(19—20):2634—2645.)*

on the inherent mechanical forces experienced by that particular valve. When grown in a 3D collagen gel, for example, aortic VICs will contract the collagen gel at a faster rate than will pulmonary VICs, suggesting that the aortic VICs are able to remodel their environment to a greater degree as compared with pulmonary VICs, and, moreover, that significant mechanobiological differences exist between these two distinct VIC types. This study illustrates how inherent valve environment, and the nuanced mechanical forces that accompany that environment, drastically influence VIC remodeling behavior, which is an important consideration in studying healthy and pathological valve progression.

An additional consideration is that all heart valves are comprise heterogeneous ECM, as will be described later in this chapter, which provides the resident VICs with a varied range of pericellular stiffnesses. This heterogeneity of composition, which contributes to the anisotropy of valve tissues, results in the cells within the leaflet being exposed to varied magnitudes of stretch based on their precise location. This organized structure is subjected to specific applied strain fields that vary over the course of a lifetime as one's heart develops and, in turn, affect matrix remodeling [16]. To mimic these complex strain fields, a study by Gould et al. subjected VICs embedded in 3D collagen gels to biaxial stretch. From this, a relationship between increased degree of anisotropy and VIC activation, as well as proliferation and apoptosis, was established. Other in vitro experiments have been utilized to observe how mechanical stimuli promote or suppress cell behavior. Studies using the Flexcell bioreactor system, which operates via stretching of cells cultured on an elastomeric substrate, have shown that VICs will change their protein expression in response to differing stretch conditions. For instance, in one study, 10% distension of human VICs over the course of 72 h produced an upregulation of osteogenic markers (osteopontin, osteocalcin, and Alkaline phosphatase (ALP)) [17].

2.2.3 Influence of sex and age

Another important factor that remains to be studied further is the influence of subject sex and age on VIC phenotype. Although it has long been established that biological sex and patient age are two of the strongest clinical risk factors for calcific aortic valve disease (CAVD), researchers are just beginning to dig deeper into exploring specific, cellular-level differences based on these inherent traits. In one study, production of hyaluronan (HA) was shown to be significantly downregulated in older porcine valves as compared with younger valves [18]. Biological sex, including associated hormones, has also been shown to have potentially profound effects on VIC and cardiac cell functioning on a whole. The presence of sex steroid receptors in the heart could point to differences in cardiac development for males and females. In VICs specifically, it has been shown that male- and female-derived aortic VICs exhibit differential gene expression governing important biological processes such as proliferation, migration, and remodeling. Specifically, male VICs demonstrate higher expression of genes that dictate proliferation and apoptosis [19]. Furthermore, an interesting but understudied trend is the difference in diseased aortic valve phenotypes between males and females affected by CAVD: females exhibit increased valvular fibrosis and decreased calcification in diseased valves as compared with their male counterparts. This distinctive form of leaflet remodeling then affects ventricle adaptation, with female left ventricles becoming hypertrophic and male left ventricles dilating [20]. These trends are important because developing effective treatments for male and female patients will require understanding the causes of these fundamental differences.

2.3 Cell signaling and microenvironment

VICs do not exist in isolation in the interior of heart valve leaflets; the leaflet surfaces are covered with VECs. In many ways, VECs are similar to endothelial cells from blood vessels. Both valvular and vascular endothelial cells express platelet endothelial cell adhesion molecule (PECAM-1, also known as CD31), vascular endothelial cadherin (VE-cadherin), and von Willebrand factor (vWF) and are negative for αSMA expression [21]. However, there are a number of key characteristics that distinguish VECs from their vascular counterparts. For example, in the presence of fluid flow, vascular endothelial cells rearrange their cytoskeletal fibers to align parallel to the direction of shear stress. In contrast, VECs have been shown to align perpendicular to the direction of shear stress, likely due to the unique mechanical environment in heart valves [22].

Various biochemical and mechanical stimuli in the valvular microenvironment influence the phenotypic characteristics and differentiation status of both VICs and VECs. These stimuli serve either to maintain valvular homeostasis under normal conditions or to induce remodeling under injured or diseased conditions. As previously stated, VICs can exist in five distinct phenotypic states—embryonic progenitor, adult progenitor, quiescent, activated myofibroblast-like, and osteoblastic—depending on the signals present in their microenvironment. When considering VECs, a particularly important phenotypic change is the EndMT, which can be considered to be a special case of the better-characterized epithelial–mesenchymal transition. EndMT is marked by a loss of expression of endothelial-specific markers, such as CD31 and vWF, and a gain of mesenchymal-like characteristics, such as the presence of αSMA stress fibers [23,24].

In heart valves, EndMT plays an important role during both development and wound healing by generating or replenishing the population of VICs in the valve. During development, endothelial progenitor cells within endocardial cushions in the developing atrioventricular canal and ventricular outflow tract undergo EndMT, transdifferentiating into VICs and giving rise to the atrioventricular and semilunar valve leaflets, respectively [25]. Similarly, during wound healing, EndMT contributes to fibrosis by enabling endothelial cells to transdifferentiate into activated myofibroblasts [26]. This next section of the chapter will review how these phenotypic changes in VICs and VECs can be induced through several important biochemical and mechanical signaling pathways, including TNFα, TGFβ, nitric oxide, and ECM stiffness.

2.3.1 Tumor necrosis factor alpha

TNFα is a cytokine that is secreted primarily by macrophages and T cells. Originally identified for its cytotoxic effects on cancer cells, TNFα has since been implicated as a potent proinflammatory factor with a broad range of effects on multiple tissue types [27]. TNFα is produced as a transmembrane protein that is proteolytically cleaved and secreted as a soluble homotrimer. In its soluble form, this protein binds to its receptors, TNFR1 and TNFR2, initiating an intracellular signaling cascade that ultimately activates a number of transcription factors including nuclear factor κB (NFκB) [28].

Inflammation has been shown to be a major mechanism for tissue remodeling in valvular disease [29]. In aortic valve stenosis, proinflammatory cytokines such as TNFα are produced by macrophages that have infiltrated the valve leaflet. TNFα promotes calcification by increasing VIC proliferation and upregulating the activity of ALPL, which is responsible for mineralization [8,30]. In addition to its effects on VICs, TNFα also induces EndMT in VECs in an NFκB-dependent manner, suggesting a parallel VEC-mediated mechanism for the role of TNFα in valvular disease [31]. Interestingly, TNFα-induced EndMT is highly heterogeneous, with some VECs displaying increased susceptibility to EndMT compared to others [32].

2.3.2 Transforming growth factor beta

Transforming growth factor beta (TGFβ) is a growth factor that regulates a multitude of cellular processes throughout the body. TGFβ initiates its signaling effects by binding to its receptor, TGFβRII, causing it to recruit and phosphorylate TGFβRI. The heterodimerized TGFβR complex subsequently phosphorylates downstream proteins known as receptor Smads (R-Smads), which translocate to the nucleus and regulate gene transcription [33]. During wound healing, TGFβ is produced abundantly by infiltrating leukocytes, and it drives fibrotic remodeling by inducing excessive production of ECM components and inhibiting ECM degradation [34]. In the myocardium, elevated TGFβ production is associated with pathological states such as hypertrophic cardiomyopathy and myocardial infarction [35].

A number of recent studies have investigated the role of TGFβ signaling in heart valves. Using a scratch assay model (an in vitro model of wound repair), one group showed that TGFβ is produced by VICs after injury and is required for VIC activation and migration into the wound [36]. In VECs, EndMT, and αSMA expression can be induced by the addition of TGFβ [23], but this phenotypic shift is inhibited by soluble paracrine signals secreted by VICs [7]. Furthermore, several groups have begun to explore the interaction between TGFβ signaling and heart valve mechanobiology by using bioreactors such as the Flexcell system. Using such a system, it was shown that TGFβ and cyclic strain together induce maladaptive HA remodeling in the aortic valve, pointing to a potential mechanism of ECM dysregulation in CAVD [37]. In the mitral valve, TGFβ induces VICs to adopt a myofibroblastic phenotype, but this effect is partially reversed by cyclic strain [38]. Taken together, these results indicate an important role of TGFβ and mechanical forces in regulating tissue homeostasis in the heart valve.

2.3.3 Nitric oxide

Nitric oxide is a free-radical compound that is biosynthesized from the amino acid L-arginine by nitric oxide synthase enzymes. Its high diffusivity and short half-life of several seconds make nitric oxide an ideal signaling molecule for eliciting transient responses in many tissues and organs [39]. In the cardiovascular system, nitric oxide is produced by endothelial cells via endothelial nitric oxide synthase (eNOS), and it exerts various effects such as vasodilation and inhibition of platelet aggregation [40].

Dysfunctional eNOS activity and nitric oxide signaling can lead to various cardiovascular disease states. For example, aging induces a phenomenon known as eNOS uncoupling in which reduced bioavailability of tetrahydrobiopterin, a cofactor for eNOS, causes the enzyme to produce superoxide anions instead of nitric oxide. These superoxide anions increase the oxidative stress in the blood vessel. As such, eNOS uncoupling has been implicated as a possible mechanism for age-related atherosclerosis [41]. There is some evidence that superoxide production from eNOS uncoupling plays a role in aortic valve calcification as well [42]. Conversely, nitric oxide secreted from VECs appears to protect against VIC activation and valve calcification [43]. Notably, the magnitude of this protective effect depends on the stiffness of the microenvironment. In one study, VICs were indirectly cocultured with VECs on substrates of various stiffnesses. VICs grown on stiffer substrates displayed increased activation and myofibroblastic properties, an effect that was reversed by paracrine nitric oxide from VECs. Furthermore, the VICs on the soft substrates were significantly more responsive to nitric oxide than the VICs on the stiff substrates [44]. Because of this intersection of mechanical and chemical signaling, the effects of microenvironmental mechanics on valve cell phenotypes will be discussed in more detail in the following paragraphs.

2.3.4 Substrate composition

Proteins and other bioactive factors present on a substrate surface have the potential to influence VIC behavior and differentiation. This influence has been studied extensively in 2D cultures by seeding VICs on surfaces coated with a variety of peptides and relevant extracellular proteins. For instance, a study comparing peptide coatings on the degree to which they influence calcific nodule formation indicated that arginylglycylaspartic acid

(RGD)-containing proteins (including fibrin, collagen, fibronectin, and laminin) are the most procalcific [45]. In contrast, VICs cultured on HA demonstrated a reduction in calcific nodule formation [46]. Furthermore, in 2D in vitro studies, fibrin, laminin, and heparin coatings on tissue culture plastic spurred increased VIC calcific nodule production as compared with collagen and fibronectin, on which VICs produced few nodules [47,48].

2.3.5 Microenvironmental mechanics and geometry

Substrate stiffness and geometry have been identified as factors that significantly influence aspects of VIC phenotypic behavior, including cell activation, calcification potential, and proliferation. To investigate this phenomenon, substrates for VIC culture have been prepared using hydrogel polymers in differing molecular weights and weight fractions. Several such investigations have demonstrated that VIC αSMA expression was upregulated and that VICs displayed an altered, more branching structure on increasingly stiff substrates [49–51]. With respect to specific ranges of substrate stiffness, VICs grown in 2D atop substrates stiffer than 15 kPa became activated with a sprouting morphology and many αSMA-positive stress fibers, whereas VICs grown on soft substrates (<15 kPa) exhibited a more cuboidal morphology with few to no stress fibers [52].

VICs grown on relatively compliant 2D polyethylene glycol (PEG) hydrogels, both coated with ECM proteins and uncoated, demonstrated decreased activation and calcification capabilities as compared with VICs grown on traditional tissue culture plastic, indicating that stiffness of a substrate positively correlates with VIC osteogenic potential [50]. Substrate stiffness also altered the proliferation capabilities of VICs [49]. More recently, substrates with real time stiffness tunability capabilities have been utilized to more closely examine VIC response and differentiation plasticity in response to varying mechanical environments. Using this approach, VICs demonstrated increased osteogenic potential when cultured in 2D on a gradually stiffening environment, with reduced osteogenic capabilities when the substrate was temporally brought back to baseline stiffness [53].

The geometry of the cellular environment is another factor that influences VIC phenotype. VICs cultured within relatively stiff 3D HA/gelatin composite hydrogels demonstrate a more quiescent phenotype as compared with those in more compliant 3D hydrogels. This result is particularly interesting because it is the opposite trend from what has been observed in 2D cultures, in which VICs were more activated on stiffer substrates [52]. In a different study where VICs were either encapsulated within collagen I gels or grown atop collagen I–coated tissue culture plastic, the VICs cultured in the 3D environment were less likely to demonstrate activation [54]. Taken together, it is clear that pericellular composition, mechanics, and adhesion (2D vs. 3D) work in concert to govern VIC function, although there is not yet consensus about the effects of specific 3D microenvironmental cues.

External stretch and shear mechanics are also highly influential factors in VIC behavior. Stretch alone has been shown to trigger calcification in VICs in vitro. Cyclic stretch bioreactors have been used to determine how different stretch magnitudes affect the calcification potential of VICs, with 10% stretch shown to promote calcification the most as compared with static and 20% stretch conditions [55]. The anisotropy of the valve is another important consideration. In a custom-built bioreactor, VICs were embedded within a round gel, containing a spiral spring around the outer perimeter to allow for anisotropic stretch [16]. Within this system, VICs were shown to orient along the long axis of anisotropy and remodel accordingly. Compared with isotropic molds in this system, VICs cultured in the anisotropic molds demonstrated an increase in apoptotic and proliferative capabilities. In another bioreactor system, VICs and VECs were cocultured in tubular molds exposed to shear stress. In response to this shear, VECs downregulated αSMA expression and dampened proliferation by the VICs [1]. In addition, VICs have demonstrated pronounced differences in behavior when seeded within substrates experiencing compressive versus tensile loading [56]. Notably, VICs in the tensile loading hydrogel formed a monolayer of cells on the substrate surfaces, whereas VICs in the compressive loading hydrogel remained quiescent and did not form a monolayer (Fig. 2.2). Overall, these studies illustrate how mechanics of the microenvironment vastly influence VIC behavior.

2.4 Role of extracellular matrix in heart valve biomechanics

To ensure unidirectional blood flow, heart valve leaflets must open and close completely. To promote this functional behavior, leaflets possess specific mechanical properties that allow for the appropriate amount of tensile strength, compression resistance, and elasticity. These essential mechanical properties are made possible by the ECM structure of the valve leaflets.

Each of the three leaflets of the aortic valve has a distinct, trilaminar ECM structure, comprising the fibrosa, spongiosa, and ventricularis layers. The fibrosa faces the aortic side of the valve and predominantly consists of bundles of fibrillar collagen, aligned in the circumferential direction; the fibrosa serves as the tensile strength—bearing component of the leaflet. The fibrosa also has a corrugated structure, which allows it to unfold during valve opening. The spongiosa is the middle layer of the valve leaflet and is rich in glycosaminoglycans (GAGs) and proteoglycans (PGs), which impart compression resistance and flexibility. The ventricularis is the thinnest layer and comprises mainly of radially aligned elastin interspersed with a small amount of collagen; this layer provides the leaflet with the ability to recoil. Overall, collagen and elastin, respectively, comprise 50% and 13% of the leaflet in terms of dry weight [57]. Collagen, elastin, and GAGs/PGs function together to withstand the repetitive forces and pressures that the valve is subjected to through each cardiac cycle [58].

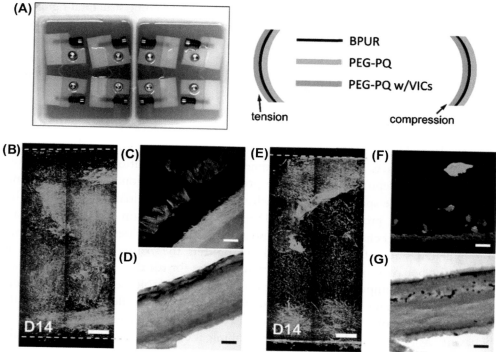

Figure 2.2 Aortic valvular interstitial cells (VICs) on tensile and compressive surfaces exhibit differential behavior. (A) VICs were seeded atop constructs that consisted of a biodegradable poly(ether ester urethane) urea (BPUR) embedded within polyethylene glycol (PEG) gels conjugated to peptide PQ allowing for cellular degradation (PEG-PQ). Cells were seeded on the side of bent constructs corresponding to tension or compression forces. Holders were used to secure gel in bent position while culturing. (B, C, D) Analysis of VICs seeded on tensile side. (B) Staining with f-actin (green) illustrates confluence of VICs after 14 days of culture on the surface (C) Cross-section demonstrating thick layer of cells atop the construct. (D) Hematoxylin and eosin (H&E) staining shows ECM production and cellular proliferation atop the BPUR construct. (E, F, G) Analysis of VICs seeded on compressive side. (E) Incomplete monolayer formation evidenced by f-actin staining. (F) Cross-section of gel demonstrating lack of VICs and ECM production. (G) H&E stain demonstrates low levels of cellular proliferation and ECM production as compared with the tensile surface. *(Credit: Adapted with permission from Puperi, Daniel S, et al. Electrospun polyurethane and hydrogel composite scaffolds as biomechanical mimics for aortic valve tissue engineering. ACS Biomaterials Science and Engineering 2016;2.9:1546—58.)*

Bundles of crimped collagen fibers are aligned in a network that is predominantly oriented to provide strength along the circumferential direction [59]. Collagen type I is the most abundant, comprising 74% of valvular collagen, whereas collagen type III comprises 24%. Collagen type I is predominantly localized in the fibrosa, whereas collagen type III is distributed throughout the three layers of the leaflet. The overall collagen network serves as the primary strength-bearing component of the leaflet, distributing critical stresses during both systole and diastole. This fibrous network is able to reduce peak stresses during diastole, as well as maintaining structural rigidity (and therefore limiting fluttering) of the leaflets during systole, as observed via computational modeling [60]. As pressure causes leaflets to coapt during diastole, the leaflet expands in the circumferential direction, leading to the collagen fibers becoming uncrimped and highly aligned. This directional alignment allows for increased stiffness and rigidity of the closed valve during diastole.

The spongiosa layer contains the majority of the highly hydrated GAGs and PGs within the valve. The most abundant GAG is HA, which is 60% of the total GAG content [61]. HA is a very large GAG and is strongly negatively charged, which consequently attracts a substantial volume of water. In addition to this biophysical attribute, HA is reported to influence proliferation, migration, and ECM assembly by VICs [62]. The spongiosa is also rich in the large interstitial PG versican, which consists of a core protein bound to 15—20 long chondroitin sulfate GAG chains. Like HA, the GAG chains on versican also attract substantial water content. The resulting hydrated state of the spongiosa layer bolsters the compressive resistance and viscoelastic behavior of the overall valve and allows for rearrangement of the collagenous and elastic components during the cardiac cycle [63]. Numerous other PGs, most notably biglycan and decorin, are present throughout the leaflet layers, where they perform diverse roles such as coordinating collagen fibrillogenesis and directing cell behavior. More comprehensive data describing the GAGs and PGs in heart valves are reported in work from the Grande-Allen group [61—64].

Elastin, in the form of elastic fibers, makes up 13% of the dry weight of the aortic valve and is found in the ventricularis and spongiosa of the leaflet [57,65]. In the ventricularis layer, elastic fibers are found in aligned sheets. These elastic fibers form a network that takes the form of a lattice structure in the spongiosa, which aids in connecting the ventricularis layer with the fibrosa. This elastic coupling of the leaflet layers provides the valve with a high degree of extensibility when it stretches radially during valve opening, as well as recoil capabilities for when the valve is closing.

The mitral valve leaflet has a similar trilaminar ECM structure to the aortic valve leaflet, with an elastin–rich atrialis layer (analogous to the ventricularis of the aortic valve) on the atrial side, a GAG–rich spongiosa layer in the interior, and a collagen–rich fibrosa layer on the ventricular side. That resemblance aside, there are several important structural differences between aortic and mitral valves. In contrast to the three symmetrical leaflets of the aortic valve, the mitral valve has two asymmetrical leaflets: the anterior leaflet and the posterior leaflet. Furthermore, to prevent the leaflets from prolapsing into the atrium due to the high-pressure environment of the ventricle, the leaflets of the mitral valve are tethered to the ventricular wall via chordae tendineae, tendon-like structures consisting of a collagen–rich core surrounded by an elastin sheath [64] (Fig. 2.3).

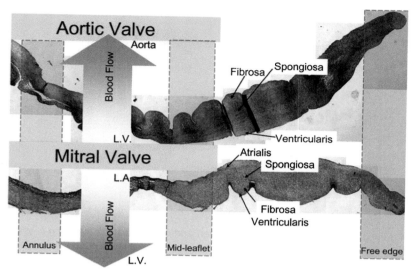

Figure 2.3 Histological anatomy of the aortic and mitral valve. Movat's pentachrome stain was used to distinguish collagen (saffron yellow), elastin (hematoxylin black), and PGs/GAGs (alcian blue) within the valves. *(Credit: Adapted with permission from Stephens, Elizabeth H., Chia-Kai Chu K., Jane Grande-Allen. Valve proteoglycan content and glycosaminoglycan fine structure are unique to microstructure, mechanical load and age: Relevance to an age-specific tissue-engineered heart valve. Acta Biomaterialia 2008;4.5:1148–1160.)*

2.5 Extracellular matrix remodeling in heart valve disease

The two most common valvular diseases are CAVD and mitral valve regurgitation. Over the course of both diseases, a dramatic disruption in cellular function and ECM integrity occurs. Though the precise mechanisms that cause these changes are still being investigated, it is known that a bidirectional cell-matrix relationship exists within the valves and dictates homeostasis and pathobiology. This relationship is superimposed on the biomechanical role of the ECM in guiding valve tissue mechanics as described in the previous section. The ECM plays just as important a role as a scaffold that transduces mechanical signals and influences cellular processes by controlling the availability of exogenous factors. These activities contribute to regulating how valvular cells develop and respond within the tissue, whether the valve is healthy or diseased.

2.5.1 Calcific aortic valve disease

CAVD impairs the ability of the aortic valve to open fully during ventricular systole. This disease is characterized by fibrotic thickening, angiogenesis, and development of mineralized lesions throughout the bulk of the leaflet (Fig. 2.4). Over the course of

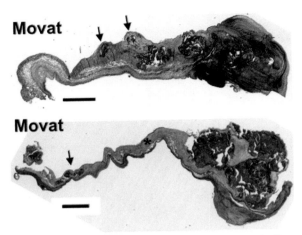

Figure 2.4 Movat's pentachrome stain on calcified aortic valve leaflets (collagen = yellow, elastin = black, and PGs/GAGs = blue). Relative to healthy aortic valve leaflets, CAVD-affected leaflets demonstrate increased fibrotic remodeling, increased cellular proliferation, increased GAG/PG deposition, and development of mineralized nodules throughout the bulk of the leaflet (nodules indicated by arrows). *(Credit: Adapted with permission from Stephens, Elizabeth H, et al. Differential proteoglycan and hyaluronan distribution in calcified aortic valves. Cardiovascular Pathology 2011;20.6:334–342.)*

the disease onset, the structured ECM of the leaflet is increasingly compromised. As described above, fibrillar collagen (namely types I and III) is highly aligned and is primarily localized to the fibrosa layer in normal valves. In CAVD-affected valves, these fibers become highly disorganized, and the collagen content of valves increases due to increased activity in matrix metalloproteinases, which spur fibrotic remodeling [60,66]. Together with the calcific mineral deposition, this fibrosis increases the overall stiffness of the leaflets and the pericellular VIC microenvironment. In addition, certain types of collagen and collagen fragments have been shown to influence activation in VICs, which could contribute to progression of CAVD within the tissue [59].

The quantity of PGs and GAGs (including versican, biglycan, decorin, and HA) increases among all three layers of CAVD-affected valves [61]. In late-stage CAVD, GAGs and PGs often localize around calcific nodules that form throughout the bulk of the leaflets. These components could play a role in the pathological remodeling of the valve via accumulation of lipoproteins and inflammatory cells, caused by the relatively highly negative charge of GAGs. PGs are also capable of sequestering specific soluble proteins, which can influence spatial concentrations of specific growth factors and cytokines. In addition, PGs and GAGs can interact directly with VICs to influence their behavior, for instance, treating VICs with decorin was shown to activate the TGFβ pathway [67].

Relative to healthy valves, the elastin content in CAVD-affected valves is reduced, and the elastic fibers are fragmented. These structural alterations also contribute to an increase in overall tissue stiffness. Elastin fragments have been implicated in promoting VIC activation and calcification via induction of ALPL activity and expression of RunX2, an osteogenic transcription factor [65].

2.5.2 Mitral valve regurgitation

Mitral valve regurgitation occurs when the two leaflets of the mitral valve cannot properly coapt to close the valve during ventricular systole, leading to blood leaking backward into the atrium. Broadly, mitral regurgitation can be categorized into degenerative mitral regurgitation, which is caused by myxomatous remodeling of the valve tissue, and functional mitral regurgitation, in which fibrotic remodeling of the mitral valve is induced by pathological remodeling of the surrounding heart muscle, which distorts the geometry of the mitral valve so that it cannot close properly.

In degenerative mitral regurgitation, the mitral valve undergoes myxomatous (myxoid) remodeling, which thickens and enlarges the tissue while decreasing the stiffness and increasing the extensibility of both the leaflets and the chordae tendineae [68]. Various changes in cell phenotype and ECM composition contribute to these mechanical changes during myxomatous remodeling (Fig. 2.5). Myxoid leaflets display enlargement of the PG- and GAG-rich spongiosa layer, along with disruption of both the collagen-rich fibrosa layer and the elastin-rich atrialis layer arising from the overexpression of collagen-degrading enzymes such as MMP-1 and MMP-13 by aVICs in the leaflet [69]. Compared with normal leaflets, myxoid leaflets contain significantly higher levels of PGs, particularly decorin, biglycan, and versican [64]. Interestingly, myxomatous remodeling induces significant changes in the GAG composition of the mitral valve. In myxoid chordae, the proportions of chondroitin-6-sulfate and HA are elevated and the proportion of dermatan-4-sulfate is reduced compared with normal chordae, with similar trends occurring to a lesser extent in myxoid leaflets [70]. Overall, these ECM changes weaken the mitral valve and allow it to prolapse into the left atrium, leading to regurgitation.

Functional mitral regurgitation is initially caused by left ventricular conditions such as myocardial infarction, congestive heart failure, and dilated cardiomyopathy. In these conditions, the geometry of the heart changes in a manner that stretches the annulus and displaces the papillary muscles (and attached chordae tendineae) apically and laterally, preventing the mitral valve from closing properly [71]. In contrast with degenerative mitral regurgitation (which is marked by weakening of the valve tissue), functional mitral regurgitation increases the stiffness and decreases the extensibility of both the leaflets and the chordae of the mitral valve [72]. These changes in valvular mechanical properties closely parallel the fibrotic ECM remodeling that is observed in the mitral valve during end-stage congestive heart failure. Specifically, mitral valve leaflets from heart failure patients display higher levels of collagen and GAGs and increased cellularity compared with leaflets from age-matched normal patients [73].

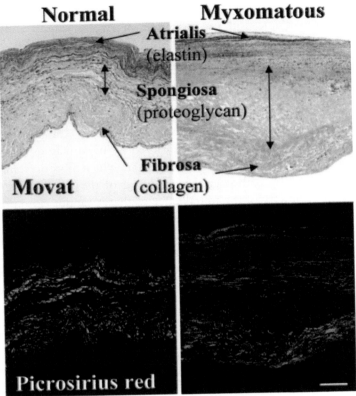

Figure 2.5 Histological analysis of normal and myxomatous mitral valve leaflets. Movat's pentachrome staining (collagen = yellow, elastin = black, and PGs/GAGs = blue) shows disrupted layered architecture in myxomatous valves, with increased fibrotic remodeling and expanded spongiosa layer. Picrosirius red stain allows for imaging of collagen fiber arrangement in mitral valves, with myxomatous valves demonstrating lower birefringence, indicating overall disruption in collagen fiber alignment and density. *(Credit: Adapted with permission from Rabkin, Elena, et al. Activated interstitial myofibroblasts express catabolic enzymes and mediate matrix remodeling in myxomatous heart valves. Circulation 2001; 104.21:2525—2532.)*

Recently, novel bioreactors have been developed to isolate the effects of mechanical forces during tissue remodeling in both degenerative and functional mitral regurgitation. One such system is the Rice University Flow Loop System, which was designed by our group to perform long-term dynamic organ culture of living heart valves while enabling the hemodynamics of mitral regurgitation to be recapitulated by independently manipulating the position of the papillary muscles and valve annulus. Using this system, it was demonstrated that the mechanical changes associated with both degenerative and functional mitral regurgitation are sufficient to induce the characteristic ECM remodeling of both conditions [74]. The same system was used to investigate whether the fibrotic remodeling that occurs during functional mitral regurgitation could potentially be reversed by restoring the valve to a nonregurgitant geometry. The results indicated that reversal of fibrotic remodeling occurs in a heterogeneous manner. Although the levels of fibrotic markers decreased in the posterior leaflet after the valve was returned to its preregurgitant geometry, the anterior leaflet continued to undergo fibrotic remodeling [75]. Together, these studies show that functional mitral regurgitation is an active process involving fibrotic ECM remodeling, rather than being a purely passive consequence of altered ventricular geometry as was historically believed.

2.6 Mechanobiology considerations for tissue engineering atrioventricular and semilunar valves

2.6.1 Tissue engineered heart valve replacements

Tissue engineering is an emerging strategy for treating atrioventricular and semilunar valve diseases such as CAVD and mitral regurgitation. Currently, severe heart valve disease is treated by replacing the valve with a mechanical or bioprosthetic heart valve. However, both of these classes of devices can be problematic. Mechanical heart valves, which are made of plastic or metal, can elicit thrombogenic responses that must be managed through lifelong administration of anticoagulative medication. Bioprosthetic valves, which are derived from decellularized animal tissues, are less durable than native valves as they do not contain living cells and are incapable of self-repair. Over time, the repetitive forces in the environment cause cumulative damage to the replacement valve, limiting its useful life span. Moreover, neither mechanical nor bioprosthetic valves experience growth or remodeling, which is a major problem for pediatric patients who receive valve replacements in infancy, because they must undergo multiple valve replacement surgeries as their hearts grow. Therefore, to sidestep the issues with both mechanical and bioprosthetic heart valves, living tissue-engineered heart valves (TEHVs) are emerging as a promising new therapeutic strategy for diseases that require valve replacement [76].

Although there is a great variety of natural and synthetic scaffold materials for heart valve tissue engineering, all TEHV replacements share similar design goals. The ideal TEHV replacement should be biocompatible, nonimmunogenic, and nonthrombotic. Its mechanical properties should be comparable with native valve tissue, to ensure long-term functionality and durability. Finally, the scaffold should provide appropriate adhesion sites for cells to populate the material and transform it into a living structure.

Decellularized allograft and xenograft heart valves are popular scaffolds for TEHVs because they contain bioactive ligands that can support cell adhesion (and ideally replicate native valve cell mechanobiology) and because their mechanical properties are similar to native tissue. The goal of decellularization is to maintain the integrity of the ECM while removing donor cells and antigenic material. This goal can be accomplished by treating the tissue with cocktails of detergents and/or enzymes [77]. Once the valve is decellularized, it can be repopulated by cells in vitro by seeding cells onto the scaffold in a bioreactor [78]. Alternatively, the decellularized valve can be chemically modified to enable in situ recellularization by endogenous cells [79].

Despite their advantages, TEHVs prepared from decellularized scaffolds have drawbacks as well. Because they are derived from animal or human tissues, there can be variability between valves. Furthermore, allograft-derived scaffolds are limited by the finite supply of cadaver tissue. Xenograft-derived scaffolds would avoid this constraint because they can be harvested from animals, but a significant barrier to clinical transplantation of xenografts is the presence of the α-gal epitope. This epitope is an oligosaccharide moiety that is abundantly present in the glycoproteins and glycolipids of nonprimate tissues. However, in primates, numerous antibodies against α-gal are present in the circulation, leading to hyperacute immune rejection of grafts containing this epitope [80]. Although it is possible to denature the α-gal epitope through detergent-based and enzymatic washes, it remains difficult to do so in a way that preserves the bioactivity of desirable ligands and retains the recellularization capacity of the scaffold. Furthermore, decellularization dramatically affects the mechanical properties of the valve leaflet, which can be problematic for TEHV durability. In one study of aortic valves that were decellularized using various detergent-based and enzymatic methods, the decellularized leaflets displayed an 80% decrease in stiffness and a 100% increase in extensibility compared with native leaflets [81].

Alternatively, rationally designed synthetic materials can be used as TEHV scaffolds. In particular, several groups are developing TEHV scaffolds consisting of textiles embedded within bioactive hydrogels. In these composite scaffolds, the hydrogel component provides adhesion sites for cells, while the textile component provides mechanical strength to the construct. For example, one group embedded an anisotropic polyacrylonitrile mesh inside a VIC-laden gelatin-HA hydrogel, obtaining a construct that displayed good cell morphology and comparable mechanical properties to the native aortic valve leaflet [82]. Similar studies have been conducted using bioresorbable materials such as poly(ether ester urethane) encapsulated within a VIC-laden PEG hydrogel [56].

Both in vitro and in situ cellularization methods have advantages and disadvantages. In vitro cell seeding methods allow TEHVs to be preconditioned with appropriate ECM, enabling them to function immediately on implantation. However, these cell-laden constructs are less shelf-stable than are polymer-only materials that are designed to promote cellularization in situ. To overcome the limitations of both methods, some investigators are combining in vitro ECM conditioning with in situ cell seeding in a three-step workflow. First, cells are cultured on a fast-degrading polymer scaffold in a bioreactor, to allow them to synthesize ECM. Then, the valve is decellularized to create an off-the-shelf, stable product. Finally, the TEHVs is implanted and recellularized in situ. A consortium of European researchers has used this strategy to develop a pulmonary TEHV replacement—the LifeValve—that has yielded promising results in preliminary benchtop and animal trials [83].

Percutaneous valve replacement is another surgical strategy that can reduce the risks of valve replacement for patients who cannot tolerate open heart surgery. To date, traditional bioprosthetic valves such as the Edwards SAPIEN valve and the Medtronic CoreValve have been implanted percutaneously in over 100,000 patients [84]. Recently, percutaneous delivery methods have been developed for TEHVs as well. For example, the LifeValve group has reported that their tissue-engineered construct is compatible with existing percutaneous surgical methods such as transapical and transvenous insertion [85]. As TEHVs continue to improve, they should be designed with percutaneous deliverability in mind. It will also be important to assess how the crimping process required to insert the TEHV into the delivery catheter impacts the resulting biomechanics and mechanobiology of the newly grown valve.

2.6.2 Innovative in vitro models

As previously discussed, gene expression related to cell structure and motility is influenced by the dimensionality of the substrate. In the context of valves, VICs grown within 3D matrices demonstrated decreased activation as compared with VICs grown on tissue culture plastic. Moreover, VICs grown in 2D on tissue culture plastic demonstrate significant differences in genetic expression compared with native VICs. Because of these differences, investigators are moving toward engineering models that better mimic the native tissue environment. Furthermore, bioreactors that subject samples to physiologically relevant forces are critical toward constructing a more accurate understanding of valvular response. To this end, bioreactors and 3D in vitro model systems that incorporate relevant mechanical and biochemical cues are currently being developed to elucidate specific targets for novel treatment strategies.

The most widely used medium for development of these 3D systems is polymer hydrogels, both synthetic and natural. Natural hydrogels used in these studies include collagen, gelatin, and modified HA. A widely used synthetic hydrogel is PEG, which lends itself to cell culture applications due to the fact that it is easily modifiable with peptide sequences and exogenous factors. Additionally, PEG is bioinert, making it easy to discern influential components on cellular behavior. A heterogeneous coculture system to model the aortic valve was developed using modified PEG hydrogels. This system was designed such that VICs were embedded in 3D in a PEG hydrogel modified with peptide sequences that encouraged VIC attachment and remodeling of the matrix. VECs were then seeded on top of the construct, mimicking the endothelial layer present on the native aortic leaflet [86]. This model allows for easy customization in terms of ligand conjugation and ability to tune bulk mechanical properties of the system to accurately recapitulate the cellular heterogeneity of the valve leaflet.

To understand the underpinnings of valvular disease, it is important to elucidate VIC–VEC interactions. To assess these interactions without the potentially confounding influence of a scaffold material, a method was developed to create a layered, VIC–VEC coculture in a scaffold-free system. Using magnetic nanoparticles to aggregate individual populations of VICs or VECs, cells were grown in specific clusters. Homogenous clusters of VICs and VECs were then stacked using a customized magnetic pen to create a spatially organized aggregate. In the 3D layered VIC–VEC aggregate, VICs demonstrated a downregulation in αSMA compared with VICs alone, suggesting that the VECs aid in maintaining the quiescent state of the VICs [87]. This model was useful in parsing out cellular-specific behavior that occurs absent of matrix.

Paper has recently become a more widely used material in the context of cell culture, for a variety of reasons. Paper is a readily available and relatively inexpensive material that has the capacity for directing fluid movement. In microfluidics applications, cellulose has been used extensively because of its porosity and inertness. The three-dimensional, fibrous structure of the paper has the capability to mimic the cellular microenvironment. Additionally, because paper can be easily chemically and physically modified, there is a great potential for forming a variety of culture conditions [88]. A layered filter paper model has the capacity to model substantial variability in 3D structure. This platform was originally developed to use paper as a mechanical support for cells cultured in ECM gels, ultimately utilizing this stacked cell culture platform to analyze hypoxia's effect on cancer cells [89]. This idea was built on for use in culturing VICs, by seeding VICs in collagen gels in this system and then analyzing migration and expression patterns in response to hypoxia [90]. This system is advantageous in that it can recapitulate the heterogeneity of the valve environment via precise control over geometry, matrix composition, and cellular concentrations within each discrete layer of a multilayered tissue mimic (Fig. 2.6).

Moreover, investigators have developed bioreactors for mechanobiology organ culture studies on whole valve tissues and engineered valve tissue mimics. These tissue-based approaches offer a complementary strategy that can be applied in conjunction with the 3D models described above to examine how mechanical forces are related to valvular response. A range of mechanical stimuli can be applied to valve tissues using various custom-designed systems; for example, one report applied pathophysiologically relevant shear stresses to mimic the side-specific shears experienced by VECs [91]. In another study, mitral valves were mounted in a flow bioreactor system that could manipulate valve positioning to mimic normal, prolapsed, or functional regurgitation geometry (Fig. 2.7). From this, it was possible to analyze ECM remodeling that occurs in mitral valve disease [74]. Additionally, this system could be used to assess whether surgically altering the mitral valve geometry would result in reversal of the pathological ECM remodeling [75]. By culturing valve cells within a 3D scaffold, it is also possible to test the resulting neotissue within bioreactor systems and thereby gain insight into the

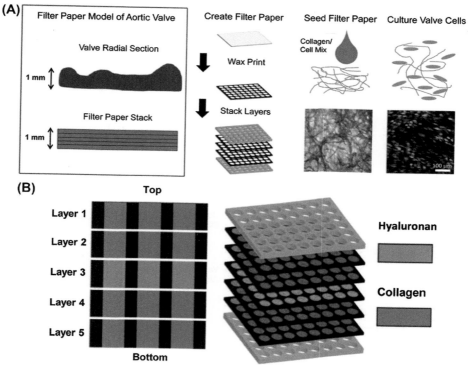

Figure 2.6 A layered, in vitro filter paper cell culture system has been used toward modeling the heterogeneous valvular ECM environment. (A) Valve cells mixed with specific prepolymers are seeded on sample regions delineated by wax on individual filter paper sheets. After the prepolymer solidifies, paper layers can be stacked and compressed to create monolithic gel constructs. Additionally, individual sheets can be stacked to mimic the thickness of the native valve leaflet. (B) Discrete layers within constructs can contain differing gel types (i.e., collagen, HA), allowing for recapitulation of native or diseased valve ECM. *(Credit: Adapted with permission from Sapp, Matthew C., et al. Multilayer three-dimensional filter paper constructs for the culture and analysis of aortic VICs. Acta Biomaterialia 2015; 13:199–206.)*

mechanical behavior of VICs under different stimuli. One elegant system cast VIC–laden collagen gels within a spring coiled around the periphery of an elastomeric membrane, allowing the entire system to be stretched once the gel had set [92]. In this scheme, the scaffold serves predominantly as a means of transducing mechanical stretch to the cells.

2.7 Future directions

Although the last two decades have witnessed an explosion in knowledge of heart valve mechanobiology, much remains to be learned. As described above, there is need for clarity in the broad range of results that have been shown when substrate stiffness is varied. It will be important to reconcile the findings from 2D and 3D studies and to

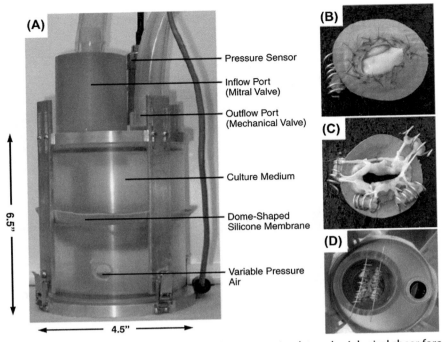

Figure 2.7 The Rice University Flow Loop System bioreactor simulates physiological shear forces and hydrostatic pressure pulses, with high tunability in regard to valve geometry. (A) Ventricular chamber of the system operates via pulses of pressurized air on a silicon membrane. This drives media through the mitral valve samples and a mechanical aortic valve mimic. (B–D) Attachment configuration for mitral valves. (B) Atrial side of a whole porcine mitral valve is sutured onto a ring to secure the annulus. (C) Papillary muscles are secured to steel coils, allowing for more precise geometrical configuration in the bioreactor. (D) Geometry of papillary muscles in relation to mitral valve is secured via attachment of coils to steel rods on the ventricular chamber. Placement of papillary muscles affects alignment of valve leaflets. To allow for finer control of the papillary muscle placement, subsequent designs of the system utilized plastic holders onto which the muscles were sutured and the holders were then placed in the correct location using a screw and washer system in place of the steel rods. *(Credit: Adapted with permission from Gheewala, Nikhil K., Jane Grande-Allen. Design and mechanical evaluation of a physiological mitral valve organ culture system. Cardiovascular Engineering and Technology 2010;1.2: 123–131.)*

elucidate the influence of scaffold choice on valve cell behavior. In addition, many of these mechanobiology studies have been performed using valve cells harvested from healthy young adult animals. Pediatric-aged animals and older animals will be essential to examine for future studies of the mechanobiology of congenital heart defects and adult valve disease, respectively. There is also much work to be done understanding in vivo mechanobiology, which can be performed with novel animal technologies such as conditional knockout mice. These models will provide insight into normal valve mechanobiology, remodeling, disease, and surgical valve repair. Finally, the majority of valve mechanobiology investigations examine the effects of mechanical stimulation on a very limited set of cell phenotypic characteristics. Advances in genomic and proteomic approaches will allow for a fuller appreciation for the complex cellular responses. Together, these novel methods for studying heart valve mechanobiology will improve clinical outcomes for the millions of patients who are affected by heart valve disease worldwide.

References

[1] Rutkovskiy A, Malashicheva A, Sullivan G, Bogdanova M, Kostareva A, Stensløkken K, Fiane A, Vaage J. Valve interstitial cells: the key to understanding the pathophysiology of heart valve calcification. J Am Heart Assoc 2017;6:e006339.

[2] Liu AC, Joag VR, Gotlieb AI. The emerging role of valve interstitial cell phenotypes in regulating heart valve pathobiology. Am J Pathol 2007;171:1407—18.

[3] Liskova J, Hadraba D, Filova E, Konarik M, Pirk J, Jelen K, Bacakova L. Valve interstitial cell culture: production of mature type I collagen and precise detection. Microsc Res Tech 2017;80:936—42.

[4] Filip DA, Radu A, Simionescu M. Interstitial cells of the heart valves possess characteristics similar to smooth muscle cells. Circ Res 1986;59:310—20.

[5] Lester WM, Damji AA, Gedeon I, Tanaka M. Interstitial Cells From the Atrial and Ventricular Sides of the Bovine Mitral Valve Respond Differently to Denuding Endocardial Injury. In Vitro Cellular & Developmental Biology 29A; 1993. p. 41—50.

[6] Taylor PM, Allen SP, Yacoub MH. Phenotypic and functional characterization of interstitial cells from human heart valves, pericardium and skin. J Heart Valve Dis 2000;9:150—8.

[7] Shapero K, Wylie-Sears J, Levine RA, Mayer JE, Bischoff J. Reciprocal interactions between mitral valve endothelial and interstitial cells reduce endothelial-to-mesenchymal transition and myofibro-blastic activation. J Mol Cell Cardiol 2015;80:175—85.

[8] Kaden JJ, Dempfle C-E, Grobholz R, Fischer CS, Vocke DC, Kiliç R, Sarikoç A, Piñol R, Hagl S, Lang S, et al. Inflammatory regulation of extracellular matrix remodeling in calcific aortic valve stenosis. Cardiovasc Pathol 2005;14:80—7.

[9] Freeman RV. Spectrum of calcific aortic valve disease: pathogenesis, disease progression, and treatment strategies. Circ Res 2005;111:3316—26.

[10] Wang H, Sridhar B, Leinwand LA, Anseth KS. Characterization of cell subpopulations expressing progenitor cell markers in porcine cardiac valves. PLoS One 2013;8:e69667.

[11] Rajamannan NM, Subramaniam M, Rickard D, Stock SR, Donovan J, Springett M, Orszulak T, Fullerton DA, Tajik AJ, Bonow RO, et al. Human aortic valve calcification is associated with an oste-oblast phenotype. Circ Res 2003;107:2181—4.

[12] Hjortnaes J, Goettsch C, Hutcheson JD, Camci-Unal G, Lax L, Scherer K, Body S, Schoen FJ, Kluin J, Khademhosseini A, et al. Simulation of early calcific aortic valve disease in a 3D platform: a role for myofibroblast differentiation. J Mol Cell Cardiol 2016;94:13—20.

[13] Yu Z, Seya K, Daitoku K, Motomura S, Fukuda I, Furukawa K-I. Tumor necrosis factor- accelerates the calcification of human aortic valve interstitial cells obtained from patients with calcific aortic valve stenosis via the BMP2-Dlx5 pathway. J Pharmacol Exp Ther 2011;337:16—23.

[14] Merryman WD, Liao J, Parekh A, Candiello JE, Lin H, Sacks MS. Differences in tissue-remodeling potential of aortic and pulmonary heart valve interstitial cells. Tissue Eng 2007;13:2281—9.

[15] Tseng H, Puperi DS, Kim EJ, Ayoub S, Shah JV, Cuchiara ML, West JL, Grande-Allen KJ. Anisotropic poly(ethylene glycol)/polycaprolactone hydrogel—fiber composites for heart valve tissue engineering. Tissue Eng 2014a;20:2634—45.

[16] Gould RA, Chin K, Santisakultarm TP, Dropkin A, Richards JM, Schaffer CB, Butcher JT. Cyclic strain anisotropy regulates valvular interstitial cell phenotype and tissue remodeling in three-dimensional culture. Acta Biomater 2012;8:1710—9.

[17] Ferdous Z, Jo H, Nerem RM. Differences in valvular and vascular cell responses to strain in osteogenic media. Biomaterials 2011;32:2885—93.

[18] Prabhavathi K. Role of biological sex in normal cardiac function and in its disease outcome — a review. J Clin Diagn Res 2014;8(8):BE01.

[19] McCoy CM, Nicholas DQ, Masters KS. Sex-related differences in gene expression by porcine aortic valvular interstitial cells. PLoS One 2012;7:e39980.

[20] Blenck CL, Harvey PA, Reckelhoff JF, Leinwand LA. The importance of biological sex and estrogen in rodent models of cardiovascular health and disease. Circ Res 2016;118:1294—312.

[21] Gould RA, Butcher JT. Isolation of valvular endothelial cells. J Vis Exp 2010;46.

[22] Butcher JT, Nerem RM. Valvular endothelial cells and the mechanoregulation of valvular pathology. Philos Trans R Soc Lond B Biol Sci 2007;362:1445—57.

[23] Paranya G, Vineberg S, Dvorin E, Kaushal S, Roth SJ, Rabkin E, Schoen FJ, Bischoff J. Aortic valve endothelial cells undergo transforming growth factor-β-mediated and non-transforming growth factor-β-mediated transdifferentiation in vitro. Am J Pathol 2001;159:1335—43.

[24] Paruchuri S, Yang J-H, Aikawa E, Melero-Martin JM, Khan ZA, Loukogeorgakis S, Schoen FJ, Bischoff J. Human pulmonary valve progenitor cells exhibit endothelial/mesenchymal plasticity in response to VEGF-A and TGFβ2. Circ Res 2006;99:861—9.

[25] Monaghan MG, Linneweh M, Liebscher S, Van Handel B, Layland SL, Schenke-Layland K. Endocardial-to-mesenchymal transformation and mesenchymal cell colonization at the onset of human cardiac valve development. Development 2016;143:473—82.

[26] Pardali E, Sanchez-Duffhues G, Gomez-Puerto MC, ten Dijke P. TGF-β-induced endothelial-mesenchymal transition in fibrotic diseases. Int J Mol Sci 2017;18(10):2157.

[27] Bradley JR. TNF—mediated inflammatory disease. J Pathol 2008;214:149—60.

[28] Chen G. TNF-R1 signaling: a beautiful pathway. Science 2002;296:1634—5.

[29] Coté N, Mahmut A, Bosse Y, Couture C, Pagé S, Trahan S, Boulanger M-C, Fournier D, Pibarot P, Mathieu P. Inflammation is associated with the remodeling of calcific aortic valve disease. Inflammation 2013;36:573—81.

[30] Galeone A, Paparella D, Colucci S, Grano M, Brunetti G. The role of TNF-α and TNF superfamily members in the pathogenesis of calcific aortic valvular disease. Sci World J 2013;2013.

[31] Mahler GJ, Farrar EJ, Butcher JT. Inflammatory cytokines promote mesenchymal transformation in embryonic and adult valve endothelial cells. Arterioscler Thromb Vasc Biol 2013;33:121—30.

[32] Farrar EJ, Butcher JT. Heterogeneous susceptibility of valve endothelial cells to mesenchymal transformation in response to TNFα. Ann Biomed Eng 2014;42:149—61.

[33] Derynck R, Zhang YE. Smad-dependent and Smad-independent pathways in TGF-β family signalling. Nature 2003;425:577—84.

[34] Meng X, Nikolic-Paterson DJ, Lan HY. TGF-β: the master regulator of fibrosis. Nat Rev Nephrol 2016;12:325—38.

[35] Rosenkranz S. TGF-β1 and angiotensin networking in cardiac remodeling. Cardiovasc Res 2004;63:423—32.

[36] Liu AC, Gotlieb AI. Transforming growth factor-β regulates in vitro heart valve repair by activated valve interstitial cells. Am J Pathol 2008;173:1275—85.

[37] Krishnamurthy VK, Stout AJ, Sapp MC, Matuska B, Lauer ME, Grande-Allen KJ. Dysregulation of hyaluronan homeostasis during aortic valve disease. Matrix Biol 2017;62:40—57.

[38] Waxman AS, Kornreich BG, Gould RA, Sydney Moïse N, Butcher JT. Interactions between TGFβ1 and cyclic strain in modulation of myofibroblastic differentiation of canine mitral valve interstitial cells in 3D culture. J Vet Cardiol 2012;14:211—21.

[39] Ignarro LJ. Nitric oxide. A novel signal transduction mechanism for transcellular communication. Hypertension 1990;16:477—83.

[40] Tousoulis D, Kampoli AM, Tentolouris Nikolaos Papageorgiou C, Stefanadis C. The role of nitric oxide on endothelial function. Curr Vasc Pharmacol 2012;10(1):4—18.

[41] Yang Y-M, Huang A, Kaley G, Sun D. eNOS uncoupling and endothelial dysfunction in aged vessels. Am J Physiol Heart Circ Physiol 2009;297:H1829—36.

[42] Miller JD, Chu Y, Brooks RM, Richenbacher WE, Peña-Silva R, Heistad DD. Dysregulation of antioxidant mechanisms contributes to increased oxidative stress in calcific aortic valvular stenosis in humans. J Am Coll Cardiol 2008;52:843—50.

[43] Richards J, El-Hamamsy I, Chen S, Sarang Z, Sarathchandra P, Yacoub MH, Chester AH, Butcher JT. Side-specific endothelial-dependent regulation of aortic valve calcification: interplay of hemodynamics and nitric oxide signaling. Am J Pathol 2013;182:1922—31.

[44] Gould ST, Matherly EE, Smith JN, Heistad DD, Anseth KS. The role of valvular endothelial cell paracrine signaling and matrix elasticity on valvular interstitial cell activation. Biomaterials 2014;35: 3596—606.

[45] Wu Y, Jane Grande-Allen K, West JL. Adhesive peptide sequences regulate valve interstitial cell adhesion, phenotype and extracellular matrix deposition. Cell Mol Bioeng 2016;9:479—95.

[46] Duan B, Hockaday LA, Kapetanovic E, Kang KH, Butcher JT. Stiffness and adhesivity control aortic valve interstitial cell behavior within hyaluronic acid based hydrogels. Acta Biomater 2013;9: 7640—50.

[47] Benton JA, Kern HB, Anseth KS. Substrate properties influence calcification in valvular interstitial cell culture. J Heart Valve Dis 2008;17:689—99.

[48] Ulloa Severino L, Santoro R, Pesce M, Casalis L, Scaini D. Substrate chemistry and morphology influence the valvular interstitial cells mechanobiology. Biophys J 2017;112:437a.

[49] Coombs KE, Leonard AT, Rush MN, Santistevan DA, Hedberg-Dirk EL. Isolated effect of material stiffness on valvular interstitial cell differentiation. J Biomed Mater Res A 2017;105:51—61.

[50] Gould ST, Anseth KS. Role of cell-matrix interactions on VIC phenotype and tissue deposition in 3D PEG hydrogels: VIC phenotype and tissue deposition in 3D PEG hydrogels. J Tissue Eng Regenerat Med 2016;10:E443—53.

[51] Masters KS, Shah DN, Leinwand LA, Anseth KS. Crosslinked hyaluronan scaffolds as a biologically active carrier for valvular interstitial cells. Biomaterials 2005;26:2517—25.

[52] Throm Quinlan Angela M, Billiar Kristen L. Investigating the role of substrate stiffness in the persistence of valvular interstitial cell activation. J Biomed Mater Res Part A 2012;100A:2474—82.

[53] Ma H, Killaars AR, DelRio FW, Yang C, Anseth KS. Myofibroblastic activation of valvular interstitial cells is modulated by spatial variations in matrix elasticity and its organization. Biomaterials 2017;131: 131—44.

[54] Yip CYY, Chen J-H, Zhao R, Simmons CA. Calcification by valve interstitial cells is regulated by the stiffness of the extracellular matrix. Arterioscler Thromb Vasc Biol 2009;29:936—42.

[55] Hutcheson JD, Venkataraman R, Baudenbacher FJ, David Merryman W. Intracellular Ca^{2+} accumulation is strain-dependent and correlates with apoptosis in aortic valve fibroblasts. J Biomech 2012;45:888—94.

[56] Puperi DS, Kishan A, Punske ZE, Wu Y, Cosgriff-Hernandez E, West JL, Grande-Allen KJ. Electrospun polyurethane and hydrogel composite scaffolds as biomechanical mimics for aortic valve tissue engineering. ACS Biomater Sci Eng 2016;2:1546—58.

[57] Vesely I. The role of elastin in aortic valve mechanics. J Biomech 1997;31:115—23.

[58] Gupta V, Werdenberg JA, Lawrence BD, Mendez JS, Stephens EH, Grande-Allen KJ. Reversible secretion of glycosaminoglycans and proteoglycans by cyclically stretched valvular cells in 3D culture. Ann Biomed Eng 2008;36:1092—103.

[59] Rock CA, Han L, Doehring TC. Complex collagen fiber and membrane morphologies of the whole porcine aortic valve. PLoS One 2014;9:e86087.

[60] De Hart J, Peters GWM, Schreurs PJG, Baaijens FPT. Collagen fibers reduce stresses and stabilize motion of aortic valve leaflets during systole. J Biomech 2004;37:303—11.

[61] Stephens EH, Saltarrelli JG, Baggett LS, Nandi I, Kuo JJ, Davis AR, Olmsted-Davis EA, Reardon MJ, Morrisett JD, Grande-Allen KJ. Differential proteoglycan and hyaluronan distribution in calcified aortic valves. Cardiovasc Pathol 2011;20:334—42.

[62] Allison DD, Grande-Allen KJ. Hyaluronan: a powerful tissue engineering tool. Tissue Eng 2006;12(8): 2131—40.

[63] Stephens EH, Chu C-K, Grande-Allen KJ. Valve proteoglycan content and glycosaminoglycan fine structure are unique to microstructure, mechanical load and age: relevance to an age-specific tissue-engineered heart valve. Acta Biomater 2008;4:1148—60.

[64] Gupta V, Barzilla JE, Mendez JS, Stephens EH, Lee EL, Collard CD, Laucirica R, Weigel PH, Grande-Allen KJ. Abundance and location of proteoglycans and hyaluronan within normal and myxomatous mitral valves. Cardiovasc Pathol 2009;18:191—7.

[65] Tseng H, Grande-Allen KJ. Elastic fibers in the aortic valve spongiosa: a fresh perspective on its structure and role in overall tissue function. Acta Biomater 2011;7:2101—8.

[66] Eriksen HA, Satta J, Risteli J, Veijola M, Väre P, Soini Y. Type I and type III collagen synthesis and composition in the valve matrix in aortic valve stenosis. Atherosclerosis 2006;189:91—8.

[67] Hinton RB. Extracellular matrix remodeling and organization in developing and diseased aortic valves. Circ Res 2006;98:1431—8.

[68] Barber JE, Kasper FK, Ratliff NB, Cosgrove DM, Griffin BP, Vesely I. Mechanical properties of myxomatous mitral valves. J Thorac Cardiovasc Surg 2001;122:955—62.

[69] Rabkin E, Aikawa M, Stone JR, Fukumoto Y, Libby P, Schoen FJ. Activated interstitial myofibroblasts express catabolic enzymes and mediate matrix remodeling in myxomatous heart valves. Circ Res 2001;104:2525—32.

[70] Grande-Allen KJ, Griffin BP, Ratliff NB, Cosgrove DM, Vesely I. Glycosaminoglycan profiles of myxomatous mitral leaflets and chordae parallel the severity of mechanical alterations. J Am Coll Cardiol 2003;42:271—7.

[71] Kono T, Sabbah HN, Stein PD, Brymer JF, Khaja F. Left ventricular shape as a determinant of functional mitral regurgitation in patients with severe heart failure secondary to either coronary artery disease or idiopathic dilated cardiomyopathy. Am J Cardiol 1991;68:355—9.

[72] Grande-Allen KJ, Barber JE, Klatka KM, Houghtaling PL, Vesely I, Moravec CS, McCarthy PM. Mitral valve stiffening in end-stage heart failure: evidence of an organic contribution to functional mitral regurgitation. J Thorac Cardiovasc Surg 2005a;130:783—90.

[73] Grande-Allen KJ, Borowski AG, Troughton RW, Houghtaling PL, DiPaola NR, Moravec CS, Vesely I, Griffin BP. Apparently normal mitral valves in patients with heart failure demonstrate biochemical and structural derangements: an extracellular matrix and echocardiographic study. J Am Coll Cardiol 2005b;45:54—61.

[74] Connell PS, Azimuddin AF, Kim SE, Ramirez F, Jackson MS, Little SH, Grande-Allen KJ. Regurgitation hemodynamics alone cause mitral valve remodeling characteristic of clinical disease states in vitro. Ann Biomed Eng 2016;44:954—67.

[75] Connell PS, Vekilov DP, Diaz CM, Kim SE, Grande-Allen KJ. Eliminating regurgitation reduces fibrotic remodeling of functional mitral regurgitation conditioned valves. Ann Biomed Eng 2018; 46:670—83.

[76] Zhu AS, Grande-Allen KJ. Heart valve tissue engineering for valve replacement and disease modeling. Curr Opin Biomed Eng 2018;5:35—41.

[77] VeDepo MC, Detamore MS, Hopkins RA, Converse GL. Recellularization of decellularized heart valves: progress toward the tissue-engineered heart valve. J Tissue Eng 2017;8. 2041731417726327.

[78] Converse GL, Buse EE, Neill KR, McFall CR, Lewis HN, VeDepo MC, Quinn RW, Hopkins RA. Design and efficacy of a single-use bioreactor for heart valve tissue engineering. J Biomed Mater Res B Appl Biomater 2017;105:249—59.

[79] Jordan JE, Williams JK, Lee S-J, Raghavan D, Atala A, Yoo JJ. Bioengineered self-seeding heart valves. J Thorac Cardiovasc Surg 2012;143:201—8.

[80] Konakci KZ, Bohle B, Blumer R, Hoetzenecker W, Roth G, Moser B, Boltz-Nitulescu G, Gorlitzer M, Klepetko W, Wolner E, et al. Alpha-Gal on bioprostheses: xenograft immune response in cardiac surgery. Eur J Clin Investig 2005;35:17—23.

[81] Liao J, Joyce EM, Sacks MS. Effects of decellularization on the mechanical and structural properties of the porcine aortic valve leaflet. Biomaterials 2008;29:1065—74.

[82] Wu S, Duan B, Qin X, Butcher JT. Living nano-micro fibrous woven fabric/hydrogel composite scaffolds for heart valve engineering. Acta Biomater 2017;51:89—100.

[83] Sanders B, Loerakker S, Fioretta ES, Bax DJP, Driessen-Mol A, Hoerstrup SP, Baaijens FPT. Improved geometry of decellularized tissue engineered heart valves to prevent leaflet retraction. Ann Biomed Eng 2016;44:1061—71.

[84] Agarwal S, Tuzcu EM, Krishnaswamy A, Schoenhagen P, Stewart WJ, Svensson LG, Kapadia SR. Transcatheter aortic valve replacement: current perspectives and future implications. Heart 2015; 101:169—77.

[85] Spriestersbach H, Prudlo A, Bartosch M, Sanders B, Radtke T, Baaijens FPT, Hoerstrup SP, Berger F, Schmitt B. First percutaneous implantation of a completely tissue-engineered self-expanding pulmonary heart valve prosthesis using a newly developed delivery system: a feasibility study in sheep. Cardiovasc Interv Ther 2017;32:36—47.

[86] Puperi DS, Balaoing LR, O'Connell RW, West JL, Grande-Allen KJ. 3-Dimensional spatially organized PEG-based hydrogels for an aortic valve co-culture model. Biomaterials 2015;67:354—64.

[87] Tseng H, Balaoing LR, Grigoryan B, Raphael RM, Killian TC, Souza GR, Grande-Allen KJ. A three-dimensional co-culture model of the aortic valve using magnetic levitation. Acta Biomater 2014b;10:173—82.

[88] Ng K, Gao B, Yong KW, Li Y, Shi M, Zhao X, Li Z, Zhang X, Pingguan-Murphy B, Yang H, et al. Paper-based cell culture platform and its emerging biomedical applications. Mater Today 2017;20: 32—44.

[89] Derda R, Tang SKY, Laromaine A, Mosadegh B, Hong E, Mwangi M, Mammoto A, Ingber DE, Whitesides GM. Multizone paper platform for 3D cell cultures. PLoS One 2011;6:e18940.

[90] Sapp MC, Fares HJ, Estrada AC, Grande-Allen KJ. Multilayer three-dimensional filter paper constructs for the culture and analysis of aortic valvular interstitial cells. Acta Biomater 2015;13: 199—206.

[91] Sun L, Rajamannan NM, Sucosky P. Design and validation of a novel bioreactor to subject aortic valve leaflets to side-specific shear stress. Ann Biomed Eng 2011;39:2174—85.

[92] Farrar EJ, Pramil V, Richards JM, Mosher CZ, Butcher JT. Valve interstitial cell tensional homeostasis directs calcification and extracellular matrix remodeling processes via RhoA signaling. Biomaterials 2016;105:25—37.

CHAPTER 3

Epidemiology of heart valve disease

Geoffrey D. Huntley, Jeremy J. Thaden, Vuyisile T. Nkomo
Department of Cardiovascular Medicine, Mayo Clinic, Rochester, MN, United States

Contents

Abbreviations

AR Aortic regurgitation
AS Aortic stenosis
BAV Bicuspid aortic valve
CHD Congenital heart disease
IE Infective endocarditis
MR Mitral regurgitation
MS Mitral stenosis
RHD Rheumatic heart disease
VHD Valvular heart disease

3.1 Introduction

More people die annually from cardiovascular disease than from any other cause [1]. Although most frequently due to coronary pathology, there is a growing burden of valvular heart disease (VHD) in developed countries. Based on a large, US population-based study of 11,911 patients using systematic echocardiographic screening, the national prevalence of moderate or severe VHD, corrected for age and sex, is 2.5% and increases considerably with age [2]. With projected shifts to older and larger populations and

Principles of Heart Valve Engineering
ISBN 978-0-12-814661-3, https://doi.org/10.1016/B978-0-12-814661-3.00003-4

improvements in echocardiographic screening, VHDs are certainly an important and growing public health problem in developed countries. Worldwide, the majority of the morbidity and mortality attributable to VHD is due to rheumatic heart disease (RHD). *The 2015 Global Burden of Disease study estimates that acute rheumatic fever and subsequent RHD affects about 33.4 million people worldwide* [1,3]. However, the greatest burden of VHD in high-income countries is due to a degenerative disease process.

3.1.1 Acute rheumatic fever as a precursor to rheumatic heart disease

Acute rheumatic fever occurs weeks after a group A *Streptococcus* infection of the tonsillo-pharynx, usually in children, and molecular mimicry between the streptococcal M protein and cardiac protein leads to progressive valve inflammation and fibrosis [4]. In the past, RHD was the leading cause of VHD in all countries worldwide, and it remains the dominant cause of VHD in developing countries. However, for the past half-century, degenerative valve disease has displaced RHD as the most common cause of VHD in high-income countries. The reduced burden of RHD in economically developed nations is largely associated with improved living conditions and the effective use of antibiotics to treat rheumatic fever [5]. In 1944, Dr. TD Jones established the Jones criteria [6], and subsequent revisions have laid the groundwork for enhanced detection and subsequent treatment of the deadly disease. By the second half of the 20th century, RHD began to decline in developed countries. Incidence rates per 100,000 in Rochester, Minnesota, among 5- to 14-year olds dropped from 65 in the period from 1935 to 1949 to 41 from 1950 to 1964 and to 9 from 1965 to 1978 [7]. Aside from a few outbreaks in the 1980s [8], this trend continued into the 20th century (Fig. 3.1).

Rheumatic fever has been considered a disease of poverty, and most of the reduction in rheumatic fever in developed countries is due to improved living conditions, less

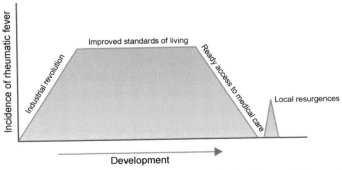

Figure 3.1 Changes in the incidence of rheumatic fever. The incidence of rheumatic fever peaked during the industrial revolution, likely due to overcrowding. However, with increasing living standards, better access to medical care, and widespread use of antibiotics, the incidence declined. Isolated outbreaks continue to occur. *(Credit: Fig. 3.1, Soler-Soler J, Galve E. Worldwide perspective of valve disease. Heart 2000;83:721—25.)*

overcrowding, and better hygiene causing less transmission of group A *Streptococci* [5]. The emergence of penicillin as primary prevention has also played a role in the reduction of RHD. Still, the prevalence of VHD has remained high in developed countries due to the increasing life span and greater incidence of degenerative valve disease later in life.

3.2 Epidemiology of heart valve disease in developed regions

It is estimated that 2.5% of the US adult population has moderate or severe VHD [2], and in Sweden, there are six new VHD diagnoses per 10,000 persons per year [9]. Many cases of VHD may be undiagnosed, as up to 50.8% of a United Kingdom population with VHD was asymptomatic, of which about 6.4% had lesions of moderate or greater in severity [10]. VHD is a serious cause of morbidity and mortality. It is present in 17% of the community with heart failure [11] and was listed on 85,000 US hospital discharges in 2014 [12]. It accounts for 10%–20% of all cardiac surgical procedures [13]. VHD was the sole contributor to 25,114 US deaths in 2014 and was mentioned in another 51,005 deaths [12]. Mitral regurgitation (MR) is the most common VHD in the community [2], while aortic stenosis (AS) is the most frequent single-valve disease referred to the hospital [9,14]. The US prevalence of moderate or greater MR is 1.7%; aortic regurgitation (AR) 0.5%; AS 0.4%; and mitral stenosis (MS) 0.1% (Table 3.1 and Fig. 3.2) [2].

The Euro Heart Survey included 5,001 patients referred to hospital or clinic and found the following distribution of VHD: AS 34%, MR 25%, AR 10%, and MS 10% (Table 3.2) [14].

Because of the improvement in living conditions and broad access to sulfonamides, the greatest burden of VHD in developed regions is now due to degenerative valve disease, rather than RHD (Fig. 3.3). A degenerative etiology contributes to 63% of VHD referred to the hospital [14]. However, RHD is still present in 22% of patients, the second most frequent etiology [14]. Degenerative valve disease comprises a heterogeneous group of lesions characterized by different pathophysiologic processes and may involve one or multiple valves. Degenerative valve disease is most classically represented by calcific AS, which involves progressive thickening, fibrosis, and calcification of the aortic valve leaflets over years that impedes outflow [15,16]. Degenerative processes also contribute to diseases of the other valves. For example, myxomatous infiltration and fibroelastic deficiency of the mitral valve leads to progressive prolapse resulting in regurgitation [17]. Thus, prevalence of any valve disease increases with age; one out of eight persons aged 75 years or older will have some form of VHD [2]. There is no difference in the total prevalence of valve disease based on sex [2]. As life expectancy continues to increase in industrialized nations [18], the prevalence of VHD is expected to rise concomitantly. Some estimate the prevalence of any VHD will increase by 122% by 2056 [10] and the number of patients with severe AS will double by 2040 and triple by 2060 [19].

Table 3.1 Prevalence of valvular heart disease in population-based studies.

	Age (years)					p-value for trend	Frequency adjusted to 2000 US adult population
	18–44	45–54	55–64	65–74	≥75		
Participants (n) Male, n (%)	4,351 1,959 (45%)	696 258 (37%)	1,240 415 (33%)	3,879 1,586 (41%)	1,745 826 (47%)		209,128,094 100,994,367 (48%)
Mitral regurgitation (n=449)	23, 0.5% (0.3–0.8)	1, 0.1% (0–0.8)	12, 1.0% (0.5–1.8)	250, 6.4% (5.7–7.3)	163, 9.3% (8.1–10.9)	<0.0001	1.7% (1.5–1.9)
Mitral stenosis (n=15)	0, 0% (0–0.1)	1, 0.1% (0–0.8)	3, 0.2% (0.1–0.7)	7, 0.2% (0.1–0.4)	4, 0.2% (0.1–0.6)	0.006	0.1% (0.02–0.2)
Aortic regurgitation (n=90)	10, 0.2% (0.1–0.4)	1, 0.1% (0–0.8)	8, 0.7% (0.3–1.3)	37, 1.0% (0.7–1.3)	34, 2.0% (1.4–2.7)	<0.0001	0.5% (0.3–0.6)
Aortic stenosis (n=102)	1, 0.02% (0–0.1)	1, 0.1% (0–0.8)	2, 0.2% (0.6–1.9)	50, 1.3% (1.0–1.7)	48, 2.8% (2.1–3.7)	<0.0001	0.4% (0.3–0.5)
Any valve disease							
Overall (n=615)	31, 0.7% (0.5–1.0)	3, 0.4% (0.1–1.3)	23, 1.9% (1.2–2.8)	328, 8.5% (7.6–9.4)	230, 13.2% (11.7–15.0)	<0.0001	2.5% (2.2–2.7)
Women (n=356)	19, 0.8% (0.5–1.3)	1, 0.2% (0.01–1.3)	13, 1.6% (0.9–2.7)	208, 9.1% (8.0–10.4)	115, 12.6% (10.6–15.0)	<0.0001	2.4% (2.1–2.8)
Men (n=259)	12, 0.6% (0.3–1.1)	2, 0.8% (0.1–2.8)	10, 2.4% (1.2–4.4)	120, 7.6% (6.3–9.0)	115, 14.0% (11.7–16.6)	<0.0001	2.5% (2.1–2.9)

Prevalence data are n, % (95% CI). Percentages are rounded to one decimal place.
From Vuyisile T Nkomo,Julius M Gardin,Thomas N Skelton,John S Gottdiener,Christopher G Scott,Maurice Enriquez-Sarano, Burden of valvular heart diseases: a population-based study. Reprinted with permission from Elsevier (The Lancet, 2006, 368(9540), Pages 1005–1011)

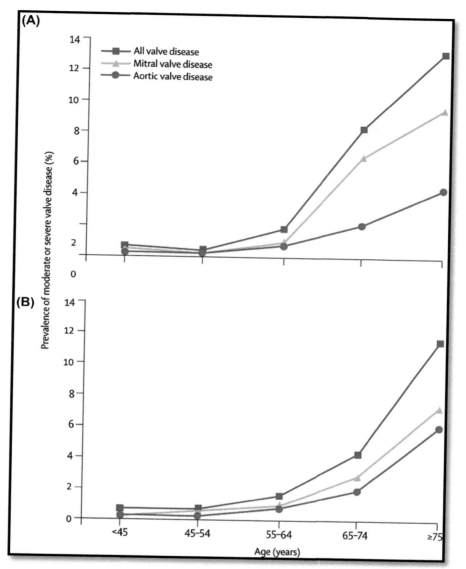

Figure 3.2 Prevalence of valvular heart disease by age in (A) population-based studies and (B) in the Olmsted County, Minnesota, USA community. *(Credit: From Vuyisile T Nkomo,Julius M Gardin,Thomas N Skelton,John S Gottdiener,Christopher G Scott,Maurice Enriquez-Sarano, Burden of valvular heart diseases: a population-based study. Reprinted with permission from Elsevier (The Lancet, 2006, 368(9540), Pages 1005-1011).)*

Table 3.2 Prevalence of valvular heart disease referred to hospital or clinic.

	Total population, n = 5,001	Patients with intervention, n = 1,269
Native valve disease (%)	71.9	87.0
Aortic (% native)	44.3	57.4
Aortic stenosis (%)	33.9	46.6
Aortic regurgitation (%)	10.4	10.8
Mitral (% native)	34.3	24.3
Mitral stenosis (%)	9.5	10.2
Mitral regurgitation (%)	24.8	14.1
Multiple (% native)	20.2	16.8
Right (% native)	1.2	1.5
Previous cardiac intervention (%)	28.1	13.0
Conservative surgery (%)	18.4	28.7
Valve replacement (%)	81.6	71.3

Credit: Table 3.2, Iung B, Baron G, Butchart EG, Delahaye F, Gohlke-Barwolf C, Levang OW, et al., A prospective survey of patients with valvular heart disease in Europe: the euro heart survey on valvular heart disease. Eur Heart J, 2003; 24:1231—43.

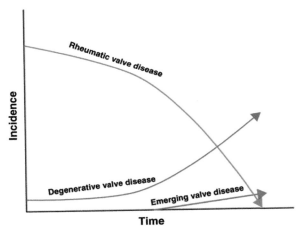

Figure 3.3 Evolution of valvular heart disease in developed regions. The incidence of rheumatic heart disease has declined, whereas the incidence of degenerative valve disease is increasing due to increased aging. New types of valve disease consistently emerge, although they account for a minority. (Credit: Fig. 3.2, Soler-Soler J, Galve E. Worldwide perspective of valve disease. Heart 2000;83:721—25.)

3.2.1 Aortic stenosis

Degenerative, calcific aortic valve disease is the most common cause of AS in developed countries. Aortic valve calcification is an active process similar to atherosclerosis, involving lipoprotein accumulation, cellular infiltration, and extracellular matrix formation that leads to progressive valve calcification and consequential cusp immobility and outflow obstruction [15,16]. Degenerative aortic valve disease is the cause of 81.9% of AS presenting to the hospital or outpatient clinic, whereas RHD is the cause in 11.2% and congenital heart disease (CHD) in 5.4% [14].

The prevalence of AS increases with age. The US prevalence of moderate or more severity AS is 0.4% [2]. Before 65 years of age, the prevalence is less than 0.2% and increases to 1.3% between 65 and 74 years and to 2.8% after age 75 [2]. Other population-based studies have found similar trends and prevalence. In a study which included institutionalized patients in old-age homes or hospitals aged 55 years or older, the prevalence of at least moderate AS was 4.8% (no cases of AS were found in the age group < 71 years) and increased with age from 2.5% in those 75–76 years to 8.1% in those 85–86 years, and 2.9% of those aged 75–86 years had at least critical AS [20]. In another study, any degree of AS was found in 2.4% in those aged 25–84 years and 4% in those greater than 85 years [21]. Another population-based study in Northern Norway (the Tromso study) also reported increasing prevalence of AS of any severity with age: 0.2% for 50–59 years, 1.3% for 60–69 years, 3.9% for 70–79 years, and 9.8% for 80–89 years [22]. The annual incidence of AS is estimated to be between 0.24 and 4.9 new cases per 1,000. The Tromso study [22] reported an annual incidence rate of 4.9 new cases of AS of any severity per 1,000, while a Swedish hospital based study [9] reported an annual incidence of primarily moderate to severe AS of 0.38 per 1,000 for males and 0.24 per 1,000 for females. A Canadian study of 1.12 million individuals older than 65 years found an AS incidence of 1.7 and 1.3 per 1,000 person-years for males and females, respectively [23]. AS is the most frequent VHD seen in the hospital or clinic, contributing to 33.9% of all native VHDs [14]. As the population ages, so, too, will AS. The number of patients aged 75 or older with AS is expected to double in the next two to three decades [19,24].

Although the association between AS and age is clear, the gender association is inconsistent. Some report higher rates of AS in males [2,9,21,23], while others report no difference according to gender [20,22]. As the pathophysiology of AS shares mechanisms with atherosclerosis, determinants of AS closely follow those of atherosclerosis. Taken together, age, hyperlipidemia, hypertension, diabetes, and prior history of smoking are all associated with the development of AS. In addition, the presence of a bicuspid aortic valve (BAV) increases the risk for development of AS. Pathologic review of operatively excised stenotic aortic valves demonstrate that 49% have BAV [25] and 24% of those with a BAV will undergo aortic valve replacement within 20 years of diagnosis, with AS as the indication in about 70% [26].

3.2.2 Aortic regurgitation

The US prevalence of moderate or more severity AR is 0.5% [2]. The prevalence increases with age and is unrelated to gender: less than 0.7% before 65 years of age, 1.0% aged 65–74 years, and 2.0% after 75 years [2]. The Framingham Heart Study found at least mild AR in 13.0% of males and 8.5% of females and similar rates of moderate or more severity that increased with age [27]. In another population-based study including a younger cohort of 21–35-year-old US patients, the prevalence of any severity of AR was 0.8%, of which 62% was moderate severity or greater, and there was no difference by gender or race [28]. In Sweden, the hospital incidence is estimated to be 20.2 per 10,000 person-years for males and 10.8 per 10,000 person-years for females [9]. AR contributes to 10.4% of all native VHD presenting to the hospital or outpatient clinic [14].

Degenerative aortic root disease is the most frequent etiology of AR in developed countries. Similar to AS, the prevalence of AR increases with age. Degenerative disease is the cause in 50.3% of AR presenting to the hospital or clinic, while 15.2% is due to RHD, 15.2% is due to congenital lesions, and 7.5% is due to infective endocarditis (IE) [14]. AR can also be characterized into primary or secondary. Primary AR includes those with pure valve dysfunction, such as a bicuspid, rheumatic, or infected valve. Secondary AR includes aortic root dilation, aortitis, acute dissection, systemic hypertension, or systemic diseases such as Marfan, Ehlers-Danlos, and osteogenesis imperfecta. Pathologic review of excised valves after undergoing aortic valve replacement for pure AR demonstrates that the proportion of primary (valvular dysfunction) to secondary (ascending aorta dysfunction) etiologies is essentially equal [29]. BAV is found in 22%, IE in 17%, aortic dissection in 10%, and unclear cause (hypertension suspected) in 34% [29]. Acute AR occurs in 18%, of which 56% is due to IE and the rest acute aortic dissection [29]. Lastly, valve calcification, particularly calcific deposits located at the valve coaptation or on the raphe of a BAV, hinders proper valve closure and leads to regurgitation [20,29].

3.2.3 Mitral stenosis

MS primarily occurs due to RHD, and thus, it is the least common VHD in developed countries; the prevalence in the US is 0.1% [2]. Prevalence increases with age and typically has a female predominance [30]. The Sweden hospital incidence was 1.5 per 10,000 person-years for males and 2.3 per 10,000 person-years for females [9]. MS accounts for 9.5% of all native VHD presenting to the hospital or outpatient clinic [14]. MS is most commonly caused by RHD, but it also occurs secondary to degenerative calcification of the mitral annulus in elderly people. Mitral annular calcification is present in about 10% of the population [31,32] but rarely results in mitral valve disease. As a chronic degenerative process of the fibrous base of the mitral valve, mitral annular calcification has been found to be associated with increasing age, female gender, chronic kidney disease, and cardiovascular risk factors [31–33] and previous chest irradiation. The Euro Heart Survey found that RHD was the cause of MS 85% of the time, while 13% was due degenerative processes [14].

3.2.4 Mitral regurgitation

In contrast to MS, MR is the most frequent heart valve disease. The Euro Heart Survey found that MR was the consequence of a degenerative valve disease in 61.3%, of RHD in 14.2%, of ischemia in 7.3%, and of congenital lesions in 4.8% [14]. Similar to AR, MR can be characterized into primary or secondary etiologies. Primary, or organic, MR is due to dysfunction of the mitral leaflets, whereas secondary, or functional, MR is the result of an abnormal mitral valve apparatus from a diseased left ventricle or enlarged left atrium, which ultimately hinder proper coaptation of the mitral leaflets [34]. Mitral valve prolapse (also known as click–murmur syndrome, Barlow's syndrome, or floppy valve syndrome) is the most common cause of primary MR. Mitral valve prolapse is considered to be a degenerative process where myxomatous infiltration and fibroelastic deficiency of the mitral tissue allow systolic displacement of the leaflets into the left atrium [17,35]. However, in rare instances, mitral valve prolapse is associated with some syndromic connective tissue disorders such as Marfan, Ehlers-Danlos, and osteogenesis imperfecta. The Framingham Heart Study reported a mitral valve prolapse prevalence of 2.4%, of which only 3.5% had severe MR [36]. A population-based study of young adults aged 21–35 years reported that mitral valve prolapse was suspected in 13% of cases of MR of any severity [28]. Other causes of primary MR include RHD, IE, mitral annulus calcification, and rupture of the chordae tendineae or papillary muscle.

Secondary, or functional, MR is most commonly a consequence of cardiomyopathy, either ischemic or dilated, where ventricular remodeling and enlargement causes tethering of the mitral leaflets and mitral annular dilatation with subsequent disruption of normal leaflet coaptation. Functional MR is present in approximately 50% of patients after a myocardial infarction and is moderate or severe in 12% [11]. It is estimated that 30%–50% of those with systolic dysfunction present with MR [37–41]. Multiple studies have shown functional MR to be an independent predictor of poor outcome for patients with heart failure and left ventricular dysfunction [38,40,42,43].

The US prevalence of moderate to severe MR is 1.7% [2]. The prevalence increases considerably with age; before 65 years of age, the prevalence is less than 1.0% but increases to 6.4% between 65 and 74 years and 9.3% after 75 years of age. In the Framingham Heart Study, the prevalence of at least mild MR was 19.0% and of moderate to severe MR was 1.6%, and it increased with age [27]. For example, in males before 60 years of age, the prevalence of moderate or more MR was less than 1.6%, but the prevalence was 2.4% and 11.2% in males aged 60–69 years and 70–83 years, respectively. In young adults, isolated MR of any severity was found in 10.4% and moderate to severe MR was found in 1.0% [28]. The hospital incidence of MR in Sweden was 21.3 per 10,000 person-years for males and 16.0 per 10,000 person-years for females [9]. Although MR is the most common VHD in the population, it is the second most common VHD presenting to the hospital or clinic after AS. MR contributes to 24.8% of all native VHD in the hospital or clinic, according to the Euro Heart Survey [14]. Aside from increasing age and systolic dysfunction, MR has been shown to be associated with systemic hypertension and lower BMI, but not gender [27].

3.2.5 Right-sided valvular heart disease

Right-sided VHD comprised 1.2% of all native lesions in the Euro Heart Survey presenting to the hospital or clinic [14]. Tricuspid regurgitation (TR) is the most frequent right-sided lesion. In the Framingham Heart Study, TR of any degree was detectable in 82% of males and 86% of females [27]. However, the prevalence of moderate or more TR was only 0.8%. The prevalence increased with age: for 70–83 year olds, the prevalence of at least moderate TR was 1.5% and 5.6% for males and females, respectively. Most TRs, and all right-sided VHD, are more frequently associated with left-sided VHD [44]. RV enlargement secondary to pulmonary hypertension is a frequent cause of functional TR, so MR often occurs concomitantly. RHD is also a frequent cause of both right- and left-sided VHD. Isolated tricuspid valve disease is rare but etiologies include congenital (Ebstein anomaly) and IE. There is increasing recognition of patients with TR caused by endomyocardial biopsy or intracardiac leads [45]. Isolated pulmonary valve disease is even more uncommon, and congenital lesions (pulmonary stenosis, tetralogy of Fallot) are the most common causes.

3.2.6 Infective endocarditis

IE is a rare disease that involves adhesion and survival of circulating bacteria onto the fibrin–platelet matrix of a traumatized valvular surface [46]. This can manifest as bacterial vegetations and valve destruction that can result in valvular regurgitation, paravalvular abscess and pseudoaneurysm, cardiac conduction disturbances, and embolization. IE contributes to 3% of all VHD: 0.8% of AS, 7.5% of AR, 0.6% of MS, and 3.5% of MR [14,47]. IE directly contributed to over 6,500 US deaths in 2014 [12]. Viridans group (oral) *Streptococci* are a common pathogen that affects patients with predisposing valvular abnormalities, such as RHD. However, developed countries have seen a shift in the epidemiology of IE in recent years. There has been an increase in the proportion of acute IE caused by staphylococci, with a concomitant decrease in the number of cases caused by streptococci [48–53]. In addition, there has been a decrease in cases due to RHD and an increase in cases associated with prosthetic valves and devices [49,54]. Lastly, studies have found that the age of onset and the proportion of males affected has increased over time [48,49,54]. Table 3.3 demonstrates recent studies of patients with IE in developed countries.

The incidence of IE has remained stable at the rate of 3–7 cases per 100,000 person-years [49,50,52,53,55]. Males greater than the age of 65 years have the highest rates of IE. Although the number of prosthetic valves affected have increased in recent years, IE still affects a native valve 70%–80% of the time [47,48,54–56]. The aortic and mitral valves are the most common native valves to be infected, constituting 30% and 50% of all cases, respectively. IE affects multiple valves (mitral and aortic) in around 10%, and the tricuspid valve is affected in about 5%. Devices, such as a pacemaker or ICD, are implicated in

Table 3.3 Studies of patients with infective endocarditis in developed countries.

	Range	Hoen et al.[a]	Murdoch et al.[b]	Tornos et al.[c]	Sy et al.[d]	Fedeli et al.[e]	Selton-suty et al.[f]	DeSimone et al.[g]	Tleyjeh et al.[h]
Age, mean, or median (years)	56–69	60	57	56	62	68	62	69	62
Male (%)	62–74	71	68	70	64	62	74	59	73
Intravenous drug use (%)	3–10	6	10	5	3		6	10	3
Prior infective endocarditis (%)	6–8		8				6	6	7
Native valve (%)	69–79	79	72	74				69	79
Prosthetic valve (%)	13–29	16	21	26	13		21	29	21
Device-related (%)	5–13	5	7				13		
Health care–associated (%)	27–30				30		27		
Early valve surgery (%)	15–52	49	48	52	20			16	15
In-hospital mortality (%)	13–23	16	18	13	14	19	23		
Incidence per 100,000 person-years	3–7.4	3.0			4.7	4.4		7.4	6.1
Microbiology(%)									
Streptococci	23–54	48	39	42	23	36	36	16	54
Viridans	13–33	17	17	13			19		33
Pyogenes	5–11	6	11				5		
Group D	13–25	25					13		
Staphylococci	29–44	29	42	34	44	41	36	33	33
Staphylococcus aureus	23–34	23	31		32	31	27	22	26
Enterococci	6–22	8	10	14	9	7	11		6
Culture negative	1–12	9	10	12			5		1

[a][55].
[b][56].
[c][47].
[d][53].
[e][50].
[f][52].
[g][48].
[h][54].

5%—7% of cases [55,56]; however, recent estimates report that it may be as high as high as 15% [49]. Intravenous drug use is associated with 3%—10% of cases [47,48,54—56]. IE remains an exceptionally deadly disease, and in-hospital mortality is approximately 15%—20%, which has remained stable over the past two decades [49].

The number of cases caused by *Staphylococcus aureus* has increased in recent years, contributing to 20%—40% of cases, and is now a more common cause of IE than viridans group *Streptococci*, which contributes to 10%—20% of cases [47,48,50,52,53,55,56]. Recent changes in the microbiological profile are partially explained by the increasing number of cases of health care—associated IE, which accounts for greater than 25% of cases, and leads to more cases of *S. aureus* [52,53,57]. Furthermore, the high rates of health care—associated IE likely contribute to the static and large in-hospital mortality. Approximately 50% of patients undergo valvular surgery during the initial hospitalization [47,55,56], and patients with health care—associated IE have a large burden of comorbidities, including old age, diabetes, renal failure, cancer, and immunosuppression [52,53,57].

Guidelines on antibiotic prophylaxis to prevent IE are a point of controversy and are subject to reviews and revisions. In the past two decades, major changes to guidelines have recommended simpler and shorter antibiotic regimens and have restricted the number of individuals and procedures where antibiotic prophylaxis is recommended [58]. Most of the world has adopted the 2007 American Heart Association or the 2015 European Society of Cardiology guidelines, which recommend prophylaxis for high-risk patients with one dose of Amoxicillin orally before high-risk procedures [59,60]. High-risk patients are defined as those with prosthetic valves, previous IE, cyanotic CHD, and cardiac transplant recipients who develop valvulopathy. Both the European Society of Cardiology and American Heart Association define dental procedures requiring manipulation of the gingiva or perforation of the oral mucosa high risk, while the American Heart Association also includes procedures of the respiratory tract and infected skin or musculoskeletal tissue. There are a few countries, such as Sweden, that have adopted the 2015 National Institute for Health and Care Excellence (NICE) guidelines, which recommends against antibiotic prophylaxis for people undergoing dental procedures [61]. Of those adopting the American Heart Association or the European Society of Cardiology guidelines, most studies report no change in the incidence of IE [48,49,62—64]. Of those adopting the NICE guidelines, studies have shown that prescriptions for antibiotic prophylaxis have fallen and the incidence of IE has increased significantly [65]. The NICE guidelines were amended in 2016 to include the recommendation that patients at high risk undergoing high-risk dental procedures should receive antibiotic prophylaxis [61]. There have been no randomized controlled studies to elucidate the effectiveness of antibiotic prophylaxis on the incidence of IE.

3.3 Epidemiology of heart valve disease in developing regions

RHD is the leading cause of VHD in developing countries Based on estimates from the 2015 Global Burden of Disease study, 33.4 million people have RHD worldwide, of which 73% come from India (13.17 million cases), China (7.07 million), Pakistan (2.25 million), Indonesia (1.18 million), and the Democratic Republic of the Congo (805,000)(1). The age-standardized prevalence of RHD is 444 cases per 100,000 population for countries with an endemic pattern and 3.4 cases per 100,000 population for countries with a nonendemic pattern. There were 319,400 deaths attributable to RHD in 2015, yielding an age-standardized mortality of 4.8 deaths per 100,000. The highest age-standardized death rates occurred in the regions of Oceania (14 deaths per 100,000), South Asia (12 deaths per 100,000), and central sub-Saharan Africa (8 deaths per 100,000). In 2005, it was estimated that there were 471,000 new cases of acute rheumatic fever each year (95% occur in developing regions) [3]. Unfortunately, limited access to healthcare and proper antibiotic prophylaxis, poor living conditions, and overcrowding allow RHD to remain a significant burden in developing countries [5,66,67]. For example, of 3,343 patients with RHD in 12 African countries, India, and Yemen, only 55% were on proper prophylaxis, only 28.3% of those on oral anticoagulants had therapeutic INRs, and only 3.6% of females of childbearing age were on contraception [67]. The incidence of acute rheumatic fever remains relatively elevated, estimated to be between 1 and 5 cases per 1,000 per year in countries of Eastern Europe, Asia, Australasia, and the Middle East [68], and about 60% of all acute rheumatic fever cases develop RHD each year [3].

Based on clinical screening, the prevalence of RHD in school-aged children is estimated between 1 and 7 cases per 1,000 population (Fig. 3.4) [3,66,69—71]. There are an estimated two million children aged 5—14 years living with RHD in Asia [69]. However, systematic screening using echocardiography reveals up to a 10-fold higher prevalence of RHD [70]. The estimated incidence of RHD in an urban South African population was 2.4 cases per 1,000 per year for those older than 14 years of age [72]. RHD accounted for 72% of VHD presenting to a South African center [72] and 64% of newly diagnosed VHD in a South India center [73].

The mitral valve is most commonly diseased in RHD, with MR predominating during the first and second decades of life and MS predominating during the third decade of life and after [74]. Mixed aortic and mitral valve involvement is present in a third of RHD cases [72,73]. Females are almost twice as likely to be affected than males [67,70,72,73,75]. RHD typically manifests at a younger age in developing regions. About 25% of patients presenting with RHD in an Eastern Nepal center were younger than 20 years of age [75], and the median age of 3,343 patients with RHD in 12 African countries, India, and Yemen was 28 years [67]. RHD complicates many pregnancies among young women. For example, rheumatic mitral valve disease accounts for over

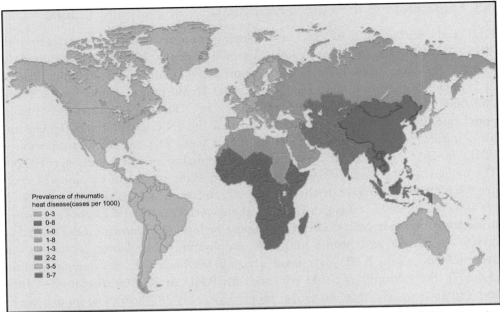

Figure 3.4 Prevalence of rheumatic heart disease in children aged 5–14 years. *(Credit: From Jonathan R Carapetis,Andrew C Steer,E Kim Mulholland,Martin Weber, The global burden of group A streptococcal diseases. Reprinted with permission from Elsevier (The Lancet, 2005, 5(11), Pages 685-694).)*

75% of cardiac diseases associated with pregnancy in Africa and may account for up to 35% of maternal mortality [71,76–78].

RHD is the main predisposing factor for IE in developing regions. Studies indicate that at least a half of all cases of IE are in patients with underlying RHD [69,71,79–82]. IE is a disease of the young in these countries, with peak incidences in the third to fourth decades of life [69,71,79–82]. Streptococci (most commonly viridans-group) is the most common microbiologic cause, followed by staphylococci, including *S. aureus*. The mortality of IE is considerable: in-hospital mortality is around 25%. CHD is also an important risk factor for IE. The epidemiology of IE in developing countries appears similar to that of developed countries decades before: it is a disease of the young with RHD and is most commonly caused by streptococci that results in substantial morbidity and mortality [71,83].

Because of the high rates of RHD and low life expectancy in developing regions compared with developed regions, degenerative valve disease, such as calcific AS or myxomatous MR, is much less common. The median life expectancies in 2015 of Mozambique, South Africa, Mali, and Afghanistan were 57.6, 62.9, 58.2, and 60.5 years, respectively, which is years before the severity of calcific AS becomes clinically significant [1].

In a South African center, 21% of newly diagnosed VHD were due to degenerative causes [72]. Degenerative valve disease still causes the majority of AS and more than a third of MR [73]. With increasing life expectancy in developing countries, the burden of degenerative valve disease in developing countries is likely to increase [84].

3.4 Epidemiology of congenital heart valve disease

Historically, most patients with CHD died in early childhood. However, because of extraordinary advances in cardiology and cardiovascular surgical procedures, the number of adults living with CHD has increased significantly in recent decades [85–87]. It is now estimated that about 85% of patients with CHD will survive to adulthood, and if trends continue, one out of every 150 young adults will have some form of CHD [88,89]. For example, in Quebec, Canada, the median age of severe CHD was 11 years in 1985 and increased to 17 years in 2000. In the same population, the prevalence of severe CHD in adults increased by 85%, whereas for children, there was a 22% increase [90].

The overall prevalence of CHD is between 6 and 10 per 1,000 live births, 4–6 per 1,000 adults, and 6 per 1,000 general population [86,90–92]. Based on this prevalence, 0.85 to 1.3 million people with CHD are living in the United States, of which 80,000 to 180,000 have complex lesions [86,90,93]. Left to right shunts are the most common types of CHD: ventricular septal defects and atrial septal defects occur in 17.6–44.8 and 3.7–10.6 infants per 10,000 live births, respectively. Tetralogy of Fallot and transposition of the great arteries are the most common cyanotic CHDs. The congenital lesions that affect the valvular apparatus have the following prevalences per 10,000 live births (Table 3.4): tetralogy of Fallot 2.9–5.8; pulmonary valve stenosis 3.6–8.4; congenital AS 1.1–3.9; coarctation of aorta 2.9–4.9; tricuspid atresia 0.2–1.2; Ebstein's anomaly 0.4–1.6; and pulmonary atresia 0.4–1.5 [91,92]. Low birth weight, low gestational age, older maternal age, and a multiple gestation pregnancy are all associated with the presence of CHD [92]. Congenital lesions are not a common cause of isolated VHD, and in the Euro Heart Survey, they made up 5.4% of all AS, 15.3% of all AR, 0.6% of all MS, and 4.8% of all MR [14].

Technically, the most common congenital heart valve malformation is the BAV. Although it is associated with other congenital heart syndromes, it is not typically considered an isolated CHD in most epidemiologic studies because clinically relevant disease of the BAV, such as AS or AR, does not typically manifest until the fourth decade of life or later [26]. BAV has been identified in 4.6 out of 1,000 live births, with a 3:1 male to female predominance [94]. In the general population, it is widely accepted that the prevalence is 1%–2%, and there are about 3 million people alive with BAV in the United States [93,95]. The presence of two leaflets, instead of three, predisposes the cusps to high mechanical and functional stress that results in degenerative AS [96,97]. AR is also a common finding of a BAV, secondary to valve prolapse, myxoid degeneration, or

Table 3.4 Prevalence of congenital valvular heart disease.

Congenital heart disease	Prevalence per 10,000 live births interquartile range[a,b]
Total valvular CHD (excluding BAV)	11.5—27.3
Tetralogy of Fallot	2.9—5.8
Pulmonary valve stenosis	3.6—8.4
Congenital aortic stenosis	1.1—3.9
Coarctation of aorta	2.9—4.9
Tricuspid atresia	0.2—1.2
Ebstein's anomaly	0.4—1.6
Pulmonary atresia	0.4—1.5
Bicuspid aortic valve[c]	46
Ventricular septal defects	17.6—44.8
Atrial septal defects	3.7—10.6
All CHD (excluding BAV)	60.2—105.7

BAV, bicuspid aortic valve; CHD, congenital heart disease.
[a][91].
[b][92].
[c][94].

dilation of the ascending aorta [98]. It is estimated that 12%—37% develop moderate-to-severe AS and 13%—32% develop moderate-to-severe AR [99]. The abnormal cusp geometry provides a substrate for IE, and the lifetime risk is predicted to be between 2% and 5% [99,100]. BAV is also associated with aortopathies in up to 40% of patients, characterized by ascending aorta dilatation that may develop into aneurysm, dissection, or rupture [99]. The incidence of aneurysm and dissection are 84.9 and 3.1 per 10,000 patient-years, respectively, and the age-adjusted relative risk versus the general population is 8.4 for aneurysm and 6.2 for dissection [101]. The risk of aneurysm and dissection increases substantially after the age of 50 years [101]. It is estimated that 25% of patients with BAV will need aorta surgery within 25 years of their diagnosis [101].

References

[1] WHO. World health statistics 2016. Wolrd Health Organization; 2016.
[2] Nkomo VT, Gardin JM, Skelton TN, Gottdiener JS, Scott CG, Enriquez-Sarano M. Burden of valvular heart diseases: a population-based study. Lancet 2006;368:1005—11.
[3] Carapetis JR, Steer AC, Mulholland EK, Weber M. The global burden of group a streptococcal diseases. Lancet Infect Dis 2005;5:685—94.
[4] Tandon R, Sharma M, Chandrashekhar Y, Kotb M, Yacoub MH, Narula J. Revisiting the pathogenesis of rheumatic fever and carditis. Nat Rev Cardiol 2013;10:171—7.
[5] Soler-Soler J, Galve E. Worldwide perspective of valve disease. Heart 2000;83:721—5.
[6] Jones TD. Diagnosis of rheumatic fever. JAMA Cardiol 1944;126:481—4.
[7] Annegers JF, Pillman NL, Weidman WH, Kurland LT. Rheumatic fever in Rochester, Minnesota, 1935-1978. Mayo Clin Proc 1982;57:753—7.

[8]   Bisno AL. The resurgence of acute rheumatic fever in the United States. Annu Rev Med 1990;41: 319—29.

[9]   Andell P, Li X, Martinsson A, Andersson C, Stagmo M, Zoller B, Sundquist K, Smith JG. Epidemiology of valvular heart disease in a Swedish nationwide hospital-based register study. Heart 2017;103(21):1696—703.

[10]   D'arcy JL, Coffey S, Loudon MA, Kennedy A, Pearson-Stuttard J, Birks J, Frangou E, Farmer AJ, Mant D, Wilson J, Myerson SG, Prendergast BD. Large-scale community echocardiographic screening reveals a major burden of undiagnosed valvular heart disease in older people: the OxVALVE Population Cohort Study. Eur Heart J 2016;37:3515—22.

[11]   Bursi F, Weston SA, Redfield MM, Jacobsen SJ, Pakhomov S, Nkomo VT, Meverden RA, Roger VL. Systolic and diastolic heart failure in the community. J Am Med Assoc 2006;296: 2209—16.

[12]   Benjamin EJ, Blaha MJ, Chiuve SE, Cushman M, Das SR, Deo R, DE Ferranti SD, Floyd J, Fornage M, Gillespie C, Isasi CR, Jimenez MC, Jordan LC, Judd SE, Lackland D, Lichtman JH, Lisabeth L, Liu S, Longenecker CT, Mackey RH, Matsushita K, Mozaffarian D, Mussolino ME, Nasir K, Neumar RW, Palaniappan L, Pandey DK, Thiagarajan RR, Reeves MJ, Ritchey M, Rodriguez CJ, Roth GA, Rosamond WD, Sasson C, Towfighi A, Tsao CW, Turner MB, Virani SS, Voeks JH, Willey JZ, Wilkins JT, Wu JH, Alger HM, Wong SS, Muntner P. Heart disease and stroke statistics-2017 update: a report from the American heart association. Circulation 2017;135: e146—603.

[13]   Mann DLE, Zipes DPE, Libby PE, Bonow ROE, Braunwald EE. Braunwald's heart disease : a textbook of cardiovascular medicine. 2014.

[14]   Iung B, Baron G, Butchart EG, Delahaye F, Gohlke-Barwolf C, Levang OW, Tornos P, Vanoverschelde JL, Vermeer F, Boersma E, Ravaud P, Vahanian A. A prospective survey of patients with valvular heart disease in Europe: the euro heart survey on valvular heart disease. Eur Heart J 2003;24:1231—43.

[15]   Pawade TA, Newby DE, Dweck MR. Calcification in aortic stenosis: the skeleton key. J Am Coll Cardiol 2015;66:561—77.

[16]   Rajamannan NM, Evans FJ, Aikawa E, Grande-Allen KJ, Demer LL, Heistad DD, Simmons CA, Masters KS, Mathieu P, O'brien KD, Schoen FJ, Towler DA, Yoganathan AP, Otto CM. Calcific aortic valve disease: not simply a degenerative process: a review and agenda for research from the National Heart and Lung and Blood Institute Aortic Stenosis Working Group. Executive summary: calcific aortic valve disease-2011 update. Circulation 2011;124:1783—91.

[17]   Pellerin D, Brecker S, Veyrat C. Degenerative mitral valve disease with emphasis on mitral valve prolapse. Heart 2002;88(Suppl. 4):iv20—i28.

[18]   Kontis V, Bennett JE, Mathers CD, Li G, Foreman K, Ezzati M. Future life expectancy in 35 industrialised countries: projections with a Bayesian model ensemble. Lancet 2017;389:1323—35.

[19]   Danielsen R, Aspelund T, Harris TB, Gudnason V. The prevalence of aortic stenosis in the elderly in Iceland and predictions for the coming decades: the AGES-Reykjavik study. Int J Cardiol 2014;176: 916—22.

[20]   Lindroos M, Kupari M, Heikkila J, Tilvis R. Prevalence of aortic valve abnormalities in the elderly: an echocardiographic study of a random population sample. J Am Coll Cardiol 1993;21:1220—5.

[21]   Stewart BF, Siscovick D, Lind BK, Gardin JM, Gottdiener JS, Smith VE, Kitzman DW, Otto CM. Clinical factors associated with calcific aortic valve disease. Cardiovascular health study. J Am Coll Cardiol 1997;29:630—4.

[22]   Eveborn GW, Schirmer H, Heggelund G, Lunde P, Rasmussen K. The evolving epidemiology of valvular aortic stenosis. the Tromso study. Heart 2013;99:396—400.

[23]   Yan AT, Koh M, Chan KK, Guo H, Alter DA, Austin PC, Tu JV, Wijeysundera HC, Ko DT. Association between cardiovascular risk factors and aortic stenosis: the CANHEART aortic stenosis study. J Am Coll Cardiol 2017;69:1523—32.

[24]   Iung B, Vahanian A. Degenerative calcific aortic stenosis: a natural history. Heart 2012;98(Suppl. 4): iv7—13.

[25] Roberts WC, Ko JM. Frequency by decades of unicuspid, bicuspid, and tricuspid aortic valves in adults having isolated aortic valve replacement for aortic stenosis, with or without associated aortic regurgitation. Circulation 2005;111:920—5.

[26] Michelena HI, Desjardins VA, Avierinos JF, Russo A, Nkomo VT, Sundt TM, Pellikka PA, Tajik AJ, Enriquez-Sarano M. Natural history of asymptomatic patients with normally functioning or minimally dysfunctional bicuspid aortic valve in the community. Circulation 2008;117:2776—84.

[27] Singh JP, Evans JC, Levy D, Larson MG, Freed LA, Fuller DL, Lehman B, Benjamin EJ. Prevalence and clinical determinants of mitral, tricuspid, and aortic regurgitation (the Framingham heart study). Am J Cardiol 1999;83:897—902.

[28] Reid CL, Anton-Culver H, Yunis C, Gardin JM. Prevalence and clinical correlates of isolated mitral, isolated aortic regurgitation, and both in adults aged 21 to 35 years (from the CARDIA study). Am J Cardiol 2007;99:830—4.

[29] Roberts WC, Ko JM, Moore TR, Jones 3rd WH. Causes of pure aortic regurgitation in patients having isolated aortic valve replacement at a single US tertiary hospital (1993 to 2005). Circulation 2006; 114:422—9.

[30] Movahed MR, Ahmadi-Kashani M, Kasravi B, Saito Y. Increased prevalence of mitral stenosis in women. J Am Soc Echocardiogr 2006;19:911—3.

[31] Abramowitz Y, Jilaihawi H, Chakravarty T, Mack MJ, Makkar RR. Mitral annulus calcification. J Am Coll Cardiol 2015;66:1934—41.

[32] Kanjanauthai S, Nasir K, Katz R, Rivera JJ, Takasu J, Blumenthal RS, Eng J, Budoff MJ. Relationships of mitral annular calcification to cardiovascular risk factors: the Multi-Ethnic Study of Atherosclerosis (MESA). Atherosclerosis 2010;213:558—62.

[33] Akram MR, Chan T, Mcauliffe S, Chenzbraun A. Non-rheumatic annular mitral stenosis: prevalence and characteristics. Eur J Echocardiogr 2009;10:103—5.

[34] Nishimura RA, Otto CM, Bonow RO, Carabello BA, Erwin 3rd JP, Guyton RA, O'gara PT, Ruiz CE, Skubas NJ, Sorajja P, Sundt 3rd TM, Thomas JD. 2014 AHA/ACC guideline for the management of patients with valvular heart disease: a report of the American college of cardiology/American heart association task force on practice guidelines. Circulation 2014;129:e521—643.

[35] Delling FN, Vasan RS. Epidemiology and pathophysiology of mitral valve prolapse: new insights into disease progression, genetics, and molecular basis. Circulation 2014;129:2158—70.

[36] Freed LA, Levy D, Levine RA, Larson MG, Evans JC, Fuller DL, Lehman B, Benjamin EJ. Prevalence and clinical outcome of mitral-valve prolapse. N Engl J Med 1999;341:1—7.

[37] Agricola E, Stella S, Figini F, Piraino D, Oppizzi M, D'amato R, Slavich M, Ancona MB, Margonato A. Non-ischemic dilated cardiopathy: prognostic value of functional mitral regurgitation. Int J Cardiol 2011;146:426—8.

[38] Bursi F, Barbieri A, Grigioni F, Reggianini L, Zanasi V, Leuzzi C, Ricci C, Piovaccari G, Branzi A, Modena MG. Prognostic implications of functional mitral regurgitation according to the severity of the underlying chronic heart failure: a long-term outcome study. Eur J Heart Fail 2010;12:382—8.

[39] Deja MA, Grayburn PA, Sun B, Rao V, She L, Krejca M, Jain AR, Leng Chua Y, Daly R, Senni M, Mokrzycki K, Menicanti L, Oh JK, Michler R, Wrobel K, Lamy A, Velazquez EJ, Lee KL, Jones RH. Influence of mitral regurgitation repair on survival in the surgical treatment for ischemic heart failure trial. Circulation 2012;125:2639—48.

[40] Rossi A, Dini FL, Faggiano P, Agricola E, Cicoira M, Frattini S, Simioniuc A, Gullace M, Ghio S, Enriquez-Sarano M, Temporelli PL. Independent prognostic value of functional mitral regurgitation in patients with heart failure. A quantitative analysis of 1256 patients with ischaemic and non-ischaemic dilated cardiomyopathy. Heart 2011;97:1675—80.

[41] Varadarajan P, Sharma S, Heywood JT, Pai RG. High prevalence of clinically silent severe mitral regurgitation in patients with heart failure: role for echocardiography. J Am Soc Echocardiogr 2006;19:1458—61.

[42] Bursi F, Enriquez-Sarano M, Nkomo VT, Jacobsen SJ, Weston SA, Meverden RA, Roger VL. Heart failure and death after myocardial infarction in the community: the emerging role of mitral regurgitation. Circulation 2005;111:295—301.

[43] Grigioni F, Enriquez-Sarano M, Zehr KJ, Bailey KR, Tajik AJ. Ischemic mitral regurgitation: long-term outcome and prognostic implications with quantitative Doppler assessment. Circulation 2001;103:1759—64.

[44] Iung B, Baron G, Tornos P, Gohlke-Barwolf C, Butchart EG, Vahanian A. Valvular heart disease in the community: a European experience. Curr Probl Cardiol 2007;32:609—61.

[45] Fender EA, Zack CJ, Nishimura RA. Isolated tricuspid regurgitation: outcomes and therapeutic interventions. Heart 2017;104(10):798—806.

[46] Sullam PM, Drake TA, Sande MA. Pathogenesis of endocarditis. Am J Med 1985;78:110—5.

[47] Tornos P, Iung B, Permanyer-Miralda G, Baron G, Delahaye F, Gohlke-Barwolf C, Butchart EG, Ravaud P, Vahanian A. Infective endocarditis in Europe: lessons from the euro heart survey. Heart 2005;91:571—5.

[48] Desimone DC, Tleyjeh IM, Correa de Sa DD, Anavekar NS, Lahr BD, Sohail MR, Steckelberg JM, Wilson WR, Baddour LM. Temporal trends in infective endocarditis epidemiology from 2007 to 2013 in Olmsted County, MN. Am Heart J 2015;170:830—6.

[49] Duval X, Delahaye F, Alla F, Tattevin P, Obadia JF, Le Moing V, Doco-Lecompte T, Celard M, Poyart C, Strady C, Chirouze C, Bes M, Cambau E, Iung B, Selton-Suty C, Hoen B. Temporal trends in infective endocarditis in the context of prophylaxis guideline modifications: three successive population-based surveys. J Am Coll Cardiol 2012;59:1968—76.

[50] Fedeli U, Schievano E, Buonfrate D, Pellizzer G, Spolaore P. Increasing incidence and mortality of infective endocarditis: a population-based study through a record-linkage system. BMC Infect Dis 2011;11:48.

[51] Pant S, Patel NJ, Deshmukh A, Golwala H, Patel N, Badheka A, Hirsch GA, Mehta JL. Trends in infective endocarditis incidence, microbiology, and valve replacement in the United States from 2000 to 2011. J Am Coll Cardiol 2015;65:2070—6.

[52] Selton-Suty C, Celard M, Le Moing V, Doco-Lecompte T, Chirouze C, Iung B, Strady C, Revest M, Vandenesch F, Bouvet A, Delahaye F, Alla F, Duval X, Hoen B. Preeminence of Staphylococcus aureus in infective endocarditis: a 1-year population-based survey. Clin Infect Dis 2012;54: 1230—9.

[53] Sy RW, Kritharides L. Health care exposure and age in infective endocarditis: results of a contemporary population-based profile of 1536 patients in Australia. Eur Heart J 2010;31:1890—7.

[54] Tleyjeh IM, Steckelberg JM, Murad HS, Anavekar NS, Ghomrawi HM, Mirzoyev Z, Moustafa SE, Hoskin TL, Mandrekar JN, Wilson WR, Baddour LM. Temporal trends in infective endocarditis: a population-based study in Olmsted County, Minnesota. Jama 2005;293:3022—8.

[55] Hoen B, Alla F, Selton-Suty C, Beguinot I, Bouvet A, Briancon S, Casalta JP, Danchin N, Delahaye F, Etienne J, Le Moing V, Leport C, Mainardi JL, Ruimy R, Vandenesch F. Changing profile of infective endocarditis: results of a 1-year survey in France. J Am Med Assoc 2002;288: 75—81.

[56] Murdoch DR, Corey GR, Hoen B, Miro JM, Fowler JR VG, Bayer AS, Karchmer AW, Olaison L, Pappas PA, Moreillon P, Chambers ST, Chu VH, Falco V, Holland DJ, Jones P, Klein JL, Raymond NJ, Read KM, Tripodi MF, Utili R, Wang A, Woods CW, Cabell CH. Clinical presentation, etiology, and outcome of infective endocarditis in the 21st century: the international collaboration on endocarditis-prospective cohort study. Arch Intern Med 2009;169:463—73.

[57] Fowler Jr VG, Miro JM, Hoen B, Cabell CH, Abrutyn E, Rubinstein E, Corey GR, Spelman D, Bradley SF, Barsic B, Pappas PA, Anstrom KJ, Wray D, Fortes CQ, Anguera I, Athan E, Jones P, Van der Meer JT, Elliott TS, Levine DP, Bayer AS. Staphylococcus aureus endocarditis: a consequence of medical progress. J Am Med Assoc 2005;293:3012—21.

[58] Thornhill MH, Dayer M, Lockhart PB, Prendergast B. Antibiotic prophylaxis of infective endocarditis. Curr Infect Dis Rep 2017;19:9.

[59] Habib G, Lancellotti P, Antunes MJ, Bongiorni MG, Casalta JP, Del Zotti F, Dulgheru R, El Khoury G, Erba PA, Iung B, Miro JM, Mulder BJ, Plonska-Gosciniak E, Price S, Roos-Hesselink J, Snygg-Martin U, Thuny F, Tornos Mas P, Vilacosta I, Zamorano JL, Erol C, Nihoyannopoulos P, Aboyans V, Agewall S, Athanassopoulos G, Aytekin S, Benzer W, Bueno H, Broekhuizen L, Carerj S, Cosyns B, De Backer J, De Bonis M, Dimopoulos K, Donal E, Drexel H, Flachskampf FA,

Hall R, Halvorsen S, Hoen B, Kirchhof P, Lainscak M, Leite-Moreira AF, Lip GY, Mestres CA, Piepoli MF, Punjabi PP, Rapezzi C, Rosenhek R, Siebens K, Tamargo J, Walker DM. 2015 ESC guidelines for the management of infective endocarditis: the task force for the management of infective endocarditis of the European Society of Cardiology (ESC). Endorsed by: European association for cardiothoracic surgery (EACTS), the European Association of Nuclear Medicine (EANM). Eur Heart J 2015; 36:3075—128.

[60] Wilson W, Taubert KA, Gewitz M, Lockhart PB, Baddour LM, Levison M, Bolger A, Cabell CH, Takahashi M, Baltimore RS, Newburger JW, Strom BL, Tani LY, Gerber M, Bonow RO, Pallasch T, Shulman ST, Rowley AH, Burns JC, Ferrieri P, Gardner T, Goff D, Durack DT. Prevention of infective endocarditis: guidelines from the American heart association: a guideline from the American heart association rheumatic fever, endocarditis, and kawasaki disease committee, council on cardiovascular disease in the young, and the council on clinical cardiology, council on cardiovascular surgery and anesthesia, and the quality of care and outcomes research interdisciplinary working group. Circulation 2007;116:1736—54.

[61] Thornhill MH, Chambers JB, Dayer M, Shanson D. A change in the NICE guidelines on antibiotic prophylaxis for dental procedures. Br J Gen Pract 2016;66:460—1.

[62] Bikdeli B, Wang Y, Kim N, Desai MM, Quagliarello V, Krumholz HM. Trends in hospitalization rates and outcomes of endocarditis among Medicare beneficiaries. J Am Coll Cardiol 2013;62: 2217—26.

[63] Pasquali SK, He X, Mohamad Z, Mccrindle BW, Newburger JW, Li JS, Shah SS. Trends in endocarditis hospitalizations at US children's hospitals: impact of the 2007 American heart association antibiotic prophylaxis guidelines. Am Heart J 2012;163:894—9.

[64] Rogers AM, Schiller NB. Impact of the first nine months of revised infective endocarditis prophylaxis guidelines at a university hospital: so far so good. J Am Soc Echocardiogr 2008;21:775.

[65] Dayer MJ, Jones S, Prendergast B, Baddour LM, Lockhart PB, Thornhill MH. Incidence of infective endocarditis in England, 2000-13: a secular trend, interrupted time-series analysis. Lancet 2015;385: 1219—28.

[66] Rizvi SF, Khan MA, Kundi A, Marsh DR, Samad A, Pasha O. Status of rheumatic heart disease in rural Pakistan. Heart 2004;90:394—9.

[67] Zuhlke L, Engel ME, Karthikeyan G, Rangarajan S, Mackie P, Cupido B, Mauff K, Islam S, Joachim A, Daniels R, Francis V, Ogendo S, Gitura B, Mondo C, Okello E, Lwabi P, AL-Kebsi MM, Hugo-Hamman C, Sheta SS, Haileamlak A, Daniel W, Goshu DY, Abdissa SG, Desta AG, Shasho BA, Begna DM, Elsayed A, Ibrahim AS, Musuku J, Bode-Thomas F, Okeahialam BN, Ige O, Sutton C, Misra R, Abul Fadl A, Kennedy N, Damasceno A, Sani M, Ogah OS, Olunuga T, Elhassan HH, Mocumbi AO, Adeoye AM, Mntla P, Ojji D, Mucumbitsi J, Teo K, Yusuf S, Mayosi BM. Characteristics, complications, and gaps in evidence-based interventions in rheumatic heart disease: the global rheumatic heart disease registry (the REMEDY study). Eur Heart J 2015;36:1115. 22a.

[68] Tibazarwa KB, Volmink JA, Mayosi BM. Incidence of acute rheumatic fever in the world: a systematic review of population-based studies. Heart 2008;94:1534—40.

[69] Carapetis JR. Rheumatic heart disease in Asia. Circulation 2008;118:2748—53.

[70] Marijon E, Ou P, Celermajer DS, Ferreira B, Mocumbi AO, Jani D, Paquet C, Jacob S, Sidi D, Jouven X. Prevalence of rheumatic heart disease detected by echocardiographic screening. N Engl J Med 2007;357:470—6.

[71] Nkomo VT. Epidemiology and prevention of valvular heart diseases and infective endocarditis in Africa. Heart 2007;93:1510—9.

[72] Sliwa K, Carrington M, Mayosi BM, Zigiriadis E, Mvungi R, Stewart S. Incidence and characteristics of newly diagnosed rheumatic heart disease in urban African adults: insights from the heart of Soweto study. Eur Heart J 2010;31:719—27.

[73] Manjunath CN, Srinivas P, Ravindranath KS, Dhanalakshmi C. Incidence and patterns of valvular heart disease in a tertiary care high-volume cardiac center: a single center experience. Indian Heart J 2014;66:320—6.

[74] Marcus RH, Sareli P, Pocock WA, Barlow JB. The spectrum of severe rheumatic mitral valve disease in a developing country. Correlations among clinical presentation, surgical pathologic findings, and hemodynamic sequelae. Ann Intern Med 1994;120:177—83.

[75] Shrestha NR, Pilgrim T, Karki P, Bhandari R, Basnet S, Tiwari S, Dhakal SS, Urban P. Rheumatic heart disease revisited: patterns of valvular involvement from a consecutive cohort in eastern Nepal. J Cardiovasc Med 2012;13:755—9.

[76] Abdel-Hady ES, El-Shamy M, El-Rifai AA, Goda H, Abdel-Samad A, Moussa S. Maternal and peri-natal outcome of pregnancies complicated by cardiac disease. Int J Gynaecol Obstet 2005;90:21—5.

[77] El Kady AA, Saleh S, Gadalla S, Fortney J, Bayoumi H. Obstetric deaths in menoufia governorate, Egypt. Br J Obstet Gynaecol 1989;96:9—14.

[78] Naidoo DP, Desai DK, Moodley J. Maternal deaths due to pre-existing cardiac disease. Cardiovas J S Afr 2002;13:17—20.

[79] Agarwal R, Bahl VK, Malaviya AN. Changing spectrum of clinical and laboratory profile of infective endocarditis. J Assoc Phys India 1992;40:721—3.

[80] Garg N, Kandpal B, Tewari S, Kapoor A, Goel P, Sinha N. Characteristics of infective endocarditis in a developing country-clinical profile and outcome in 192 Indian patients, 1992-2001. Int J Cardiol 2005;98:253—60.

[81] Kanafani ZA, Mahfouz TH, Kanj SS. Infective endocarditis at a tertiary care centre in Lebanon: predominance of streptococcal infection. J Infect 2002;45:152—9.

[82] Koegelenberg CF, Doubell AF, Orth H, Reuter H. Infective endocarditis in the Western Cape Province of South Africa: a three-year prospective study. QJM 2003;96:217—25.

[83] Essop MR, Nkomo VT. Rheumatic and nonrheumatic valvular heart disease: epidemiology, management, and prevention in Africa. Circulation 2005;112:3584—91.

[84] Shetty P. Grey matter: ageing in developing countries. Lancet 2012;379:1285—7.

[85] Khairy P, Ionescu-Ittu R, Mackie AS, Abrahamowicz M, Pilote L, Marelli AJ. Changing mortality in congenital heart disease. J Am Coll Cardiol 2010;56:1149—57.

[86] Marelli AJ, Ionescu-Ittu R, Mackie AS, Guo L, Dendukuri N, Kaouache M. Lifetime prevalence of congenital heart disease in the general population from 2000 to 2010. Circulation 2014;130:749—56.

[87] Warnes CA, Williams RG, Bashore TM, Child JS, Connolly HM, Dearani JA, Del Nido P, Fasules JW, Graham JR TP, Hijazi ZM, Hunt SA, King ME, Landzberg MJ, Miner PD, Radford MJ, Walsh EP, Webb GD. ACC/AHA 2008 guidelines for the management of adults with congenital heart disease: executive summary: a report of the American college of cardiology/American heart association task force on practice guidelines (writing committee to develop guidelines for the management of adults with congenital heart disease). Circulation 2008;118:2395—451.

[88] Wren C, O'sullivan JJ. Survival with congenital heart disease and need for follow up in adult life. Heart 2001;85:438—43.

[89] Best KE, Rankin J. Long-Term survival of individuals born with congenital heart disease: a systematic review and meta-analysis. J Am Heart Assoc 2016;5.

[90] Marelli AJ, Mackie AS, Ionescu-Ittu R, Rahme E, Pilote L. Congenital heart disease in the general population: changing prevalence and age distribution. Circulation 2007;115:163—72.

[91] Hoffman JI, Kaplan S. The incidence of congenital heart disease. J Am Coll Cardiol 2002;39:1890—900.

[92] Reller MD, Strickland MJ, Riehle-Colarusso T, Mahle WT, Correa A. Prevalence of congenital heart defects in metropolitan Atlanta, 1998-2005. J Pediatr 2008;153:807—13.

[93] Hoffman JI, Kaplan S, Liberthson RR. Prevalence of congenital heart disease. Am Heart J 2004;147:425—39.

[94] Tutar E, Ekici F, Atalay S, Nacar N. The prevalence of bicuspid aortic valve in newborns by echocardiographic screening. Am Heart J 2005;150:513—5.

[95] Ward C. Clinical significance of the bicuspid aortic valve. Heart 2000;83:81—5.

[96] Beppu S, Suzuki S, Matsuda H, Ohmori F, Nagata S, Miyatake K. Rapidity of progression of aortic stenosis in patients with congenital bicuspid aortic valves. Am J Cardiol 1993;71:322—7.

[97] Robicsek F, Thubrikar MJ, Cook JW, Fowler B. The congenitally bicuspid aortic valve: how does it function? Why does it fail? Ann Thorac Surg 2004;77:177—85.

[98] Sadee AS, Becker AE, Verheul HA, Bouma B, Hoedemaker G. Aortic valve regurgitation and the congenitally bicuspid aortic valve: a clinico-pathological correlation. Br Heart J 1992;67:439—41.

[99]   Masri A, Svensson LG, Griffin BP, Desai MY. Contemporary natural history of bicuspid aortic valve disease: a systematic review. Heart 2017;103:1323—30.

[100]   Michelena HI, Katan O, Suri RM, Baddour LM, Enriquez-Sarano M. Incidence of infective endocarditis in patients with bicuspid aortic valves in the community. Mayo Clin Proc 2016;91:122—3.

[101]   Michelena HI, Khanna AD, Mahoney D, Margaryan E, Topilsky Y, Suri RM, Eidem B, Edwards WD, Sundt 3rd TM, Enriquez-Sarano M. Incidence of aortic complications in patients with bicuspid aortic valves. J Am Med Assoc 2011;306:1104—12.

Further reading

[1]   Prophylaxis against infective endocarditis: antimicrobial prophylaxis against infective endocarditis in adults and children undergoing interventional procedures. 2008 [London].

CHAPTER 4

Surgical heart valves

Evan H. Phillips, Craig J. Goergen
Weldon School of Biomedical Engineering, Purdue University, West Lafayette, IN, United States

Contents

4.1 Introduction: history of surgical heart valves

The creation of artificial valves has revolutionized the surgical treatment of structural heart disease. Based on bottle stoppers from the 1800s, the first mechanical heart valves created in the 1950s had a ball-and-cage design [1]. Albert Starr, a cardiac surgeon, and Miles Lowell Edwards, an engineer with interest in entrepreneurship, met in 1957 and by extending principles from hydraulics and fuel pump operations [2] created a novel mechanical valve prosthesis that was first implanted in a patient in 1960 [3]. This artificial valve represented a life-saving technology for many patients who had no other option in the following decade. Unfortunately, this valve also created high transvalvular gradients that disrupted hemodynamics and required lifetime anticoagulation [4]. Although

Principles of Heart Valve Engineering
ISBN 978-0-12-814661-3, https://doi.org/10.1016/B978-0-12-814661-3.00004-6

the Starr–Edwards valve is no longer used today because of these drawbacks, artificial valve technology has advanced substantially over the past 60+ years with a variety of mechanical and bioprosthetic options. Today, more than 80,000 valve replacements utilizing mechanical or bioprosthetic valves occur per year in the United States, and more than 300,000 valve replacements occur worldwide [5,6]. Although transcatheter valve technology, covered in greater detail in Chapter 5, has also become popular in the past 20 years in high-risk surgical patients [7], surgical valve replacement is still the gold standard for treatment of many forms of structural heart disease. Today, mechanical aortic valve replacement (AVR) is associated with a substantial survival benefit, an early perioperative mortality less than 4%, and late-term mortality less than 2% per year [8]. Here, we describe a brief history of surgical valve development, with further details on the current use of both mechanical and bioprosthetic surgical valves. Future directions for improving surgical heart valve replacement will also be highlighted.

4.2 Mechanical valves

Several different designs in mechanical heart valves have been used since their original development more than 50 years ago. Current mechanical valve manufacturers include Edwards Lifesciences, Medtronic, St. Jude Medical, Livanova, and others (Fig. 4.1). Ongoing development continues to focus on improving valve hemodynamics and durability to reduce hemolysis and thrombus formation and ingrowth of native tissue called pannus that can hamper leaflet motion. However, current clinical recommendations still include lifetime anticoagulant to reduce the risk of thromboembolism in patients with mechanical heart valves, limiting their usage in some patients [9].

The Magovern–Cromie valve, a sutureless valve with a ball-and-cage design, was the first implanted mechanical heart valve [10]. Implantations were performed into the 1980s in thousands of patients with operative mortality dropping to 4.9% [11]. The Starr–Edwards valve, another ball-and-cage design, created lateralization of

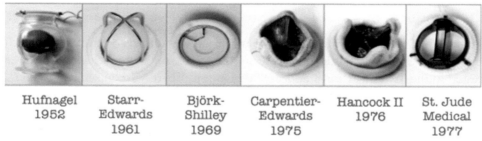

| Hufnagel 1952 | Starr- Edwards 1961 | Björk- Shilley 1969 | Carpentier- Edwards 1975 | Hancock II 1976 | St. Jude Medical 1977 |

Figure 4.1 *Early evolution of heart valves since 1950s* (https://cardiovascmed.ch/en/article/doi/cvm.2017.00532/). *(Reproduced from Naef AP and von Segesser LK. The development of open valve surgery. In: Thoracic and Cardiovascular Surgery. From the magic mountain to rocket)*

forward flow leading to complex flow patterns, and a large sewing ring restricted the size of the effective orifice area. Mitral valve usage was also limited due to the high profile [12,13]. To increase the effective valve opening area, the Björk—Shiley valve was developed with a tilting disk design where a central disk is held in place by two struts [14]. This valve creates two orifices, with a higher flow resistance in the small opening. The design was ultimately modified into a convexo-concave shape where the disk could move roughly 2 mm during each cycle. In a small but not insignificant percent of patients, long-term wear issues led to release of small metal fragments and fracture of the valve struts [15]. According to a New Your Times article, the FDA ultimately withdrew approval in 1986 after the valve was "implicated in about 300 deaths," highlighting the need for disclosing design changes and focusing on patient safety [16].

Rigid bileaflet mechanical valves create three outflow areas with more uniform central flow and better hemodynamics as compared with previous designs (Fig. 4.2). Additionally, the lower profile of the valve as compared with the ball-and-cage design made implantation easier. The first of these valves, a St. Jude Medical cardiac valve prosthesis, showed favorable results for AVR, mitral valve replacement (MVR), and double valve replacement (DVR) with "no cases of mechanical failure" [17]. With these advantages, a rigid bileaflet valve is the most common mechanical valve design used today.

4.2.1 Mechanical aortic valves

The St. Jude Medical bileaflet mechanical valve (now St. Jude Medical Regent valve) continues to be implanted in many younger patients undergoing AVR. The St. Jude valve is perhaps the most commonly used of all valves, and its design remains similar to the original design from the late 1970s. Review of the clinical outcomes at 3 years from a study published in 1981 revealed a low rate of thromboembolism (0.6/1000 patient-months) and no valve failures for 114 patients receiving AVR [17]. Harada et al. conducted a follow-up of 50 children between 1 and 10 years old in 1990 and determined that there was an actuarial survival rate of 91% [18]. These valves are now made in multiple different sizes to maximize the effective orifice area and diminish the risk for blood stagnation [19].

| Medtronic
Open Pivot | St Jude Medical
Regent | Sorin Group
Bicarbon Slimline | Sorin Group
Carbomedics | On-X Life
Technologies |

Figure 4.2 *Example mechanical valves from a variety of major manufactures.* *(From Arash Kheradvar, Elliott M. Groves, Craig A. Simmons et al, Emerging Trends in Heart Valve Engineering: Part III. Novel Technologies for Mitral Valve Repair and Replacement,Annals of Biomedical Engineering, 43(4),2018)*

Korteland et al. [8] have recently reported on long-term clinical outcomes after meta-analysis of more than 5000 patients (48 years of age and followed for 5.7 years on average) receiving AVR with bileaflet mechanical valves. They found that this nonelderly population had "suboptimal survival and considerable lifetime risk of anticoagulation-related complications, but also reoperation" [8]. Although the authors confirmed that structural valve deterioration (SVD) was not a major concern for these valves, their analysis showed that the youngest patients below 25 years of age at surgery had a high lifetime risk for reintervention (15%) and an elevated combined risk for thromboembolism, valve thrombosis, and bleeding (53%) [8].

To mimic a native valve and normal blood flow through a central orifice, there have been further research efforts dedicated to developing trileaflet valves. Li and Lu [20] compared a trileaflet valve to the St. Jude Medical bileaflet valve using simulations and found that the trileaflet design closed more slowly, which could reduce the occurrence of cavitation and risk of hemolysis. Kuan et al. [21] compared downstream blood flow profiles of trileaflet and bileaflet valve models using computational fluid dynamic modeling and observed the expected central region of flow and higher wall shear stresses with the trileaflet model. Gallegos et al. [22] tested the Triflo valve (Triflo Medical; Fig. 4.3) in sheep and observed good hemodynamics and a reduction in myocardial hypertrophy as compared with sheep implanted with the St. Jude Medical bileaflet valve. Unfortunately, the Triflo valve has not entered clinical trials despite these very promising experimental results in sheep [22] and calves [23].

To date, only preclinical studies have investigated the performance of sutureless mechanical valves. Berreklouw [24] implanted suture-denuded St. Jude Medical mechanical aortic valves with nitinol attachment rings in pigs. The device stayed "well attached at its annular position without migration or paradevice leakage"; however, tissue overgrowth was an issue in multiple animals [24]. While potentially advantageous by reducing paravalvular leak and overall surgical time, sutureless mechanical valves will need to be studied further before assessing the feasibility of this approach in humans.

Figure 4.3 Triflo valve (http://www.cardiovascular-engineering.com), Institute of Applied Medical Engineering, RWTH Aachen University.

4.2.2 Mitral position

The native mitral valve has two leaflets and is connected to the aortic valve via a fibrous tract called the aortomitral continuity. A bileaflet valve, rather than a trileaflet valve, enables proper closure and prevents potential obstruction to the left ventricular outflow tract to the aorta [25]. Mechanical surgical valves are now commonly used in the mitral position as well. Although the bileaflet design is common for both locations, the age threshold for MVR with either a bioprosthetic or mechanical valve remains controversial. Woo and colleagues found from a population-level analysis that 8027 patients between 50 and 69 years of age who received MVR had decreased survival (1.16 hazard ratio with a 95% confidence interval between 1.04 and 1.30) when implanted with a bioprosthesis [26]. They maintain that setting an age threshold younger than 70 years of age is not advisable; however, changes in practice over the course of this 18-year-long study may have biased these conclusions. Conversely, Chikwe et al. suggested from their retrospective analysis of 3433 MVR patients between 50 and 69 years of age that there was no significant difference in survival at 15 years between patients receiving mechanical and bioprosthetic valves [27]. Clearly, more prospective studies on the lifetime risks for MVR with either a bioprosthetic or mechanical valve are needed to improve the guidelines for age recommendations.

4.2.3 Other positions

The use of mechanical valves in the pulmonary position is relatively rare, but these valves are occasionally used to treat adult and congenital disease. Although a bioprosthetic valve is preferred in this position, mechanical pulmonary valves have been implanted in patients who are already taking anticoagulants or do not prefer a bioprosthetic [28]. Dehaki and colleagues retrospectively analyzed the clinical outcomes of congenital heart disease patients receiving mechanical pulmonary valves, primarily for Tetralogy of Fallot, and found low rates of reoperation (4% at 10 years) but an increased risk with "younger age, [and] longer interval between the repair of congenital defect and mPVR [mechanical pulmonary valve replacement]" [28,29]. Stulak et al. have recommended that patients with an International Normalized Ratio (INR) of 3 (described further below) and who are taking warfarin are suitable for mechanical pulmonary valves [30]. Their analysis of 59 patients who received a mechanical pulmonary valve at the Mayo Clinic showed no valve thrombosis and "no reoperations … for pannus formation, paravalvular leak, endocarditis, valve thrombosis or prosthesis dysfunction" [30]. Taherkhani et al. also retrospectively analyzed 16 patients receiving thrombolytic therapy for mechanical pulmonary or tricuspid valve thrombosis and found alteplase to be effective antithrombotic therapy in children [31]. Thus, select patients are recommended for mechanical pulmonary valve replacement when a bioprosthetic option is not favorable and anticoagulant administration can be administered.

Mechanical tricuspid valve replacement has been performed on a very small number of patients. Rossello et al. recently reported on 62 patients, primarily older females with rheumatic fever or functional regurgitation, who received mechanical tricuspid valves [32]. Their results showed high mortality rates postsurgery (18%) and in the long-term (34%), particularly for those with comorbidities, although surviving patients experienced freedom from symptoms associated with right heart failure [32]. The outlook for the use of mechanical tricuspid valves in adult patients is uncertain, given the greater use of bioprosthetic valves and transcatheter valves. However, young patients who have had a previous surgery could benefit from tricuspid valve replacement with a mechanical valve due to the increased longevity.

4.2.4 Need for anticoagulation

Anticoagulant therapy is required after heart valve replacement with mechanical prosthesis to reduce the risk of thrombosis formation. Warfarin is the standard of care oral anticoagulant for patients after aortic and mitral mechanical valve surgery [33]. Patients with a mechanical mitral valve are also advised to take aspirin [34]. Level IIa recommendations for anticoagulant therapy after bioprosthetic valve surgery are varied depending on the valve position and the patient. For example, low-dose aspirin (75—100 mg per day [34]) for the 3 months after implantation of an aortic bioprosthesis is recommended, while other indications for anticoagulation, such as venous thromboembolism, require lifelong warfarin [35].

A prothrombin time (PT) assay can be used to provide a patient-specific value that represents how quickly blood clots. PT results are normalized to a control sample, standardized according to the specific assay used, and usually provided as a ratio known as the INR [36,37]. A high INR suggests a greater likelihood of bleeding, whereas a low INR suggests a greater likelihood of thrombosis. Warfarin dosing should be tailored to a patient's target INR, so that the risks of bleeding and thrombosis are minimized. Postoperative patients have their INR monitored regularly until it is within the therapeutic range, and 5 mg of warfarin is typically recommended as a maintenance dosage [38]. Patients with mechanical valves in some European countries can self-manage their INR using portable devices and are trained to adjust their medication dosage. Koerfer et al. found from their review of two prospective studies that patients who trained and performed self-management used lower doses of anticoagulant and had a "high percentage of their measured INR values lying within the predetermined therapeutic range, thus resulting in a low rate of complications such as bleeding and thromboembolism" [39]. INR self-management has promising outcomes for many patients [40], and more studies should reveal which subgroups of patients require more training or medical oversight to minimize risk of complications.

Pregnancy raises the risk for thrombosis, and pregnant women with mechanical heart valves should seek medical advice about how to minimize clotting and bleeding risks for themselves and the fetus. Valve thrombosis has been shown to occur most frequently during the first trimester [41]. The primary issue is that heparin is less effective than warfarin at preventing thromboembolism in expecting mothers, but warfarin is potentially teratogenic [42]. Whether a low dosage of warfarin (less than 5 mg) may provide a significant improvement to fetal health over other anticoagulants is still not clear owing to the small number of pregnancies and lack of outcomes to be studied to date [43]. During labor, delivery, and after delivery, the risk of bleeding complications rises. Unfractionated heparin is typically administered closer to the time of delivery so that an epidural may be used; however, this form of the anticoagulant may result in a higher risk for bleeding complications than low molecular weight heparin [42]. Thus, the use of mechanical valves in young female patients is limited as many of these patients hope to become pregnant at some point in the future.

4.2.5 Evaluation techniques

The manufacturing, testing, and quality inspection of mechanical and bioprosthetic heart valves are typically conducted in a mostly manual fashion. Manual processes including assembly, microscopic inspection, and quality control documentation are rigorous and time-consuming. Benchtop testing of a valve typically includes measuring how well a valve closes, checking whether any leaks are found in the presence of physiological flow, and simulating fatigue and failure of the valve. Advanced computational techniques can now improve the capabilities of valve manufacturers to model valve geometry for simulating physiologically relevant flow. For example, Siemens has developed Simcenter to simulate the performance of valves through various models, including dynamic fluid–body interaction. With this approach, Khalili has shown that peak velocity values could be useful for identifying a dysfunctional valve based on a model of the St. Jude Regent Medical valve [44]. Overall, these computational approaches help to reduce the amount and length of benchtop testing, providing further confidence in the effects of minor design improvements.

4.3 Bioprosthetic valves

Bioprosthetic valves are typically constructed from porcine valve leaflets or bovine pericardium. A common design for bioprosthetic valves is to mount the xenograft tissue on a self-expanding nitinol stent. These valves are primarily designed as trileaflet aortic valves and can be implanted surgically by AVR or by a transcatheter approach (Chapter 5). More recent alternative bioprosthetic designs include stentless (e.g., Edwards Prima

Plus [45]) and sutureless (e.g., Sorin Perceval S [46]) valves. The primary advantages of bioprosthetic valves compared with mechanical valves are more ideal hemodynamics and a reduced need for anticoagulant therapy. Although improvements in valve design are extending the durability and lifetime of bioprosthetic valves, SVD still remains a key complication.

Surgeons now implant bioprosthetic valves in roughly 80% of all patients receiving a surgical implant, particularly older patients [47]. The 2017 AHA/ACC guideline for the management of patients with valvular disease recommends that bioprosthetic valves be used in older patients above 70 years of age when there is a higher risk of anticoagulation complications, and the issues associated with SVD are less important given the expected life expectancy [34]. Patients between 50 and 70 years of age need to weigh multiple factors, such as valve durability, bleeding risk, and the need for reintervention, as well as personal preference in deciding between a mechanical and bioprosthetic valve. Mechanical valves are generally favored for patients less than 50 years of age because the need for reoperation is higher in younger patients whose bioprosthetic valves undergo SVD [34]. Still, the next-generation of bioprosthetic valves may have improved durability, diminishing the need for reintervention. In this section, we highlight recent studies demonstrating the positive and negative clinical and preclinical findings of bioprosthetic aortic valves, while also discussing the outlook for bioprosthetic valves and their use in other cardiac positions.

4.3.1 Aortic bioprosthetic valves

4.3.1.1 Porcine aortic xenografts

In the industrialized world, older patients with aortic stenosis are routinely implanted with porcine xenograft valves. Multiple stented designs have been released by Medtronic (e.g., Mosaic bioprosthesis and Freestyle root) and St. Jude Medical (e.g., Epic xenograft and Toronto root) (Fig. 4.4A−D). Riess and colleagues recently showed that the Mosaic bioprosthesis led to an improved mortality but lower freedom from SVD at 17 years in patients less than 60 years of age at implant when compared with patients older than 60 [48]. They found, however, that the younger patients who underwent reoperations due to SVD had a very low mortality rate, providing some support for bioprosthetic implantation in younger patients. In a larger single-center study, Lehmann et al. evaluated the long-term clinical outcomes for patients with aortic stenosis or regurgitation who received AVR with the Epic xenograft (St. Jude Medical) [49]. After 10 years, 1843 out of 1920 patients with an average age of 77 years were free from structural valve disease.

For patients implanted with a Medtronic Freestyle aortic root bioprosthesis or a St. Jude Medical Toronto bioprosthesis, a comparative retrospective study to mechanical valves concluded that "the age threshold for the use of new-generation xenoroots may need to be lowered" [50]. The authors examined the long-term outcomes of 50- to 60-year-old patients receiving aortic root replacement with either a stentless porcine xenograft or mechanical composite valve and found similar 10-year survival (76% overall) and freedom from reoperation after 12 years (95%) [50].

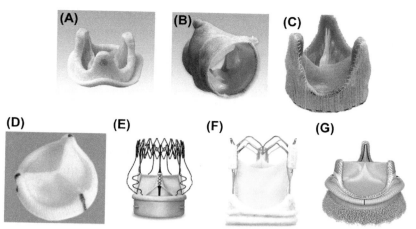

Figure 4.4 *Bioprosthetic surgical heart valves.* (A—C) Stented porcine bioprostheses. (A) Medtronic Mosaic, (B) Freestyle root (medtronic.com), (C) Toronto SPV (cthsurgery.com), (D) Premium bovine pericardium valve *Fábio Rocha Farias et al. Aortic valve replacement with the Cardioprotese Premium bovine pericardium bioprosthesis: four-year clinical results, Interactive CardioVascular and Thoracic Surgery, Volume 15, Issue 2, August 2012, Pages 229—234, https://academic.oup.com/icvts/article/15/2/229/804702.* (E—G) Sutureless bioprosthetic valves. (E) LivaNova Perceval S (livanova.sorin.com), (F) ATS 3f Enable (Eur J Cardio Thorac Surg (10.1016/j.ejcts.2010.12.068)), and (G) Edwards Intuity (edwards.com). *((A,B) Reproduced with permission of Medtronic, Inc.; (D) From Fábio Rocha Farias et al. Aortic valve replacement with the Cardioprotese Premium bovine pericardium bioprosthesis: four-year clinical results, Interactive CardioVascular and Thoracic Surgery, Volume 15, Issue 2, August 2012, Pages 229—234; (F) Perceval® Sutureless Aortic Heart Valve (LivaNova). Used with permission from LivaNova; (H) with permission from Edwards Lifesciences.)*

Stentless valve designs have shown to preserve hemodynamics and to have improved resistance under conditions of rest as well as exercise. Khoo et al. [51] compared five different valve designs, including stentless porcine and bovine aortic prostheses. Their study demonstrated that "stentless bioprostheses have superior hemodynamics to stented and mechanical valves at rest, which become more significant during stress" [51]. These findings suggest that stentless valves have favorable designs to mimic physiologic blood flow. Future studies will need to compare the long-term durability of these valves.

4.3.1.2 Bovine pericardium

In the 1980s, Cosgrove et al. compared bovine pericardial valves with the standard Carpentier–Edwards porcine valve during operations and found that the former had a superior hemodynamic profile [52]. Patients with a small aortic root are especially likely to experience higher pressure gradients after aortic valve surgery. Modi et al. [53] determined that the St. Jude Medical Trifecta valve, which was "designed to produce less impedance to transvalvular flow compared to other stented bioprostheses," had low transvalvular gradients approximately 1 year after surgery. Goldman et al. [54] showed that the Trifecta bioprosthesis yielded "excellent hemodynamic performance, particularly in the smaller valve sizes, that is maintained through 6 years of follow-up." Comparing their results to those from studies investigating the PERIMOUNT valve from Edwards

Lifesciences after 7 years [55], they noted similar freedom from SVD (97%). Overall, bovine pericardial valves exhibit good hemodynamic performance leading to several advantageous properties, especially in patients with small aortic roots.

4.3.1.3 Sutureless bioprostheses

Bioprostheses are now available as sutureless valves without a sewing ring. This design offers potential advantages for reducing operation times, enabling less invasive surgical approaches while improving outcomes for higher risk patients. The Perceval S is a widely used sutureless valve that is commercially available in Europe. Two other commonly used sutureless valves are the ATS 3f Enable and the Edwards Intuity valves (Fig. 4.4E−G). The first 5-year follow-up study for the Perceval S had "excellent clinical and hemodynamic outcomes" for 30 patients who were 75−88 years of age at surgery [56]. Other studies on the long-term performance of the Perceval S are limited, but the early results are positive for both minimally invasive and standard AVR [57]. The recognition that standard size valves can result in "patient-prosthesis mismatch" is a primary motivation for choosing sutureless valves [58]. That is, patients with a small and/or calcified aortic root could benefit from valves without a sewing ring, allowing for a larger effective orifice area [59]. Indeed, an in vitro study by Tasca et al. [58] provided preliminary evidence that the Perceval bioprosthesis has "good fluid dynamic performance" in small aortic annuli as compared with two standard pericardial stented bioprostheses (Crown and Magna).

Another option for high-risk patients requiring an aortic valve is to undergo surgery through a minimally invasive approach (i.e., transcatheter aortic valve implantation [TAVI]; see Chapter 5). A recent meta-analysis of studies comparing sutureless AVR to TAVI revealed that sutureless valves led to a reduction in early mortality and a lower incidence of paravalvular leakage compared with transcatheter valves [60]. This report recognized, however, that data are lacking on long-term outcomes, the performance of newer generation valves, and differences in costs between these procedures. Other case studies have described the use of a Perceval S bioprosthesis in difficult cases where patients required reoperation after previously receiving a stentless Freestyle root. Valve–in–valve procedures with a sutureless valve are typically required in instances when a patient has high operative mortality risk and a lengthy operation is a concern [61].

4.3.1.4 Outlook for bioprosthetic valves

Bioprosthetic valves are favorable in many moderate- to high-risk patients of advanced age. Current areas of design improvements include better tissue fixation to extend valve durability and the addition of outer material, such as wraps, which allow for better valve sealing. Patients undergoing a redo procedure [46] or those with an extremely calcified (porcelain) aorta [62], a degenerated aortic root [63], or a small aortic root [59] have all had positive outcomes after receiving sutureless bioprosthetic valves. For patients who currently undergo conventional surgery rather than minimally invasive AVR, both reintervention rates and overall procedure costs are lower.

As sutureless valves are self-expanding, the sizing required to fit the native aortic annulus in vivo is an important consideration during surgery. Cerillo et al. [64] conducted a clinical study where they calculated the degree of oversizing of the Perceval valve and found that oversizing was associated with an increased postoperative gradient. They suggested that developing a "different sizing algorithm" with a more conservative sizing strategy could improve the use of sutureless valves [64]. Further reports by Muneretto et al. [65] and Biancari et al. [66] discovered that postoperative pacemaker implantations were less frequent with the use of sutureless valves (2% and 10%, respectively) as compared with TAVI (26% and 17%). Future work will be needed to update guidelines and strategies associated with sutureless bioprosthetic aortic valve implantation as experience is gained with these devices.

4.3.2 Other locations

4.3.2.1 Mitral position

MVR using a bioprosthesis is often performed with a stented valve. Commonly used models include the Medtronic Mosaic, St. Jude Medical Epic, and Carpentier-Edwards Perimount bioprostheses. Riess et al. [48] recently reported on the 17-year outcomes for the Mosaic mitral bioprosthesis and concluded that patients younger than 60 years old at implant did not have a statistically significant different freedom from explant due to SVD when compared with older patients. Celiento et al. [67] found that the Mosaic provided an "actuarial freedom at 15 years" of more than 90% from thromboembolic episodes and SVD, suggesting it to be suitable for older MVR patients. Patients receiving the Epic xenograft in the mitral position have also been followed over the long term. As part of Lehmann and colleagues' single-center trial [49], they compared clinical outcomes for AVR, MVR, and DVR patients and found similar levels for "acceptable survival, freedom from valve-related complications, and freedom from valve reintervention" between the AVR and MVR cohorts. For the Carpentier-Edwards Perimount mitral bioprosthesis, a recent study by Bourguignon et al. [68] demonstrated that it has an "expected valve durability of 14.2 years" when implanted in MVR patients at 65 years of age and younger. Recently, a case of leaflet detachment was reported in a woman who underwent MVR 9 years ago [69]. Despite the likelihood of requiring a reoperation, the advantage for the majority of younger patients who elect for a bioprosthetic MVR is the increased quality of life from not having to take anticoagulation medication indefinitely [68,70].

Other bioprosthetic mitral valve designs have also been developed to mitigate the deficits of the conventional bileaflet design. In the early 90s, a quadrileaflet stentless mitral valve (SMV) incorporating bovine pericardium was designed by Dr. Robert Frater [71]. More recently, Kasegawa et al. [72] designed and performed in vitro testing on a novel bileaflet SMV with a flexible ring and two "legs" (i.e., papillary flaps) and observed high transvalvular flow, large effective orifice area, and "excellent closing function." Preliminary in vivo testing in pigs yielded positive results in terms of mitral regurgitation [73], and two human subjects who required redo procedures were "free from reintervention and readmission for heart failure during 12 months postoperatively" [74]. This valve design could potentially allow for implantation in a small mitral annulus and "good coaptation postimplantation, even when the anchoring of the SMV legs is less than optimal" [72]. Despite this initial success, these alternative designs have yet to be used in a large number of patients. Overall, SMVs represent a promising design but will require further clinical assessment on long-term durability.

4.3.2.2 Pulmonary position

Bioprosthetic valves are not often implanted in the pulmonary position. Long-term durability and utility of these valves have yet to be demonstrated for patients with congenital heart defects or those requiring reoperation. Lodge and colleagues retrospectively analyzed 170 cases of patients between 0.5 and 72 years of age receiving either a porcine or pericardial valve and found that the patients younger than 15 were at a greater risk for reintervention, pulmonic insufficiency, or pulmonic stenosis [75]. The results of this single-center study suggest that both valves "provide a durable pulmonary valve substitute and good quality of life, with most patients free from significant pulmonic stenosis or pulmonic insufficiency at intermediate follow-up" [75]. Further work will be needed to fully optimize bioprosthetic replacement valves for the relatively high velocities and low pressure associated with pulmonary valve hemodynamics.

4.3.2.3 Tricuspid position

Tricuspid valve replacement is reserved for only select high-risk surgical patients. Nevertheless, tricuspid valve repair and valve replacement surgeries are being performed more regularly now than in the past decade [76]. Bioprosthetic valves are routinely used and have been suggested to have better durability as compared with valves implanted on the left side of the heart [77,78]. Still, it remains a challenge to reduce the mortality rate for patients undergoing tricuspid valve replacement, suggesting further improvements are needed.

4.3.3 Longevity issues for bioprosthetic valves

SVD is the primary issue with bioprosthetic valves. Several examples of SVD have been reported in the literature, including calcification [79] and tearing [54]. For sutureless valves such as the Perceval S, there are also newly identified issues arising from the design and surgical placement. Dalén et al. [80] found that patients receiving the Perceval S and taking anticoagulant medication were likely to develop hypoattenuated leaflet thickening and reduced leaflet motion. Aljalloud et al. [81] described a cusp-fluttering effect that occurs because of the flexible design of the Perceval S. Lee et al. [82] highlighted the case of a patient who developed exacerbated mitral regurgitation after AVR with the Perceval S but no aortic regurgitation or paravalvular leakage. They attributed this to the low placement required for the bioprosthesis and impingement on a mitral valve leaflet. Given the variety of longevity issues, protecting leaflets from significant degeneration in vivo is a clear unmet clinical need.

4.3.4 Summary

Positive postprocedural outcomes and reasonable costs are observed in patients receiving bioprosthetic valves. Further development in bioprosthetic valve technology could lead to increased use in elderly and middle-aged patients, as well as device approval in low- and middle-income countries.

4.4 Prosthetic heart valve selection and development

The decision about which heart valve to implant is multifaceted and requires advanced medical expertise, awareness, and access to different prosthetic valves. Understanding both the risks associated with the procedure and required postoperative care is key. Current medical guidelines based on clinical trial evidence are extensive, providing age recommendations and contraindications, among others. It can thus be difficult for some patients to decide between a mechanical or bioprosthetic valve. Ideally, the selected valve should provide an optimal balance of improved survival with minimal risk of complications.

There are multiple criteria to consider when developing new valves that improve on current well-established designs: (1) embolism prevention, (2) valve durability, (3) ease and security of attachment to native tissue, (4) preservation of surrounding tissue function, (5) reduction of turbulence, (6) reduction of blood trauma, (7) minimal noise, (8) use of materials compatible with blood, and (9) development of improved methods for sterilization and storage. Many of the valves detailed above meet only some of these requirements. In combination with more advanced technology for testing and validation, it is anticipated that newer generation mechanical and bioprosthetic valves can better address all of these criteria.

4.5 Unmet clinical needs and future areas of development

Work is still needed to create long-lasting replacement surgical heart valves that can be easily implanted, have improved hemodynamics, and are not susceptible to pannus that can obstruct leaflet motion. Indeed, the ideal valve would not be associated with complications that require pacemakers and would completely eliminate the need for anticoagulants. Although a large amount of effort was initially made to improve the overall design of mechanical heart valves, the bileaflet design offered by most medical device manufacturers has been around since the 1970s. This leaves open the opportunity to use advanced computational and imaging approaches to optimize valve hemodynamics in both normal and dysfunctional states [44,83,84]. Computational fluid dynamic techniques are now available to simulate blood flow through surgical heart valves using a rigid body assumption of leaflet motion and a generalized Newtonian model of blood that can capture its shear-thinning properties [85−89]. More recently, fluid−structure interaction (FSI) approaches have been developed, which can take into account deformations of cardiac tissue and arteries [90−93]. Improvement in hemodynamics could ultimately lead to reduced need for blood thinning medication, such as warfarin, that is administered to reduce the risk of thrombus formation and stroke [94,95]. Current clinical guidelines suggest that patients with mechanical heart valves maintain a target INR level of 2.5 [96]. Yet in 2015, FDA expanded the on-label use of On-X aortic heart valve to patients with a target INR level of 1.5−2.0 because of the improved hemodynamics and thromboresistant surface of this valve [97]. Indeed, this lower INR level is closer to an unmedicated state, reduces the risk of hemorrhagic complications, and increases the number of activities patients can participate in. Despite these recent advancements, continued work is still needed to further optimize mechanical valve hemodynamics and completely eliminate the need for anticoagulation therapy.

Anticoagulation therapy is even more complicated for pregnant women. Unfortunately, the current recommendations for pregnant women with mechanical valves require careful management of the anticoagulation that still brings about tradeoffs in risk for the mother and fetus. An optimized dosing strategy or an anticoagulant drug that sufficiently reduces risk of thrombosis while posing no risk to the fetus remains an unmet need.

A clear benefit of bioprosthetic valves is the reduced need for anticoagulation therapy. Unfortunately, tissue valves typically wear out due to SVD and typically fail within 10−15 years in many patients. Thus, a bioprosthetic valve with improved durability that lasts longer would open up the possibility of implanting tissue valves in younger patients who often prefer not to be on warfarin or other anticoagulants indefinitely. Extending tissue valve durability is an active area of development for many of the major medical device manufacturers, as previous storage in tissue fixative is now being replaced by more robust dry storage methods [98,99]. Expandable bioprosthetic valves that are suitable for a later valve-in-valve TAVI are another potentially encouraging option for patients who may want to avoid invasive reintervention. Furthermore, tissue-engineered heart valves are an exciting field that shows promise for young growing patients and for regenerative medicine (see Chapter 6). Polymeric heart valves (see Chapter 14) are also being developed to overcome the greatest limitations of both mechanical and bioprosthetic valves [100]. These growing fields could have a profound effect for patients through improvements to the longevity and durability of surgically implanted valves.

Future development for mitral valves must consider the position-specific factors and circumstances that make MVR difficult. In particular, patients who have undergone MVR and require reoperation are more likely to have complications. Therefore, valve longevity and durability must be prioritized in the design of new-generation mitral valves. As compared to the aortic position, there is also a greater risk for thromboembolic events to occur with mechanical mitral valves. On the right side of the heart where lower pressure conditions exist, surgeons are now often performing initial testing of tricuspid valve replacement as patients can survive for many years without a functioning valve in this position.

FSI models of left ventricular flow can reveal differences among aortic and mitral replacement valves. Modeling cardiac flow is particularly helpful for identifying areas of stagnated flow, which are particularly problematic with bileaflet mechanical mitral valves (Fig. 4.5A−C) [101,102]. Moreover, implementing an accurate tissue constitutive model from various diseased aorta and valve anatomies will clearly impact FSI simulations (Fig. 4.5D−E) [103,104]. Echocardiographic assessment of patients should also be used in helping to predict outcomes. For example, assessing which valves lead to greater regression in left ventricular hypertrophy may be a strong indicator for which valves have a greater likelihood of long-term survival [105].

Ultimately, the ideal surgical prosthesis will be a long-lasting, hemodynamically optimized valve that can be implanted through less invasive techniques and requires no anticoagulation, pacemakers, or other devices. Furthermore, the chosen surgical approach should have minimal complications. In low- and middle-income countries where access to clinical care is limited, there is a pressing need for cardiac valves. In this setting, rheumatic fever leads to significant valvular disease, which is less common in the developed world. However, the cost for valve replacement surgery is prohibitive in many settings [106]. TAVI, however, shows promise for resource-limited locations [107]. Additionally, it has already been shown that handheld ultrasound devices perform well and can be used to predict the development of rheumatic heart disease [108]. Cardiac care programs could be established in countries with limited resources to identify and treat urgent cases of severe rheumatic fever before structural heart disease has developed. Meanwhile, minimally invasive approaches such as TAVI and portable ultrasound devices could be solutions to providing more accessible options.

Mechanical and bioprosthetic heart valves are now being implanted, which are suitable for patients of different ages and with differently sized valves. Novel valve designs, such as sutureless valves, can provide an individualized fit with improved hemodynamics for patients. Future recommendation guidelines will need to weigh the evidence from prospective clinical trials to determine whether age limits can be lowered or other clinical factors should be adjusted for the next generation of surgical valves. With continued advancements, surgical heart valves will likely remain in clinical use for decades to come, helping to improve the lives of patients with structural heart disease.

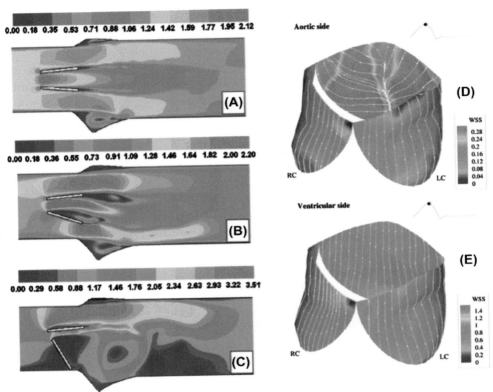

Figure 4.5 *Computational fluid dynamic modeling of surgical valves.* Cardiac flow with a bileaflet mechanical valve (left column). FSI simulation using a tissue constitutive model (right column). (A), Normally functioning bileaflet prosthesis. The flow velocity within the central orifice is higher than that in the lateral orifices. Accordingly, the peak gradient across the central orifice is 19 mm Hg, which is higher than the actual peak transprosthetic gradient (10 mm Hg). (B), Mild prosthesis dysfunction with 25% restriction in the opening of one leaflet. The peak gradient is 20 mm Hg. (C), Severe prosthesis dysfunction with one leaflet blocked in the closed position. The peak gradient is 50 mm Hg [102]. Contours of wall shear stresses (WSS, in Pa) and limiting streamlines (LSL, white lines with arrows, indicating direction of a blood flow) are shown on leaflet surfaces from the aortic side (D) and from the ventricular side (E) at peak systole (t = 0.23 s) [103]. ((A,B,C) Courtesy of Drs Othman Smadi and Lyes Kadem, Concordia University, Montreal, Québec, Canada.; D,E) Reproduced from "Flow—Structure Interaction Simulations of the Aortic Heart Valve at Physiologic Conditions: The Role of Tissue Constitutive Model," by Anvar Gilmanov; Henryk Stolarski; Fotis Sotiropoulos, Journal of Biomech Eng. 2018; 140(4), with permission from ASME.)

References

[1] Gott VL, Alejo DE, Cameron DE. Mechanical heart valves: 50 years of evolution. Ann Thorac Surg 2003;76:S2230—9.

[2] Marchand MA, Aupart MR, Norton R, Goldsmith IRA, Pelletier LC, Pellerin M, Dubiel T, Daenen WJ, Herijgers P, Casselman FP, Holden MP, David TE. Fifteen-year experience with the mitral Carpentier-Edwards PERIMOUNT pericardial bioprosthesis. Ann Thorac Surg 2001;71: S236—9.

[3] Starr A, Edwards ML. Mitral replacement: clinical experience with a ball-valve prosthesis. Ann Surg 1961;154:726—40.

[4] Offstad J, Andersen K, Paulsson P, Andreasson J, Kjellman U, Lundblad O, Engstrom KG, Haaverstad R, Svennevig JL. The Scandinavian multicenter hemodynamic evaluation of the SJM regent aortic valve. J Cardiothorac Surg 2011;6:163.

[5] Rahimtoola SH. Choice of prosthetic heart valve in adults an update. J Am Coll Cardiol 2010;55: 2413—26.

[6] Sato M, Harasaki H, Wika KE, Soloviev MV, Lee AS. Blood compatibility of a newly developed trileaflet mechanical heart valve. Am Soc Artif Intern Organs J 2003;49:117—22.

[7] Leon MB, Smith CR, Mack M, Miller DC, Moses JW, Svensson LG, Tuzcu EM, Webb JG, Fontana GP, Makkar RR, Brown DL, Block PC, Guyton RA, Pichard AD, Bavaria JE, Herrmann HC, Douglas PS, Petersen JL, Akin JJ, Anderson WN, Wang D, Pocock S, Investigators PT. Transcatheter aortic-valve implantation for aortic stenosis in patients who cannot undergo surgery. N Engl J Med 2010;363:1597—607.

[8] Korteland NM, Etnel JRG, Arabkhani B, Mokhles MM, Mohamad A, Roos-Hesselink JW, Bogers A, Takkenberg JJM. Mechanical aortic valve replacement in non-elderly adults: meta-analysis and microsimulation. Eur Heart J 2017;38:3370—7.

[9] Kheradvar A, Groves EM, Goergen CJ, Alavi SH, Tranquillo R, Simmons CA, Dasi LP, Grande-Allen KJ, Mofrad MR, Falahatpisheh A, Griffith B, Baaijens F, Little SH, Canic S. Emerging trends in heart valve engineering: part II. Novel and standard technologies for aortic valve replacement. Ann Biomed Eng 2015;43:844—57.

[10] Roman TP, Schattenberg TT, Oxman HA, Wallace RB. Magovern-Cromie ball valve prosthesis for aortic valve replacement. Mayo Clin Proc 1973;48:537—40.

[11] Magovern GJ, Liebler GA, Park SB, Burkholder JA, Sakert T, Simpson KA. Twenty-five-year review of the Magovern-Cromie sutureless aortic valve. Ann Thorac Surg 1989;48:S33—4.

[12] Cohn LH. Cardiac surgery in the adult. 3rd ed. New York: McGraw Hill Professional; 2008.

[13] Chaikof EL. The development of prosthetic heart valves—lessons in form and function. N Engl J Med 2007;357:1368—71.

[14] Lindblom D, Rodriguez L, Björk VO. Mechanical failure of the Björk-Shiley valve. Updated follow-up and considerations on prophylactic rereplacement. J Thorac Cardiovasc Surg 1989;97:95.

[15] Blot WJ, Ibrahim MA, Ivey TD, Acheson DE, Brookmeyer R, Weyman A, Defauw J, Smith JK, Harrison D. Twenty-five-year experience with the Bjork-Shiley convexoconcave heart valve: a continuing clinical concern. Circulation 2005;111:2850—7.

[16] T. A. Press, in the New York times. 1992. p. 001009.

[17] Nicoloff DM, Emery RW, Arom KV, Northrup WF, Jorgensen CR, Wang Y, Lindsay WG. Clinical and hemodynamic results with the St. Jude Medical cardiac valve prosthesis. A three-year experience. J Thorac Cardiovasc Surg 1981;82:674.

[18] Harada Y, Imai Y, Kurosawa H, Ishihara K, Kawada M, Fukuchi S. Ten-year follow-up after valve replacement with the St. Jude Medical prosthesis in children. J Thorac Cardiovasc Surg 1990;100: 175—80.

[19] Czer LSC, Chaux A, Matloff JM, DeRobertis MA, Nessim SA, Scarlata D, Khan SS, Kass RM, Po Tsai T, Blanche C, Gray RJ. Ten-year experience with the St. Jude Medical valve for primary valve replacement. J Thorac Cardiovasc Surg 1990;100:44—55.

[20] Li CP, Lu PC. Numerical comparison of the closing dynamics of a new trileaflet and a bileaflet mechanical aortic heart valve. J Artif Organs 2012;15:364—74.

[21] Kuan YH, Nguyen VT, Kabinejadian F, Leo HL. Computational hemodynamic investigation of bileaflet and trileaflet mechanical heart valves. J Heart Valve Dis 2015;24:393—403.

[22] Gallegos RP, Rivard AL, Suwan PT, Black S, Bertog S, Steinseifer U, Armien A, Lahti M, Bianco RW. In-vivo experience with the Triflo trileaflet mechanical heart valve. J Heart Valve Dis 2006;15:791—9.

[23] Gregoric ID, Eya K, Tamez D, Cervera R, Byler D, Conger J, Tuzun E, Chee HK, Clubb FJ, Kadipasaoglu K, Frazier OH. Preclinical hemodynamic assessment of a new trileaflet mechanical valve in the aortic position in a bovine model. J Heart Valve Dis 2004;13:254—9.

[24] Berreklouw E, Koene B, De Somer F, Bouchez S, Chiers K, Taeymans Y, Van Nooten GJ. Sutureless replacement of aortic valves with St Jude Medical mechanical valve prostheses and Nitinol attachment rings: feasibility in long-term (90-day) pig experiments. J Thorac Cardiovasc Surg 2011;141:1231–7. e1231.

[25] D'Ancona G, Neuhausen-Abramkina A, Atmowihardjo I, Kische S, Ince H. Tri-leaflet mitral valve anatomy: a rare occurrence leading to severe mitral valve regurgitation. Eur Heart J 2015;36:1697.

[26] Goldstone AB, Chiu P, Baiocchi M, Lingala B, Patrick WL, Fischbein MP, Woo YJ. Mechanical or biologic prostheses for aortic-valve and mitral-valve replacement. N Engl J Med 2017;377:1847–57.

[27] Chikwe J, Chiang YP, Egorova NN, Itagaki S, Adams DH. Survival and outcomes following bioprosthetic vs mechanical mitral valve replacement in patients aged 50 to 69 years. J Am Med Assoc 2015;313:1435–42.

[28] Dehaki MG, Al-Dairy A, Rezaei Y, Omrani G, Jalali AH, Javadikasgari H, Dehaki MG. Mid-term outcomes of mechanical pulmonary valve replacement: a single-institutional experience of 396 patients. Gen Thorac Cardiovasc Surg 2019;67(3):289–96. https://www.ncbi.nlm.nih.gov/pubmed/30209777.

[29] Dehaki MG, Ghavidel AA, Omrani G, Javadikasgari H. Long-term outcome of mechanical pulmonary valve replacement in 121 patients with congenital heart disease. Thorac Cardiovasc Surg 2015;63:367–72.

[30] Stulak JM, Mora BN, Said SM, Schaff HV, Dearani JA. Mechanical pulmonary valve replacement. Semin Thorac Cardiovasc Surg Pediatr Card Surg Annu 2016;19:82–9.

[31] Taherkhani M, Hashemi SR, Hekmat M, Safi M, Taherkhani A, Movahed MR. Thrombolytic therapy for right-sided mechanical pulmonic and tricuspid valves: the largest survival analysis to date. Tex Heart Inst J 2015;42:543–7.

[32] Rossello X, Munoz-Guijosa C, Mena E, Camprecios M, Mendez AB, Borras X, Padro JM. Tricuspid valve replacement with mechanical prostheses: short and long-term outcomes. J Card Surg 2017;32:542–9.

[33] Baumann Kreuziger L, Karkouti K, Tweddell J, Massicotte MP. Antithrombotic therapy management of adult and pediatric cardiac surgery patients. J Thromb Haemost 2018;16(11):2133–46. https://www.ncbi.nlm.nih.gov/pubmed/30153372.

[34] Nishimura RA, Otto CM, Bonow RO, Carabello BA, Erwin 3rd JP, Fleisher LA, Jneid H, Mack MJ, McLeod CJ, O'Gara PT, Rigolin VH, Sundt 3rd TM, Thompson A. AHA/ACC focused update of the 2014 AHA/ACC guideline for the management of patients with valvular heart disease: a report of the American college of cardiology/American heart association task force on clinical practice guidelines. J Am Coll Cardiol 2017;70:252–89.

[35] Falk V, Baumgartner H, Bax JJ, De Bonis M, Hamm C, Holm PJ, Iung B, Lancellotti P, Lansac E, Munoz DR, Rosenhek R, Sjogren J, Tornos Mas P, Vahanian A, Walther T, Wendler O, Windecker S, Zamorano JL, Group ESCSD. ESC/EACTS Guidelines for the management of valvular heart disease. Eur J Cardiothorac Surg 2017;52:616–64.

[36] Poller L. Therapeutic ranges in anticoagulant administration. Br Med J (Clin Res Ed) 1985;290:1683–6.

[37] Edmunds Jr LH. Thrombotic and bleeding complications of prosthetic heart valves. Ann Thorac Surg 1987;44:430–45.

[38] Wigle P, Hein B, Bloomfield HE, Tubb M, Doherty M. Updated guidelines on outpatient anticoagulation. Am Fam Physician 2013;87:556–66.

[39] Koerfer R, Reiss N, Koertke H. International normalized ratio patient self-management for mechanical valves: is it safe enough? Curr Opin Cardiol 2009;24:130–5.

[40] Pozzi M, Mitchell J, Henaine AM, Hanna N, Safi O, Henaine R. International normalized ratio self-testing and self-management: improving patient outcomes. Vasc Health Risk Manag 2016;12:387–92.

[41] van Hagen IM, Roos-Hesselink JW, Ruys TP, Merz WM, Goland S, Gabriel H, Lelonek M, Trojnarska O, Al Mahmeed WA, Balint HO, Ashour Z, Baumgartner H, Boersma E, Johnson MR, Hall R, ROPAC Investigators and the EURObservational Research Programme (EORP) Team, Pregnancy in women with a mechanical heart valve: data of the European society of cardiology registry of pregnancy and cardiac disease (ROPAC). Circulation 2015;132:132–42.

[42] Fogerty AE. Challenges of anticoagulation therapy in pregnancy. Curr Treat Options Cardiovasc Med 2017;19:76.

[43] D'Souza R, Ostro J, Shah PS, Silversides CK, Malinowski A, Murphy KE, Sermer M, Shehata N. Anticoagulation for pregnant women with mechanical heart valves: a systematic review and meta-analysis. Eur Heart J 2017;38:1509—16.

[44] Khalili F, Gamage P, Mansy H. Hemodynamics of a bileaflet mechanical heart valve with different levels of dysfunction. J Appl Biotechnol Bioeng 2017;2:00044.

[45] Auriemma S, D'Onofrio A, Brunelli M, Magagna P, Paccanaro M, Rulfo F, Fabbri A. Long-term results of aortic valve replacement with Edwards prima plus stentless bioprosthesis: eleven years' follow up. J Heart Valve Dis 2006;15:691—5. discussion 695.

[46] Santarpino G, Pfeiffer S, Concistre G, Fischlein T. REDO aortic valve replacement: the sutureless approach. J Heart Valve Dis 2013;22:615—20.

[47] Antunes MJ. Porcine or bovine: does it really matter? Eur J Cardiothorac Surg 2015;47:1075—6.

[48] Riess FC, Fradet G, Lavoie A, Legget M. Long-term outcomes of the Mosaic bioprosthesis. Ann Thorac Surg 2018;105:763—9.

[49] Lehmann S, Merk DR, Etz CD, Oberbach A, Uhlemann M, Emrich F, Funkat AK, Meyer A, Garbade J, Bakhtiary F, Misfeld M, Mohr FW. Porcine xenograft for aortic, mitral and double valve replacement: long-term results of 2544 consecutive patients. Eur J Cardiothorac Surg 2016;49: 1150—6.

[50] Etz CD, Girrbach FF, von Aspern K, Battellini R, Dohmen P, Hoyer A, Luehr M, Misfeld M, Borger MA, Mohr FW. Longevity after aortic root replacement: is the mechanically valved conduit really the gold standard for quinquagenarians? Circulation 2013;128:S253—62.

[51] Khoo JP, Davies JE, Ang KL, Galinanes M, Chin DT. Differences in performance of five types of aortic valve prostheses: haemodynamic assessment by dobutamine stress echocardiography. Heart 2013;99:41—7.

[52] Cosgrove DM, Lytle BW, Gill CC, Golding LA, Stewart RW, Loop FD, Williams GW. In vivo hemodynamic comparison of porcine and pericardial valves. J Thorac Cardiovasc Surg 1985;89: 358—68.

[53] Modi A, Budra M, Miskolczi S, Velissaris T, Kaarne M, Barlow CW, Livesey SA, Ohri SK, Tsang GM. Hemodynamic performance of Trifecta: single-center experience of 400 patients. Asian Cardiovasc Thorac Ann 2015;23:140—5.

[54] Goldman S, Cheung A, Bavaria JE, Petracek MR, Groh MA, Schaff HV. Midterm, multicenter clinical and hemodynamic results for the Trifecta aortic pericardial valve. J Thorac Cardiovasc Surg 2017; 153:561—569 e562.

[55] Frater RW, Salomon NW, Rainer WG, Cosgrove 3rd DM, Wickham E. The Carpentier-Edwards pericardial aortic valve: intermediate results. Ann Thorac Surg 1992;53:764—71.

[56] Meuris B, Flameng WJ, Laborde F, Folliguet TA, Haverich A, Shrestha M. Five-year results of the pilot trial of a sutureless valve. J Thorac Cardiovasc Surg 2015;150:84—8.

[57] Sian K, Li S, Selvakumar D, Mejia R. Early results of the Sorin((R)) Perceval S sutureless valve: systematic review and meta-analysis. J Thorac Dis 2017;9:711—24.

[58] Tasca G, Vismara R, Mangini A, Romagnoni C, Contino M, Redaelli A, Fiore GB, Antona C. Comparison of the performance of a sutureless bioprosthesis with two pericardial stented valves on small annuli: an in vitro study. Ann Thorac Surg 2017;103:139—44.

[59] Shrestha M, Maeding I, Hoffler K, Koigeldiyev N, Marsch G, Siemeni T, Fleissner F, Haverich A. Aortic valve replacement in geriatric patients with small aortic roots: are sutureless valves the future? Interact Cardiovasc Thorac Surg 2013;17:778—82. discussion 782.

[60] Shinn SH, Altarabsheh SE, Deo SV, Sabik JH, Markowitz AH, Park SJ. A systemic review and meta-analysis of sutureless aortic valve replacement versus transcatheter aortic valve implantation. Ann Thorac Surg 2018;106:924—9.

[61] Lio A, Miceli A, Ferrarini M, Glauber M. Perceval S valve solution for degenerated Freestyle root in the presence of chronic aortic dissection. Ann Thorac Surg 2016;101:2365—7.

[62] Gatti G, Benussi B, Camerini F, Pappalardo A. Aortic valve replacement within an unexpected porcelain aorta: the sutureless valve option. Interact Cardiovasc Thorac Surg 2014;18:396—8.

[63] Folliguet TA, Laborde F. Sutureless Perceval aortic valve replacement in aortic homograft. Ann Thorac Surg 2013;96:1866—8.

[64] Cerillo AG, Amoretti F, Mariani M, Cigala E, Murzi M, Gasbarri T, Solinas M, Chiappino D. Increased gradients after aortic valve replacement with the perceval valve: the role of oversizing. Ann Thorac Surg 2018;106:121—8.

[65] Muneretto C, Bisleri G, Moggi A, Di Bacco L, Tespili M, Repossini A, Rambaldini M. Treating the patients in the 'grey-zone' with aortic valve disease: a comparison among conventional surgery, sutureless valves and transcatheter aortic valve replacement. Interact Cardiovasc Thorac Surg 2015; 20:90—5.

[66] Biancari F, Barbanti M, Santarpino G, Deste W, Tamburino C, Gulino S, Imme S, Di Simone E, Todaro D, Pollari F, Fischlein T, Kasama K, Meuris B, Dalen M, Sartipy U, Svenarud P, Lahtinen J, Heikkinen J, Juvonen T, Gatti G, Pappalardo A, Mignosa C, Rubino AS. Immediate outcome after sutureless versus transcatheter aortic valve replacement. Heart Vessel 2016;31:427—33.

[67] Celiento M, Blasi S, De Martino A, Pratali S, Milano AD, Bortolotti U. The Mosaic mitral valve bioprosthesis: a long-term clinical and hemodynamic follow-up. Tex Heart Inst J 2016;43:13—9.

[68] Bourguignon T, Espitalier F, Pantaleon C, Vermes E, El-Arid JM, Loardi C, Karam E, Candolfi P, Ivanes F, Aupart M. Bioprosthetic mitral valve replacement in patients aged 65 years or younger: long-term outcomes with the Carpentier-Edwards PERIMOUNT pericardial valve. Eur J Cardiothorac Surg 2018;54:302—9.

[69] Arnaiz-Garcia ME, Gonzalez-Santos JM, Iscar-Galan A, Lopez-Rodriguez FJ. Leaflet detachment in a Carpentier-Edwards Perimount Magna mitral pericardial valve bioprosthesis. Eur J Cardiothorac Surg 2018;53:1293.

[70] Ruel M, Kulik A, Lam BK, Rubens FD, Hendry PJ, Masters RG, Bedard P, Mesana TG. Long-term outcomes of valve replacement with modern prostheses in young adults. Eur J Cardiothorac Surg 2005;27:425—33. discussion 433.

[71] Frater R, Liao K, Seifter E. In: Gabbay S, Frater R, editors. New horizons and the future of heart valve bioprostheses. Austin: Silent Partners, Inc.; 1994. p. 103—19. chap. 8.

[72] Kasegawa H, Iwasaki K, Kusunose S, Tatusta R, Doi T, Yasuda H, Umezu M. Assessment of a novel stentless mitral valve using a pulsatile mitral valve simulator. J Heart Valve Dis 2012;21:71—5.

[73] Kainuma S, Kasegawa H, Miyagawa S, Nishi H, Yaku H, Takanashi S, Hashimoto K, Okada Y, Nakatani S, Umezu M, Daimon T, Sakaguchi T, Toda K, Sawa Y, V. Committee of the Japanese Society of Stentless Mitral. In vivo assessment of novel stentless valve in the mitral position. Circ J 2015;79:553—9.

[74] Nishida H, Kasegawa H, Kin H, Takanashi S. Early clinical outcome of mitral valve replacement using a newly designed stentless mitral valve for failure of initial mitral valve repair. Heart Surg Forum 2016;19:E306—7.

[75] Chen XJ, Smith PB, Jaggers J, Lodge AJ. Bioprosthetic pulmonary valve replacement: contemporary analysis of a large, single-center series of 170 cases. J Thorac Cardiovasc Surg 2013;146:1461—6.

[76] Zack CJ, Fender EA, Chandrashekar P, Reddy YNV, Bennett CE, Stulak JM, Miller VM, Nishimura RA. National trends and outcomes in isolated tricuspid valve surgery. J Am Coll Cardiol 2017;70:2953—60.

[77] Kawano H, Oda T, Fukunaga S, Tayama E, Kawara T, Oryoji A, Aoyagi S. Tricuspid valve replacement with the St. Jude medical valve: 19 years of experience. Eur J Cardiothorac Surg 2000;18: 565—9.

[78] Altaani HA, Jaber S. Tricuspid valve replacement, mechanical vs. biological valve, which is better? Int Cardiovasc Res J 2013;7:71—4.

[79] Cunanan CM, Cabiling CM, Dinh TT, Shen SH, Tran-Hata P, Rutledge 3rd JH, Fishbein MC. Tissue characterization and calcification potential of commercial bioprosthetic heart valves. Ann Thorac Surg 2001;71:S417—421.

[80] Dalen M, Sartipy U, Cederlund K, Franco-Cereceda A, Svensson A, Themudo R, Svenarud P, Bacsovics Brolin E. Hypo-attenuated leaflet thickening and reduced leaflet motion in sutureless bioprosthetic aortic valves. J Am Heart Assoc 2017;6.

[81] Aljalloud A, Shoaib M, Egron S, Arias J, Tewarie L, Schnoering H, Lotfi S, Goetzenich A, Hatam N, Pott D, Zhong Z, Steinseifer U, Zayat R, Autschbach R. The flutter-by effect: a comprehensive study of the fluttering cusps of the Perceval heart valve prosthesis. Interact Cardiovasc Thorac Surg 2018;27(5):664—70. https://www.ncbi.nlm.nih.gov/pubmed/29788476.

[82] Lee T, Mittnacht AJC, Itagaki S, Stewart A. Mitral regurgitation exacerbation due to sutureless aortic valve replacement. Ann Thorac Surg 2018;105:e103—5.

[83] Morbiducci U, D Avenio G, Del Gaudio C, Grigioni M. Testing requirements for steroscopic particle image velocimetry measurements of mechanical heart valves fluid dynamics. Rapporti ISTISAN 2005;46.

[84] Amatya D, Troolin D, Longmire E. 3D3C velocity measurements downstream of artificial heart valves. Methods 2009;7.

[85] Siemens. Siemens, vol. 2018; 2018. https://www.plm.automation.siemens.com/global/en/webinar/mechanical-heart-valves-cfd-simulation/34664.

[86] Alemu Y, Bluestein D. Flow-induced platelet activation and damage accumulation in a mechanical heart valve: numerical studies. Artif Organs 2007;31:677—88.

[87] Yin W, Alemu Y, Affeld K, Jesty J, Bluestein D. Flow-induced platelet activation in bileaflet and monoleaflet mechanical heart valves. Ann Biomed Eng 2004;32:1058—66.

[88] Midha PA, Raghav V, Condado JF, Okafor IU, Lerakis S, Thourani VH, Babaliaros V, Yoganathan AP. Valve type, size, and deployment location affect hemodynamics in an in vitro valve-in-valve model. JACC Cardiovasc Interv 2016;9:1618—28.

[89] Zakerzadeh R, Hsu MC, Sacks MS. Computational methods for the aortic heart valve and its replacements. Expert Rev Med Devices 2017;14:849—66.

[90] Cheng R, Lai YG, Chandran KB. Three-dimensional fluid-structure interaction simulation of bileaflet mechanical heart valve flow dynamics. Ann Biomed Eng 2004;32:1471—83.

[91] Dasi LP, Ge L, Simon HA, Sotiropoulos F, Yoganathan AP. Vorticity dynamics of a bileaflet mechanical heart valve in an axisymmetric aorta. Phys Fluids 2007;19.

[92] Quaini A, Canic S, Glowinski R, Igo S, Hartley CJ, Zoghbi W, Little S. Validation of a 3D computational fluid-structure interaction model simulating flow through an elastic aperture. J Biomech 2012;45:310—8.

[93] Stijnen JMA, de Hart J, Bovendeerd PHM, van de Vosse FN. Evaluation of a fictitious domain method for predicting dynamic response of mechanical heart valves. J Fluids Struct 2004;19:835—50.

[94] Le Tourneau T, Lim V, Inamo J, Miller FA, Mahoney DW, Schaff HV, Enriquez-Sarano M. Achieved anticoagulation vs prosthesis selection for mitral mechanical valve replacement: a population-based outcome study. Chest 2009;136:1503—13.

[95] Heras M, Chesebro JH, Fuster V, Penny WJ, Grill DE, Bailey KR, Danielson GK, Orszulak TA, Pluth JR, Puga FJ, et al. High risk of thromboemboli early after bioprosthetic cardiac valve replacement. J Am Coll Cardiol 1995;25:1111—9.

[96] Whitlock RP, Sun JC, Fremes SE, Rubens FD, Teoh KH. Antithrombotic and thrombolytic therapy for valvular disease: antithrombotic therapy and prevention of thrombosis. In: American college of chest physicians evidence-based clinical practice guidelines. 9th ed., vol. 141; 2012. e576S—600S. Chest.

[97] Puskas J, Gerdisch M, Nichols D, Quinn R, Anderson C, Rhenman B, Fermin L, McGrath M, Kong B, Hughes C, Sethi G, Wait M, Martin T, Graeve A, Investigators P. Reduced anticoagulation after mechanical aortic valve replacement: interim results from the prospective randomized on-X valve anticoagulation clinical trial randomized Food and Drug Administration investigational device exemption trial. J Thorac Cardiovasc Surg 2014;147:1202—10. discussion 1210-1201.

[98] Fumoto H, Chen JF, Zhou Q, Massiello AL, Dessoffy R, Fukamachi K, Navia JL. Performance of bioprosthetic valves after glycerol dehydration, ethylene oxide sterilization, and rehydration. Innovations (Phila) 2011;6:32—6.

[99] Puskas JD, Bavaria JE, Svensson LG, Blackstone EH, Griffith B, Gammie JS, Heimansohn DA, Sadowski J, Bartus K, Johnston DR, Rozanski J, Rosengart T, Girardi LN, Klodell CT, Mumtaz MA, Takayama H, Halkos M, Starnes V, Boateng P, Timek TA, Ryan W, Omer S, Smith CR, Investigators CT. The COMMENCE trial: 2-year outcomes with an aortic bioprosthesis with RESILIA tissue. Eur J Cardiothorac Surg 2017;52:432—9.

[100] Piatti F, Sturla F, Marom G, Sheriff J, Claiborne TE, Slepian MJ, Redaelli A, Bluestein D. Hemo-dynamic and thrombogenic analysis of a trileaflet polymeric valve using a fluid-structure interaction approach. J Biomech 2015;48:3641—9.

[101] Meschini V, de Tullio M, Querzoli G, Verzicco R. 55th annual technical meeting of the society of engineering science (SES2018), vol. 7. Madrid, Spain: Leganés; 2018.

[102] Pibarot P, Dumesnil JG. Prosthetic heart valves: selection of the optimal prosthesis and long-term management. Circulation 2009;119:1034—48. PMID 19237674.

[103] Gilmanov A, Stolarski H, Sotiropoulos F. Flow-structure interaction simulations of the aortic heart valve at physiologic conditions: the role of tissue constitutive model. J Biomech Eng 2018;140. PMID 29305610.

[104] Sotiropoulos F, Gilmanov A. 55th annual technical meeting of the society of engineering science (SES2018). Madrid, Spain: Leganés; 2018.

[105] Ali A, Patel A, Ali Z, Abu-Omar Y, Saeed A, Athanasiou T, Pepper J. Enhanced left ventricular mass regression after aortic valve replacement in patients with aortic stenosis is associated with improved long-term survival. J Thorac Cardiovasc Surg 2011;142:285—91.

[106] Nkoke C, Dzudie A, Makoge C, Luchuo EB, Jingi AM, Kingue S. Rheumatic heart disease in the South West region of Cameroon: a hospital based echocardiographic study. BMC Res Notes 2018; 11:221.

[107] Saidi T, Douglas TS. Minimally invasive transcatheter aortic valve implantation for the treatment of rheumatic heart disease in developing countries. Expert Rev Med Devices 2016;13:979—85.

[108] Beaton A, Aliku T, Okello E, Lubega S, McCarter R, Lwabi P, Sable C. The utility of handheld echocardiography for early diagnosis of rheumatic heart disease. J Am Soc Echocardiogr 2014;27: 42—9.

CHAPTER 5

Transcatheter heart valves

Mohammad Sarraf[1], Elliott M. Groves[2], Arash Kheradvar[3,4,5]

[1]Division of Cardiovascular Disease, School of Medicine, University of Alabama at Birmingham, Birmingham, AL, United States; [2]Division of Cardiology, College of Medicine, University of Illinois at Chicago, Chicago, IL, United States; [3]Department of Biomedical Engineering, University of California Irvine, Irvine, CA, United States; [4]Department of Medicine, Division of Cardiology, University of California Irvine, CA, United States; [5]The Edwards Lifesciences Center for Advanced Cardiovascular Technology, University of California Irvine, CA, United States

Contents

5.1 History of transcatheter heart valves

For about 40 years, open heart surgery was the only means to replace or repair severe aortic stenosis, up until early 2000s when transcatheter aortic valve replacement (TAVR) was introduced in its current form. TAVR was an emerging technology with great promise for revolutionizing the treatment of valvular heart disease, particularly in elderly patients who were deemed unfit for cardiac surgery given the operative morbidity and mortality.

According to John G. Harold [1], the very first catheter-based approach of balloon aortic valvuloplasty (BAV) was developed by Dr. Alain Cribier, who, in 1985, completed a BAV on a 77-year-old woman with inoperable severe aortic stenosis. BAV leads to restenosis in most treated patients within the first year after intervention. In 1989, with a hypothesis that a balloon-expandable valve can be implanted in a manner similar to stenting a coronary artery, Dr. Henning R. Andersen developed a handmade metal stent with porcine aortic valves sewn into it along with a deflated

Principles of Heart Valve Engineering
ISBN 978-0-12-814661-3, https://doi.org/10.1016/B978-0-12-814661-3.00005-8

balloon within the apparatus. He then crimped the device and implanted that across a live pig's native aortic valve. However, his efforts did not result in translation of this concept to patients due to lack of acceptance in the medical community. In the meantime, Dr. Cribier persisted his efforts on the concept of percutaneous heart valve replacement. He considered that a valvular structure could be inserted within a stenotic valve using a balloon-expandable stent to mitigate the native valve structure and function. Based on his idea, and the purchase of the patent from Dr. Anderson, a startup company—Percutaneous Valve Technology (PVT)—was formed by Dr. Cribier, Dr. Martin Leon, Stanley Rabinovich, and Stanton J. Rowe to develop and commercialize transcatheter aortic valves.

The very first TAVR in its modern form was performed on a 57-year-old inoperable patient with severe calcific aortic stenosis who was presented in cardiogenic shock with an ejection fraction of 12% and multiple comorbidities, which prevented him undergoing open heart surgery. His clinical status significantly improved post-TAVR, but he died 4 months later due to complications not related to his TAVR procedure [2]. Since then, TAVR technology has been rapidly expanded, and subsequent clinical trials have validated the clinical use of TAVR in a variety of patient groups [1]. It is now expected that TAVR procedures have been performed in well over 100,000 patients around the world [3,4] and will exceed 300,000 per year by 2025 [1].

Considering the success of TAVR in older and high-risk patients, many investigators have been trying to develop sophisticated catheter-based interventional procedures for other heart valves such as mitral valve. Nevertheless, when compared with the aortic valve replacement, many more challenges involve successful transcatheter mitral valve replacement (TMVR). The mitral valve is more heterogenous in shape and anatomy as it is asymmetric with a significantly larger orifice area compared with the aortic valve. Furthermore, its dynamic saddle annulus, which is usually noncalcified in case of mitral insufficiency, makes it more challenging for proper anchoring of a prosthetic apparatus to be implanted via transcatheter means.

This chapter describes the current state of the art related to the transcatheter heart valves and discusses the engineering challenges to overcome the clinical unmet needs.

5.2 Transcatheter aortic valves

Since its inception in 2002, TAVR has been performed globally in well over 100,000 patients, and its indication is growing among groups of patients at lower risk for surgery [3,4]. Short-term data (between 1.6 and 2.8 years) suggest that transcatheter aortic valve replacement in patients with severe aortic stenosis who were unsuitable candidates for surgery significantly reduces all-cause mortality, the composite endpoint of all-cause death, or repeat hospitalization and cardiac symptoms, compared with standard therapy, with a comparable incidence of stroke and major vascular events [5]. More recent studies suggest that TAVR will provide patients with intermediate risk for traditional surgery a similar outcome with respect to the primary endpoint of death or disabling stroke at 2 years [6]. Fig. 5.1 shows all the TAVR systems currently approved or in clinical and preclinical stages.

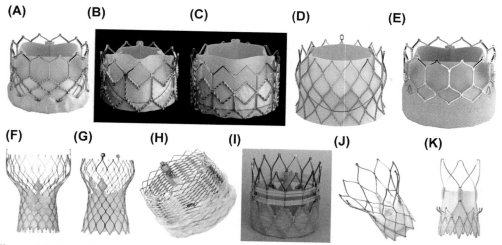

Figure 5.1 Currently-available transcatheter aortic valve technologies either clinically available or at different preclinical stages. (A) Edwards SAPIEN 3 Valve by Edwards Lifesciences; (B) and (C) Edwards SAPIEN XT Valve by Edwards Lifesciences; Sapien XT by Edwards Lifesciences; (D) Edwards CENTERA Valve (self-expanding valve) by Edwards Lifesciences; (E) Edwards SAPIEN 3 Ultra Valve by Edwards Lifesciences; (F) CoreValve by Medtronic; (G) CoreValve Evolut R by Medtronic; (H) LOTUS *Edge*^TM Aortic Valve System by Boston Scientific; (I) FoldaValve by ValVention Inc.; (J) Portico by Abbott; (K) ACURATE neo by Boston Scientific. *((F,G) Reproduced with permission of Medtronic, Inc.; (H,K) Image provided courtesy of Boston Scientific. © 2019 Boston Scientific Corporation or its affiliates. All rights reserved; (J) Portico is a trademark of Abbott or its related companies. Reproduced with permission of Abbott, © 2019. All rights reserved;)*

5.2.1 Transfemoral approach

From the onset of TAVR, using transfemoral (TF) access was preferred to the other alternative access points. There are several reasons that operators prefer the TF approach. To begin with, interventional cardiologists, as important members of the heart team, are usually quite facile with femoral artery access. Given the long history of coronary angiography performed through the common femoral artery (CFA), cardiologists generally feel comfortable with percutaneous as opposed to cut down access in the CFA. An additional and critically important reason for the use of TF access is the fact that the CFA either partially or fully lies over the bony anterior pelvis, and thus the CFA is compressible, and hemostasis can be achieved even if there is failure of a closure device [7,8].

Patient Selection: Selecting the appropriate patients for TF access is critical to the success of the procedure. In general, the vast majority of patients would have appropriate TF access [8]. While not always the case, ideally, there should be little to no calcification or tortuosity of the iliofemoral system on the side of preferred access. In addition, and perhaps most importantly, the vessels must be of adequate size to accommodate the delivery system chosen by the operators. This sizing varies by the device manufacturer and by the size of the device within each manufacturer. Currently in the United States, Three devices are approved for commercial use by the Food and Drug Administration (FDA). The approved devices are the Sapien valve family (Edwards Lifesciences, Irvine, CA, USA), the CoreValve family (Medtronic, Minneapolis, MN, USA) and the LOTUS *Edge*™ Aortic Valve System (Boston Scientific, MA, USA) [8,9]. Sapien and CoreValve systems require a minimum vessel diameter of 5—5.5 mm or greater depending on the valve size while Lotus valve and the Portico valve (Abbott Vascular, IL, USA; currently under clinical trial) require a slightly larger vessel diameter [10,11].

Access Techniques: There are numerous techniques for femoral vascular access before large bore femoral sheath insertion in TAVR. Currently, most operators use one of the following three techniques or a combination of those. The first technique is a combination of ultrasound and fluoroscopy. Using an anterior—posterior projection, an ultrasound probe is positioned over the anterior pelvis and then passed along the course of the artery distally to identify the CFA bifurcation. Following this, the probe is proximally passed a safe distance from the bifurcation. Fluoroscopy is then again used to ensure that the chosen site of puncture is over the anterior pelvis [7,12]. Using either a micropuncture technique or modified Seldinger technique, a 6F sheath is inserted into the CFA. Proper position is confirmed either with an iliofemoral angiogram through the 6F sheath or the micropuncture sheath before insertion of the larger sheath. This technique ensures CFA puncture through the anterior wall and away from any calcium, which can be detected on ultrasound images. A drawback is given shifts in the ultrasound probe and the difficulty in using concurrent fluoroscopy and ultrasound, which makes extremely precise puncture difficult. The benefits of ultrasound can also be mitigated by obesity.

Another slightly more advanced technique is a puncture under direct fluoroscopic and angiographic guidance. This technique is accomplished by obtaining access in the CFA not being used for the TAVR, which is needed regardless for a pigtail catheter. Through this access site, a catheter is taken over the iliac bifurcation and into the contralateral external iliac, which is the chosen side for the large sheath. Through the catheter over the bifurcation, an angiogram is taken from the TAVR side CFA. Then using small contrast injections, a needle is used to puncture the CFA in the exact desired location. This technique does require more radiation and may result in puncture through calcium, not seen on fluoroscopy. However, the CT scan before the procedure can help identify where the calcium is located. Variations of this technique using a wire in the artery have been used; however, given that the operator cannot be certain where the wire is located in the artery, a side wall puncture is always a possibility.

Finally, the easiest technique to execute is simply puncturing the TAVR sheath side of the CFA using a combination of palpation to locate the artery and fluoroscopy to locate the pelvis. Once the artery is accessed either with micropuncture or standard techniques, an iliofemoral angiogram is completed, and if the puncture is acceptable, then the operator proceeds with the large sheath insertion.

When the access is secured with a short 6F sheath, a wire is passed into the aorta, and two Perclose closure devices are deployed to "preclose" the arteriotomy. Once the preclose devices have deployed, a stiff wire is passed into the aorta over which the main transcatheter heart valve delivery sheath is placed. The procedure is then completed in the standard manner. Following valve deployment and assessment, the large bore sheath is removed followed by deployment of the Perclose device [12]. A digitally subtracted angiogram can be completed at the conclusion of the procedure to ensure that no contrast extravasation is present.

Iliofemoral Challenges: Many challenges are anticipated when using iliofemoral access [12]. One challenge is excessive tortuosity, which can be overcome by using a soft steerable wire to pass an exchange catheter into the aorta. The wire can be then exchanged for a stiff wire that will straighten the tortuosity. A sheath can usually then be passed over the stiff wire and into the aorta, past the tortuosity. If the sheath still cannot be passed, a balloon–assisted tracking (BAT) technique can be used.

A second significant challenge is calcification and iliofemoral disease. Before TAVR, the patient can be taken to the peripheral angiography suite for angioplasty, atherectomy, or stenting of the iliac arteries. However, this approach becomes more complicated with significant CFA disease. BAT can be used to overcome this challenge as well. However, this can lead to flow-limiting dissections and needs to be monitored after removal of the sheath.

The PARTNER trial had two separate studies, PARTNER A, which enrolled patients with higher risk for surgical aortic valve replacement (SAVR). These patients could either have received a standard TF approach or transapical approach (TA) for TAVR. PARTNER B enrolled patients who were not eligible for SAVR and as such were randomized to the combination of BAV and medical therapy or TAVR via the TF approach. PARTNER B demonstrated a significantly lower all-cause mortality at 1 year that sustained up to 5 years [5]. However, PARTNER A resulted in noninferiority of TAVR versus SAVR due to all-cause mortality in 1 year. Among the patients who were randomized to TAVR, TF approach had lower all-cause mortality and lower risk of stroke, albeit statistically nonsignificant [8].

The benefits of TF access was then again demonstrated in the PARTNER II trial [6]. As-treated population analysis and intention to treat analysis showed that TAVR was noninferior to SAVR with a trend toward superiority. In a subanalysis of the TF cohort, statistical superiority was demonstrated in the as-treated population [6]. This was suggesting that TF access is superior to alternative access.

5.2.2 Alternative access

Patients are always evaluated for TF access when TAVR is being considered due to its superior outcomes as noted earlier. However, nearly a quarter of patients in PARTNER

II trial were not eligible to undergo TF TAVR procedure due to the condition of the femoral access site [6]. Alternative access consists of transthoracic access (Transapical, Transaortic), subclavian/axillary artery access, carotid artery access and transcaval access [13]. These techniques are used to a varying degree depending on the familiarity of the operator, availability of surgical expertise and access to new technologies.

Patient Selection: Selection for alternative access begins with the exclusion of TF access. Following this, the analysis of the CT will allow the operators to assess alternative options. Currently, most institutions consider the subclavian/axillary approach to be the first option if TF access is unavailable [13]. This approach depends on the presence of a vessel that is large enough and other variables such as the location of vessel insertion into the aortic arch, and the angle of the ascending aorta to the horizontal plane. Left subclavian/axillary artery access is generally preferred, however, a grafted left internal mammary artery (LIMA) may sometimes preclude this approach and is another consideration.

If subclavian/axillary access is not a suitable option, other approaches are available that depends on the institution and operator's preference. Most operators will use a transthoracic approach although this approach has become less popular over the last few years due to longer hospitalization, post-procedure care and cost [14]. During the early days of TAVR, this was almost exclusively limited to transapical access [8]. However, transaortic access has more recently become a popular point of access due to the concerns for myocardial puncture and scar or apical aneurysm formation in the transapical access. Transaortic access is indeed not feasible in patients with a porcelain aorta [14]. Both of these methods must be performed under general anesthesia.

In general, a transcaval approach has not been widely used, but in several high-volume institutions it has been regularly used with excellent outcomes, comparable or even better than conventional alternative access [15]. The major drawback of transcaval access is the cost. Transcaval access starts with the common femoral vein access. After the operator reaches the inferior vena cava (IVC) below the renal artery/venous system, a puncture is performed between the IVC and abdominal aorta. Subsequently, a wire will be placed from the common femoral vein and cross the retroperitoneal space to the abdominal aorta over which the TAVR sheath will be placed. After completion of the procedure, this connection is closed by a vascular plug [15].

Finally, transcarotid access has been used in select patients with limited other options. If the contralateral carotid from the access vessel is diseased, then transcarotid approach may not be a good option.

Access Techniques: Subclavian/axillary access is generally approached in one of the two ways. The most common approach is a surgical cut down which allows for surgical control of a non-compressible artery [13]. Following surgical exposure, the artery is accessed and dilated in the normal manner. Once the procedure is completed, the artery is repaired and the cut down is closed. Recently, a percutaneous approach has become more popular [16]. There are multiple techniques for access, however, the majority of them include a combination of ultrasound and fluoroscopy. An additional useful technique is balloon tamponade from a femoral access point during pre-closure to aid in hemostasis at the procedure's conclusion. These techniques can also be applied to carotid access.

Transapical (TA) access was the preferred alternative access approach in the early experience. However, more recently, transaortic (TAo) access has become preferred due to the lack of direct left ventricular puncture. TA access is accomplished with an intercostal approach and rib retraction [14]. A needle is then used to directly puncture the LV apex. TAo is completed using an upper ministernotomy, followed by direct puncture of the aorta with a needle [14]. The TA approach can be complicated by the ventricular repair. By using TAo and thus avoiding the intercostal approach, there is less pain associated with the surgery, reduced impact on respiratory function, and a faster postoperative recovery with reduced length of stay.

Transcaval approach is not routinely used except for highly specialized centers. Briefly, a sheath is passed into the cava and at a pre-identified position, an 0.014″ wire is used under cautery to pass into the aorta with the position marked by a snare [15]. The wire is then snared and over the wire an 0.035″ exchange catheter is used to pass a stiff wire into the aorta. A sheath is then passed over the wire into the aorta and TAVR is completed in the typically retrograde fashion. Following valve deployment and assessment, the sheath is removed and a vascular plug or ventricular septal defect occluder is used. Bailout covered stents can be used if needed, but this was uncommon in the published experience [15].

Challenges: Alternative access is alternative for several reasons, chiefly amongst them is the increased complexity of the procedure [14]. Any transthoracic access significantly increases the procedure's morbidity and the hospital stay. In many centers, TF patients are routinely being discharged in 24—48 h. However, TA and TAo patients require longer stays and have much more post-operative pain. Subclavian/axillary access is associated with a higher risk of bleeding as the artery is not compressible and over the thoracic cavity (although it is compressible more distally), and the approach is not something familiar to most interventional cardiologists. There is also some concern for a higher stroke risk, although this has not been adjudicated. Carotid access runs the risk of stroke if any prolonged ischemia is necessary. Finally, transcaval access, has a higher risk of catastrophic consequences in the hands of those less experienced than the experts [15].

The data on transthoracic versus TF access have already been reviewed. Insufficient use of transcarotid access has resulted in no significant data sets being available. There is a recent publication on 100 patients who underwent transcaval access in very experienced hands [15]. 99 of the 100 patients had successful caval to aortic access, with ultimately only one patient needing a covered stent following the procedure. Major vascular complications were 13%, and 30-day survival was 92% with a median length of stay of 4 days [15].

Based on the recent studies, subclavian/axillary access has become second line in the eyes of many operators [16]. A study of 202 patients demonstrated 30-day survival of 94.6%, with similar rates of stroke and major vascular complications [13]. The difference regarding bleeding, acute kidney injury and need for a pacemaker were also not statistically significant. Worth mentioning, it is generally felt that procedural time is longer with any alternative access, including subclavian.

5.2.3 Valve design principles

Design principles for transcatheter heart valves are quite complex. Given the need to implant a new valve inside the native anatomy without distorting the numerous components of the cardiovascular system that coalesce in the aortic root, many design principles must be taken into account. An ideal transcatheter valve would have little to no residual gradient, excellent coaptation of the leaflets, no paravalvular/central leak, with a minimal rate of coronary obstruction, root rupture and need for a permanent pacemaker [6,8,9]. These characteristics should also be coupled with a design that would allow for recapture/repositioning and a low delivery profile to minimize vascular complications along with allowing TF access in those with small/diseased femoral/iliac arteries [9].

Leaving no residual gradient is challenging in the smaller valve sizes. This challenge is frequently seen in surgical valves, which in small sizes typically have a residual gradient postoperatively [17]. One effective solution to this issue is to place the valve in a supra-annular position [9]. A supra-annular position ensures excellent coaptation, which in turn eliminates the risk of central regurgitation. The reason for this, is the need to ensure the valve is circular. The native aortic annulus is not circular, and thus with a circular valve, the annulus must be forced to conform to the valve as to avoid poor coaptation and central regurgitation. The outward force is available with a balloon-expandable valve or a self-expandable valve with higher radial force, and thus more conformational potential with the native annulus to ensure optimal valve function. However, this indeed increases the risk of root rupture. Early reconstitution of laminar flow distal to the valve is advantageous to minimization of a post-operative gradient [18].

Paravalvular leak that is more than mild has been consistently associated with poor outcomes in TAVR patients (Fig. 5.2) [19]. One common design method is to include a sealing skirt on the outer frame of the valve's inflow. A sealing skirt fills the gaps between the frame and the annulus to seal any potential leaks. Downsides of a sealing skirt are that they make the outer diameter of the valve larger and can be associated with increased rates of permanent pacemakers [9]. Rigid balloons in a balloon expandable valve can be used to seal the valve and avoid the leak. Self-expanding valves tend to continue to safely expand over time and seal leaks that are present immediately after implantation.

Complications can be avoided with valve design in addition to proper procedural technique. A low-profile valve is much less likely to cause coronary obstruction [6].

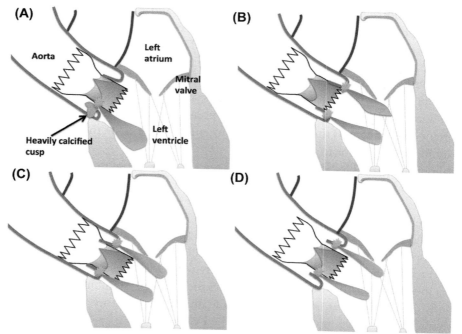

Figure 5.2 Paravalvular aortic leak due to improper positioning of transcatheter aortic valve. Paravalvular leaks with consecutive peri-prosthetic aortic regurgitation result from improper positioning of the stent frame, usually caused by (A) calcifications of the annulus or the cusps of the native valve; (B) valve malposition with too high; (C) or too low; (D) implantation depth of the prosthesis, and/or annulus-prosthesis-size mismatch. *(The image is adapted from Sinning J-M, Hammerstingl C, Vasa-Nicotera M, Adenauer V, Lema Cachiguango SJ, Scheer A-C, Hausen S, Sedaghat A, Ghanem A, Müller C, Grube E, Nickenig G, Werner N. Aortic regurgitation index defines severity of peri-prosthetic regurgitation and predicts outcome in patients after transcatheter aortic valve implantation. J Am Coll Cardiol 2012;59:1134–1141.)*

If the top of the valve does not reach the sinotubular junction, then in most cases, with the exception of subjects with effaced sinuses, there is very little chance of coronary obstruction.

Coronary obstruction is a dreaded potential complication of TAVR procedure. The most common reason for coronary obstruction is the coronary height. Generally, if the height of the coronary artery is more than 10 mm, the risk of the coronary obstruction is low. Other variables that may lead to coronary obstruction are the diameter of the sinuses of Valsalva (SOV), the density/volume of the calcification of the aortic leaflets and the effacement of the sinotubular junction. All these variables can be predetermined by gated cardiac CT scan before the procedure.

Root rupture may occur due to balloon dilatation inside the annulus. This might be reduced by using a self-expanding design. The majority of self-expandable valves lack the sudden radial force required to rupture the annulus and can always be post dilated if needed [9]. Permanent pacemaker implantation is one of the complications that is far more frequent with TAVR than SAVR [8,9,11]. This can be avoided with a low profile valve that is precisely implanted just below the native annulus and above the membranous septum. The more a valve comes in contact with the outflow tract, the more likely that compression of the conduction system arises, and a pacemaker is needed. Avoiding a sealing skirt can also help minimize the rate of post-operative pacemaker need. The most important determinant of the pacemaker requirement seems to be the depth of the valve positioning from the aortic annulus.

Balloon expandable valves are not recaptured or repositioned [8]. However, self-expanding valves can accomplish both of these useful characteristics [9]. On the other hand, self-expanding valves also tend to have larger frames, which can lead to challenges in valve delivery. Allowing for a low profile when fully crimped is critical to the progression of TAVR given the superiority of TF access. Self-expanding valves, due to the lack of a balloon, can typically be crimped to a smaller diameter [9]. The balloon can be loaded in line with the valve in the delivery system with assembly inside the body to make a low profile. Other design characteristics such as even assembling the leaflets inside the body, and leaflets that sit outside the frame during crimping are in development and can drastically lower the profile of the crimped valve [20].

As is clear from the many ideal characteristics, there is no optimal valve in the clinical trials nor in the commercial use. Some valves have optimal hemodynamics, but a high rate of pacemakers, and some minimize leak, but put the patient at risk of other complications and increase delivery system size.

5.2.4 Long-term outcomes and durability concerns

Two independent consensus articles have recently published on structural valve degeneration of SAVR and TAVR from the European Association of Percutaneous Cardiovascular Interventions (EAPCI) [21], in December 2017 and from the Valve-in-Valve International Data Investigators (VIVID) in January 2018 [22]. These two consensus articles from expert groups from Europe and North America independently emphasized the significance of dysfunction and failure in both TAVR and SAVR, according to valve design, delivery and implantation, individual patient characteristics, and the premise for understanding the failure mechanisms. According to the VIVID group, there have been some concerning reports about long-term TAVR performance [23]. For example, the follow-up of 704 patients (mean age 82 years) who underwent TAVR at St. Paul's Hospital in Vancouver, Canada, and Hôpital Charles Nicolle in Rouen, France, for up to 10 years indicates that among the 100 patients survived at least 5 years post-TAVR, 35 showed signs of valve degeneration, with median time to degeneration reported as 61 months [24]. Accordingly, based on Kaplan—Meier estimation [25], the curve for freedom from degeneration drops from 94% at 4 years to 82% at 6 years and to approximately 50% at 8 years among surviving patients, per Dvir et al. [24].

Concerns on thrombosis: According to recent studies, subclinical leaflet thrombosis has been frequently observed after both SAVR and TAVR and seems to be more common post-TAVR [26,27]. Chakravarty et al. [28] reported in the RESOLVE[1] and SAVORY[2] registries that out of 931 patients who had CT imaging done (657 [71%] in the RESOLVE registry and 274 [29%] in the SAVORY registry), 890 [96%] had interpretable CT scans (626 [70%] in the RESOLVE registry and 264 [30%] in the SAVORY registry). Of the 890 patients, 106 (12%) had subclinical leaflet thrombosis, including 5 (4%) of 138 with thrombosis of surgical valves versus 101 (13%) of 752 with thrombosis of transcatheter valves ($P = .001$). These large studies clearly show that subclinical leaflet thrombosis incidence in TAVR is over three times more likely than in SAVR (4% vs. 13% [$P = .001$], even considering the most recent TAV generation (e.g., Edwards' Sapien three and Medtronic's Evolut-R) [28]. Subclinical leaflet thrombosis has reported to be associated with increased rates of transient ischemic attacks (TIAs) and strokes.

5.2.5 Delivery systems

TAVR delivery systems are greatly influenced by the valve which they are delivering. This has been discussed somewhat regarding the valve design. Currently, there are three types of systems commonly used. The first and most straightforward is the fully assembled valve is crimped at the Cath lab and loaded over the delivery system that is passed through a fixed diameter delivery sheath [11,29]. Once the valve is deployed, the delivery system is then removed through the same sheath and closure is completed. Another approach is the one currently used in the Edwards' Sapien three valve [30]. This system involves two unique components, an expandable sheath and a delivery system that requires assembly of the valve/balloon system in the aorta. An expandable sheath can temporarily enlarge by unfolding to allow entry of the crimped valve and successively recoils to its reduced profile as the stent-crimped heart valve passes through [30]. This expansion occurs when the valve is inserted and a seam in the sheath allows it to stretch and expand. Occasionally, this can lead to a complete separation of the sheath and bleeding on removal. Once the valve is in the aorta, the balloon and valve apparatus are assembled using external controls, then advanced and deployed. The final currently available alternative is the inline sheath, as is used in Medtronic Evolut-R or Evolut-Pro TAVR systems [31]. In this design, the valve and the sheath are all part of the delivery system. A placeholder sheath is inserted

---

[1] Assessment of Transcatheter and Surgical Aortic Bioprosthetic Valve Thrombosis and its Treatment with Anticoagulation (RESOLVE) registry.
[2] Subclinical Aortic Valve Bioprosthesis Thrombosis Assessed with Four-Dimensional Computed Tomography (SAVORY) registry.

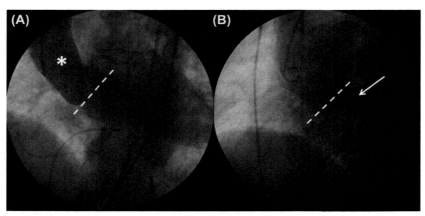

Figure 5.3 (A) Angiography shows high position of CoreValve prosthesis resulting in severe perivalvular leak. (B) Angiographic confirmation of deployment fix using second CoreValve device resulting in valve-in-valve. *(Figure from Ussia GP, Barbanti M, Imme S, Scarabelli M, Mule M, Cammalleri V, Aruta P, Pistritto AM, Capodanno D, Deste W, Di Pasqua MC, Tamburino C. Management of implant failure during transcatheter aortic valve implantation. Cathet Cardiovasc Interv 2010;76:440—449)*

and then once a stiff wire is in place in the ventricle, the placeholder is removed, and the entire delivery system is inserted followed by valve deployment. Once the valve is deployed, the entire system is removed, and the placeholder is reinserted. This system has the advantage of providing a known OD while providing the lowest profile delivery capability.

Repositioning and retrieval: Studies have shown that the TAVR procedure results in a 20% reduced death rate after 1-year when compared to the standard surgical therapy [5]. Despite these results, the main challenge for a successful TAVR procedure is optimal positioning of the transcatheter aortic prosthesis [32]. Approximately 25% of TAVR failures are due to the device being implanted abnormally low or high within the aortic root, requiring a bailout procedure to correct the implantation error (Fig. 5.3) [33]. If valve deployment is too high within the native valve annulus or even beyond the aortic annulus, there is a risk of aortic injury, valvular regurgitation, and aortic embolization (Fig. 5.4). Valve deployment that is extending too low into the ventricle can cause valvular dysfunction, heart block, regurgitation and embolization into the left ventricle [7,34,35]. Overall, prosthesis embolization that occurs immediately after deployment is due to a serious error in positioning of the device and a subsequent effective ventricular contraction during deployment (Fig. 5.4) [35].

To remedy potential malposition, acceptable bailout procedures involve utilizing a snare to reposition the valve, or deploying a second TAVR device within the first TAVR leaving one operational valve [36—38]. The two-TAVR valve situation is generally referred to as "valve-in-valve (V-in-V)" (Fig. 5.5) [33]. In the worst case scenario, surgical bailout is required to retrieve a damaged or malfunctioning TAVR device.

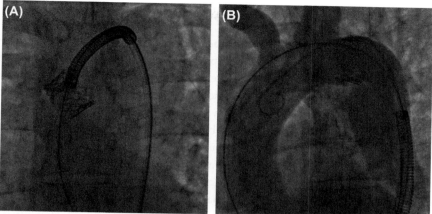

Figure 5.4 (A) The embolized valve orientation is maintained by the wire position. (B) The stented valve is secured in the aorta with no detectable gradient across it. *(From Masson J-B, Kovac J, Schuler G, Ye J, Cheung A, Kapadia S, Tuzcu ME, Kodali S, Leon MB, Webb JG. Transcatheter aortic valve implantation review of the nature, management, and avoidance of procedural complications. JACC Cardiovasc Interv 2009;2:811−820.)*

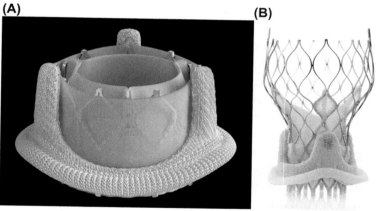

Figure 5.5 Valve-in-valve (V-in-V). V-in-V is a procedure through which a new transcatheter valve is tightly placed into the orifice of a failed centrally-opening bioprosthetic valve, pushing the old valve leaflets aside. (A) Edwards' Sapien valve implanted in a bioprosthetic surgical valve; (B) Medtronic's CoreValve implanted in a surgical bioprosthetic valve.

Valve–in-valve has become an acceptable technique to correct for early implant failure [37−39]. However, as the second valve compresses and deforms the leaflets of the first valve, this can significantly affect the flow field downstream of the valve [40]. Additionally, intermediate and long-term consequences of V–in–V implantation have not yet been extensively studied [41]. Even minimally invasive bailout procedures, such as the use of a repositioning snare, have resulted in an increased risk of embolization [42].

It is recommended that the delivery catheters smaller than 15 French (Fr) provide a seemingly smooth transition and delivery [42,43]. Relatively larger diameter of the delivery catheter (greater than 22Fr) has been a major limitation of TF TAVR with high incidence of arterial dissection and perforation. Dissection of the ascending or descending aorta can occur due to catheter trauma [35,44]. Vascular perforation can lead to retroperitoneal hemorrhage, which is a serious potential complication of TAVR. The complication of vascular injury has been consistently associated with higher mortality in clinical trials. These issues can be resolved with the introduction of a repositionable TAVR system deliverable equipped with an imaging modality that can convey the local position of the valve within the vascular structure.

Application of imaging modalities in TAVR: As mentioned earlier, repositionability is an important option to have in case of initial malpositioning. However, an efficient imaging modality that is integrated into the valve delivery system should facilitate accurate positioning and result in dramatically increased procedural success without the need to reposition. Patients suffering from highly calcified aortic valve leaflets can experience significant valvular regurgitation with a replacement valve implanted in place of the native valve. Therefore, in addition to the need for repositionability, accurate image-guidance is a necessity in endovascular procedures where visualization of the root and the leaflet anatomy is a must. Currently, TAVR procedures are guided by X–ray fluoroscopy and angiography. However, X–ray includes several limitations such as the need for nephrotoxic contrast agents during the procedure, and radiation exposure to the patient and the interventional team [45,46]. Incorrect valve sizing and positioning are shown to occur, due to difficulties that exist in imaging the optimal view of the native valve and annulus [47]. The limited 2D-projection in X–ray can underestimate the size and shape of the aortic valve when the imaging plane is oriented obliquely to the valve [48,49]. Therefore, three-dimensional imaging modalities such as CT and MRI are used for screening and follow-up in TAVR but cannot be used for intra-procedural imaging due to their relatively slow acquisition speed [50–52]. Furthermore, although CT and MRI can visualize the aortic root and aortic arch, both modalities have difficulty imaging leaflet calcification [53]. Leaflet imaging is necessary to determine the level of calcification of the native valve. Despite its shortcomings, angiography is still used for real–time assessment during TAVR procedures. For valve sizing, currently CT is primarily used. While intravascular ultrasound (IVUS) has not been directly compared to CT in aortic valve imaging, it has been compared to CT in aortic imaging and guidance of aortic endograft placement; in human and animal studies, IVUS has been shown to be as reliable as CT in the measurement of aortic luminal diameter [54,55]. This reliability in imaging and procedural guidance will likely extend to imaging and procedures involving the aortic valve as well.

Ultrasound has been utilized as a procedural imaging technique for valve implantation based on its real-time capabilities and non-ionizing modality. TEE and transthoracic echocardiography (TTE) have been used as imaging tools during TAVR procedures [56,57]. Both TEE and TTE can visualize the aortic root and ventricular portions of the anatomy, as well as providing other anatomical references that might support more accurate positioning [58,59]. TEE is currently used to image leaflet calcification based on the superior ability of ultrasound to resolve calcium deposits in tissues. Several studies have shown that TEE-guided TAVR allows for shorter procedural duration and requires less use of ionizing radiation [45]. TEE is limited by the resolution in which it can resolve TAVR deployment targets within the anatomy, and unfavorable imaging windows can render TEE useless, a problem not shared by IVUS. Increasing the center frequency of the transducer can create higher ultrasound resolutions, albeit at the expense of imaging depth. However, imaging depth cannot be reduced in TEE due to the fixed distance between the ultrasound transducer and the aortic anatomy. TEE is still utilized as a complement to current imaging in TAVR, and still requires procedural image guidance by fluoroscopy and/or angiography. The shortcomings of the current procedural technologies may be resolved by using an IVUS system incorporated into the TAVR delivery catheter. A similar use of ultrasound in TAVR procedures is known as intracardiac echocardiography (ICE), which utilizes a lower resolution transducer to visualize the entire heart within the imaging plane. In addition to the assessment of coronary arteries, IVUS has been used for full evaluation of the aorta due to its small catheter size [60]. In fact, vascular surgeons have called the use of aortic imaging using IVUS "essential" during aortic endografting despite aortic motion [61]. IVUS has also been utilized to determine the optimal placement of coronary stents based on the high specular reflection of the acoustic beam off of stents' metal surfaces [62,63]. The higher resolution of IVUS compared to TEE or ICE provides detailed evaluation of the size, tortuosity and presence of calcification [64], but at the moment cannot be used simultaneously during the procedure since only one catheter (TAVR or IVUS) can occupy the aortic root region at any time-point. The IVUS catheter, if used separately, can be physically damaged once the stent frame expands from crimped size to a full 23 mm or greater diameter. Therefore, it is essential to develop a TAVR delivery system with an integral IVUS component that inherently provides a real-time imaging ultrasound modality. This delivery system also allows for repositioning and retrieval of the valve to effectively mitigate implantation errors and establish optimal deployment targets within the aortic root. Utilizing IVUS during a TAVR would also involve no additional morbidity or mortality to the patient as it would be integrated into the delivery system. IVUS requires no additional sedation or mechanical ventilation. TEE while safe does carry the risk of devastating consequences such as esophageal perforation, gastrointestinal bleeding, pharyngeal hematoma and methemoglobinemia [65–68].

An ideal delivery system should be low profile, trackable through tortuosity and calcification, and easy to use. Currently as discussed there are three available systems with numerous more in development. As TAVR progresses there will be a push for image-guided systems with lower profiles and the ability to repositions and retrieve the valve.

5.3 Transcatheter mitral valve repair and replacement

Since the FDA approval of Melody valve (Melody, Medtronic, Fridley, MN) for pulmonic valve disease, there has been an explosive interest in percutaneous approaches to other valvular heart diseases. The introduction of TAVR in 2002 [2], opened a new horizon to a new era of transcatheter heart valve (THV) therapies, including mitral valve disorders. As mentioned in the previous section, the successful randomized control trials (RCTs) of TAVR led to approval of two different TAVR systems, i.e., Edwards' Sapien THV and Medtronic's CoreValve THV, in high and intermediate risk patients [5,6,8,69−71]. However, the approach to THV therapies for mitral valve disease is far more challenging due to the complexity of the anatomy as well as the disease states of the mitral valve disorders. To date, the only FDA approved device is MitraClip (Abbott Vascular, Abbott Park, IL) in patients with high risk or prohibitive risk undergoing mitral valve repair or replacement [72].

In this section, we focus on the current available THV systems for mitral valve repair (MVr) and mitral valve replacement (MVR).

5.3.1 Mitral valve apparatus function

TMVR is one of the most challenging areas for cardiac device innovation because of the complexity of mitral valve anatomy, function and the different pathologies that can lead to mitral valve disease. As such even in the surgical field, there are numerous surgical corrective therapies pending on the pathology requiring treatment.

Mitral valve apparatus naturally involves a complex interaction among multiple components; namely mitral valve leaflets, mitral annulus, chordae tendineae, papillary muscles, left ventricle, and left atrium. The mitral valve apparatus has two fundamental roles in circulation. First, it functions as a hemodynamic valve to control inflow and outflow of the blood from left atrium toward the aortic valve. Second, it plays an important role in the structural and functional integrity of the left ventricle (LV). Any disruption to the mitral-ventricular continuity leads to remodeling and reduced LV performance (Fig. 5.6). One of the fundamental roles of the mitral valve apparatus with respect to LV function is the annulus dynamic and its contribution to the cardiac output [73]. Therefore, prosthetic mitral valve surgery, which fixes the prosthesis to the annulus may lead to a global reduction in contractility. Non-randomized studies have shown the long-term outcomes of patients who undergo MVR is worse than MVr in patients with degenerative mitral regurgitation (MR) [74,75]. This complex interaction between the mitral valve and LV indicates that the hemodynamic valve function of the mitral valve is only a tiny proportion of the its function to maintain normal cardiac output.

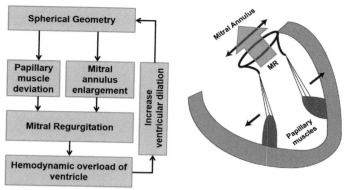

Figure 5.6 Both functional mitral regurgitation (FMR) or degenerative mitral regurgitation (DMR) contribute to LV remodeling and dilatation through a vicious cycle that ultimately develops annular dilatation of the mitral annulus as well as lateral displacement of the papillary muscles that results in further worsening of MR.

5.3.2 TMVR and TMVr technologies

A spectrum of TMVR and TMVr technologies to treat degenerative mitral regurgitation (DMR) and functional mitral regurgitation (FMR) has been developed very rapidly over the last 2 decades. These technologies have addressed separate yet pertinent component of mitral valve apparatus for treatment of MR. These technologies are mainly classified into replacement strategy (TMVR) or repair technologies (TMVr).

TMVR and TMVr each comes with advantages and challenges. The TMVR's main advantages are versatility, mitigating MR regardless of the etiology of the mitral valve disease, and its predictably (fairly). However, TMVR is also associated with some major challenges such as, paravalvular leak, obstruction of the left circumflex artery, possible obstruction the LV outflow tract (LVOT), potential need for long-term anticoagulation, and most importantly, anchoring and sealing of the mitral annulus. Table 5.1 describes the ideal characteristics of TMVR [76].

TMVr has the advantage of maintaining the native mitral valve's hemodynamics and relatively lower profile of delivery that makes the procedure safer with no need for chronic anticoagulation. One crucial challenge of TMVr is that different etiologies of MR may require different approaches/devices. Furthermore, the TMVr procedures heavy depend on imaging. Regardless of the method, a residual MR usually remains in almost all the currently available TMVr technologies.

The lessons learned from the surgical literature may also have a significant impact on the future of TMVR(r). Patients with DMR have a primary mitral valve disease, which if left untreated would lead to LV remodeling and finally LV failure. Several studies have shown that surgical repair of the mitral valve results in significant improvement in the patients' outcomes, even in the elderly patients [8,74,77]. The guidelines in the Europe

Table 5.1 The ideal TMVR system.

Simple and reproducible implant
Absence of transvalvular gradient
Absence of peri-valvular regurgitation
No LVOT, coronary sinus, or LCX obstruction
Preservation of LV contractility, hemodynamics, and blood flow pattern
Non-thrombogenic (no need for anti-coagulation)
Low infection rates No acute or delayed embolization
Durable

and U.S. have emphasized on the significance of mitral valve repair technologies [78,79], albeit they are contingent to the surgeon's expertise, the institution as well as the pathology of the underlying disease.

The results of the clinical studies and in particular a large randomized controlled trial in FMR has been less convincing for surgical repair than replacement. In that study, patients with severe ischemic MR were randomized to chorda valve sparing mitral valve repair or mitral valve replacement. Accordingly, no significant difference was found between the two groups up to 2 years after the procedure [80,81]. Other non-randomized studies have also shown no significant difference in the outcomes of patients who undergo MVR or MVr for FMR [82–85].

In summary, the underlying pathology of the mitral valve disease may have a distinct role in the strategy of the clinician and the engineers to choose TMVR versus TMVr in a specific pathology leading to MR.

TMVr: The design and development of TMVr devices should be focused to the anatomical features of the mitral valve. As noted, the complexity of the mitral valve apparatus prohibits a particular repair system mitigating all mitral valve's pathologies. Thus, each device may only address a particular disease at the time, depending on the mitral valve's underlying pathology. The current TMVr technologies are categorized as below:

- Leaflet technology
- Chordal technology
- Annuloplasty technology (Direct and Indirect Annuloplasty)
- Ventricular reshaping technology

5.3.2.1 TMVr's leaflet technologies

1. *MitraClip system:* In 1991, Alfieri et al. introduced an "edge-to-edge" approximation of the mitral leaflets as a strategy for mitral valve repair, with excellent performance in short- and long-term [86]. The surgery was initially exclusively developed for DMR patients with redundant mitral valve leaflets. MitraClip system follows the

Alfieri stitch technique by a percutaneous approach via femoral vein with a trans-septal approach [87]. The MitraClip system brings the anterior and posterior leaflets at A2 − P2 by a "gripper" and "clip" combination. By closure of the device, the operator generates a double orifice mitral valve. The bridge formed between the anterior and posterior leaflets reduces the MR. It is, however, important to mention that the clip placement must be precisely on the location originating the MR; i.e., if the MR is on the mitral valve's medial side, then clipping of the middle portion of the valve will not reduce the MR. Currently, the MitraClip system by Abbott Vascular (Abbott Park, IL) is the only FDA approved system for commercial use in patients with DMR.

The EVEREST II (Endovascular Valve Edge-to-Edge Repair) randomized controlled trial evaluated the safety and effectiveness of the MitraClip compared to mitral valve surgery (MVR or MVr) [88]. The primary efficacy end point was freedom from death, from surgery for mitral-valve dysfunction, and from grade 3 + or 4 + MR at 1 year. The study included DMR (74%) and FMR (26%) patients. The EVEREST II trial met its primary endpoint of non-inferiority of MitraClip to surgery and superiority of MitraClip safety compared to surgery. The follow up study up to 4 years also demonstrated that the results lasted between the two groups with MitraClip having a better safety profile, while the surgical approach had superior efficacy in reduction of MR [89].

FMR is a ventricular disease and many surgical studies have shown repair or replacement of the mitral regurgitation may reduce the patient's symptoms, however, it does not lead to lower mortality. Virtually all the literature on treatment of FMR have been non-randomized and frequently single center studies. The COAPT trial (Cardiovascular Outcomes Assessment of the MitraClip Percutaneous Therapy for Heart Failure Patients with Functional Mitral Regurgitation) [90] randomized patient with 3/4 + MR with LVEF between 20% and 50% into guideline-directed medical therapy or guideline-directed medical therapy and MitraClip. The primary endpoint of the study was a reduction of HF admission, and two of the main 10 key secondary endpoints were all-cause mortality and cardiovascular death. For the first time, the COAPT investigators demonstrated that MitraClip has led to a reduction of HF admissions to the hospital, as well as all-cause and cardiovascular mortality [90]. Although COAPT was a strongly positive trial in favor of MitraClip, another randomized controlled trial MITRA-FR trial investigated a similar patient population and failed to show any benefit in patients with FMR who were randomized to MitraClip [91], unlike the COAPT trial [90]. This sharp discordance has generated a heated discussion in the field of structural heart disease. However, there have been some major challenges in the MITRA-FR trial that were not present in the COAPT trial. MITRA-FR trial is a smaller study with 304 patients compared to 614 patients in the COAPT trial.

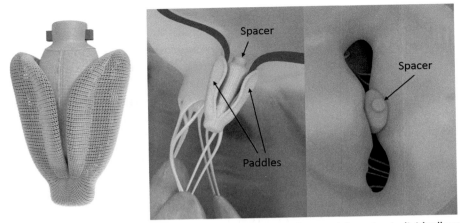

Figure 5.7 The PASCAL Repair System enables transcatheter valve repair by using individually adjustable clasps to place a spacer between the native valve leaflets. *(Courtesy of Edwards Lifesciences, Irvine, CA, USA.)*

The LV end-diastolic volume was significantly larger in the MITRA-FR trial suggesting more remodeled LV compared to COPAT trial's and the effective orifice area was smaller in the MITRA-FR trial. The acute failure rate was higher in the MITRA-FR trial compared to COAPT trial (9% vs. 5%). One of the most important differences between the two studies are the management of the guideline-directed medical therapy in the COAPT trial that was not present in the MITRA-FR trial. Therefore, the medical management of the MITRA-FR patients were subject to change throughout the study, unlike COPAT trial. Finally, nearly 30% of the patients who were randomized to medical therapy arm in MITRA-FR trial showed improvement in the severity of FMR. This underlies the lack of appropriate medical therapy before randomization of the patients to the study.

2. *Pascal System:* The PASCAL Transcatheter Valve Repair System: The PASCAL Repair System (Edwards Lifesciences, Irvine, CA, USA) is an alternative leaflet-focused technology. This device requires a femoral vein, transseptal approach to access the left atrium and uses clasps and paddles to place a spacer between the native valve leaflets (Fig. 5.7). The clasps can be independently adjusted to optimize leaflet capture and fine tune leaflet position. The complementary contoured paddles and spacer design minimize stress concentration on the native leaflets. Furthermore, the spacer fills the regurgitant orifice area and reduces MR. In a compassionate use setting, the PASCAL repair system showed feasibility with 96% technical success, reduction of MR severity, and significant clinical improvements [92]. A multicenter, prospective, single-arm CLASP study is currently enrolling patients with clinically significant mitral regurgitation to assess the safety, performance, and clinical outcomes of the PASCAL repair system (clinicaltrials.gov ID: NCT03170349).

3. *Mitra-Spacer by Cardiosolutions:* The Mitra-Spacer system (Cardiosolutions Inc., Stoughton, MA) is a percutaneously delivered balloon-shaped. The spacer occupies the space between the mitral leaflets and increases the coaptation area between the leaflets and the device, and anchors to the LV apex. The spacer can "self-center" within the regurgitant orifice. The balloon's volume can be adjusted in real time by the operator and the residual leak can be minimized by inflating or deflating the balloon. There is currently no available published data on this device except for one case report that has demonstrated a successful bridge of a failing LV to successful MVR [93].

4. *Middle Peak Medical:* Middle Peak Medical (Palo Alto, CA) developed a device to simulate the shape of the posterior leaflet to stop mitral regurgitation as a posterior neoleaflet. The goal of any corrective surgery of the mitral valve is to improve the coaptation of the mitral valve leaflets. In a normal heart, anterior leaflet occupies about one-third of the valve circumference and controls the blood flow during diastole. Posterior leaflet has less prominent role in the overall valve function, and in particular coaptation during systole. Having said that, the MR pathology frequently involves the posterior leaflet, often with restricted motion. Thus, it is reasonable to consider the posterior leaflet as a target to improve MR by correction of the mitral valve apparatus. The Mitral Peak Medical device is designed for surgical or percutaneous insertion of the posterior leaflet. This device can be considered for DMR and FMR. There is currently no available data about the device. Middle Peak Medical was acquired by Symetis in 2017 and is now called Polares Medical.

5.3.2.2 TMVr — chordal implantation

Synthetic chords have been implanted via transapical or transseptal approach, and anchored between the LV myocardium at the apex and mitral leaflet. By adjusting the length of the chord, the MR can be reduced or eliminated. Currently, there are three different systems in the preclinical or clinical stage; the transapically delivered MitraFlex (TransCardiac Therapeutics, Atlanta, GA), NeoChord (Neochord, Inc., Minnetonka, Minnesota) devices, and the V-Chordal (Valtech Cardio Ltd., Or-Yehuda, Israel).

1. *Neochord DS 1000 System:* The NeoChord DS1000 system is designed to transapically approach to capture a flail leaflet segment and pierce it with a needle to attach a polytetrafluoroethylene (PTFE) artificial chord. The NeoChord is subsequently anchored to the apical entry site with a pledgeted suture.

2. *Mitraflex System:* The MitraFlex (TransCardiac Therapeutics, Atlanta, GA, USA) places an anchor on the LV and another on the leaflet via a thoracoscopic TA and connects the two with a synthetic chord. This device can also simultaneously perform an edge-to-edge repair, if/when indicated. There is limited number of cases performed to date. The patient can be on the normothermic cardiopulmonary bypass for the entire length of the procedure.

3. *V-Chordal System:* The V-Chordal (Valtech Cardio Ltd., Or-Yehuda, Israel) system uses a slow rotation, semi-automatic insertion of a helical element to fixate the chordae to the apex of the papillary muscle. The procedure is performed via transatrial approach or robotic, and the chords are subsequently placed through the apex of the papillary muscle. Subsequently, while the sutures are still attached to the system, the heart is resuscitated, and under echocardiography guidance in the beating heart, the suture would be tightened to the desired length before finally released from the system. TF and transseptal approaches are in development. Valtech Cardio Ltd. was acquired by Edwards Lifesciences in 2016.

5.3.2.3 TMVr — annuloplasty

Annular dilatation occurs invariably in patients with MR, especially in patients with FMR. Annular band or ring placement is one of the fundamental aspects of surgical correction of MVr to reduce the septo-lateral dimension of the mitral annulus. Different technologies have been developed to reduce and reshape the mitral annulus that are at different stages of research and development. They all aim to address different anatomical and pathophysiological concepts.

1. *Direct Annuloplasty:* This technology approaches the mitral annulus directly without using the juxtaposed structures such as coronary sinus or great cardiac vein. The access to the mitral annulus can be either via transseptal puncture or retrograde through the aortic valve and LV. Suture-based or suture-less devices are implanted onto the annulus and subsequently used to directly reduce the mitral annulus size. These technologies would be most useful for FMR although they have the potential to mitigate DMR as well.

   • **Mitralign**

   The Mitralign Percutaneous Annuloplasty System (Mitralign Inc., Tewksbury, MA) is a suture-based annuloplasty system for mitral valve. The operator retrogradely places a deflectable and steerable catheter through the aortic valve. The steerable catheter is advanced to the LV and delivers pledgeted anchors through the posterior annulus. These pledgeted sutures can be pulled together to shorten the annulus up to 17 mm, resulting in a segmental posterior mitral annuloplasty, as shown in Fig. 5.8 [94].

   • **Cardioband**

   The Edwards Cardioband^{TM} Mitral Reconstruction System (Edwards Lifesciences, Irvine, CA, USA), is a transcatheter device designed to reduce mitral regurgitation via annular reduction. The Cardioband implant consists of a contraction wire and polyester fabric with small radiopaque markers (Fig. 5.10E). The implant is placed in a step-wise fashion along the annulus of the mitral valve with guidance of transesophageal echocardiogram (TEE) and fluoroscopy. Once placed, implant is contracted to reduce mitral regurgitation, allowing real-time adjustment and confirmation of results via TEE. The CE Mark trial has recently been presented

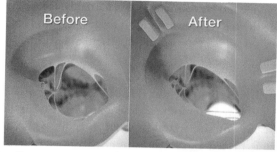

Figure 5.8 The Mitralign Percutaneous Annuloplasty System is a suture-based annuloplasty system for mitral valve. The steerable catheter is advanced to the LV and delivers pledgeted anchors through the posterior annulus. These pledgeted sutures can be pulled together to shorten the annulus up to 17 mm, resulting in a segmental posterior mitral annuloplasty.

at TVT 2019 held in Chicago, IL, USA showing the 61 patient cohort technical success rate approached 80% with 3% procedural mortality unrelated to the device or procedure. The post market study, MiBAND (NCT03600688) is currently recruiting.

- **Millipede**

 Millipede IRIS system (Boston Scientific, Marlborough, MA) is a complete ring annuloplasty system and since it is positioned supra-annularly, like the other supra-annular systems, it does not have any interaction with the subvalvular apparatus. Millipede IRIS system is deliverable transfemorally and transseptally, is fully repositionable and retrievable, and is totally customizable to the patient's anatomy. The system is comprised of eight helical anchors that can be individually advanced and retracted. Presently, the device is placed by TEE. No available published data are currently available on this device. There are anecdotes with successful outcomes, yet there is no peer-reviewed data available.

2. *Indirect Annuloplasty:* Great cardiac vein and coronary sinus are juxtaposed to the mitral annulus; therefore, this relationship has given an opportunity for percutaneous approaches to mitral annuloplasty. Originally, the first attempts to reduce MR without surgery mimicked surgical ring annuloplasty through placement of devices in the coronary sinus.

- **Carillon**

  The Carillon Mitral Contour System (Cardiac Dimension, Inc., Kirkland, WA) consists of self-expandable nitinol semi-helical, distal and proximal anchors connected by a nitinol bridge placed in the great cardiac vein and proximal coronary sinus [95]. The tension generated by the Carillon Mitral Contour system results in cinching of the posterior mitral annulus tissue anteriorly. The device is placed percutaneously via the right internal jugular vein approach and can be easily retrieved if MR reduction is not favorable or if a coronary artery is compromised. Several clinical studies have been performed with reasonably acceptable results with the first generation of the device [96]. A limitation to the device is that in some patients, the circumflex artery's flow is compromised particularly at the crossing of the coronary sinus with circumflex artery [97]

- **Accucinch**

  The Accucinch system (Anchora Heart, Santa Clara, CA) is based on suture annuloplasty. A retrograde femoral arterial approach to gain access to the ventricular/subannular aspect of the mitral annulus is required. A module guide tunnel (MGT) catheter is placed under the posterior mitral leaflet next to the anterior trigone. Between nine to 12 anchors can be placed from the anterior to posterior commissures along the posterior mitral annulus. These anchors are connected by a suture that can cinch the annulus to reduce the mitral annular circumference and FMR under direct TEE.

5.3.2.4 TMVr — ventricular reshaping

Whether the mechanism of MR is FMR or DMR, the end result is LV remodeling and dilatation. This indeed leads to a vicious cycle to develop annular dilatation of the mitral annulus as well as lateral displacement of the papillary muscles that results in further worsening of MR (Fig. 5.6). Some technologies have mainly focused on the prevention of further LV remodeling. These technologies aim to use intra- or extra-cardiac devices that attenuates or reverses the LV remodeling. By reducing the anteroposterior diameter, the septo-lateral distance will also decrease in size.

1. *iCoapsys System:* This device has two extracardiac epicardial pads connected by a flexible tether running through transventricular subvalvular space and anchored on the other side of epicardial surface. The chord can be shortened to reduce the MR during the operation, as shown in Fig. 5.9. In the Randomized Evaluation of a Surgical Treatment for Off-Pump Repair of the Mitral Valve (RESTORE-MV) Trial, 165 patients were randomly assigned to undergo CABG with or without iCoapsys ventricular reshaping. Patients treated with the device had a remarkable reverse LV remodeling, lower MR grades, and lower mortality at 2 years, compared to the control arm. A sub-xiphoid percutaneous approach known as Myocor i-Coapsys

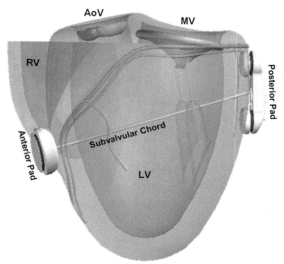

Figure 5.9 Schematic of an implanted iCoapsys device. (AoV, aortic valve; MV, mitral valve; RV, right ventricle; LV, left ventricle). The iCoapsys device has two extracardiac epicardial pads connected by a flexible tether running through transventricular subvalvular space and anchored on the other side of epicardial surface. The chord can be shortened to reduce the MR during the operation. *(The image is from Pedersen WR, Block P, Leon M, Kramer P, Kapadia S, Babaliaros V, Kodali S, Tuzcu EM, Feldman T. Icoapsys mitral valve repair system: percutaneous implantation in an animal model. Cathet Cardiovasc Interv 2008;72:125−131.)*

(Edwards Lifesciences, Inc., Irvine, CA) was also developed with successful animal model, yet the device is no longer manufactured due to financial challenges [98].

2. *Mardil BACE System:* The Mardil BACE (Basal Annuloplasty of the Cardia Externally; Mardil, Inc., Morrisville, NC) consists of a strip of mesh implanted at the base of the heart and positioned at the level of the AV groove [99]. The system is composed of inflatable chambers that are slipped externally around the base of the beating heart. The chambers can be inflated by saline through subcutaneous ports, and their volume can be adjusted intra- and postoperatively, thus remodeling the mitral valve annulus and sub-valvular apparatus. A percutaneous iteration has been developed known as VenTouch device. The first in man has been performed but no published literature or case report is currently available on this device.

Some of the current TMVr technologies are shown in Fig. 5.10.

5.3.2.5 TMVR systems

In the surgical space, mitral valve repair continues to be the leading strategy for treatment of patients with MR. Yet, there are some advantages of the mitral valve replacement (MVR). The same principle applies to TMVR in carefully selected patients. The main advantages of TMVR are reproducibility, applicability to majority of the patients and

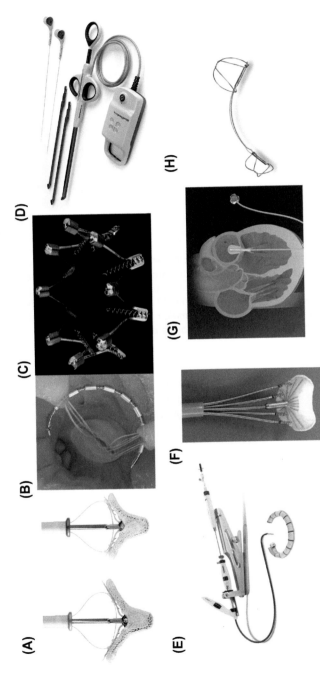

Figure 5.10 Some of the currently-available transcatheter mitral valve repair technologies that are either clinically available or at different pre-clinical stages are summarized here. (A) MitraClip by Abbott; (B) Accucinch by AncoraHeart; (C) Millipede by Boston Scientific; (D) NeoChord DS 1000; (E) Edwards Cardioband™ Transcatheter Reconstruction System by Edwards Lifesciences; (F) Posterior Neoleaflet by Middle Peak Medical; (G) MitraSpacer by Cardiosolutions Inc; (H) Carillon Mitral Contour System ® by Cardiac Dimensions. ((A) MitraClip is a trademark of Abbott or its related companies. Reproduced with permission of Abbott, © 2019. All rights reserved; (B) Used with permission from Ancora Heart, Inc.; (C) Image provided courtesy of Boston Scientific. © 2019 Boston Scientific Corporation or its affiliates. All rights reserved; (D) Used with permission from NeoChord, Inc.; (E) Used with permission from Edwards Lifesciences; (F) From Andrea Colli et al., Transcatheter Chordal Repair for Degenerative Mitral Regurgitation, Vol. 12, No. 4 July/August 2018 Cardiac Interventions Today; (G) Used with permission from Cardiosolutions Inc.; (H) Used with permission from Cardiac Dimensions, Inc.)

predictability of the results. Furthermore, in patients with mitral stenosis (calcific or rheumatic), none of the TMVr systems apply. Virtually all TMVr systems are designed on MR pathologies other than the stenotic pathologies.

There are many different TMVR systems worldwide in different stages of development. All TMVR systems currently use self-expanding nitinol frames. Valve anchoring is either by axial fixation, outward radial force, and sometimes a combination of both mechanisms. Some have features that capture the native mitral leaflets and secure them to the prosthetic stent. Figure 5.11 summarizes few TMVR systems in the pre-clinical and clinical stages.

Medical and Biomedical Design Challenges to TMVR Systems: The concept of TMVR is assumed to be an ultimate challenge in medical device. This is because there are many major challenges related to design and delivery of the TMVR system. these challenges preclude a clinically-approved system to be available for patients yet. Some of these challenges are further described here.

Access site is a major challenge for any TMVR designs. Almost all TAVR systems are first designed based on a TA. The advantage of this route is the short distance and direct access from the apex toward the left atrium. However, most patients with MR have chronic systolic LV dysfunction, with thin walled LV. Therefore, accessing the apex of a chronically dysfunctional LV put the patient at risk of arrhythmias, bleeding and, in the long-term, pseudoaneurysm, as has been the case in many cases of transapical TAVR. Furthermore, the hemodynamics of chronic MR patients is such that the LV is always exposed to a lower afterload due to lower left atrial pressure. Conceptually, since TMVR systems eliminate the MR immediately, the afterload may acutely rise after the procedure. Transapical access may hamper the myocardial reserve to overcome this immediate increase in afterload. As such, some operators have advocated for direct transatrial access. Although transatrial access does not have the above-mentioned issues for TA, the alignment of the TMVR for valve deployment is not ideal in this approach. Moreover, many MR patients already have other comorbidities such as chronic lung disease or have been exposed to prior cardiac or chest surgeries that may hamper transatrial access. Hence, most TMVR designs have now focused on designs that accommodate the transseptal approach. The main advantage of the transseptal approach is its minimally-invasive nature with no chest wall invasion. However, since most TMVR systems are bulky (over 24 French for delivery), there is always a concern as to whether the residual interatrial iatrogenic septal defect would be closed or not after the procedure. Nevertheless, it is currently believed that the transseptal approach would be the future of all competitive TMVR systems. Some of TMVR systems currently under clinical and pre-clinical studies are shown in Fig. 5.11.

The acute hemodynamic changes immediately post-MVR may become critical peri-operatively. Acute rise in afterload has already been discussed. Another major challenge in all TMVR systems is the prevention of LV outflow tract obstruction (LVOT).

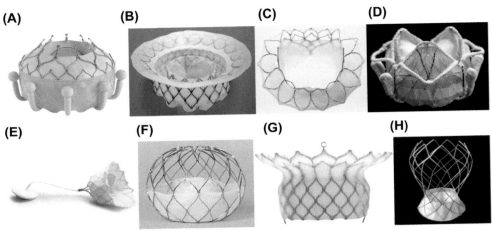

Figure 5.11 Some of the currently-available transcatheter mitral valve replacement technologies that are either clinically available or at different preclinical stages are summarized here: (A) Edwards EVO-QUE™ Mitral Valve Replacement System by Edwards Lifesciences; (B) Intrepid by Medtronic; (C) Tiara by Neovasc; (D) Caisson by LivaNova; (E) Tendyne by Abbott; (F) AltaValve by 4C Medical; (G) HighLife TMVR by HighLife Medical Inc; (H) AValve by ValVention Inc. *((A) Used with permission from Edwards Lifesciences; (B) Reproduced with permission of Medtronic, Inc.; (C) © Neovasc Inc. The Tiara TM System is a trademark of Neovasc Inc. All rights reserved; (E) Tendyne is a trademark of Abbott or its related companies. Reproduced with permission of Abbott, © 2019. All rights reserved; (F) Used with permission from 4C Medical Technologies, Inc.)*

Due to the common pathway between the anterior mitral leaflet and LVOT, the risk of LVOT obstruction is not trivial. Successful TMVR designs should be able to prevent LVOT obstruction at any cost.

Calcification of the mitral annulus is present only in some MR pathologies. As such, reliable anchoring is challenging. Alternatively, fixation methods that primarily rely on radial force do not seem to be successful. Tremendous radial force, even if possible, increases the risk of compression and may potentially damage the adjacent structures e.g., the LVOT, cardiac conduction system, coronary sinus, left circumflex artery, and aortic root. Hence, novel innovative fixation methods are needed to ensure proper fixation to the LV or to other components of the sub-valvular apparatus.

Thrombus formation on the metal stent or even the valve's leaflets are quite likely since all TMVR designs have a higher burden of nitinol. This problem is particularly more important in patients with atrial fibrillation who have a stagnant or low atrial blood flow velocity at the atrial side that can lead to clot formation. Although, anticoagulation can always be prescribed, it must be considered that many of these patients are elderly with increased risk of bleeding.

Durability is yet another important aspect that must be considered in TMVR designs. TMVR systems have large profile and as such crimping of the valve may have an early damage to the leaflets. The early leaflet damage may result in short term durability that becomes a major challenge if these devices are used in younger patient population.

Conclusion and the TMVR Future Direction: Presently, there is a significant momentum to develop safe, durable and reliably-reproducible TMVR and repair systems. All TMVR and TMVr systems come with individual challenges. While the result for MR reduction may be reproducible, the durability and potential damage to the adjacent structures by TMVR will continue to be the short-term unanswered questions. TMVR systems are currently being developed regardless of the etiology of the mitral valve disease, namely mitral regurgitation or stenosis. TMVr systems have steeper learning curve, but the results may be more permanent. Contrary to TMVR, TMVr systems are more specific to the underlying pathology of the MR. They have less hemodynamic compromise and are more likely to be consistent with the normal physiology of the heart and circulation.

The role of team approach of the clinicians, scientists, imaging specialists and biomedical engineers cannot be overemphasized in innovating TMVR and TMVr systems.

5.4 Pediatric transcatheter heart valves

The very first successful transcatheter heart valve was indeed a valve developed for pediatric purpose. In 2000, Bonhoeffer et al. [100] reported the successful transcatheter implantation of a pulmonary valve in a 12-year-old patient with Tetralogy of Fallot (TOF). Contrary to heart valve disease in adults, which is mainly degenerative, in pediatric age group, congenital heart defects (CHDs) are responsible for most interventions involving heart valves. CHDs occur in at least 1% of newborns in the U.S. and are responsible for more than 24% of infant deaths per year [101–103]. CHDs often involve pulmonary or aortic valves that are malformed. These valves may not have enough tissue leaflets, may have the wrong size or shape, or they may be stenotic. For example, TOF, a common form of cyanotic CHD with a prevalence of 3.4 per 10,000 live births in the United States [104], accounts for approximately 6.8% of live-born patients with CHD [105]. TOF presents as a heterogeneous range of phenotypes. The major anatomical features of TOF are pulmonary and subpulmonary stenosis or atresia, a subaortic ventricular septal defect (VSD), and RV hypertrophy. Most patients born with a dysfunction in their right ventricle outflow tract (RVOT) need serial pulmonary valve replacements throughout their life. Most studies suggest that pulmonary valve replacement for chronic pulmonary valve regurgitation (PVR) should be considered to avoid RV dysfunction [106–109].

In 2010, Medtronic, Inc. announced that its Melody valve (Fig. 5.12) for transcatheter pulmonary valve replacement (TPVR), originally developed by Bonhoeffer et al. in 2000 [100], has received U.S. FDA approval under a Humanitarian Device Exemption (HDE). Considering the early safety and efficacy, the Melody valve has shown excellent results so far according to multiple single and multicenter studies in Europe, Canada, and the United States [109–112]. Now that over a decade has passed since the early introduction of the Melody valve for transcatheter pulmonary valve replacement, the technology is mature and proven to be effective for mitigating regurgitation and relieving obstruction [113].

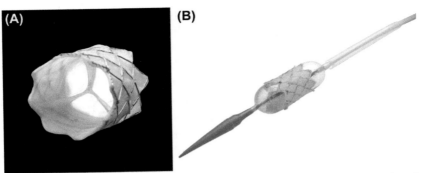

Figure 5.12 Melody valve by Medtronic. (A) The fully assembled valve is shown in a closed position. Melody comes with the bovine jugular venous segment sewn to a gold-brazed platinum–iridium stent. (B) The Melody valve is shown over the delivery system with the outer balloon of the balloon-in-balloon (BiB) fully inflated. *(Reproduced with permission of Medtronic, Inc.)*

Every success comes with challenges as well. in a study by Cheatham et al., 14 out of the 148 patients experienced endocarditis of the Melody valve or persistent bacteremia [114]. Accordingly, this study reported an unadjusted rate of infectious endocarditis (IE) post-TPVR of 9.5%, or ≈2% per year [113,114]. However, endocarditis post heart valve implantation is not unique to the Melody valve. Endocarditis rate in surgical valves are similar to those reported by Cheatham et al. [114], particularly when the surgical valve is the Contegra bovine jugular vein [115] as bovine jugular vein (Fig. 5.12) seems to be particularly predisposing to infective endocarditis [113,116,117].

Overall, over the past 15 years, the Melody valve and its delivery system have not received any significant modification compared to TAVR systems whose third and fourth generations are either currently approved for clinical use or in later stages of pre-clinical studies. One potential reason behind this lack of progress in pulmonary valve development is the wide range of anatomic complexity observed in patients with CHDs in addition to the overall smaller market for TPVR compared to TAVR and TMVR. Future advancements in pediatric valve implantation are expected to utilize the technologies developed for TAVR, TMVR and their delivery systems. Furthermore, a living tissue-engineered valve that can grow and remodel with the child has been the much-loved unmet clinical need for CHD patients, which still seems farfetched despite advancement in tissue engineering and heart valve science.

References

[1]   Harrold JG. The evolution of transcatheter aortic valve replacement. Cardiology 2017;46:38.
[2]   Cribier A, Eltchaninoff H, Bash A, Borenstein N, Tron C, Bauer F, Derumeaux G, Anselme F, Laborde F, Leon MB. Percutaneous transcatheter implantation of an aortic valve prosthesis for calcific aortic stenosis: first human case description. Circulation 2002;106:3006–8.

[3]   Hamm CW, Arsalan M, Mack MJ. The future of transcatheter aortic valve implantation. Eur Heart J 2016;37:803—10.

[4]   Kappetein AP, Osnabrugge RL, Head SJ. Patient selection for tavi in 2014: is there a justification for treating low- or intermediate-risk patients? The surgeon's view. EuroIntervention 2014;(10 Suppl. 1 U):U11—5.

[5]   Leon MB, Smith CR, Mack M, Miller DC, Moses JW, Svensson LG, Tuzcu EM, Webb JG, Fontana GP, Makkar RR, Brown DL, Block PC, Guyton RA, Pichard AD, Bavaria JE, Herrmann HC, Douglas PS, Petersen JL, Akin JJ, Anderson WN, Wang D, Pocock S. Transcatheter aortic-valve implantation for aortic stenosis in patients who cannot undergo surgery. N Engl J Med 2010;363:1597—607.

[6]   Leon MB, Smith CR, Mack MJ, Makkar RR, Svensson LG, Kodali SK, Thourani VH, Tuzcu EM, Miller DC, Herrmann HC, Doshi D, Cohen DJ, Pichard AD, Kapadia S, Dewey T, Babaliaros V, Szeto WY, Williams MR, Kereiakes D, Zajarias A, Greason KL, Whisenant BK, Hodson RW, Moses JW, Trento A, Brown DL, Fearon WF, Pibarot P, Hahn RT, Jaber WA, Anderson WN, Alu MC, Webb JG. Transcatheter or surgical aortic-valve replacement in intermediate-risk patients. N Engl J Med 2016;374:1609—20.

[7]   Seto AH, Abu-Fadel MS, Sparling JM, Zacharias SJ, Daly TS, Harrison AT, Suh WM, Vera JA, Aston CE, Winters RJ, Patel PM, Hennebry TA, Kern MJ. Real-time ultrasound guidance facilitates femoral arterial access and reduces vascular complications: faust (femoral arterial access with ultrasound trial). JACC Cardiovasc Interv 2010;3:751—8.

[8]   Smith CR, Leon MB, Mack MJ, Miller DC, Moses JW, Svensson LG, Tuzcu EM, Webb JG, Fontana GP, Makkar RR, Williams M, Dewey T, Kapadia S, Babaliaros V, Thourani VH, Corso P, Pichard AD, Bavaria JE, Herrmann HC, Akin JJ, Anderson WN, Wang D, Pocock SJ. Transcatheter versus surgical aortic-valve replacement in high-risk patients. N Engl J Med 2011; 364:2187—98.

[9]   Reardon MJ, Van Mieghem NM, Popma JJ, Kleiman NS, Søndergaard L, Mumtaz M, Adams DH, Deeb GM, Maini B, Gada H, Chetcuti S, Gleason T, Heiser J, Lange R, Merhi W, Oh JK, Olsen PS, Piazza N, Williams M, Windecker S, Yakubov SJ, Grube E, Makkar R, Lee JS, Conte J, Vang E, Nguyen H, Chang Y, Mugglin AS, Serruys PWJC, Kappetein AP. Surgical or transcatheter aortic-valve replacement in intermediate-risk patients. N Engl J Med 2017;376: 1321—31.

[10]   Denegri A, Nietlispach F, Kottwitz J, Suetsch G, Haager P, Rodriguez H, Taramasso M, Obeid S, Maisano F. Real-world procedural and 30-day outcome using the portico transcatheter aortic valve prosthesis: a large single center cohort. Int J Cardiol 2018;253:40—4.

[11]   Meredith IT, Walters DL, Dumonteil N, Worthley SG, Tchétché D, Manoharan G, Blackman DJ, Rioufol G, Hildick-Smith D, Whitbourn RJ, Lefèvre T, Lange R, Müller R, Redwood S, Feldman TE, Allocco DJ, Dawkins KD. 1-year outcomes with the fully repositionable and retrievable lotus transcatheter aortic replacement valve in 120 high-risk surgical patients with severe aortic stenosis: results of the reprise ii study. JACC Cardiovasc Interv 2016;9:376—84.

[12]   Toggweiler S, Gurvitch R, Leipsic J, Wood DA, Willson AB, Binder RK, Cheung A, Ye J, Webb JG. Percutaneous aortic valve replacement: vascular outcomes with a fully percutaneous procedure. J Am Coll Cardiol 2012;59:113—8.

[13]   Gleason TG, Schindler JT, Hagberg RC, Deeb GM, Adams DH, Conte JV, Zorn III GL, Hughes GC, Guo J, Popma JJ, Reardon MJ. Subclavian/axillary access for self-expanding transcatheter aortic valve replacement renders equivalent outcomes as transfemoral. Ann Thorac Surg 2018; 105:477—83.

[14]   Dunne B, Tan D, Chu D, Yau V, Xiao J, Ho KM, Yong G, Larbalestier R. Transapical versus trans-aortic transcatheter aortic valve implantation: a systematic review. Ann Thorac Surg 2015;100: 354—61.

[15]   Greenbaum AB, Babaliaros VC, Chen MY, Stine AM, Rogers T, O'Neill WW, Paone G, Thourani VH, Muhammad KI, Leonardi RA, Ramee S, Troendle JF, Lederman RJ. Transcaval access and closure for transcatheter aortic valve replacement: a prospective investigation. J Am Coll Cardiol 2017;69:511—21.

[16] Mathur M, Krishnan SK, Levin D, Aldea G, Reisman M, McCabe JM. A step-by-step guide to fully percutaneous transaxillary transcatheter aortic valve replacement. Structural Heart 2017;1:209—15.

[17] Koene BM, Soliman Hamad MA, Bouma W, Mariani MA, Peels KC, van Dantzig J-M, van Straten AH. Can postoperative mean transprosthetic pressure gradient predict survival after aortic valve replacement? Clin Res Cardiol 2014;103:133—40.

[18] Sengupta PP, Pedrizzetti G, Kilner PJ, Kheradvar A, Ebbers T, Tonti G, Fraser AG, Narula J. Emerging trends in cv flow visualization. JACC Cardiovasc Imaging 2012;5:305—16.

[19] Daubert MA, Weissman NJ, Hahn RT, Pibarot P, Parvataneni R, Mack MJ, Svensson LG, Gopal D, Kapadia S, Siegel RJ, Kodali SK, Szeto WY, Makkar R, Leon MB, Douglas PS. Long-term valve performance of tavr and savr: a report from the partner i trial. JACC Cardiovasc Imaging 2017;10: 15—25.

[20] Kheradvar A, Groves EL, Tseng EE. Proof of concept of foldavalve a novel 14fr totally repositionable and retrievable transcatheter aortic valve. Euro Intervention 2015;10:591—6.

[21] Capodanno D, Petronio AS, Prendergast B, Eltchaninoff H, Vahanian A, Modine T, Lancellotti P, Sondergaard L, Ludman PF, Tamburino C, Piazza N, Hancock J, Mehilli J, Byrne RA, Baumbach A, Kappetein AP, Windecker S, Bax J, Haude M. Standardized definitions of structural deterioration and valve failure in assessing long-term durability of transcatheter and surgical aortic bioprosthetic valves: a consensus statement from the european association of percutaneous cardiovascular interventions (eapci) endorsed by the european society of cardiology (esc) and the european association for cardio-thoracic surgery (eacts). Eur Heart J 2017;38:3382—90.

[22] Dvir D, Bourguignon T, Otto CM, Hahn RT, Rosenhek R, Webb JG, Treede H, Sarano ME, Feldman T, Wijeysundera HC, Topilsky Y, Aupart M, Reardon MJ, Mackensen GB, Szeto WY, Kornowski R, Gammie JS, Yoganathan AP, Arbel Y, Borger MA, Simonato M, Reisman M, Makkar RR, Abizaid A, McCabe JM, Dahle G, Aldea GS, Leipsic J, Pibarot P, Moat NE, Mack MJ, Kappetein AP, Leon MB. Standardized definition of structural valve degeneration for surgical and transcatheter bioprosthetic aortic valves. Circulation 2018;137:388.

[23] O'Riordan M. Structural valve degeneration: vivid group proposes new, graduated definition. 2018.

[24] Dvir D, al e. First look at long-term durability of transcatheter heart valves: assessment of valve function up to 10-years after implantation. Paris, France: Presented at EuroPCR; May 2016.

[25] Kaplan EL, Meier P. Nonparametric estimation from incomplete observations. J Am Stat Assoc 1958; 53:457—81.

[26] Makkar RR, Chakravarty T. Transcatheter aortic valve thrombosis: new problem, new insights*. JACC Cardiovasc Interv 2017;10:698—700.

[27] Makkar RR, Fontana G, Jilaihawi H, Chakravarty T, Kofoed KF, de Backer O, Asch FM, Ruiz CE, Olsen NT, Trento A, Friedman J, Berman D, Cheng W, Kashif M, Jelnin V, Kliger CA, Guo H, Pichard AD, Weissman NJ, Kapadia S, Manasse E, Bhatt DL, Leon MB, Søndergaard L. Possible subclinical leaflet thrombosis in bioprosthetic aortic valves. N Engl J Med 2015;0.

[28] Chakravarty T, Søndergaard L, Friedman J, De Backer O, Berman D, Kofoed KF, Jilaihawi H, Shiota T, Abramowitz Y, Jørgensen TH, Rami T, Israr S, Fontana G, de Knegt M, Fuchs A, Lyden P, Trento A, Bhatt DL, Leon MB, Makkar RR. Subclinical leaflet thrombosis in surgical and transcatheter bioprosthetic aortic valves: an observational study. The Lancet 2017;389:2383—92.

[29] Tzikas A, Amrane H, Bedogni F, Brambilla N, Kefer J, Manoharan G, Makkar R, Möllman H, Rodés-Cabau J, Schäfer U, Settergren M, Spargias K, van Boven A, Walther T, Worthley SG, Sondergaard L. Transcatheter aortic valve replacement using the portico system: 10 things to remember. J Interv Cardiol 2016;29:523—9.

[30] Binder RK, Rodés-Cabau J, Wood DA, Mok M, Leipsic J, De Larochellière R, Toggweiler S, Dumont E, Freeman M, Willson AB, Webb JG. Transcatheter aortic valve replacement with the sapien 3: a new balloon-expandable transcatheter heart valve. JACC Cardiovasc Interv 2013;6: 293—300.

[31] Hellhammer K, Piayda K, Afzal S, Kleinebrecht L, Makosch M, Hennig I, Quast C, Jung C, Polzin A, Westenfeld R, Kelm M, Zeus T, Veulemans V. The latest evolution of the medtronic corevalve system in the era of transcatheter aortic valve replacement: matched comparison of the evolut pro and evolut r. JACC Cardiovasc Interv 2018;11:2314—22.

[32] Geisbusch S, Bleiziffer S, Mazzitelli D, Ruge H, Bauernschmitt R, Lange R. Incidence and management of corevalve dislocation during transcatheter aortic valve implantation. Circ Cardiovasc Interv 2010;3:531—6.

[33] Ussia GP, Barbanti M, Imme S, Scarabelli M, Mule M, Cammalleri V, Aruta P, Pistritto AM, Capodanno D, Deste W, Di Pasqua MC, Tamburino C. Management of implant failure during transcatheter aortic valve implantation. Cathet Cardiovasc Interv 2010;76:440—9.

[34] Ussia GP, Barbanti M, Ramondo A, Petronio AS, Ettori F, Santoro G, Klugmann S, Bedogni F, Maisano F, Marzocchi A, Poli A, Napodano M, Tamburino C. The valve-in-valve technique for treatment of aortic bioprosthesis malposition an analysis of incidence and 1-year clinical outcomes from the Italian corevalve registry. J Am Coll Cardiol 2011;57:1062—8.

[35] Masson J-B, Kovac J, Schuler G, Ye J, Cheung A, Kapadia S, Tuzcu ME, Kodali S, Leon MB, Webb JG. Transcatheter aortic valve implantation review of the nature, management, and avoidance of procedural complications. JACC Cardiovasc Interv 2009;2:811—20.

[36] Giri J, Bortnick AE, Wallen T, Walsh E, Bannan A, Desai N, Szeto WY, Bavaria J, Herrmann HC. Procedural and clinical outcomes of the valve-in-valve technique for severe aortic insufficiency after balloon-expandable transcatheter aortic valve replacement. Cathet Cardiovasc Interv 2012.

[37] Webb JG, Wood DA, Ye J, Gurvitch R, Masson JB, Rodes-Cabau J, Osten M, Horlick E, Wendler O, Dumont E, Carere RG, Wijesinghe N, Nietlispach F, Johnson M, Thompson CR, Moss R, Leipsic J, Munt B, Lichtenstein SV, Cheung A. Transcatheter valve-in-valve implantation for failed bioprosthetic heart valves. Circulation 2010;121:1848—57.

[38] Gurvitch R, Cheung A, Ye J, Wood DA, Willson AB, Toggweiler S, Binder R, Webb JG. Transcatheter valve-in-valve implantation for failed surgical bioprosthetic valves. J Am Coll Cardiol 2011; 58:2196—209.

[39] Piazza N, Schultz C, de Jaegere PP, Serruys PW. Implantation of two self-expanding aortic bioprosthetic valves during the same procedure-insights into valve-in-valve implantation ("Russian doll concept"). Cathet Cardiovasc Interv 2009;73:530—9.

[40] Groves EM, Falahatpisheh A, Su JL, Kheradvar A. The effects of positioning of transcatheter aortic valves on fluid dynamics of the aortic root. Am Soc Artif Intern Organs J 2014;60:545—52.

[41] Ruiz CE, Laborde JC, Condado JF, Chiam PT, Condado JA. First percutaneous transcatheter aortic valve-in-valve implant with three year follow-up. Catheter Cardiovasc Interv 2008;72: 143—8.

[42] Sarkar K, Ussia G, Tamburino C. Core valve embolization: technical challenges and management. Catheter Cardiovasc Interv 2012;79:777—82.

[43] Thielmann M, Kahlert P, Konorza T, Erbel R, Jakob H, Wendt D. Current developments in transcatheter aortic valve implantation techniques. Herz 2011;36:696—704.

[44] Svensson LG, Dewey T, Kapadia S, Roselli EE, Stewart A, Williams M, Anderson WN, Brown D, Leon M, Lytle B, Moses J, Mack M, Tuzcu M, Smith C. United States feasibility study of transcatheter insertion of a stented aortic valve by the left ventricular apex. Ann Thorac Surg 2008; 86:46—54.

[45] Bagur R, Rodes-Cabau J, Doyle D, De Larochelliere R, Villeneuve J, Lemieux J, Bergeron S, Cote M, Bertrand OF, Pibarot P, Dumont E. Usefulness of tee as the primary imaging technique to guide transcatheter transapical aortic valve implantation. JACC Cardiovasc Imaging 2011;4: 115—24.

[46] Kahlert P, Parohl N, Albert J, Schafer L, Reinhardt R, Kaiser GM, McDougall I, Decker B, Plicht B, Erbel R, Eggebrecht H, Ladd ME, Quick HH. Towards real-time cardiovascular magnetic resonance guided transarterial corevalve implantation: in vivo evaluation in swine. J Cardiovasc Magn Reson 2012;14:21.

[47] Rodes-Cabau J. Progress in transcatheter aortic valve implantation. Rev Española Cardiol 2010;63: 439—50.

[48] Su JL, Wang B, Emelianov SY. Photoacoustic imaging of coronary artery stents. Optic Express 2009; 17:19894—901.

[49] Elgort DR, Hillenbrand CM, Zhang S, Wong EY, Rafie S, Lewin JS, Duerk JL. Image-guided and -monitored renal artery stenting using only MRI. J Magn Reson Imaging 2006;23:619—27.

[50] de Heer LM, Kluin J, Stella PR, Sieswerda GT, MMWP T, van Herwerden LA, Budde RP. Multi-modality imaging throughout transcatheter aortic valve implantation. Future Cardiol 2012;8: 413–24.

[51] Buzzatti N, Maisano F, Latib A, Cioni M, Taramasso M, Mussardo M, Colombo A, Alfieri O. Computed tomography-based evaluation of aortic annulus, prosthesis size and impact on early residual aortic regurgitation after transcatheter aortic valve implantation. Eur J Cardiothorac Surg 2012.

[52] Kempfert J, Van Linden A, Lehmkuhl L, Rastan AJ, Holzhey D, Blumenstein J, Mohr FW, Walther T. Aortic annulus sizing: echocardiographic vs. Computed tomography derived measurements in comparison with direct surgical sizing. Eur J Cardiothorac Surg 2012.

[53] Koos R, Kuhl HP, Muhlenbruch G, Wildberger JE, Gunther RW, Mahnken AH. Prevalence and clinical importance of aortic valve calcification detected incidentally on ct scans: comparison with echocardiography. Radiology 2006;241:76–82.

[54] Blasco A, Piazza A, Goicolea J, Hernández C, García-Montero C, Burgos R, Domínguez JR, Alonso-Pulpón L. Intravascular ultrasound measurement of the aortic lumen. Rev Española Cardiol 2010;63:598–601.

[55] White RA, Verbin C, Kopchok G, Scoccianti M, de Virgilio C, Donayre C. The role of cinefluoro-scopy and intravascular ultrasonography in evaluating the deployment of experimental endovascular prostheses. J Vasc Surg 1995;21:365–74.

[56] Moss RR, Ivens E, Pasupati S, Humphries K, Thompson CR, Munt B, Sinhal A, Webb JG. Role of echocardiography in percutaneous aortic valve implantation. JACC. Cardiovascular imaging 2008;1: 15–24.

[57] Naqvi TZ. Echocardiography in percutaneous valve therapy. JACC. Cardiovascular imaging 2009;2: 1226–37.

[58] Dumont E, Lemieux J, Doyle D, Rodes-Cabau J. Feasibility of transapical aortic valve implantation fully guided by transesophageal echocardiography. J Thorac Cardiovasc Surg 2009;138:1022–4.

[59] Janosi RA, Kahlert P, Plicht B, Bose D, Wendt D, Thielmann M, Jakob H, Eggebrecht H, Erbel R, Buck T. Guidance of percutaneous transcatheter aortic valve implantation by real-time three-dimensional transesophageal echocardiography–a single-center experience. Minim Invasive Ther Allied Technol 2009;18:142–8.

[60] Kpodonu J, Ramaiah VG, Diethrich EB. Intravascular ultrasound imaging as applied to the aorta: a new tool for the cardiovascular surgeon. Ann Thorac Surg 2008;86:1391–8.

[61] Beebe HG. Imaging modalities for aortic endografting. J Endovasc Surg 1997;4:111–23.

[62] Kawase Y, Hoshino K, Yoneyama R, McGregor J, Hajjar RJ, Jang IK, Hayase M. In vivo volumetric analysis of coronary stent using optical coherence tomography with a novel balloon occlusion-flushing catheter: a comparison with intravascular ultrasound. Ultrasound Med Biol 2005;31: 1343–9.

[63] Mintz GS, Nissen SE, Anderson WD, Bailey SR, Erbel R, Fitzgerald PJ, Pinto FJ, Rosenfield K, Siegel RJ, Tuzcu EM, Yock PG. American college of cardiology clinical expert consensus document on standards for acquisition, measurement and reporting of intravascular ultrasound studies (ivus). A report of the american college of cardiology task force on clinical expert consensus documents. J Am Coll Cardiol 2001;37:1478–92.

[64] Ferrari E, Niclauss L, Berdajs D, von Segesser LK. Imaging for trans-catheter pulmonary stent-valve implantation without angiography: role of intravascular ultrasound. Eur J Cardiothorac Surg 2011;40: 522–4.

[65] Mathur SK, Singh P. Transoesophageal echocardiography related complications. Indian J Anaesth 2009;53:567–74.

[66] Jánosi RA, Plicht B, Kahlert P, Eißmann M, Wendt D, Jakob H, Erbel R, Buck T. Quantitative analysis of aortic valve stenosis and aortic root dimensions by three-dimensional echocardiography in patients scheduled for transcutaneous aortic valve implantation. Curr Cardiovasc Imaging Rep 2014;7:9296.

[67] Maragiannis D, Little SH. Interventional imaging: the role of echocardiography. Methodist Debakey Cardiovasc J 2014;10:172–7.

[68] Klein AA, Skubas NJ, Ender J. Controversies and complications in the perioperative management of transcatheter aortic valve replacement. Anesth Analg 2014;119:784—98.

[69] Makkar RR, Fontana GP, Jilaihawi H, Kapadia S, Pichard AD, Douglas PS, Thourani VH, Babaliaros VC, Webb JG, Herrmann HC, Bavaria JE, Kodali S, Brown DL, Bowers B, Dewey TM, Svensson LG, Tuzcu M, Moses JW, Williams MR, Siegel RJ, Akin JJ, Anderson WN, Pocock S, Smith CR, Leon MB. Transcatheter aortic-valve replacement for inoperable severe aortic stenosis. N Engl J Med 2012;366:1696—704.

[70] Kodali SK, Williams MR, Smith CR, Svensson LG, Webb JG, Makkar RR, Fontana GP, Dewey TM, Thourani VH, Pichard AD, Fischbein M, Szeto WY, Lim S, Greason KL, Teirstein PS, Malaisrie SC, Douglas PS, Hahn RT, Whisenant B, Zajarias A, Wang D, Akin JJ, Anderson WN, Leon MB. Two-year outcomes after transcatheter or surgical aortic-valve replacement. N Engl J Med 2012;366:1686—95.

[71] Adams DH, Popma JJ, Reardon MJ, Yakubov SJ, Coselli JS, Deeb GM, Gleason TG, Buchbinder M, Hermiller J, Kleiman NS, Chetcuti S, Heiser J, Merhi W, Zorn G, Tadros P, Robinson N, Petrossian G, Hughes GC, Harrison JK, Conte J, Maini B, Mumtaz M, Chenoweth S, Oh JK. Transcatheter aortic-valve replacement with a self-expanding prosthesis. N Engl J Med 2014;370:1790—8.

[72] Feldman T, Foster E, Glower DD, Kar S, Rinaldi MJ, Fail PS, Smalling RW, Siegel R, Rose GA, Engeron E, Loghin C, Trento A, Skipper ER, Fudge T, Letsou GV, Massaro JM, Mauri L. Percutaneous repair or surgery for mitral regurgitation. N Engl J Med 2011;364:1395—406.

[73] Levack MM, Jassar AS, Shang EK, Vergnat M, Woo YJ, Acker MA, Jackson BM, Gorman 3rd JH, Gorman RC. Three-dimensional echocardiographic analysis of mitral annular dynamics: implication for annuloplasty selection. Circulation 2012;126:S183—8.

[74] Ling LH, Enriquez-Sarano M, Seward JB, Orszulak TA, Schaff HV, Bailey KR, Tajik AJ, Frye RL. Early surgery in patients with mitral regurgitation due to flail leaflets: a long-term outcome study. Circulation 1997;16:1819—25.

[75] Detaint D, Sundt Thoralf M, Nkomo Vuyisile T, Scott Christopher G, Tajik AJ, Schaff Hartzell V, Enriquez-Sarano M. Surgical correction of mitral regurgitation in the elderly. Circulation 2006;114: 265—72.

[76] Maisano F, Alfieri O, Banai S, Buchbinder M, Colombo A, Falk V, Feldman T, Franzen O, Herrmann H, Kar S, Kuck K-H, Lutter G, Mack M, Nickenig G, Piazza N, Reisman M, Ruiz CE, Schofer J, Søndergaard L, Stone GW, Taramasso M, Thomas M, Vahanian A, Webb J, Windecker S, Leon MB. The future of transcatheter mitral valve interventions: competitive or complementary role of repair vs. Replacement? Eur Heart J 2015;36:1651—9.

[77] Yun KL, Miller DC. Mitral valve repair versus replacement. Cardiol Clin 1991;9:315—27.

[78] Baumgartner H, Falk V, Bax JJ, De Bonis M, Hamm C, Holm PJ, Iung B, Lancellotti P, Lansac E, Rodriguez Muñoz D, Rosenhek R, Sjögren J, Tornos Mas P, Vahanian A, Walther T, Wendler O, Windecker S, Zamorano JL, Group ESCSD. 2017 esc/eacts guidelines for the management of valvular heart disease. Eur Heart J 2017;38:2739—91.

[79] Nishimura RA, Otto CM, Bonow RO, Carabello BA, Erwin JP, Guyton RA, O'Gara PT, Ruiz CE, Skubas NJ, Sorajja P, Sundt TM, Thomas JD. 2014 aha/acc guideline for the management of patients with valvular heart disease: a report of the american college of cardiology/american heart association task force on practice guidelines. J Am Coll Cardiol 2014;63:e57—185.

[80] Acker MA, Parides MK, Perrault LP, Moskowitz AJ, Gelijns AC, Voisine P, Smith PK, Hung JW, Blackstone EH, Puskas JD, Argenziano M, Gammie JS, Mack M, Ascheim DD, Bagiella E, Moquete EG, Ferguson TB, Horvath KA, Geller NL, Miller MA, Woo YJ, D'Alessandro DA, Ailawadi G, Dagenais F, Gardner TJ, O'Gara PT, Michler RE, Kron IL, CTSN. Mitral-valve repair versus replacement for severe ischemic mitral regurgitation. N Engl J Med 2014;370:23—32.

[81] Goldstein D, Moskowitz AJ, Gelijns AC, Ailawadi G, Parides MK, Perrault LP, Hung JW, Voisine P, Dagenais F, Gillinov AM, Thourani V, Argenziano M, Gammie JS, Mack M, Demers P, Atluri P, Rose EA, O'Sullivan K, Williams DL, Bagiella E, Michler RE, Weisel RD, Miller MA, Geller NL, Taddei-Peters WC, Smith PK, Moquete E, Overbey JR, Kron IL, O'Gara PT, Acker MA, CTSN. Two-year outcomes of surgical treatment of severe ischemic mitral regurgitation. N Engl J Med 2016;374:344—53.

[82]   Smith PK, Puskas JD, Ascheim DD, Voisine P, Gelijns AC, Moskowitz AJ, Hung JW, Parides MK, Ailawadi G, Perrault LP, Acker MA, Argenziano M, Thourani V, Gammie JS, Miller MA, Pagé P, Overbey JR, Bagiella E, Dagenais F, Blackstone EH, Kron IL, Goldstein DJ, Rose EA, Moquete EG, Jeffries N, Gardner TJ, O'Gara PT, Alexander JH, Michler RE. Cardiothoracic surgical trials network I. Surgical treatment of moderate ischemic mitral regurgitation. N Engl J Med 2014;371:2178–88.

[83]   Gillinov AM, Wierup PN, Blackstone EH, Bishay ES, Cosgrove DM, White J, Lytle BW, McCarthy PM. Is repair preferable to replacement for ischemic mitral regurgitation? J Thorac Cardiovasc Surg 2001;122:1125–41.

[84]   Bonis MD, Costa MAF. Educação em biossegurança e bioética: articulação necessária em biotecnologia. Ciência Saúde Coletiva 2009;14:2107–14.

[85]   Lorusso R, Gelsomino S, Vizzardi E, D'Aloia A, De Cicco G, Lucà F, Parise O, Gensini GF, Stefano P, Livi U, Vendramin I, Pacini D, Di Bartolomeo R, Miceli A, Varone E, Glauber M, Parolari A, Giuseppe Arlati F, Alamanni F, Serraino F, Renzulli A, Messina A, Troise G, Mariscalco G, Cottini M, Beghi C, Nicolini F, Gherli T, Borghetti V, Pardini A, Caimmi P-P, Micalizzi E, Fino C, Ferrazzi P, Di Mauro M, Calafiore AM. Mitral valve repair or replacement for ischemic mitral regurgitation? The Italian study on the treatment of ischemic mitral regurgitation (istimir). J Thorac Cardiovasc Surg 2013;145:128–39.

[86]   De Bonis M, Alfieri O. The edge-to-edge technique for mitral valve repair. HSR Proc Intensive Care Cardiovasc Anesth 2010;2:7–17.

[87]   Mack MJ. New techniques for percutaneous repair of the mitral valve. Heart Fail Rev 2006;11: 259–68.

[88]   Glower DD, Kar S, Trento A, Lim DS, Bajwa T, Quesada R, Whitlow PL, Rinaldi MJ, Grayburn P, Mack MJ, Mauri L, McCarthy PM, Feldman T. Percutaneous mitral valve repair for mitral regurgitation in high-risk patients: results of the everest ii study. J Am Coll Cardiol 2014;64:172–81.

[89]   Feldman T, Kar S, Elmariah S, Smart SC, Trento A, Siegel RJ, Apruzzese P, Fail P, Rinaldi MJ, Smalling RW, Hermiller JB, Heimansohn D, Gray WA, Grayburn PA, Mack MJ, Lim DS, Ailawadi G, Herrmann HC, Acker MA, Silvestry FE, Foster E, Wang A, Glower DD, Mauri L. Randomized comparison of percutaneous repair and surgery for mitral regurgitation: 5-year results of everest ii. J Am Coll Cardiol 2015;66:2844–54.

[90]   Stone GW, Lindenfeld J, Abraham WT, Kar S, Lim DS, Mishell JM, Whisenant B, Grayburn PA, Rinaldi M, Kapadia SR, Rajagopal V, Sarembock IJ, Brieke A, Marx SO, Cohen DJ, Weissman NJ, Mack MJ. Transcatheter mitral-valve repair in patients with heart failure. N Engl J Med 2018;379:2307–18.

[91]   Obadia J-F, Messika-Zeitoun D, Leurent G, Iung B, Bonnet G, Piriou N, Lefèvre T, Piot C, Rouleau F, Carrié D, Nejjari M, Ohlmann P, Leclercq F, Saint Etienne C, Teiger E, Leroux L, Karam N, Michel N, Gilard M, Donal E, Trochu J-N, Cormier B, Armoiry X, Boutitie F, Maucort-Boulch D, Barnel C, Samson G, Guerin P, Vahanian A, Mewton N. Percutaneous repair or medical treatment for secondary mitral regurgitation. N Engl J Med 2018;379:2297–306.

[92]   Praz F, Spargias K, Chrissoheris M, Büllesfeld L, Nickenig G, Deuschl F, Schueler R, Fam NP, Moss R, Makar M, Boone R, Edwards J, Moschovitis A, Kar S, Webb J, Schäfer U, Feldman T, Windecker S. Compassionate use of the pascal transcatheter mitral valve repair system for patients with severe mitral regurgitation: a multicentre, prospective, observational, first-in-man study. The Lancet 2017;390:773–80.

[93]   Silaschi M, Nicou N, Eskandari M, Aldalati O, Seguin C, Piemonte T, McDonagh T, Dworakowski R, Byrne J, Maccarthy P, Monaghan M, Wendler O. Dynamic transcatheter mitral valve repair: a new concept to treat functional mitral regurgitation using an adjustable spacer. EuroIntervention 2017;13:280–3.

[94]   Taramasso M, Latib A. Percutaneous mitral annuloplasty. Interv Cardiol Clin 2016;5:101–7.

[95]   Goldberg SL, Lipiecki J, Sievert H. The carillon mitral contour transcatheter indirect mitral valve annuloplasty system. EuroIntervention 2015;11:W64–6.

[96]   Bail DHL. Treatment of functional mitral regurgitation by percutaneous annuloplasty using the carillon mitral contour system—currently available data state. J Interv Cardiol 2017;30:156–62.

[97] Degen H, Schneider T, Wilke J, Haude M. Koronarsinus-devices zur behandlung der funktionellen mitralklappeninsuffizienz. Herz 2013;38:490—500.

[98] Pedersen WR, Block P, Leon M, Kramer P, Kapadia S, Babaliaros V, Kodali S, Tuzcu EM, Feldman T. Icoapsys mitral valve repair system: percutaneous implantation in an animal model. Cathet Cardiovasc Interv 2008;72:125—31.

[99] Chiam PTL, Ruiz CE. Percutaneous transcatheter mitral valve repair: a classification of the technology. JACC Cardiovasc Interv 2011;4:1—13.

[100] Bonhoeffer P, Boudjemline Y, Saliba Z, Merckx J, Aggoun Y, Bonnet D, Acar P, Le Bidois J, Sidi D, Kachaner J. Percutaneous replacement of pulmonary valve in a right-ventricle to pulmonary-artery prosthetic conduit with valve dysfunction. The Lancet 2000;356:1403—5.

[101] Srivastava D, Olson E. A genetic blueprint for cardiac development. Nature 2000;407:221—6.

[102] Rosamond W, Flegal K, Friday G, Furie K, Go A, Greenlund K, Haase N, Ho M, Howard V, Kissela B, Kittner S, Lloyd-Jones D, McDermott M, Meigs J, Moy C, Nichol G, O'Donnell CJ, Roger V, Rumsfeld J, Sorlie P, Steinberger J, Thom T, Wasserthiel-Smoller S, Hong YL. Amer Heart A. Heart disease and stroke statistics - 2007 update - a report from the american heart association statistics committee and stroke statistics subcommittee. Circulation 2007;115:E69—171.

[103] Roger VL, al e. Heart disease and stroke statistics—2011 update: a report from the american heart association. Circulation 2011;123:E18—1209.

[104] Parker SE, Mai CT, Canfield MA, Rickard R, Wang Y, Meyer RE, Anderson P, Mason CA, Collins JS, Kirby RS, Correa A. For the National Birth Defects Prevention N. Updated national birth prevalence estimates for selected birth defects in the United States, 2004—2006. Birth Defects Res Part A Clin Mol Teratol 2010;88:1008—16.

[105] Rauch R, Hofbeck M, Zweier C, Koch A, Zink S, Trautmann U, Hoyer J, Kaulitz R, Singer H, Rauch A. Comprehensive genotype—phenotype analysis in 230 patients with tetralogy of fallot. J Med Genet 2010;47:321—31.

[106] Therrien J, Siu SC, McLaughlin PR, Liu PP, Williams WG, Webb GD. Pulmonary valve replacement in adults late after repair of tetralogy of fallot: are we operating too late? J Am Coll Cardiol 2000;36:1670—5.

[107] Momenah TS, El Oakley R, Al Najashi K, Khoshhal S, Al Qethamy H, Bonhoeffer P. Extended application of percutaneous pulmonary valve implantation. J Am Coll Cardiol 2009;53:1859—63.

[108] Moiduddin N, Asoh K, Slorach C, Benson LN, Friedberg MK. Effect of transcatheter pulmonary valve implantation on short-term right ventricular function as determined by two-dimensional speckle tracking strain and strain rate imaging. Am J Cardiol 2009;104:862—7.

[109] McElhinney DB, Hennesen JT. The melody® valve and ensemble® delivery system for transcatheter pulmonary valve replacement. Ann N Y Acad Sci 2013;1291:77—85.

[110] Butera G, Milanesi O, Spadoni I, Piazza L, Donti A, Ricci C, Agnoletti G, Pangrazi A, Chessa M, Carminati M. Melody transcatheter pulmonary valve implantation. Results from the registry of the Italian society of pediatric cardiology. Cathet Cardiovasc Interv 2013;81:310—6.

[111] Eicken A, Ewert P, Hager A, Peters B, Fratz S, Kuehne T, Busch R, Hess J, Berger F. Percutaneous pulmonary valve implantation: two-centre experience with more than 100 patients. Eur Heart J 2011;32:1260—5.

[112] Vezmar M, Chaturvedi R, Lee K-J, Almeida C, Manlhiot C, McCrindle BW, Horlick EM, Benson LN. Percutaneous pulmonary valve implantation in the young: 2-year follow-up. JACC Cardiovasc Interv 2010;3:439—48.

[113] Petit Christopher J. Pediatric transcatheter valve replacement. Circulation 2015;131:1943—5.

[114] Cheatham John P, Hellenbrand William E, Zahn Evan M, Jones Thomas K, Berman Darren P, Vincent Julie A, McElhinney Doff B. Clinical and hemodynamic outcomes up to 7 years after transcatheter pulmonary valve replacement in the us melody valve investigational device exemption trial. Circulation 2015;131:1960—70.

[115] Purohit M, Kitchiner D, Pozzi M. Contegra bovine jugular vein right ventricle to pulmonary artery conduit in ross procedure. Ann Thorac Surg 2004;77:1707—10.

[116] Patel M, Malekzadeh-Milani S, Ladouceur M, Iserin L, Boudjemline Y. Percutaneous pulmonary valve endocarditis: incidence, prevention and management. Arch Cardiovasc Dis 2014;107:615—24.

[117] Van Dijck I, Budts W, Cools B, Eyskens B, Boshoff DE, Heying R, Frerich S, Vanagt WY, Troost E, Gewillig M. Infective endocarditis of a transcatheter pulmonary valve in comparison with surgical implants. Heart 2015;101:788.

[118] Sinning J-M, Hammerstingl C, Vasa-Nicotera M, Adenauer V, Lema Cachiguango SJ, Scheer A-C, Hausen S, Sedaghat A, Ghanem A, Müller C, Grube E, Nickenig G, Werner N. Aortic regurgitation index defines severity of peri-prosthetic regurgitation and predicts outcome in patients after transcatheter aortic valve implantation. J Am Coll Cardiol 2012;59:1134−41.

CHAPTER 6

Tissue-engineered heart valves

Petra Mela[1,5], Svenja Hinderer[2], Harkamaljot S. Kandail[3], Carlijn V.C. Bouten[3,4], Anthal I.P.M. Smits[3,4]

[1]Department of Biohybrid & Medical Textiles (BioTex), AME — Institute of Applied Medical Engineering, Helmholtz Institute, RWTH Aachen University, Aachen, Germany; [2]Natural and Medical Sciences Institute (NMI), University of Tübingen, Reutlingen, Germany; [3]Department of Biomedical Engineering, Eindhoven University of Technology, Eindhoven, the Netherlands; [4]Institute for Complex Molecular Systems (ICMS), Eindhoven University of Technology, Eindhoven, the Netherlands; [5]Medical Materials and Implants, Department of Mechanical Engineering, Technical University of Münich, Münich, Germany

Contents

Principles of Heart Valve Engineering
ISBN 978-0-12-814661-3, https://doi.org/10.1016/B978-0-12-814661-3.00006-X

6.1 Introduction

Heart valve replacement is a common clinical procedure. Since decades, the clinical golden standards in heart valve replacement are chemically cross-linked bioprosthetic valves and mechanical valves. Although such valves improve quality of life, both valve types are suboptimal for large cohorts of patients. The common limitation of both these types of replacements is that they are nonliving tissues, which are inherently incapable of adaptation and growth in response to environmental changes. It is this ability to adapt to changes in the hemodynamic environment that is pivotal for the long-term functionality of the native valve [1]. This has inspired the quest for a living replacement valve by tissue engineers over the last two decades [2]. This chapter deals with the requirements and potential benefits of tissue-engineered heart valves (TEHVs) (Section 6.2) and the various technological concepts that are being pursued to achieve this (Section 6.3). Sections 6.4 and 6.5 deal with the various cell sources and scaffold types that are used for heart valve tissue engineering (HVTE), respectively. As for the native valve, biomechanics play a central role in HVTE, which is reflected in the elaborate use of bioreactors (Section 6.6) and numerical models to optimize valve design and predict tissue growth and remodeling (Section 6.7). Section 6.8 concisely reflects on the compatibility of TEHV with minimally invasive delivery methods, with an emphasis on the mechanical aspects in that context. To conclude, Section 6.9 provides a perspective on some of the most imminent current challenges for HVTE to overcome, to progress from promising technology toward a robust clinical treatment option. Given the extensive body of literature on the multidisciplinary aspects of HVTE, this chapter is by no means complete, but rather it is intended to give the reader an overview of the various building blocks and concepts involved in HVTE.

6.2 The living heart valve—taking inspiration from nature

Although the mechanical function of heart valves is purely passive, driven by the pressure differences over the valve, it is crucial to realize that the native human heart valves are living dynamic tissues that grow and remodel throughout postnatal and adult life (Fig. 6.1). In their landmark study, Aikawa et al. demonstrated the development of human pulmonary and aortic valves from fetuses at various stages of gestation, neonates, children, and adults, in terms of cellular phenotype and extracellular matrix (ECM) composition [3]. Histological characterization of the valves revealed a transition from an immature, activated myofibroblast state of the valvular interstitial cells (VICs) during fetal development into a quiescent fibroblast-like phenotype in adults (Fig. 6.1A). This was accompanied by collagen maturation, with thicker and more aligned collagen fibers in adult valves compared with prenatal and pediatric valves. Building on this, Van Geemen et al. studied the age-specific structural and functional characteristics of pairs of pulmonary and aortic valves of human donors from the fetus to adult [4]. Focusing

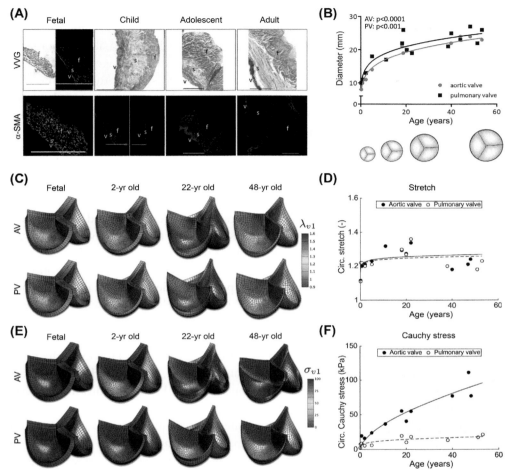

Figure 6.1 *Native human valve development throughout life.* (A) Human aortic valve leaflets stained with Verhoeff—Van Gieson (VVG) staining for collagen (red) and elastin (black), elastin antibody (red fluorescence), and α-smooth muscle actin (α-SMA; green fluorescent) showing development of the tissue architecture and the transition from activated to quiescent valvular interstitial cells from fetus to adult. Scale bars, 500 μm f, fibrosa; s, spongiosa; v, ventricularis. (B) Human semilunar valve growth throughout postnatal life. (C, D) Circumferential stretch distribution in pairs of human aortic valves (AV) and pulmonary valves (PV) revealing that valves remodel throughout life to maintain a constant circumferential stretch homeostasis, while, in contrast, (E, F) this does not apply to the circumferential stresses, which increase with age in both valves. *((A, B) Adapted from van Geemen D, Soares ALF, Oomen PJA, Driessen-Mol A, Janssen-van den Broek MWJT, van den Bogaerdt AJ, Bogers AJJC, Goumans MJTH, Baaijens FPT, Bouten CVC. Age-dependent changes in geometry, tissue composition and mechanical properties of fetal to adult cryopreserved human heart valves. PLoS One 2016;11:e0149020. https://doi.org/ 10.1371/journal.pone.0149020. and (C—F) Adapted from Oomen PJA, Loerakker S, van Geemen D, Neggers J, Goumans MJTH, van den Bogaerdt AJ, Bogers AJJC, Bouten CVC, Baaijens FPT. Age-dependent changes of stress and strain in the human heart valve and their relation with collagen remodeling. Acta Biomater 2016; 29:161—9. https://doi.org/10.1016/j.actbio.2015.10.044. reprinted with permission from the respective publishers.)*

on the ECM composition, this study revealed a progressive decrease in glycosaminogly-cans (GAGs) and an increase in collagen content and cross-links with age, which was reflected by an increase in leaflet stiffness. Importantly, this study also demonstrated that valvular growth is not restricted to prenatal and early postnatal stages but that human valves grow continuously throughout life (Fig. 6.1B). In a follow-up study, Oomen et al. postulated that these changes are likely to be attributable to continuous changes in hemodynamics and specifically to a progressive increase in the diastolic blood pressure with age [5]. Numerical modeling of the valves revealed a striking resemblance in the stretch distribution—but not in the stress distribution—in pairs of pulmonary and aortic valves for each donor (Fig. 6.1C–F). This suggests that the human semilunar valves grow and remodel throughout life to maintain a state of stretch homeostasis, presumably governed by the VICs. Although the native valve can adapt to physiological changes in hemodynamic loads, pathological hemodynamic conditions (e.g., increased blood pressure, disturbed flow-induced shear stresses) can adversely activate the VICs, giving rise to valve dysfunction and pathologies such as calcification, as reviewed in detail by Gould et al. [6].

The remarkable capacity of human valves to adapt and remodel to environmental changes is the incentive for tissue engineers to pursue the development of a living heart valve replacement. The idea of a living TEHV is particularly exciting when considering children with congenital valvular defects. A living tissue-engineered replacement valve would be able to somatically grow with the child and therefore reduce the need for repetitive reoperations, which is inherent to nonliving prosthetics. However, the notion that native human valves continuously grow and remodel, also throughout adulthood, has important implications for the target populations for HVTE, as this implies that a living replacement valve would also benefit adult patients. Indeed, several reports have described a reduced life expectancy for patients receiving either a mechanical or bioprosthetic replacement valve, when compared with age-matched controls (6–8). This has become even more relevant with increasing general life expectancy and patients "outliving" the lifetime of their prosthetic valve. The benefit of a living replacement valve in adults is perhaps best illustrated by the positive clinical outcomes associated with the Ross procedure, in which the damaged aortic valve is replaced by the patient's own pulmonary valve. A randomized controlled trial by El-Hamamsy et al. with a follow-up time of up to 13 years revealed an improved survival and improved freedom from reoperations for patients receiving a pulmonary autograft in the aortic position (i.e., the Ross procedure), when compared with patients receiving a cryopreserved homograft, underlining the importance of having a living replacement valve [7].

6.3 Heart valve tissue engineering paradigms

Within the HVTE field, there are two conceptually different strategies, which we will refer to as (1) in vitro tissue engineering and (2) in situ tissue engineering (Fig. 6.2) [8,9].

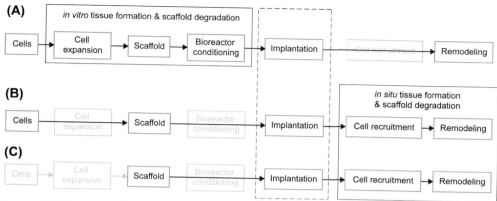

Figure 6.2 *The various heart valve tissue engineering (HVTE) paradigms.* (A) *In vitro* HVTE uses autologously harvested cells seeded onto a scaffold, followed by dynamic in vitro culturing to induce tissue formation. (B, C) *In situ* HVTE strategies, in which the in vitro culturing phases are omitted by directly implanting functional heart valve scaffolds that are designed for endogenous regeneration, either on-the-fly preseeded (B) or acellular (C).

Each of these strategies has its advantages and disadvantages; however, common goal is to provide an autologous, living replacement valve. In vitro tissue engineering is aimed at creating an autologous cellularized valve in the laboratory, before implantation (Fig. 6.2A). In situ tissue engineering employs acellular resorbable scaffolds (either biological or synthetic), which are designed to induce endogenous regeneration directly in the valve's functional site (Fig. 6.2C) [10]. A variation to the in situ tissue engineering approach is the use of on-the-fly preseeding of resorbable scaffolds with patient cells in a single operation, without in vitro culturing (Fig. 6.2B) [11,12]. By avoiding lengthy and costly in vitro culturing, in situ tissue engineering has several advantages over in vitro tissue engineering in that the former is more cost-effective and logistically less demanding. Most importantly, in situ tissue-engineered valves are off-the-shelf available. Moreover, when using a purely synthetic scaffold for in situ tissue engineering, the prosthesis is considered a medical device in terms of regulatory issues, which greatly ameliorates clinical translation [13]. The main advantage of in vitro tissue engineering, on the other hand, is that the culturing in a controlled setting allows for a relatively high level of control over the newly formed tissue composition and architecture. In contrast, in situ tissue engineering is heavily reliant on the in vivo remodeling processes, which are difficult to steer. Moreover, for in situ tissue engineering, the scaffold has to instantaneously take over the valve's function on implantation and has to maintain functionality, while the initial scaffold is gradually resorbed and replaced by newly formed tissue. To add to the challenge, this process is dependent on the patient's intrinsic regenerative capacity and immunological state, which are prone to strong interpatient variabilities [14].

6.3.1 In vitro heart valve tissue engineering

In vitro tissue engineering comprises the traditional tissue engineering paradigm using cells, scaffolds, and stimuli, as originally defined by Langer and Vacanti [15]. For this, cells are first isolated from the patient and expanded via in vitro culturing. After expansion, the cells are then seeded onto a scaffold, which acts as a temporary structural framework for the cells to produce new ECM. The cell-seeded scaffolds are exposed to various biochemical and/or biomechanical cues to induce tissue formation and remodeling. Given that the tissue organization of the native semilunar valves is heavily dictated by the hemodynamic loads to which the valve is exposed, mechanical conditioning using bioreactors plays an important role in the in vitro culturing of heart valves (see Section 6.6). In general, culturing conditions are optimized as such to engineer a valve that mimics the native valve as closely as possible, in terms of mechanical properties, tissue architecture, and cellular composition. However, it should be noted that lab-grown heart valves do not necessarily need to perfectly resemble their native counterpart, as they are anticipated to undergo functional tissue remodeling once implanted.

Following this approach, among the first reports on HVTE is the work by Shinoka et al. in which the successful replacement of a single heart valve leaflet of the ovine pulmonary valve by an in vitro tissue-engineered leaflet is described [16,17]. The proof of principle for in vitro HVTE was demonstrated in 2000, by several studies describing the complete valve replacement with an in vitro engineered trileaflet valve in the pulmonary position in lambs [18,19]. Since then, numerous research groups have engineered valves using a variety of cell sources, ranging from mature venous fibroblasts to mesenchymal stem cells (MSCs) and circulatory progenitor cells (reviewed in Ref. [20]; see Section 6.4), as well as different scaffolds materials and processing methods (reviewed in Ref. [21]; see Section 6.5). These efforts have led to increased levels of refinement of microstructural and mechanical properties of TEHVs, aiming to mimic those of native valves. Moreover, the compatibility of TEHVs with minimally invasive delivery has been demonstrated [22–25]. However, from a translational point of view, in vitro HVTE is logistically challenging and costly because of the lengthy in vitro culturing phases, hampering the progression into clinical trials to date [8]. In addition, the emphasis has laid on the in vitro recreation of the native human valve as closely as possible. The in vivo response to TEHVs upon implantation, on the other hand, has been largely underexposed, resulting in adverse in vivo remodeling leading to leaflet retraction and valve failure in preclinical models [11,18,23,26–30].

6.3.2 In situ heart valve tissue engineering

To overcome the inherent limitations of in vitro cultured valves, in situ HVTE has risen as an alternative approach to obtain autologous, living replacement valves. In situ HVTE is characterized by the use of off-the-shelf available acellular scaffolds, which rely on in vivo recellularization and tissue remodeling by endogenous cells after implantation [10,31]. The main advantage of this strategy is that it employs readily available and relatively simple scaffolds, which can compete with the clinically available valvular prostheses in terms of costs. Similar to in vitro HVTE, the first reports on in situ tissue engineering date back to nearly two decades ago. These pioneering studies employed decellularized xenografts and allografts as the scaffold materials (see also Section 6.5.1). Preclinical success of these approaches has led to numerous clinical trials, which, in particular for decellularized allografts, have demonstrated that these valves are viable alternatives to currently available bioprosthetic valves in terms of midterm functionality [32]). Notwithstanding their potential, decellularized native valves are intrinsically limited by requiring a donor valve. This is a particularly relevant consideration with allografts, given the limited availability of human donor valves. To overcome this necessity, decellularization of de novo engineered valves has emerged as a potential alternative strategy, which has demonstrated promising preclinical results [33—35]. The use of natural scaffold materials is described in more detail in Section 6.5.1.

More recently, acellular resorbable synthetic scaffolds have emerged as candidate materials for in situ HVTE (see also Section 6.5.2). Building on the exponentially growing body of data on immunomodulatory materials, this technology is inspired by the notion that the inevitable host inflammatory response to a foreign body is not only destructive to a synthetic implant but that it can also be the initiator of tissue regeneration when harnessed properly [10,36]. Although the production of scaffolds for in situ HVTE is relatively simple, controlling the in vivo response evoked by such a scaffold is highly complex [13] and requires multifactorial control over the scaffold parameters, such as material choice, microarchitecture, degradation rate, and mechanical properties [10]. Recent preclinical studies have demonstrated the proof of principle for resorbable synthetic scaffolds for in situ HVTE [37,38], which have led to the first ongoing clinical trials, with positive short-term results to date (Xeltis, Xplore-I, and Xplore-II trials).

Despite the relatively fast progress of in situ HVTE and the successful preclinical and clinical results, both for natural and synthetic scaffolds, many questions concerning in situ recellularization and the pathophysiological remodeling processes remain unanswered. Hence, gaining a mechanistic understanding of the in vivo inflammatory and regenerative processes and the variety of these over various patient populations [8] is imminent to propel the field to robust clinical application.

6.4 The cellular players in heart valve tissue engineering

The cellular population of native heart valves consists predominantly of VICs, which are responsible for tissue remodeling and adaptation during valve growth and in response to injury or changes in the hemodynamic environment. In addition to VICs, a monolayer of valvular endothelial cells (VECs) covers the valvular leaflets as a barrier between the blood and the underlying matrix. During valve development, endothelial-to-mesenchymal trans-differentiation (EndMT) of VECs to VICs is a physiological process, governed by hemo-dynamic loads [39,40]. Moreover, during adulthood, EndMT has been correlated to pathological fibrosis, as transdifferentiating VECs can give rise to activated myofibroblast-like and, potentially, osteogenic VICs. The latter was recently demonstrated in vitro by Hjortnaes et al. who showed that EndMT is inhibited by VICs during normal valve homeostasis, but that disturbance of the homeostasis during valve calcification and the osteogenic differentiation of VICs can lead to EndMT of VECs, which may further differentiate into osteogenic VICs, thereby contributing to disease progression [41]. Zhong et al. recently demonstrated that the stiffness of the ECM also is a key regulator for EndMT in this context, with increased occurrence of EndoMT with increasing matrix stiffness [42]. The mechanosensitivity of VECs is further underlined by the notion that, in native adult aortic valves, the VECs on the aortic side of the leaflet display distinct phenotypical differences from the VECs on the ventricular side, which is attributed to the differences in shear stresses on either side of the leaflet [43]. VECs actively signal to VICs to transmit mechanical cues on changes in hemodynamics [44], as well the ECM stiffness [45]. Together, the VECs and the VICs are essential for maintaining the valve's dynamic and adaptive remodeling potential.

6.4.1 Cell sources for in vitro heart valve tissue engineering

For the in vitro creation of TEHVs, numerous cell types from various sources have been investigated, as recently reviewed in more detail by Jana et al. [20]. A much used cell source for HVTE is venous cells, for example, isolated from leftover saphenous vein after coronary bypass surgery. These cells are characterized by a myofibroblast-like phenotype, which can have an activated contractile state or a quiescent fibroblast-like state depending on environmental cues, thereby resembling the VIC phenotype. Apart from phenotypical potential, an important consideration with respect to cell types for HVTE is the clinical ease of access to obtain the cells, which are ideally sourced autologously. From that perspective, MSCs make ideal candidates. Various studies have successfully employed MSCs from different sources (e.g., bone marrow, adipose tissue) to create TEHVs in vitro, as reviewed by Weber et al. [46]. For example, bone marrow—derived MSCs were seeded onto nonwoven melt extruded polyglycolic acid (PGA)/PLLA scaffolds. After a 4-week in vitro culture, the constructs were implanted as pulmonary valves in sheep. A successful proof of principle was demonstrated; the valves underwent extensive remodeling and adopted cellular phenotypes [47]. Notably, in a series of studies, Schmidt et al. demonstrated the potential of using fresh or cryopreserved prenatally harvested MSCs (e.g., from amniotic fluid) as an interesting cell source for TEHV, particularly aimed at pediatric patients with congenital defects [48,49]. Moreover, in addition to giving rise to VIC-like cells, MSCs have been demonstrated to be capable of differentiation into mature antithrombogenic VEC-like cells [50,51]. Another easily accessible source of stem and progenitor cells is the peripheral blood. For example, TEHVs seeded with peripheral blood mononuclear cells on a PGA-based scaffold and cultured in vitro were transapically implanted as pulmonary valve replacement in sheep. On implantation, the tissue-engineered constructs showed ECM formation but also massive leaflet thickening. Long-term studies are mandatory to verify cell fate and function [23]. Sales et al. demonstrated the feasibility to use peripheral blood—derived endothelial progenitor cells as a suitable cell source for TEHV creation, capable of differentiation into both VIC- and VEC-like cells [52].

6.4.2 In situ cellularization

In contrast to in vitro cultured TEHVs, in situ tissue engineering approaches fully rely on endogenous (re)population of decellularized natural scaffolds or acellular synthetic scaffolds with host cells. Although the feasibility of such in situ cellularization processes has been proven, the underlying mechanisms and the sources of the colonizing host cells are largely unknown. Potential cellularization routes include the infiltration of circulatory immune and progenitor cells, either directly from the blood stream or via the formation of a microvasculature, the transanastomotic and transmural ingrowth of cells from the native artery, and the homing of (stem) cells from the bone marrow [8]; the latter being a naturally occurring phenomenon during normal valve homeostasis [53].

Given that complete cellular colonization is critical for the regenerative and remodeling potentials of a valve, numerous studies have explored scaffold functionalization strategies to improve the homing of stem or immune cells or the infiltration of recruited cells into the scaffold material [54—57]. For example, Flameng et al. applied a combinatorial coating of stromal cell—derived factor-1α with its natural linker fibronectin to target the homing of stem cells toward acellular vascular and valvular scaffolds [55,58]. Interestingly, when applying this dual coating on decellularized allograft valves, the coated valves displayed a less strong immunogenic response and less calcification when implanted as right ventricular outflow tracts (RVOTs) in sheep [55]. Notably, however, studies on acellular vascular grafts have demonstrated that the infiltration of innate immune cells, and macrophages in particular, is essential to ignite the regenerative cascade [59,60].

Although fallout healing has been demonstrated to be possible in vascular grafts [61,62], the main route of cellularization of acellular vascular grafts was recently pinpointed to be transmural migration of adventitial cells toward the lumen [63]. Several reports on acellular synthetic TEHVs have demonstrated extensive cellularization of valvular leaflets as soon as 1 day after implantation [27,64]. As cellular ingrowth throughout the entire leaflet is unlikely in such a time frame, this strongly suggests a rapid influx of circulatory cells into the leaflet. Similarly, a recent study by Kluin et al. in which the long-term regenerative potential of an acellular synthetic TEHV as pulmonary valve replacement in sheep was examined, explanted valves revealed extensive cellularization throughout the valvular leaflets. Longitudinal follow-up up to 12 months follow-up suggested a secondary wave of progressive ingrowth of tissue-forming cells from the hinge region toward the free edge of the leaflet [37]. Similar findings were reported by Syedain et al. regarding the recellularization patterns of decellularized de novo engineered valves, demonstrating progressive colonization of vimentin-positive and α-SMA—negative cells from the hinge toward the leaflet-free edge, strongly suggesting that these cells originate from the neighboring arterial tissue [34]. Taken together, these findings suggest combinatorial cellularization routes analogue to the data reported on vascular grafts, namely an initial rapid influx of circulatory (immune) cells, followed by a secondary wave of gradual ingrowth of arterial cells to colonize the valve leaflets; the latter are potentially attracted by signaling factors secreted by the initially infiltrating cells [10]. However, these observations remain speculative, and mechanistic studies investigating the exact cellular phenotypes of colonizing host cells using dedicated antibody panels [65,66] and tracing studies are required to unravel the spatiotemporal in vivo cellularization processes and sources of colonizing cells.

An alternative to the implantation of completely acellular grafts is the use of on-the-fly preseeding with bone marrow—derived cells [11,12,27,33] (Fig. 6.3). These cells are harvested, isolated, and reseeded on cardiovascular scaffolds within the same intervention as the valve replacement. As this approach omits any in vitro culture phase, the advantage of off-the-shelf availability is maintained. The original rationale behind such on-the-fly preseeding strategies was that the preseeded bone marrow—derived cells would contain a proportion of stem cells that could subsequently differentiate into mature valvular cells. However, tracing studies using vascular grafts suggest that preseeded cells are removed relatively rapidly after implantation, suggesting that they predominantly exert paracrine effects, which may or may not ameliorate downstream cellularization and tissue regeneration [60].

6.5 Scaffolds for heart valve tissue engineering

A plethora of different scaffold types is being explored for the engineering of heart valves, both in vitro and in situ (Fig. 6.4). Generally, these scaffolds can be classified as being of natural or synthetic origin, or a combination of both (hybrid scaffolds). In addition to being the structural framework, the scaffold provides a local niche for the cells to differentiate and form and remodel the valvular matrix. As such, scaffolds transduce structural and mechanical cues to the cells, as well as biological cues that can be incorporated into the scaffold, either intrinsically or exogenously. Apart from the scaffold material source and composition, the processing of the base materials into the three-dimensional valve with the ideal microstructure and macroscopic geometry is an active field of research, which will be elaborated on in the following sections.

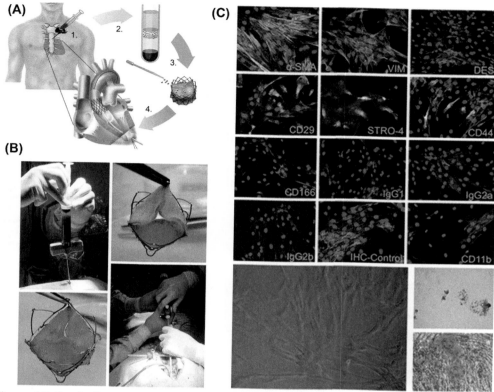

Figure 6.3 *On-the-fly preseeding of acellular heart valve scaffolds.* (A) Schematic representation illustrating the concept of on-the-fly preseeding of scaffold with bone marrow—derived cells, for which a bone marrow biopsy is collected from the sternum (1); the mononuclear cell fraction is isolated via density gradient centrifugation (2); the cells are seeded onto an acellular scaffold (3), after which the valve is directly implanted (4). (B) Photographs of the preseeding procedure applied onto polyglycolic acid—based composite scaffolds, before transapical delivery in sheep. (C) Phenotypical characterization of ovine bone marrow—derived cells, revealing the presence of mesenchymal stem cells with differentiation capacity within the derived cell fraction. *((A) Adapted from Emmert MY, Weber B, Wolint P, Behr L, Sammut S, Frauenfelder T, Frese L, Scherman J, Brokopp CE, Templin C, Grünenfelder J, Zünd G, Falk V, Hoerstrup SP. Stem cell-based transcatheter aortic valve implantation: first experiences in a pre-clinical model. JACC Cardiovasc Interv 2012;5:874—83. https://doi.org/10.1016/j.jcin. 2012.04.010. (B, C) Adapted from Emmert MY, Weber B, Behr L, Sammut S, Frauenfelder T, Wolint P, Scherman J, Bettex D, Grünenfelder J, Falk V, Hoerstrup SP. Transcatheter aortic valve implantation using anatomically oriented, marrow stromal cell-based, stented, tissue-engineered heart valves: technical considerations and implications for translational cell-based heart valve concepts. Eur J Cardiothorac Surg 2014;45:61—8. https://doi.org/10.1093/ejcts/ezt243. with permission from the respective publishers.)*

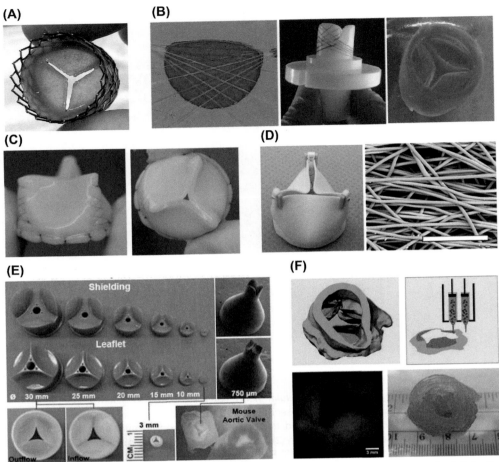

Figure 6.4 *Examples of tissue-engineered heart valve (TEHV) types and fabrication methods.* (A) Decellularized in vitro TEHV on nitinol stent for minimally invasive delivery. (B) Composite "BioTexValve"; created by arranging large diameter multifilament fibers onto a frame, fixed by a thin layer of electrospun polylactide-co-glycolide, and seeded with a cell-laden fibrin gel. (C) Tissue-engineered aortic valve mounted on Sorin's Mitroflow Frame; the valves were created by culturing human dermal fibroblasts in fibrin, after which the deposited matrix was decellularized, before implantation. (D) Electrospun valve from supramolecular elastomer on polyetheretherketone (PEEK) supportive frame with a highly porous microfibrous architecture; scale bar, 50 μm. (E) Acellular fibrous heart valves (JetValve) fabricated via an automated rotary jet spinning process, which allows for large-range scalability of valve size. (F) 3D printed aortic valve based on native (porcine) valve anatomy as obtained from micro-CT scanning, composed of a combination of smooth muscle cell— (labeled green, wall) and valvular interstitial cell (labeled red, leaflets)—laden alginate–gelatin hydrogel. *((A) Adapted from Weber B, Dijkman PE, Scherman J, Sanders B, Emmert MY, Grünenfelder J, Verbeek R, Bracher M, Black M, Franz T, Kortsmit J, Modregger P, Peter S, Stampanoni M, Robert J, Kehl D, Van Doeselaar M, Schweiger M, Brokopp CE, Walchli T, Falk V, Zilla P, Driessen-Mol A, Baaijens FPT, Hoerstrup SP. Off-the-shelf human decellularized tissue-engineered heart valves in a non-human primate model. Biomaterials 2013;34:7269—80. https://doi.org/10.1016/j.biomaterials.2013.04.059. (B) Adapted from*

6.5.1 Natural scaffolds

Natural scaffolds used for HVTE can be divided in natural polymer-based hydrogels and ECM-based scaffolds obtained by decellularization of either native or bioengineered tissues. We will mainly restrict this section to reports of complete heart valves, not including the growing number of studies on planar substrates intended for HVTE and disease modeling, which, although promising, have not been translated into functional heart valves.

6.5.1.1 Natural polymer—based hydrogels

Natural polymers offer the advantage of inherent bioactivity and enzymatic degradation when compared with their synthetic counterparts. However, they suffer from batch-to-batch reproducibility and are, to some extent, more limited in the manufacturability, as the use of harsh solvents and large temperature ranges normally applied to synthetic polymers may not be compatible with maintaining their bioactivity and functionality. Natural polymers have been extensively applied for the fabrication of TEHVs. Mainly, purified collagen and fibrinogen have been used, with only few examples involving GAGs, alginate, and gelatin being reported.

Collagen is a main component of the ECM of native cardiovascular structures responsible for their mechanical strength. Purified collagen hydrogels have been applied to engineer matrices and complete valvular conduits, alone or in combination with other valvular ECM components [69—74]. Neidert and colleagues fabricated a bileaflet heart valve by molding bovine collagen entrapping dermal neonatal human dermal fibroblasts. The authors exploited the cell-induced gel compaction to obtain biomimetic commissure-to-commissure fibril alignment in the leaflets and circumferential alignment in the root. However, the synthesized ECM was mechanically weaker than the native one and was almost entirely composed of collagen, lacking other components of

Moreira R, Neusser C, Kruse M, Mulderrig S, Wolf F, Spillner J, Schmitz-Rode T, Jockenhoevel S, Mela P. Tissue-engineered fibrin-based heart valve with bio-inspired textile reinforcement. Adv Healthc Mater 2016;5:2113—21. https://doi.org/10.1002/adhm.201600300. (C) Adapted from Syedain Z, Reimer J, Schmidt J, Lahti M, Berry J, Bianco R, Tranquillo RT, 6-Month aortic valve implantation of an off-the-shelf tissue-engineered valve in sheep. Biomaterials 2015;73:175—84. https://doi.org/10.1016/j.biomaterials. 2015.09.016. (D) Adapted from Kluin J, Talacua H, Smits AIPM, Emmert MY, Brugmans MCP, Fioretta ES, Dijkman PE, Söntjens SHM, Duijvelshoff R, Dekker S, Janssen-van den Broek MWJT, Lintas V, Vink A, Hoerstrup SP, Janssen HM, Dankers PYW, Baaijens FPT, Bouten CVC. In situ heart valve tissue engineering using a bioresorbable elastomeric implant - From material design to 12 months follow-up in sheep. Biomaterials 2017;125:101—17. https://doi.org/10.1016/j.biomaterials.2017.02.007. (E) Adapted from Capulli AK, Emmert MY, Pasqualini FS, Kehl D, Caliskan E, Lind JU, Sheehy SP, Park SJ, Ahn S, Weber B, Goss JA, Hoerstrup SP, Parker KK. JetValve: Rapid manufacturing of biohybrid scaffolds for biomimetic heart valve replacement. Biomaterials 2017;133:229—41. https://doi.org/10.1016/j.biomaterials.2017.04.033. (F) Adapted from Duan B, Hockaday LA, Kang KH, Butcher JT. 3D Bioprinting of heterogeneous aortic valve conduits with alginate/gelatin hydrogels. J Biomed Mater Res A 2013;101:1255—64. https://doi.org/10. 1002/jbm.a.34420; with permission from the respective publishers.)

fundamental importance for the valve functionality [73]. With the rationale of improving the bioactivity, Brougham et al. reported the fabrication of a semilunar heart valve by freeze-drying a collagen—GAG copolymer suspension in a custom-designed mold. They were able to obtain the complex 3D geometry with homogeneous, interconnected porosity, showing the potential of the established method for complex 3D constructs [74]. Chen and colleagues introduced another fundamental protein for heart valves and fabricated a bilayered collagen—elastin scaffold with well-connected interface and anisotropic bending moduli depending on the loading directions, which recapitulated the behavior of the native heart valves. Furthermore, asymmetric distribution of cardiosphere-derived cells within the scaffold was shown [69].

Fibrin has also demonstrated excellent properties as heart valve matrix material. It offers many advantages, such as its autologous origin [75], the rapid polymerization, the tuneable degradation via protease inhibitors [76], the autologous release of growth factors [77], and the manufacturability into complex 3D geometries with homogeneous cell seeding [78]. Fibrin is known to facilitate ECM synthesis by seeded cells in higher amounts than in collagen gels and, importantly for cardiovascular tissue engineering, elastin production in fibrin gels is also clearly enhanced [79]. Fibrin has been employed as a cell carrier in combination with another scaffold material [23,80] or as main scaffold material [26,78,81—84] to engineer heart valves with good functionality. Similar to what has been achieved with a collagen-based valve, the compaction of the fibrils that occurs in the protein-based gels under the contraction forces of the smooth muscle cells could be beneficially exploited to create ECM alignment in the constructs [83]. However, cell-mediated tissue compaction has been proven as one of the major drawbacks of heart valves, including those employing fibrin, resulting in valvular insufficiency in vitro and in vivo [11,18,23,26—30]. A further disadvantage is the weak mechanical properties that have made the implantation of such valves in the systemic circulation rather challenging. Different strategies have been adopted by the various groups to overcome these limitations, including the removal of the cells through decellularization processes (treated later on in this section) to avoid tissue shrinkage, the adoption of a tubular leaflet design, and the reinforcement of the scaffolds with macroporous textiles to contrast tissue shrinkage and increase the mechanical properties. Fiber reinforcement of fibrin-based constructs resulted in mitral valves, aortic valve conduits, and aortic valves with tubular design leaflets able to withstand the systemic circulation in vitro, to be implanted either surgically or by minimally invasive delivery (Fig. 6.4B) [25,67,85,86]. Although the textile reinforcement has the potential of preventing tissue contraction, further knowledge will be gained by the in vivo evaluation of these valves.

A multistep molding technique [81] and 3D bioprinting (e.g., Fig. 6.4F) [68] of hydrogel-based heart valves have also proven adequate for the realization of complete valved conduits with spatially controlled material and/or cells distribution contrary to most studies relying on just one material and one ECM synthesizing cell type. Duan and colleagues fabricated living alginate/gelatin hydrogel valve conduits with anatomical architecture and direct encapsulation of smooth muscle cells in the valve root and VIC in the leaflets [68].

Elastin-like recombinamers were added to fibrin-based construct with the rational of providing a functional elastic network to overcome the challenge of generating mature elastin in vitro in cardiovascular constructs [81,87]. Elastin is a fundamental component of heart valves as it provides elasticity and resilience, besides modulating important cellular processes [88—90]. The mechanisms leading to a lack of elastogenesis in vitro is still not understood, and groups that were able to induce elastin formation in cell-laden gels in vitro were, however, not able to transfer the results to TEHVs fabricated with the same materials [79,91,92].

6.5.1.2 Extracellular matrix—based scaffolds

Decellularization of native or bioengineered tissues represents a promising strategy to obtain off-the-shelf constructs with the potential of promoting in situ cellular ingrowth and tissue formation by endogenous cells. Decellularization aims at removing cellular and nuclear material from a tissue while minimizing any adverse effect on the composition, biological activity, and mechanical integrity of the remaining ECM [93]. The elimination of endogenous cell components is mandatory to avoid an adverse host response, which could result in inflammation or rejection.

Decellularization processes are tailored to the tissues or organs to be treated. In general, they consist of combination of physical, chemical, and enzymatic approaches to achieve lysis of the cell membrane, enzymatic separation of cellular components from the ECM, detergent solubilization of cytoplasmic and nuclear cellular components, and thorough removal of cellular debris and residual chemicals from the tissue. The elimination of the alpha-gal xenoantigen is of particular importance when animal tissues are used. The xenogenic epitopes are present on the surface of vascular endothelial and stromal cells in all mammals with the exceptions of apes, Old World monkeys, and humans. These species are unable to metabolize alpha-gal, and an immune response to the antigen has been indicated as mechanism for calcification in heart valve bioprostheses [94].

6.5.1.2.1 Decellularized native tissues

A straightforward approach to guide tissue regeneration is to use native heart valves (allografts) after effective removal of endogenous cellular elements. In this way, a natural scaffold that has already reached the state of maturation of its ECM fibers can be implemented as a starter matrix for in situ cell colonization. The main concerns about decellularized constructs are the immunological reaction due to the incomplete removal of the cellular component, calcification following implantation, compromised mechanical properties, and insufficient recellularization [95—97].

Decellularized allografts and xenografts have been tested in clinical studies. The groups of Haverich [98—100], Dohmen [101,102] and da Costa [103—106] reported the implantation of decellularized allografts in adult and pediatric patients, with promising results. Complete recellularization of the valve leaflets, however, has not been reported. Therefore, the capability of these valves to accommodate the somatic growth of the pediatric patients still remains debatable. Only trivial medial repopulation was reported by da Costa et al. in the aortic conduit of a decellularized heart valve implanted in the aortic position [106]. An ongoing clinical evaluation of allogeneic aortic heart valves treated with TRICOL decellularization technology is taking place in Italy. The TRICOL method was shown to be the sole method able to manufacture valve xenoscaffolds completely deprived of alpha-gal [94]. It was successfully applied in a long-term preclinical evaluation of decellularized porcine aortic valves as RVOT replacements in the Vietnamese pig model, showing favourable hemodynamic assessments up to 15 months and self-regeneration potential [107,108].

Allografts are of limited availability, and data on long-term clinical evaluation are incomplete to consider decellularized allografts as a tissue-engineered replacement option. However, even without recellularization, they seem to provide satisfactory functional performance.

With respect to xenografts, so far, only two decellularized glutaraldehyde–devoid xenogenic (porcine) matrices have been evaluated in patients, for RVOT reconstruction. CryoLife's SynerGraft valves were implanted in four children in 2001 with catastrophic early term failure [109]; however, a more recent study with the new CryoValve SynerGraft technology showed favourable results [110]. Controversial immunological findings were disclosed for AutoTissue GmbH's Matrix P valves, used since 2002 [95,111–116]. The controversial results obtained with the SynerGraft and the Matrix P suggest that more thorough preclinical studies (in non-human primates) are mandatory before proceeding to the clinical stage [94]. Moreover, as with decellularized allografts, recellularization of decellularized xenografts is highly limited, leading to the development of various bioactive and synthetic coatings aimed to improve recellularization potential and structural properties of decellularized xenografts [54,56,57,117,118]. A potential alternative to decellularized xenograft valves is the use of decellularized porcine small intestinal submucosa (SIS), also known as the FDA-approved CorMatrix technology. Ropcke et al. recently reported on the extensive recellularization, remodeling, and growth potential of decellularized SIS valves when implanted as tricuspid valves in a porcine model [119,120]. However, so far, the clinical use of CorMatrix for various valvular applications has led to mixed clinical results, with reports of severe valvular insufficiency, degeneration, and a consistent lack of tissue remodeling into the characteristic native three-layered valve structure [121–123].

6.5.1.2.2 Decellularized tissue-engineered valves

A de novo engineered construct, obtained by the classical tissue-engineering approach, can be decellularized and used as scaffold for in situ tissue engineering. This approach has the advantage of producing off-the-shelf, size-defined, homologous substitutes with minimal risk for immunogenic reactions or disease transmission. A further advantage is given by the fact that the heart valve does not need to be engineered with the patient's own cells, eliminating crucial issues on cells availability. Importantly, decellularization was proposed also as a strategy to contrast tissue compaction attributed to the contractile cells utilized to produce the ECM during TEHV culture.

Decellularized TEHV leaflets and complete heart valves have been first realized starting from classical TEHV based on fibrin by the group of Tranquillo and coworkers

[124,125] (Fig. 6.4C), or biodegradable synthetic materials, as pioneered by the groups of Baaijens and Hoerstrup [126] (Fig. 6.4A). The applied protocols were able to remove the cellular content while maintaining the collagen content and the tensile properties of the original construct. An important difference between the fibrin- and the synthetic polymer—derived valves is the fact that, in contrast to the fibrin-derived TEHVs, the latter still contains fractions of remnant polymer particles after culturing and decellularization. Sanders et al. demonstrated that these polymer remnants play an important role in the inflammatory response to such grafts, which is postulated to have a beneficial effect on in situ recellularization, acting as a kick-starter for subsequent tissue remodeling [127]. Overall, decellularized TEHVs have shown better recellularization potential with respect to native counterparts in vitro [124,126] and in vivo [27] (Fig. 6.5). This can be attributed to a less dense ECM and, therefore, higher porosity.

Despite these very promising results, increased valvular insufficiency developing over time, associated with leaflet shortening, was still reported when decellularized TEHVs were implanted in sheep and non–human primate models [27,33,127]. Leaflet shortening was also reported by Reimer and colleagues after implantation of decellularized valves in the growing lamb [35]. The valves were composed of two decellularized tissue tubes sutured together with degradable suture lines to form complete pulmonary valved conduits. The shortening was hypothesized to result from inadequate fusion of the two tubes at the commissures before suture degradation. Furthermore, the lamb model grows faster than humans during childhood, making the in situ recellularization process and therefore valvular growth quite challenging. However, as described in more detail in Section 6.7, using computational simulations, Loerakker et al. showed that radial leaflet compression occurred in the original valvular geometry when subjected to physiological pulmonary pressure condition and suggested an improved geometry to elicit radial leaflet extension and in this way counteract the host cell retraction forces [128]. Indeed, valves engineered as suggested showed competent hydrodynamic functionality up to 16 weeks in a fatigue test in vitro [129] and excellent functionality, with thin, pliable leaflets without any signs of retraction, up to 1 year in vivo when implanted in the pulmonary position in sheep [129].

6.5.2 Synthetic scaffolds and hybrids

Synthetic biocompatible polymers are widely used to generate scaffolds for tissue engineering applications. Material and fabrication method selection for tissue engineering depends on biocompatibility, mechanical properties, and biodegradability [130]. By applying synthetic polymers to create scaffolds, the mechanical properties and thus the stability and durability can be well adjusted, and highly reproducible scaffolds can be generated [131]. However, bare synthetic materials do not provide any intrinsic biochemical signals, and in contact with blood, complications such as thrombus formation occur [10]. To overcome this, chemical binding or physical adsorption are well-known and efficient strategies to immobilize biologically relevant proteins, protein sequences, sugars, chemokines, proteoglycans, or GAGs on the surface of an

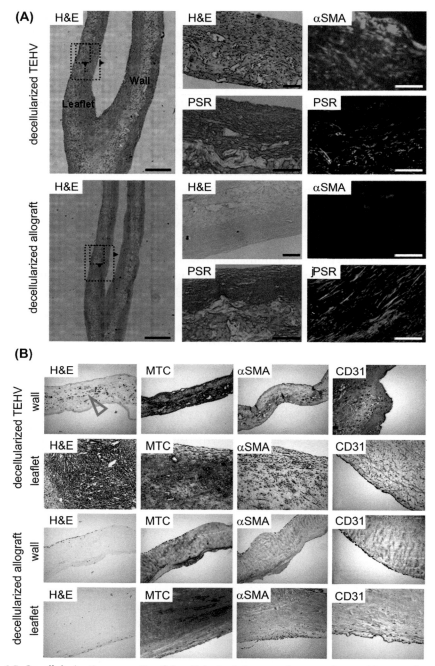

Figure 6.5 *Recellularization capacity of decellularized de novo tissue-engineered heart valves (TEHV) and decellularized allograft valves.* (A) Decellularized TEHVs demonstrate complete recellularization

implant [132,133]. Another idea is to combine synthetic polymers with natural materials during the scaffold fabrication process, which results in highly interesting biologically functional and mechanically stable hybrid scaffolds [134]. There is an array of approaches, where functionalized or structured biomaterials have been used to modulate the immune response or influence cell behavior including attachment, differentiation, migration, or proliferation [10,134,135].

6.5.2.1 Materials and functionalization

The optimal scaffold provides a microenvironment, which promotes cell adhesion and differentiation, and accommodates ECM deposition [130]. Synthetic polymers including polyglycolic acid (PGA), polylactic acid (PLA), polylactic–co–glycolic acid (PLGA), poly-ε-caprolactone (PCL), poly(ethylene glycol) (PEG) and PEG modifications, polyvinyl alcohol, polypropylene fumarate, polyacrylic acid, and many more have been intensively investigated for tissue engineering applications [136]. However, regarding HVTE, most of the studies, especially early in vivo studies, majorly focused on compliance and mechanical performance, resulting in a lack of understanding in important biological mechanisms [130].

The type of the synthetic polymer defines the mechanical characteristics of the implant, but for biological functionality, the polymer processing method for scaffold generation seems to play an equal, if not more important, role. The fabrication technique impacts the microstructure morphology of the implant, which has strong effects on the cellular response, irrespective of the polymer type used. Milleret et al. electrospun two different polymers—PLGA and Degrapol—and produced scaffolds with either comparable or different topographies and roughnesses. There were no differences regarding platelet adhesion and activation between the two polymers, but there were significant changes when considering varying topographies. Small fibers triggered low coagulation,

after in vitro reseeding with mesenchymal stromal cells, compared with minimal cell infiltration in decellularized allografts after in vitro reseeding, due to the more mature and dense ECM organization of decellularized allografts. (B) Extensive in situ recellularization of decellularized TEHVs throughout the valve's wall and leaflets after 8 weeks of implantation as pulmonary valve in a non-human primate model, compared with the limited recellularization of decellularized human valves, which is restricted to endothelial coverage of the surface. Stainings represent general histology (hematoxylin and eosin; H&E), α-smooth muscle actin (αSMA), collagen (picrosirius red; PSR, visualized without and with polarized light), general tissue composition (Masson trichrome; MTC), and endothelial cells (CD31). ((A) Adapted from Dijkman PE, Driessen-Mol A, Frese L, Hoerstrup SP, Baaijens FPT. Decellularized homologous tissue-engineered heart valves as off-the-shelf alternatives to xeno- and homografts. Biomaterials 2012;33: 4545–54. https://doi.org/10.1016/j.biomaterials.2012.03.015. (B) Adapted from Weber B, Dijkman PE, Scherman J, Sanders B, Emmert MY, Grünenfelder J, Verbeek R, Bracher M, Black M, Franz T, Kortsmit J, Modregger P, Peter S, Stampanoni M, Robert J, Kehl D, van Doeselaar M, Schweiger M, Brokopp CE, Wälchli T, Falk V, Zilla P, Driessen-Mol A, Baaijens FPT, Hoerstrup SP. Off-the-shelf human decellularized tissue-engineered heart valves in a non-human primate model. Biomaterials 2013;34 7269–80. https://doi. org/10.1016/j.biomaterials.2013.04.059. with permission from the respective publishers.)

whereas bigger fibers supported thrombin formation [137]. The impact of topography has also been demonstrated for endothelial cells and smooth muscle cells by seeding them onto flat, expanded and electrospun, polytetrafluoroethylene scaffolds [138]. These studies emphasize the importance of a proper, multifactorial scaffold design [10].

6.5.2.2 Scaffold fabrication techniques

Various scaffold fabrication techniques have been employed generate a synthetic heart valve substrate. Hydrogel molding and additional salt leaching is a frequently used method to create porous scaffolds. Polymers such as PGA, polyhydroxyalkanoate (PHA), and poly-4-hydroxybutyrate (P4HB) have been processed with this method into heart valves with defined pore sizes [139]. Alternatively, knitting has been used as a technique for HVTE. Here, fiber yarns are knitted with a hooked needle into complex three-dimensional constructs. As fiber or loop orientation can be varied and adjusted, it is also possible to recapitulate the complex three-layered architecture of heart valve leaflets [140]. PCL has been processed into a knitted heart valve, which opens and closes properly but shows leakage [141]. The combination of knitting with another fabrication technique such as electrospinning [142] or hydrogel molding [25,143,144] seems to be a promising approach.

Another interesting technology is 3D printing, where various biocompatible inks are precisely positioned through different printing mechanisms into complex and defined structures. The main printing mechanisms are summarized as inkjet, laser-assisted, and extrusion [145]. It has been nicely demonstrated by Vukicevic et al. that patient-specific constructs can be generated based on processed images obtained from high-resolution cardiac imaging [146]. First examples of printed heart valves were produced from P4HB or polyhydroxyoctanoate via stereolithography [147]. Also other polymers such as PGA/PLA copolymer [148] or poly(ethylene glycol) diacrylate [149] have been utilized for 3D printing. The latter resulted in a construct with mechanical properties similar to native aortic leaflets but softer than the aortic valve sinus [149]. Another elegant scaffold design approach has been described by Capulli et al. [61]. Their fiber-containing so-called JetValve is manufactured by collecting force-extruded fibers with a heart valve—shaped mandrel (Fig. 6.4E). By using varying ratios of P4HB and gelatin, fiber sizes of 680 nm—1.28 μm were realized. The valve demonstrated accute functionality in an ovine model for 15 h [64].

A more common method to produce nano- and microfiber-containing scaffolds is electrospinning. This method enables different scaffold microstructures, either aligned or randomly oriented. Synthetic polymers such as PCL, PLA, PGA, PEG, poly(glycerol sebacate) (PGS), PLGA, poly (ester urethane) ureas (PEUU), and many more have already been successfully used for electrospinning [134]. For heart valve engineering, polymer blends are also used to combine polymer properties and to enhance valve-specific mechanical properties of the scaffold. A blend of poly(ethylene glycol) dimethacrylate and PLA

(PEGdma—PLA) was, for example, processed via electrospinning. By that, a scaffold was created, which mimics the structural and mechanical properties of a native heart valve leaflet. Under physiological conditions, which were mimicked in vitro with a bioreactor system, the valve opened and closed properly and withstood the natural pressures [150]. Also the intrinsic anisotropy of native heart valve leaflets can be mimicked by blend electrospinning. In this case, a blend of PGS and PCL was used to fabricate a scaffold for HVTE [151]. In another study, bisurea-modified polycarbonate (PC-BU) was employed for electrospinning to generate an off-the-shelf in situ cellularizing heart valve scaffold (Fig. 6.4D). Long-term function was confirmed by implanting the material in a sheep. Furthermore, cellularization of the material and processes such as remodeling were observed in this study [37]. Neither a single fabrication technique nor a single material comprises all multifactorial properties such as structure, architecture, and function of a native heart valve leaflet. Accordingly, there are also approaches combining two or more fabrication techniques to resemble the layered native architecture and anisotropy of the heart valve leaflet [145,152].

6.5.2.3 Preseeded scaffolds

In the last decade, a massive amount of synthetic material/scaffold and cell combinations have been investigated in vitro to identify an optimal construct for HVTE. Unfortunately, only a few tissue-engineered cell-synthetic material combinations progressed from in vitro to animal studies, and so far none of them made it to the patient. The first TEHV was a PGA fiber—containing matrix seeded with autologous fibroblasts and endothelial cells. The single leaflets implanted as a pulmonary valve in a lamb showed no regurgitation or stenosis after 2 months [17]. Breuer et al. obtained similar results by seeding PLA/PGA fiber matrix with myofibroblasts and endothelial cells [153]. A few years later, Sodian et al. created a porous trileaflet heart valve from PHA and seeded vascular cells from carotid arteries onto the constructs before implantation to a lamp. Over a period of 17 weeks, no thrombus formation occurred and neotissue formation was observed. Notably, acellular scaffolds, which were used as a control, did not show tissue formation in that study [139]. To address the shortcomings of current valve options, Hoerstrup coated PGA fibers of a trileaflet heart valve substitute with P4HB. Myofibroblasts and endothelial cells were seeded and dynamically cultured in a pulse duplicator system. As before, an ovine model was used to test the preconditioned tissue-engineered constructs in vivo. Proper valve function and neither stenosis nor thrombosis were detected up to 5 months [18]. Furthermore, a vascular medial cell— and endothelial cell—seeded polyhydroxyoctanoate scaffold was successfully proven in an ovine model [154]. All of these early studies focused on varying synthetic materials seeded with differentiated primary isolated vascular cells in an ovine model. All of them describe, more or less detailed, ECM formation on the tissue-engineered matrix, but also undesired leaflet thickening was observed in some cases [22].

A conceptually different approach was introduced by Alavi and Kheradvar, who proposed the use of a very thin (thickness 25 μm), superelastic nondegradable nitinol mesh [155]. Experimental and computational assessment of hemodynamic functionality revealed important features in the design of a trileaflet valve from the nitinol mesh, such as the configuration and attachment of the leaflets to a supporting frame, which have a critical impact on valve functionality (e.g., orifice area) and durability [156]. By seeding the nitinol mesh with a combination of aortic smooth muscle cells and (myo)fibroblasts, covered by a layer of vascular ECs, using collagen gel, a hybrid valve was created consisting of natural ECM and cells, strengthened by the nitinol backbone. By seeding the nitinol meshes with cells, the inflammatory response to the valve leaflets was significantly reduced, as assessed by in vitro macrophage culture assays [157]. Although these valves intrinsically do not support somatic growth, this strategy would overcome the issues of premature scaffold degradation and tissue retraction that have been reported for other in vitro HVTE approaches.

6.5.2.4 Acellular resorbable scaffolds for in situ heart valve tissue engineering

In contrast to scaffolds for in vitro tissue engineering, acellular resorbable synthetic scaffolds for in situ tissue engineering have to immediately take over valve function on implantation, before the formation of any neotissue. Clearly, this poses stringent mechanical and functional criteria on the scaffold properties. Moreover, the degradation rate and mechanism of the scaffold material have to be in sync with the formation of new tissue. Consequently, synthetic scaffolds for in situ HVTE typically have a slower degradation profile when compared with scaffolds for in vitro tissue engineering, which are often much more rapidly degrading.

As already discussed earlier, not only surface chemistry but also mechanics and structure influence cell attachment, differentiation, survival, migration, and proliferation [132]. As soon as an acellular material is in contact with blood, proteins immediately attach to the scaffold, followed by a wide range of cells including cells of the immune system. Now, the immune system is activated; this is an important step for tissue regeneration. Although cell attachment is crucial for regeneration, there is also the risk of undesired cell agglomeration and thrombus formation. One strategy to avoid coagulation but enable binding of tissue-specific cells is to cover the synthetic surface with proteins, which serve as an attractant for specific progenitor cells. In vitro studies showed that a decorin-modified PEGdma—PLA scaffold enables progenitor cell recruitment from the circulation [133].

Cellularization and regeneration of an acellular scaffold has also been demonstrated in a sheep model by the group of Bouten et al. demonstrating the first proof of concept for this technology [37] (Fig. 6.6). Here, the electrospun synthetic PC-BU was coated with fibrin before implantation. Importantly, the authors describe that scaffold degradation was observed especially in regions of cellularization and neotissue formation, suggesting

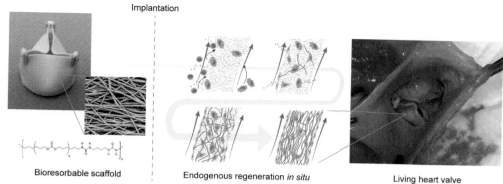

Implantation

Bioresorbable scaffold | Endogenous regeneration *in situ* | Living heart valve

Figure 6.6 *In situ heart valve tissue engineering based on an electrospun resorbable synthetic scaffold.* On implantation, the synthetic scaffold triggers an inflammatory response by the host, driving the process of endogenous regeneration to eventually form a living, autologous heart valve. The photograph represents the regenerated valve, explanted after 12 months of implantation in the pulmonary position in sheep. *(Adapted from Kluin J, Talacua H, Smits AIPM, Emmert MY, Brugmans MCP, Fioretta ES, Dijkman PE, Söntjens SHM, Duijvelshoff R, Dekker S, Janssen-van den Broek MWJT, Lintas V, Vink A, Hoerstrup SP, Janssen HM, Dankers PYW, Baaijens FPT, Bouten CVC. In situ heart valve tissue engineering using a bioresorbable elastomeric implant — from material design to 12 months follow-up in sheep. Biomaterials 2017;125:101—17. https://doi.org/10.1016/j.biomaterials.2017.02.007. with permission from the publisher.)*

that degradation is correlated to ECM formation, which is essential to maintain structural integrity and functionality of the valve throughout the regenerative process [37]. The authors employed supramolecular polymers as the scaffold material [158]. These are essentially modular polymers, which self-assemble via click chemistry, which makes them tunable in terms of degradation kinetics, mechanical properties, and incorporation of bioactive factors [159]. Using a similar class of supramolecular materials, the first clinical trials with a synthetic resorbable electrospun heart valve scaffold for endogenous tissue restoration were initiated by Xeltis. In 2016, the first patient received a Xeltis implant. Another trial called XPlore-II was approved in 2017. There are still many open questions especially regarding long-term functionality and growth potential; however, the implantation of a synthetic heart valve material into a patient is an important milestone in the field of HVTE.

6.5.3 Scaffold-free approaches

Besides scaffold-based approaches to obtain heart valves, scaffold-free strategies have been proposed, which exploit the capability of the cells to produce ECM without the initial support of a scaffold. These approaches result in completely autologous heart valves. They include the so-called in-body tissue architecture and the tissue engineering by self-assembly (TESA) method.

The in-body tissue architecture technology, also named in vivo tissue engineering, takes advantage of the inflammatory reaction to implanted nondegradable polymeric molds, resulting in their encapsulation. Once the formed tissue is released from the mold, a self-standing construct with desired geometry is obtained. This method reproposes the concept first introduced by Schilling [160] and Sparks [161] to create vascular grafts. Implanted plastic tubings subcutaneously in patients to take advantage of the consequent inflammatory reaction result in the encapsulation of the foreign body. Although highly innovative, the method was not successful in the clinical settings because of the propensity for thrombosis and aneurysm formation [162,163]. Nakayama and co-workers picked up the methods and applied it to vascular grafts [164,165] and heart valves for surgical and minimally invasive implantation [164,166–168].

The TESA concept consists in the production of ECM sheets by mesenchymal cells cultivated on flat substrates and, subsequently, the fabrication of tubular constructs by rolling one or more sheets. Also this concept was first applied for vascular grafts [169,170] and only more recently for heart valves [171–173]. Although this method eliminates the need of a scaffold and the eventual problems associated with its degradation, it requires up to 12 weeks of fabrication time, which might limit its clinical applicability. This issue could be eventually solved by decellularization of the cell sheet–based constructs resulting in off-the-shelf heart valves.

6.6 Bioreactors

Bioreactors are essential for the development of mechanically sound TEHVs, capable of functioning effectively from the moment of implantation. They have a major role in (1) the tissue maturation of cell-based heart valves; (2) testing the functionality of TEHVs under defined physiological flow and pressure conditions according to the position of implantation; and (3) advancing the development of HVTE by studies on the cell and tissue responses to controlled biological and physical factors. Concerning the latter, a wide variety of bioreactor systems have been described to study biological processes that are relevant for HVTE on a fundamental level, for example, by using small, simple-geometry samples of tissues or scaffolds in dynamic culture settings (e.g., Refs. [39,44,174–179]). Although these mechanistic studies are highly valuable to HVTE, they are beyond the scope of this chapter to discuss in great detail. Therefore, the following section is mainly focused on whole-valve bioreactors (Fig. 6.7). Besides the original therapeutic aim, TEHVs have great potential as 3D in vitro models for the investigation of cardiovascular disease, in addition to drug development and testing. These bioreactor-centered in vitro models can span the gap between 2D culture and in vivo testing, thus reducing the cost, time, and ethical burden of current approaches [180,181].

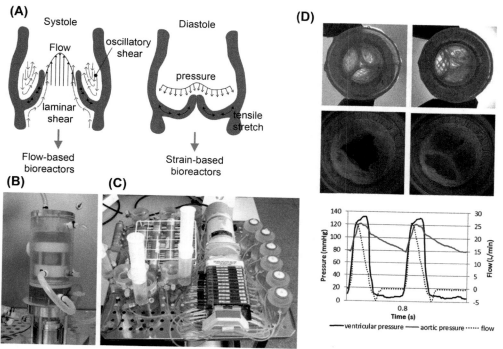

Figure 6.7 *Bioreactors for heart valve tissue engineering.* (A) Schematic of mechanical forces experienced by the aortic valve during peak systole (left) and peak diastole (right). (B, C) Bioreactors to dynamically stimulate tissue formation are indicated as flow-based when they aim at recapitulating the whole cardiac cycle (systole and diastole) (B) or strain-based if only the diastolic phase is reproduced (C). (D) Bioreactors also serve as in vitro functionality platforms, allowing the evaluation of valve opening and closing behavior under physiological hemodynamic conditions. *((A) Adapted from Balachandran K, Sucosky P, Yoganathan AP. Hemodynamics and mechanobiology of aortic valve inflammation and calcification. Int J Inflamm 2011;2011 263870. https://doi.org/10.4061/2011/263870. (C) Adapted from Mol A, Driessen NJB, Rutten MCM, Hoerstrup SP, Bouten CVC, Baaijens FPT. Tissue engineering of human heart valve leaflets: a novel bioreactor for a strain-based conditioning approach. Ann Biomed Eng 2005;33:1778–88. https://doi.org/10.1007/s10439-005-8025-4. (D) Adapted from Moreira R, Neusser C, Kruse M, Mulderrig S, Wolf F, Spillner J, Schmitz-Rode T, Jockenhoevel S, Mela P. Tissue-engineered fibrin-based heart valve with bio-inspired textile reinforcement. Adv Healthc Mater 2016;5: 2113–21. https://doi.org/10.1002/adhm.201600300. with permission from the respective publishers.)*

6.6.1 Whole-valve bioreactors for culturing and testing

It is nowadays well accepted that mechanical stimulation of cell-seeded constructs can influence the ECM synthesis and organization toward the formation of tissues with biomimetic anisotropy and improved mechanical properties in comparison with statically conditioned controls. Although the use of bioreactors to apply dynamic mechanical loads has become common practice in HVTE, there is no defined standard with respect to bioreactor design, applied mechanical forces, and stimulation protocols, in terms of duration, amplitude, cycle, frequency, etc. Moreover, apart from stimulating desirable tissue formation for HVTE purposes, mechanical conditioning typically also triggers the activation of VIC-like cells to an activated myofibroblastic phenotype, which is in principle a pathological valve phenotype that can lead to fibrosis and leaflet retraction. This seemingly paradoxical challenge of obtaining sufficient well-organized new tissue and a quiescent VIC-like phenotype requires sophisticated control of culture conditions, as excellently reviewed by Nejad et al. [182].

Heart valves can be cyclically subjected to pulsatile flow recapitulating the cardiac cycle (systole and diastole) or to the strain experienced during the diastole in so-called flow-based and strain-based bioreactors, respectively (Fig. 6.7A—C). Besides enabling the desired mechanical stimulation, bioreactors provide a sterile biochemical environment with improved nutrient and metabolites transfer, adequate gas exchange in the culture medium, controlled pH, temperature, pO_2, and pCO_2. Ideally, they are easily assembled and operated under sterile conditions, offer the possibility to refresh the medium regularly, and can be placed in standard incubators. In general, whole-valve bioreactors consist of an actuation unit generating the medium flow, a valve housing, a compliance chamber simulating the elasticity of the arteries, a flow restrictor recreating the resistance of the small diameter vessels, and a hydraulic circuit connecting all the units. The bioreactors are computer controlled and can be equipped with pressure and flow sensors which, in the most sophisticated cases, are included in a feedback loop to provide automated operation [183,184].

The more intuitive flow-based concept was the first one to be proposed for the conditioning of TEHVs [185,186] and is still largely applied by several research groups with varying levels of complexity and parameters' control [82,86,183,187—193]. In general, heart valves conditioned in flow-based bioreactors show improved tissue formation with respect to statically conditioned constructs. In the bioreactors, the valves are subjected to a combination of stimuli in the form of shear stress, pressure, cyclic stretching, and flexure as encountered in vivo and which can be set according to different rationales, from mild fetal-like conditions as indicated by Stock and Vacanti [194] to incremental stimulation toward adult physiological regimes. Normally, stroke volume, frequency, and duty cycle can be controlled. The cyclic fluid stroke is generated through a deformable membrane in contact with the medium and displaced by different systems including pneumatic systems (e.g., respirators, compressed air lines) or pistons connected to linear motors or eccentric cams actuated by a rotational motor.

Following a different approach, Mol and colleagues proposed to stimulate the TEHVs by recapitulating the condition of strain during the diastolic phase [195] based on experimental evidences on the improved tissue development in tissue-engineered vascular grafts subjected to dynamic wall straining, without the use of pulsatile flow. They developed the diastolic pulse duplicator to create a dynamic pressure difference over a closed prestrained TEHV, thereby inducing dynamic strains within the leaflets. The authors demonstrated superior tissue formation and nonlinear tissue-like mechanical properties in the conditioned valves when compared with nonloaded tissue strips. Vismara et al. adopted the pulse duplicator concept to facilitate cell migration into decellularized porcine and human valves, their integration in the existing ECM, and the production of new ECM [184]. Syedian and Tranquillo employed a different bioreactor configuration for the strain-based cultivation of TEHVs by placing the valve inside a distensible latex tube, which is cyclically pressurized with culture medium to impose controllable stretching to the root and the leaflets [196]. After 3 weeks of cyclic stretching with incrementally increasing strain amplitude, the fibrin-based TEHV possessed improved compositional and mechanical properties including native-like anisotropy when compared with the leaflets from sheep pulmonary valves.

The advances in TEHVs have led to the need to characterize them according to standardized protocol to be able to assess and compare their mechanical functionality. To this end, commercial or custom-made flow loop systems are used to create flow and pressure conditions, in some cases according to ISO 5840 guidelines, for surgical and transcatheter heart valve prostheses [34,37,64,67,85,125,197,198]. Preliminary durability testing has also been reported with frequencies up to 10 Hz, however, with a limited number of cycles [37,125,198]. The concept of accelerated testing is questionable in the case of cell-based TEHVs as prolonged stimulation at supraphysiological frequency is likely to influence the cell behavior in a drastic way. In the case of cell-free biodegradable scaffolds, while high frequencies are not a critical issue as for the clinically available prostheses, the inherent structural degradation of the scaffold will hinder its functionality and limit the duration of the test.

6.6.2 Real time noninvasive and nondestructive monitoring in bioreactors

Noninvasive and nondestructive monitoring techniques coupled to bioreactors can advance the development of heart valves throughout their life cycle by means of functionality and quality assessments. A longitudinal evaluation of the mechanical properties of a heart valve during cultivation can result, for example, in an improved conditioning protocol, while a better visualization of the 3D geometry is helpful to identify design flaws of the scaffolds at an early stage as well as to evaluate the shape's evolution in time and eventually associated malfunctions (e.g., tissue shrinkage). The need to operate under sterile conditions and to maintain the construct's structural integrity limits the monitoring possibilities; however, interesting solutions have been proposed to gain important information noninvasively and nondestructively. Kortsmit et al. developed an inverse experimental—numerical approach to assess the mechanical properties of TEHVs during cultivation in a diastolic pulse duplicator. The method relies on the measurement of the applied pressure and the induced volumetric deformation to estimate the Young's modulus using a computational model. The results were in good agreement with uniaxial tensile test data [199]. With a similar approach, Vismara and colleagues estimated the compliance of cell-seeded decellularized heart valves [184]. In a following study, Kortsmit et al. further developed the method by including a feedback controller to regulate the load-induced leaflet deformation [200]. Syedain and Tranquillo calculated the leaflets' and root's strain on the base of the circumferential displacement of the root during cyclic stretching as evaluated by video recordings [196].

The functionality of heart valves in bioreactors is typically evaluated through optical imaging from the outflow side to visualize the opening and closing (e.g., Fig. 6.7D) [201]. The application of different ultrasound modes to bioreactors for heart valves provided information on the 3D geometry otherwise not accessible by conventional optical imaging, as well as visualization and quantitative evaluation of the flow through the construct, in real time and throughout the whole in vitro fabrication phase. The ability to visualize tissue shrinkage, assess flow profiles, and analyze leaflets' behavior can further accelerate improvements in the design and conditioning of TEHV, and it gives the possibility to benefit from the vast clinical ultrasound data to advance HVTE from design to fabrication and in vitro maturation [202]. Furthermore, the increase of collagen content was correlated with the increase of the gray-scale value of ultrasound images of cell-laden fibrin gels during static cultivation, showing the potential of this technique to detect ECM development in tissue-engineered constructs in a noninvasive and longitudinal way [203]. Van Kelle and colleagues developed a bioreactor where 3D circular tissue-engineered constructs were cultivated under dynamic mechanical loading for up to 4 weeks [179]. The bioreactor offered the possibility to evaluate the changing mechanical properties over the cultivation time in a noninvasive manner. This was possible by means of a classical bulge test where ultrasound imaging was used to track the tissue displacement [204]. To achieve high spatial resolution, ultrasound transducers with high frequencies are required, which, however, have a short working distance. This might make the application of ultrasound in bioreactor systems challenging.

6.7 Computational modeling

Given the importance of biomechanics and mechanobiology in HVTE, computational modeling has been used as a valuable tool to understand valve functionality and predict growth and remodeling of TEHVs for several decades now. As expertly highlighted by Ayoub et al. [205], this requires a multiscale approach, ranging from the macroscopic valve geometry and systems hemodynamics, to the microscopic events on the cellular and tissue level in response to such loads. This section briefly touches on some recent examples from this field, including the prediction of tissue growth and remodeling and the computationally driven macroscopic design of TEHVs.

6.7.1 Predicting collagen remodeling in tissue-engineered heart valves

One of the most important factors that dictate the mechanical behavior of the TEHV is its collagen network. However, the collagen network of TEHVs is remodeled depending on the (bio)mechanical stimuli, thus leading to time-varying mechanical behavior. A thorough understanding of how collagen fibers will remodel themselves in TEHV under different physiological conditions is therefore crucial to design optimal TEHV and to enhance their in vivo functionality. One way of most effective way to quantitatively elucidate this collagen remodeling behavior in TEHV is through computational modeling. Loerakker and colleagues in 2016 combined a previously proposed numerical model from Obbink-Huizer et al. [206], which was able to predict cell traction and alignment, with laws for collagen contraction and remodeling and hence were able to predict cell-mediated tissue compaction and collagen remodeling in TEHV [207]. Using their model, they were able to gain valuable insights into underlying remodeling processes, which are responsible for leaflet retraction of TEHV, thus leading to postoperative valvular insufficiency. More specifically, their numerical model predicted that leaflets of TEHV, which were implanted in the aortic position, are not likely to retract due to high blood pressure and collagen network, which always remodels toward a circumferentially aligned network, as it is the case in native aortic valves. However, leaflets of TEHV, which were implanted in the pulmonary position, on the other hand, were prone to retraction and developing valvular insufficiency unless the cell contractility was very low. These findings, which were aided purely by the computer models, have considerable implications when it comes to designing optimal TEHV because they manifest that initial scaffold design may be less important for TEHV, which have to be implanted in the aortic position, as external hemodynamic loads will potentially lead to physiologically favorable remodeling processes. However, for TEHV to be implanted in the pulmonary position, it is important to design the initial scaffold as such that it prevents the radial alignment of the collagen fibers and induces circumferential alignment of collagen fibers via contact guidance [207].

6.7.2 Predicting growth through computational modeling

Computer models based on traditional nonlinear continuum mechanics framework cannot only provide a valuable insights in how TEHV might remodel their collagen network after implantation or how to optimize the design of TEHV to minimize the chances of leaflet retraction, but if modified adequately, such models can also be used to simulate postoperative tissue growth inside TEHV. Kuhl published a comprehensive review of computational approaches that can be used to model the phenomenon of growth in living systems [205]. It was highlighted in the aforementioned review paper that different types of microstructural growths, such as volume, area, or fiber growth, are not only driven by biochemical cues such as nutrient transport or hormones but also on the mechanical cues such as stress-, strain- or stretch-driven homeostasis. Oomen et al. analyzed collagen architecture as well as geometric and mechanical properties of native human heart valves from fetal to adult age. Based on their experimental findings, they concluded that both aortic and pulmonary valves grow and remodel to maintain stretch-driven homeostasis [5]. Equipped with this knowledge, Oomen and colleagues proposed a volume–driven growth model in 2018, which was based on the assumption that human heart valves grow and remodel as such that they would maintain a certain level of stretch. Their computational simulations elucidated that from infant to adolescence, excessive tissue stretch was decreased by tissue growth. In contrast, from adolescence to adulthood, instead of growth, tissue stretch was primarily decreased by remodeling and increased by growth [206]. It can, therefore, be concluded from their computational results that tissue growth and remodeling processes play opposing roles in preserving tissue stretch during human development from infant to adulthood and such findings are highly relevant for the optimal development of TEHV.

6.7.3 Aiding physical design via computer modeling

As the majority of the TEHV are manufactured via electrospinning, they can be sculpted into a variety of different designs. As such, apart from the level of anisotropy, the physical design or geometry of the TEHV itself has a considerable impact on the long-term outcome of the valve functionality (Fig. 6.8). In their 2013 investigation, Loerakker et al. used computer models that can predict collagen remodeling within TEHV (similar to the ones as mentioned previously) with varying geometries to delineate if, along with valve anisotropy, its geometry also has an effect on leaflet retraction [128]. They constructed various 3D computer-aided design models of TEHV according to design protocols as proposed by Hamid et al. [208] and Thubrikar [209] and observed that TEHV scaffolds created according to the design of Hamid et al. may potentially yield better TEHV than those created according to the design of Thubrikar. This was due to the fact that the curved profile of the valve leaflet due to Hamid's design increases the radial stretch of the TEHV in the closed (diastolic) position, thus leading to a configuration that mimics the physiological deformation of the native valve more closely (Fig. 6.8A). They also reported that the minimum degree of anisotropy that is required to obtain favorable radial stretch is also highly dependent on the initial geometry of the TEHV scaffold, i.e., more curved the TEHV leaflet, lesser the degree of required anisotropy [51]. Results from these computer simulations were consistent with the findings of Amoroso et al. [210], Argento et al. [211], and Fan et al. [212], leading to the conclusion that both valve geometry and degree of anisotropy are crucial to enhancing the coaptation area thus reducing chances of leaflet retraction from happening. These findings, which would have been very difficult to obtain from experimental techniques alone, advanced the design of TEHV by elucidating that valves with curved leaflets might be less prone

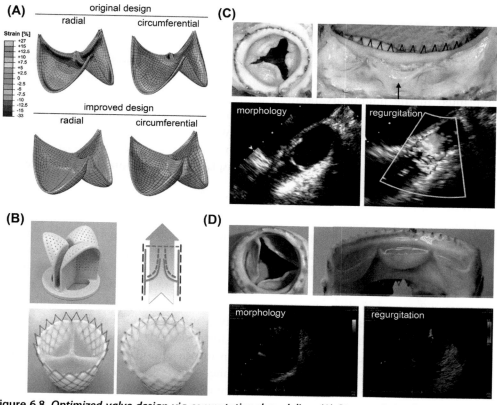

Figure 6.8 *Optimized valve design via computational modeling.* (A) Structural simulations revealing the strains in the radial and circumferential directions in tissue-engineered heart valves during diastole. While the original valve design displayed undesired compressive strains in the radial direction, this could be negated by computationally optimizing the valve geometry, characterized by more pronounced leaflet curvature and an increased coaptation area. (B) A custom bioreactor insert was designed to culture the valves according to the optimized design, as computationally established. (C) Decellularized de novo engineered tissue valves with the original valve geometry displayed adverse remodeling and leaflet contraction, leading to valve regurgitation after 24 weeks in vivo, while (D) valves cultured according to the optimized design displayed thin pliable leaflets and sustained functionality up to 1 year in vivo. *((A, B) Adapted from Sanders B, Loerakker S, Fioretta ES, Bax DJPP, Driessen-Mol A, Hoerstrup SP, Baaijens FPT. Improved geometry of decellularized tissue engineered heart valves to prevent leaflet retraction. Ann Biomed Eng 2016;44:1061–71. https://doi.org/10.1007/s10439-015-1386-4. (C) Adapted from Driessen-Mol A, Emmert MY, Dijkman PE, Frese L, Sanders B, Weber B, Cesarovic N, Sidler M, Leenders J, Jenni R, Grünenfelder J, Falk V, Baaijens FPT, Hoerstrup SP. Transcatheter implantation of homologous "off-the-shelf" tissue-engineered heart valves with self-repair capacity: long-term functionality and rapid in vivo remodeling in sheep. J Am Coll Cardiol 2014;63: 1320–29. https://doi.org/10.1016/j.jacc.2013.09.082. (D) Adapted from Emmert MY, Schmitt BA, Loerakker S, Sanders B, Spriestersbach H, Fioretta ES, Bruder L, Brakmann K, Motta SE, Lintas V, Dijkman PE, Frese L, Berger F, Baaijens FPT, Hoerstrup SP, Computational modeling guides tissue-engineered heart valve design for long-term in vivo performance in a translational sheep model. Sci Transl Med 2018;10. https://doi.org/10.1126/scitranslmed.aan4587. with permission from the respective publishers.)*

retraction as compared with their noncurved counterparts. Based on these findings, Sanders et al. proposed an off-the-shelf TEHV design with enlarged coaptation area and profound belly curvature to enable radial stretch and to prevent leaflet retraction (Fig. 6.8B). Their in vitro experiments showed the favorable hemodynamic performance of these TEHV designs under physiological pulmonary conditions [198]. When implanted in the pulmonary position in sheep, the redesigned valves were able to maintain functionality over the 12-month follow-up time, thereby avoiding any of the complications with leaflet retraction as were reported previously (Fig. 6.8C,D) [129]. Importantly, the model was also able to predict the failure of two of the outlier valves and pinpoint the reasons for failure to misplaced implantation and too thin leaflets, respectively [129].

With respect to rational scaffold design on the microscale, recent studies by the group of Humphrey and collaborators on in situ engineered synthetic vascular grafts are highly relevant. For example, Miller et al. reported on a model that allows for the tailoring of scaffold design parameters to tune tissue regeneration [213,214]. Using this model, the authors were able to attribute differences in mechanical properties of remodeled in situ tissue-engineered blood vessels, which were observed after long-term implantation in mice, to differences in the collagen type 1 and type 3 ratios [215]. Importantly, these models discriminate between inflammation-driven and mechano-driven tissue formation and remodeling, which is particularly relevant for in situ HVTE strategies. Using such a model, Szafron et al. compared the long-term in situ remodeling of vascular grafts in immunocompetent versus immunocompromised mice, revealing that excess initial inflammation-driven tissue deposition can attenuate the mechano-driven tissue remodeling in later stages [216].

From the numerical applications mentioned above, it can be concluded that when used in conjunction with experimental techniques, computational modeling for TEHV can help us to gain new in-depth insights into the postimplantation behavior of TEHV, which would have been not possible or extremely tedious to gain by traditional experimental methods alone.

6.8 Minimally invasive delivery of tissue-engineered heart valves

Surgical implantation of a replacement valve is a major surgical procedure and as such can be a traumatic experience for the patients with lengthy hospital stay and prolonged recovery time. However, with the advent of minimally invasive valve implantation techniques, such as the transfemoral or transapical approach, heart valves can now be implanted into the human body with minimum required intervention. Following the clinical progress toward minimally invasive valve replacements, various groups are aiming to merge their HVTE technology with such percutaneous implantation techniques.

One of the first reports demonstrating the feasibility of combining minimally invasive delivery technique with HVTE is that by Schmidt et al. (2010) [23]. The authors cultured living TEHVs from either venous myofibroblasts or bone marrow— or peripheral blood—derived stem cells in vitro. These TEHVs were successfully implanted transapically into the pulmonary position in sheep [23]. In the same year, Lutter and Metzner et al. reported on the feasibility and functionality of transcatheter implantation of TEHVs in the pulmonary position in sheep, using valves based on decellularized porcine xenografts, which were subsequently recellularized with autologous myofibroblasts and endothelial cells followed by dynamic in vitro culturing [217,218]. Interestingly, here the authors used a sheet of decellularized SIS to protect the TEHV from damage by the nitinol stent struts [219]. Apart from in vitro engineered valves, the percutaneous delivery of resorbable synthetic scaffolds, either acellular or on-the-fly-preseeded [11,37], and decellularized de novo engineered valvular matrices [27,33,129] has been demonstrated. Although all these studies are restricted to the pulmonary circulation, Emmert et al. demonstrated the successful percutaneous deployment of TEHVs in the systemic circulation in a series of studies using TEHVs as aortic valve replacement [12,220,221].

Minimally invasive delivery of TEHVs comes with several technical challenges in terms of the valve and stent design, as well as the delivery device, as expertly reviewed elsewhere [24]. To deliver the TEHV via the minimally invasive route, the valve is typically sutured to a balloon-expandable or self-expanding stent, following which the entire assembly is then crimped and mounted on a catheter for delivery. Dijkman et al. investigated the effect of crimping on the postoperative behavior of TEHV as compared with conventionally implanted TEHV and reported that crimping has no adverse effect on the integrity or functional outcome of TEHV [22]. Moreira et al. described the development of a fibrin-based tube-in-stent TEHV, specifically manufactured for minimally invasive delivery [25].

Along with optimally designed TEHV, it is also critical to have a properly sized stent to which the valve is sutured to in order to achieve favorable clinical results following a minimally invasive TEHV implantation. In their combined experimental and computational study, Cabrera et al. reported that considerably oversized stents will exert excessive radial forces on the arterial walls. They further hypothesized that if the radial forces exerted by the stent on the arterial walls exceed 16 N, it might lead to stent-induced vascular growth and remodeling in the arterial wall, and if these forces exceed 17.5 N, it can potentially lead to perforation of the artery [222], hence, underscoring the importance of quantifying the correct stent size. At this point, it is worth mentioning that along with correct size, the stent's materials also dictates the postoperative success of minimally delivered TEHV. This is because, contrary to bioprosthetic heart valves, the stent material has to be biodegradable so that the TEHV can grow and remodel once implanted. A proof-of-principle study by Cabrera et al. advanced the field of minimally invasive delivery of TEHV by demonstrating the feasibility to manufacture 3D printed self-expandable biodegradable stents with mechanical properties comparable with traditional non biodegradable nitinol stents along with a reasonable degree of plastic deformation [223]. The possibility of manufacturing 3D printed self-expanding biodegradable stents has the potential to revolutionize the manner in which TEHV is delivered and implanted inside the human body and has numerous clinical applications, especially in pediatric surgery.

6.9 Perspective on current challenges for heart valve tissue engineering

Since its conception, HVTE has made great progress over the last couple of decades. Particularly, the body of knowledge regarding valve biomechanics and mechanobiology and the implications for tissue engineering has grown tremendously. At the same time, HVTE has yet to be established as a routine clinical option, favorable over the current suboptimal clinical golden standards for valve replacement that have been relatively unmodified since the 1970's. Particularly, the research on in vitro engineered valves have not yet progressed into clinical trials due to the complexity of this technology. This section addresses some of the main current challenges to overcome in the translation of TEHVs toward a robust clinical treatment option for valve replacement.

6.9.1 Inducing elastogenesis

Elastin is a key matrix component of the native valves. After being stretched during diastole, the elastic fibers make sure that the collagen fibers recoil back to their original corrugated state [89]. Consequently, damage to the elastic network leads to valve failure in the long term [224]. Elastin consists of cross-linked packages of the cell-secreted tropoelastin protein, which is subsequently cross-linked to a microfibrillar network (consisting of, e.g., fibrillin-1 and fibrillin-2) via a variety of accessory proteins (e.g., fibulin-4, fibulin-5, lysyl oxidase) to form functional elastic fibers [225,226]. Apart from its well-established mechanical function, (tropo)elastin exerts important signaling functions that are relevant for cardiovascular tissue remodeling and homeostasis [227–229].

The regeneration of mature functional elastic fibers in TEHVs has been elusive to date. Although, elastin production is reported regularly, the amount and functionality thereof is marginal compared with native. The difficulty in regenerating elastin in the heart valve lies in the fact that human elastin has an extremely long half-life, and cellular production is developmentally restricted to the early fetal stages, as described by Votteler et al. [230]. During adulthood, elastin synthesis is typically only observed in certain pathological conditions [231]. However, the feasibility of inducing elastin production by adult cells in vitro has been described in numerous studies, pointing to essential roles for the mechanical cues (e.g., cyclic strain), as well as the 3D scaffold microarchitecture [91,92,232–235].

An alternative to inducing elastin production by endogenous cells is represented by the incorporation of exogenous recombinant elastin [236]. In alignment with the afore-mentioned biological signaling role of elastin, the coating of scaffolds with exogenous tropoelastin has proven to be a potent modulator of tissue regeneration as well as cell behavior in various applications [237–240].

6.9.2 Harnessing the host response and tissue homeostasis

Perhaps one of the most complex unanswered challenges for HVTE is to achieve biolog-ical and mechanical homeostasis. In a way, this is a paradoxical challenge as TEHVs—both in vitro cultured and in situ regenerated—rely on the active in vivo remodeling to recapitulate the native valve's microstructural organization and composition. VIC-like cells with an activated phenotype are needed to achieve sufficient de novo tissue formation, yet, at the same time, these cells represent a phenotype that is typically correlated to valve pathologies, as also described in previous sections. Consequently, the understanding and controlling of VIC phenotype switching is an active field of research. Maintaining a quiescent VIC phenotype in vitro is notoriously challenging. Porras et al. recently reported on a robust in vitro culturing protocol to maintain quiescent VICs that mimic a healthy VIC population [241]. As also pinpointed in Section 6.4, the stiffness of the ECM is an important regulator of pathophysiological VIC activation and related processes, such as EndMT [42,45,242]. Capitalizing on this, Hjortneas et al. described the use of a 3D hydrogel platform based on methacrylated hyaluronic acid and methacry-lated gelatin to study VIC activation [243]. Using this platform, they showed that VIC activation could indeed be directed by changing the hydrogel stiffness [243], and in follow-up work the platform was used to engineer hydrogels with layer-specific stiffness to mimic the native valve [244]. Following a similar rationale, the group of Anseth and colleagues reported on the use of PEG-based hydrogels, with which they were able to tune VIC activation and ECM deposition by dynamically varying the hydrogel stiffness [245,246] and microstructural features [246], as well as the incorporation of various ECM-derived adhesion peptides [247,248].

Apart from the biophysical matrix environment, the transient role of inflammation in the process of tissue regeneration and remodeling is being increasingly acknowledged. The notion of inflammation-driven remodeling is further propelled by the growing body of research on immunomodulatory materials designed to instruct the host response to achieve tissue regeneration, as described in more detail in numerous recent reviews, e.g., Refs. [249,250]. Nevertheless, the knowledge of the spatiotemporal inflammatory processes in TEHVs and how they may or may not drive valvular regeneration and remodeling is very limited to date. Therein, the influence of hemodynamic loads on the inflammatory response is a largely unexplored area. The relevance of this is under-lined in, for example, a series of 2D cell culture studies by Matheson et al. who demonstrated that macrophages adopt their phenotype and functional behavior in a strain-dependent way [251—253]. In line with this, Ballotta et al. demonstrated that human macrophages in 3D electrospun microfibrous scaffolds are sensitive to strain. More specifically, they demonstrated that human macrophages attain a predominantly proinflammatory phenotype at cyclic strains of 12% [254]. Correspondingly, Battiston et al. showed that human monocyte-derived macrophages govern the functionality of and ECM formation by smooth muscle cells in 3D synthetic scaffolds under the influence of cyclic strain [235,255]. Using a custom-developed mesofluidic system, Smits et al. demonstrated that specific human monocyte recruitment and infiltration into 3D electrospun scaffolds is heavily dependent on wall shear stress [175,256]. These findings might explain to the highly inhomogeneous patterns of cellularization and tissue formation that are typically observed in TEHVs after implantation in vivo, suggesting that these may indeed be inflammation-related.

Taken together, increasing evidence points to a commanding role of inflammation in the process of in situ remodeling of TEHVs and that these inflammation-driven processes are dependent on hemodynamics. Temporal control over the host inflammatory response to, eventually, achieve a state of tissue homeostasis will be essential for the success of HVTE [10].

6.9.3 Mechanistic approaches and stratification

HVTE is a highly challenging technology, which requires true interdisciplinary research, ranging from biomechanics, immunology, tissue engineering, materials science, and developmental biology. To create a heart valve, either in vitro or in vivo, a plethora of interdependent parameters has to be carefully designed and controlled. To deal with this in a systematic way, appropriate models are required, which has been a hiatus to date. Blum et al. emphasized the current lack of small animal studies for HVTE and advocated the importance of such models to bridge the gap between in vitro experiments and large animal studies [130]. An example is given by Książek et al. who demonstrated the technical feasibility of implanting a monocuspid PGA/P4HB-based TEHV in the rat systemic circulation [257]. In their recent review paper, the group of Simmons and colleagues recently called for a systems approach to tackle the complexity of HVTE, for example, by using high-throughput microsystems, rather than conventional culture methods [258]. To address this, they previously developed, for example, a small-scale high-throughput PDMS-based microfluidics device to evaluate the response of layer-specific VICs isolated from the fibrosa and ventricularis of ovine valves to dynamic strains in 2D [259]. Work by the same group describes the application of PEG-based hydrogels in a microfluidic platform to study the effect of dynamic shear stresses on VIC and VEC cultures in 3D [260], and more recently, as a screening platform to systematically investigate optimal combinations of biomaterial and environmental cues for HVTE [261]. Indeed, the application of such models would allow for a more evidence-driven design of TEHVs, thereby refining the need for large animal studies and accelerating translation. As described in previous sections, sophisticated in silico models have proven their worth in the rational design of scaffolds, being able to predict the in vivo outcome, dependent on scaffold design parameters [129].

Apart from the TEHV design parameters, the impact of interpatient variability on HVTE is an essential element to consider, which, although broadly recognized, has not received much attention to date. This is particularly relevant when considering the aforementioned role of the immune response for the integration and regeneration of TEHVs in vivo, as described in more detail in recent reviews by Smits and Bouten [14] and Wissing and Bonito et al. [10]. Patient demographics, such as age and gender, but also previous interventions, use of medication, and common systemic comorbidities (e.g., diabetes), may greatly affect the intrinsic immunological and regenerative capacity of the patient and thereby the long-term outcome after valve replacement with a TEHV. As such, these patient-specific conditions cannot be overlooked, and dedicated research is needed for the stratification of HVTE therapies. Interesting recent papers by Hjortnaes et al. and Porras et al. report on the creation of disease-inspired environments in vitro to study VIC behavior in pathological conditions [262,263]. Other relevant examples range from basic in vitro models using human cells from specific patient cohorts [264—266] to dedicated animal models to specifically investigate the effect of disease [267] or age [268] on tissue remodeling. Again, dealing with patient-to-patient variability calls for systematic approaches and dedicated models to find the clinical boundaries of the HVTE potential.

References

[1] Chester AH, El-Hamamsy I, Butcher JT, Latif N, Bertazzo S, Yacoub MH. The living aortic valve: from molecules to function. Glob Cardiol Sci Pract 2014;2014:52—77. https://doi.org/10.5339/gcsp.2014.11.

[2] Yacoub MH, Takkenberg JJM. Will heart valve tissue engineering change the world? Nat Clin Pract Cardiovasc Med 2005;2:60—1. https://doi.org/10.1038/ncpcardio0112.

[3] Aikawa E, Whittaker P, Farber M, Mendelson K, Padera RF, Aikawa M, Schoen FJ. Human semilunar cardiac valve remodeling by activated cells from fetus to adult: implications for postnatal adaptation, pathology, and tissue engineering. Circulation 2006;113:1344—52. https://doi.org/10.1161/CIRCULATIONAHA.105.591768.

[4] van Geemen D, Soares ALF, Oomen PJA, Driessen-Mol A, Janssen-van den Broek MWJT, van den Bogaerdt AJ, Bogers AJJC, Goumans M-JTH, Baaijens FPT, Bouten CVC. Age-dependent changes in geometry, tissue composition and mechanical properties of fetal to adult cryopreserved human heart valves. PLoS One 2016;11:e0149020. https://doi.org/10.1371/journal.pone.0149020.

[5] Oomen PJA, Loerakker S, van Geemen D, Neggers J, Goumans M-JTH, van den Bogaerdt AJ, Bogers AJJC, Bouten CVC, Baaijens FPT. Age-dependent changes of stress and strain in the human heart valve and their relation with collagen remodeling. Acta Biomater 2016;29:161—9. https://doi.org/10.1016/j.actbio.2015.10.044.

[6] Gould ST, Srigunapalan S, Simmons CA, Anseth KS. Hemodynamic and cellular response feedback in calcific aortic valve disease. Circ Res 2013;113:186—97. https://doi.org/10.1161/CIRCRESAHA.112.300154.

[7] El-Hamamsy I, Eryigit Z, Stevens L-M, Sarang Z, George R, Clark L, Melina G, Takkenberg JJM, Yacoub MH. Long-term outcomes after autograft versus homograft aortic root replacement in adults with aortic valve disease: a randomised controlled trial. Lancet 2010;376:524–31. https://doi.org/10.1016/S0140-6736(10)60828-8.

[8] Mol A, Smits AIPM, Bouten CVC, Baaijens FPT. Tissue engineering of heart valves: advances and current challenges. Expert Rev Med Devices 2009;6:259–75. https://doi.org/10.1586/erd.09.12.

[9] Mendelson K, Schoen FJ. Heart valve tissue engineering: concepts, approaches, progress, and challenges. Ann Biomed Eng 2006;34:1799–819. https://doi.org/10.1007/s10439-006-9163-z.

[10] Wissing TB, Bonito V, Bouten CVC, Smits AIPM. Biomaterial-driven in situ cardiovascular tissue engineering—a multi-disciplinary perspective. NPJ Regen Med 2017;2:18. https://doi.org/10.1038/s41536-017-0023-2.

[11] Weber B, Scherman J, Emmert MY, Gruenenfelder J, Verbeek R, Bracher M, Black M, Kortsmit J, Franz T, Schoenauer R, Baumgartner L, Brokopp C, Agarkova I, Wolint P, Zund G, Falk V, Zilla P, Hoerstrup SP. Injectable living marrow stromal cell-based autologous tissue engineered heart valves: first experiences with a one-step intervention in primates. Eur Heart J 2011;32:2830–40. https://doi.org/10.1093/eurheartj/ehr059.

[12] Emmert MY, Weber B, Wolint P, Behr L, Sammut S, Frauenfelder T, Frese L, Scherman J, Brokopp CE, Templin C, Grünenfelder J, Zünd G, Falk V, Hoerstrup SP. Stem cell-based transcatheter aortic valve implantation: first experiences in a pre-clinical model. JACC Cardiovasc Interv 2012;5:874–83. https://doi.org/10.1016/j.jcin.2012.04.010.

[13] Bouten CVC, Dankers PYW, Driessen-Mol A, Pedron S, a Brizard AM, Baaijens FPT. Substrates for cardiovascular tissue engineering. Adv Drug Deliv Rev 2011;63:221–41. https://doi.org/10.1016/j.addr.2011.01.007.

[14] Smits AIPM, Bouten CVC. Tissue engineering meets immunoengineering: prospective on personalized in situ tissue engineering strategies. Curr Opin Biomed Eng 6, 2018, 17–26. https://doi.org/10.1016/j.cobme.2018.02.006.

[15] Langer R, Vacanti JP. Tissue engineering. Science 1993;260:920–6. https://doi.org/10.1126/science.8493529.

[16] Shinoka T, Ma PX, Shum-Tim D, Breuer CK, Cusick RA, Zund G, Langer R, Vacanti JP, Mayer JE. Tissue-engineered heart valves. Autologous valve leaflet replacement study in a lamb model. Circulation 1996;94:II164–II168. https://doi.org/10.1016/0003-4975(95)00733-4.

[17] Shin'oka T, Breuer CK, Tanel RE, Zund G, Miura T, Ma PX, Langer R, Vacanti JP, Mayer JE. Tissue engineering heart valves: valve leaflet replacement study in a lamb model. Ann Thorac Surg 1995;60:S513–6. http://www.ncbi.nlm.nih.gov/pubmed/8604922.

[18] Hoerstrup SP, Sodian R, Daebritz S, Wang J, Bacha EA, Martin DP, Moran AM, Guleserian KJ, Sperling JS, Kaushal S, Vacanti JP, Schoen FJ, Mayer JE. Functional living trileaflet heart valves grown in vitro. Circulation 2000;102:III44–I49. https://doi.org/10.1161/01.CIR.102.suppl_3.III-44.

[19] Sodian R, Hoerstrup SP, Sperling JS, Daebritz S, Martin DP, Moran AM, Kim BS, Schoen FJ, Vacanti JP, Mayer JE. Early in vivo experience with tissue-engineered trileaflet heart valves. Circulation 2000;102:III22–I29. http://www.ncbi.nlm.nih.gov/pubmed/11082357.

[20] Jana S, Tranquillo RT, Lerman A. Cells for tissue engineering of cardiac valves. J Tissue Eng Regenerat Med 2016;10:804–24. https://doi.org/10.1002/term.2010.

[21] Capulli AK, MacQueen LA, Sheehy SP, Parker KK. Fibrous scaffolds for building hearts and heart parts. Adv Drug Deliv Rev 2016;96:83–102. https://doi.org/10.1016/j.addr.2015.11.020.

[22] Dijkman PE, Driessen-Mol A, de Heer LM, Kluin J, van Herwerden LA, Odermatt B, Baaijens FPT, Hoerstrup SP. Trans-apical versus surgical implantation of autologous ovine tissue-engineered heart valves. J Heart Valve Dis 2012;21:670–8. http://www.ncbi.nlm.nih.gov/pubmed/23167234.

[23] Schmidt D, Dijkman PE, Driessen-Mol A, Stenger R, Mariani C, Puolakka A, Rissanen M, Deichmann T, Odermatt B, Weber B, Emmert MY, Zund G, Baaijens FPT, Hoerstrup SP. Minimally-invasive implantation of living tissue engineered heart valves: a comprehensive approach from autologous vascular cells to stem cells. J Am Coll Cardiol 2010;56:510–20. https://doi.org/10.1016/j.jacc.2010.04.024.

[24] Emmert MY, Weber B, Falk V, Hoerstrup SP. Transcatheter tissue engineered heart valves. Expert Rev Med Devices 2014;11:15–21. https://doi.org/10.1586/17434440.2014.864231.

[25]   Moreira R, Velz T, Alves N, Gesche VN, Malischewski A, Schmitz-Rode T, Frese J, Jockenhoevel S, Mela P. Tissue-engineered heart valve with a tubular leaflet design for minimally invasive transcatheter implantation. Tissue Eng C Methods 2015;21:530—40. https://doi.org/10.1089/ten.TEC.2014.0214.

[26]   Flanagan TC, Sachweh JS, Frese J, Schnöring H, Gronloh N, Koch S, Tolba RH, Schmitz-Rode T, Jockenhoevel S. In vivo remodeling and structural characterization of fibrin-based tissue-engineered heart valves in the adult sheep model. Tissue Eng 2009;15:2965—76. https://doi.org/10.1089/ten.TEA.2009.0018.

[27]   Weber B, Dijkman PE, Scherman J, Sanders B, Emmert MY, Grünenfelder J, Verbeek R, Bracher M, Black M, Franz T, Kortsmit J, Modregger P, Peter S, Stampanoni M, Robert J, Kehl D, van Doeselaar M, Schweiger M, Brokopp CE, Wälchli T, Falk V, Zilla P, Driessen-Mol A, Baaijens FPT, Hoerstrup SP. Off-the-shelf human decellularized tissue-engineered heart valves in a non-human primate model. Biomaterials 2013;34:7269—80. https://doi.org/10.1016/j.biomaterials.2013.04.059.

[28]   Syedain ZH, Lahti MT, Johnson SL, Robinson PS, Ruth GR, Bianco RW, Tranquillo RT. Implantation of a tissue-engineered heart valve from human fibroblasts exhibiting short term function in the sheep pulmonary artery. Cardiovasc Eng Technol 2011;2:101—12. https://doi.org/10.1007/s13239-011-0039-5.

[29]   Shinoka T, Shum-Tim D, Ma PX, Tanel RE, Langer R, Vacanti JP, Mayer JE. Tissue-engineered heart valve leaflets: does cell origin affect outcome? Circulation 1997;96. II-102-7, http://www.ncbi.nlm.nih.gov/pubmed/9386083.

[30]   Gottlieb D, Kunal T, Emani S, Aikawa E, Brown DW, Powell AJ, Nedder A, Engelmayr GC, Melero-Martin JM, Sacks MS, Mayer JE. In vivo monitoring of function of autologous engineered pulmonary valve. J Thorac Cardiovasc Surg 2010;139:723—31. https://doi.org/10.1016/j.jtcvs.2009.11.006.

[31]   Bouten CVC, Smits AIPM, Baaijens FPT. Can we grow valves inside the heart? Perspective on material-based in situ heart valve tissue engineering. Front Cardiovasc Med 2018;5:54. https://doi.org/10.3389/fcvm.2018.00054.

[32]   Sarikouch S, Horke A, Tudorache I, Beerbaum P, Westhoff-Bleck M, Boethig D, Repin O, Maniuc L, Ciubotaru A, Haverich A, Cebotari S. Decellularized fresh homografts for pulmonary valve replacement: a decade of clinical experience. Eur J Cardiothorac Surg 2016;50:281—90. https://doi.org/10.1093/ejcts/ezw050.

[33]   Driessen-Mol A, Emmert MY, Dijkman PE, Frese L, Sanders B, Weber B, Cesarovic N, Sidler M, Leenders J, Jenni R, Grünenfelder J, Falk V, Baaijens FPT, Hoerstrup SP. Transcatheter implantation of homologous "off-the-shelf" tissue-engineered heart valves with self-repair capacity: long-term functionality and rapid in vivo remodeling in sheep. J Am Coll Cardiol 2014;63:1320—9. https://doi.org/10.1016/j.jacc.2013.09.082.

[34]   Syedain Z, Reimer J, Schmidt J, Lahti M, Berry J, Bianco R, Tranquillo RT. 6-Month aortic valve implantation of an off-the-shelf tissue-engineered valve in sheep. Biomaterials 2015;73:175—84. https://doi.org/10.1016/j.biomaterials.2015.09.016.

[35]   Reimer JM, Syedain ZH, Haynie BHTT, Tranquillo RT. Pediatric tubular pulmonary heart valve from decellularized engineered tissue tubes. Biomaterials 2015;62:88—94. https://doi.org/10.1016/j.biomaterials.2015.05.009.

[36]   van Loon SLM, Smits AIPM, Driessen-Mol A, Baaijens FPT, Bouten CVC. The immune response in in situ tissue engineering of aortic heart valves. In: Aikawa E, editor. Calcif. Aortic Dis. Rijeka: InTech; 2013. p. 207—45. https://doi.org/10.5772/54354.

[37]   Kluin J, Talacua H, Smits AIPM, Emmert MY, Brugmans MCP, Fioretta ES, Dijkman PE, Söntjens SHM, Duijvelshoff R, Dekker S, Janssen-van den Broek MWJT, Lintas V, Vink A, Hoerstrup SP, Janssen HM, Dankers PYW, Baaijens FPT, Bouten CVC. In situ heart valve tissue engineering using a bioresorbable elastomeric implant — from material design to 12 months follow-up in sheep. Biomaterials 2017;125:101—17. https://doi.org/10.1016/j.biomaterials.2017.02.007.

[38]   Bennink G, Torii S, Brugmans M, Cox M, Svanidze O, Ladich E, Carrel T, Virmani R. A novel restorative pulmonary valved conduit in a chronic sheep model: mid-term hemodynamic function and histologic assessment. J Thorac Cardiovasc Surg 2017. https://doi.org/10.1016/j.jtcvs.2017.12.046.

[39] Mahler GJ, Frendl CM, Cao Q, Butcher JT. Effects of shear stress pattern and magnitude on mesenchymal transformation and invasion of aortic valve endothelial cells. Biotechnol Bioeng 2014;111: 2326—37. https://doi.org/10.1002/bit.25291.

[40] Steed E, Boselli F, Vermot J. Hemodynamics driven cardiac valve morphogenesis. Biochim Biophys Acta Mol Cell Res 2016;1863:1760—6. https://doi.org/10.1016/j.bbamcr.2015.11.014.

[41] Hjortnaes J, Shapero K, Goettsch C, Hutcheson JD, Keegan J, Kluin J, Mayer JE, Bischoff J, Aikawa E. Valvular interstitial cells suppress calcification of valvular endothelial cells. Atherosclerosis 2015;242:251—60. https://doi.org/10.1016/j.atherosclerosis.2015.07.008.

[42] Zhong A, Mirzaei Z, Simmons CA. The roles of matrix stiffness and ß-catenin signaling in endothelial-to-mesenchymal transition of aortic valve endothelial cells. Cardiovasc Eng Technol 2018;9:158—67. https://doi.org/10.1007/s13239-018-0363-0.

[43] Simmons CA, Grant GR, Manduchi E, Davies PF. Spatial heterogeneity of endothelial phenotypes correlates with side-specific vulnerability to calcification in normal porcine aortic valves. Circ Res 2005;96:792—9. https://doi.org/10.1161/01.RES.0000161998.92009.64.

[44] Butcher JT, Nerem RM. Valvular endothelial cells regulate the phenotype of interstitial cells in co-culture: effects of steady shear stress. Tissue Eng 2006;12:905—15. https://doi.org/10.1089/ten.2006.12.905.

[45] Gould ST, Matherly EE, Smith JN, Heistad DD, Anseth KS. The role of valvular endothelial cell paracrine signaling and matrix elasticity on valvular interstitial cell activation. Biomaterials 2014; 35:3596—606. https://doi.org/10.1016/j.biomaterials.2014.01.005.

[46] Weber B, Emmert MY, Hoerstrup SP. Stem cells for heart valve regeneration. Swiss Med Wkly 2012;142:w13622. https://doi.org/10.4414/smw.2012.13622.

[47] Sutherland FWH, Perry TE, Yu Y, Sherwood MC, Rabkin E, Masuda Y, Garcia GA, McLellan DL, Engelmayr GC, Sacks MS, Schoen FJ, Mayer JE. From stem cells to viable autologous semilunar heart valve. Circulation 2005;111:2783—91. https://doi.org/10.1161/CIRCULATIONAHA.104.498378.

[48] Schmidt D, Mol A, Breymann C, Achermann J, Odermatt B, Gössi M, Neuenschwander S, Prêtre R, Genoni M, Zund G, Hoerstrup SP. Living autologous heart valves engineered from human prenatally harvested progenitors. Circulation 2006;114:I125—31. https://doi.org/10.1161/CIRCULATIONAHA.105.001040.

[49] Schmidt D, Mol A, Odermatt B, Neuenschwander S, Breymann C, Gössi M, Genoni M, Zund G, Hoerstrup SP. Engineering of biologically active living heart valve leaflets using human umbilical cord-derived progenitor cells. Tissue Eng 2006;12:3223—32. https://doi.org/10.1089/ten.2006.12.3223.

[50] Schmidt D, Achermann J, Odermatt B, Breymann C, Mol A, Genoni M, Zund G, Hoerstrup SP. Prenatally fabricated autologous human living heart valves based on amniotic fluid derived progenitor cells as single cell source. Circulation 2007;116:I64—70. https://doi.org/10.1161/CIRCULATIONAHA.106.681494.

[51] Schmidt D, Achermann J, Odermatt B, Genoni M, Zund G, Hoerstrup SP. Cryopreserved amniotic fluid-derived cells: a lifelong autologous fetal stem cell source for heart valve tissue engineering. J Heart Valve Dis 2008;17:446—55. discussion 455, http://www.ncbi.nlm.nih.gov/pubmed/18751475.

[52] Sales VL, Mettler BA, Engelmayr GC, Aikawa E, Bischoff J, Martin DP, Exarhopoulos A, Moses MA, Schoen FJ, Sacks MS, Mayer JE. Endothelial progenitor cells as a sole source for ex vivo seeding of tissue-engineered heart valves. Tissue Eng 2010;16:257—67. https://doi.org/10.1089/ten.TEA.2009.0424.

[53] Hajdu Z, Romeo SJ, a Fleming P, Markwald RR, Visconti RP, Drake CJ. Recruitment of bone marrow-derived valve interstitial cells is a normal homeostatic process. J Mol Cell Cardiol 2011; 51:955—65. https://doi.org/10.1016/j.yjmcc.2011.08.006.

[54] Ota T, Sawa Y, Iwai S, Kitajima T, Ueda Y, Coppin C, Matsuda H, Okita Y. Fibronectin-hepatocyte growth factor enhances reendothelialization in tissue-engineered heart valve. Ann Thorac Surg 2005;80:1794—801. https://doi.org/10.1016/j.athoracsur.2005.05.002.

[55] Flameng W, De Visscher G, Mesure L, Hermans H, Jashari R, Meuris B. Coating with fibronectin and stromal cell–derived factor-1α of decellularized homografts used for right ventricular outflow

tract reconstruction eliminates immune response-related degeneration. J Thorac Cardiovasc Surg 2014;147:1398—1404.e2. https://doi.org/10.1016/j.jtcvs.2013.06.022.

[56] Williams JK, Miller ES, Lane MR, Atala A, Yoo JJ, Jordan JE. Characterization of CD133 antibody-directed recellularized heart valves. J Cardiovasc Transl Res 2015;8:411—20. https://doi.org/10.1007/s12265-015-9651-3.

[57] Juthier F, Vincentelli A, Gaudric J, Corseaux D, Fouquet O, Calet C, Le Tourneau T, Soenen V, Zawadzki C, Fabre O, Susen S, Prat A, Jude B. Decellularized heart valve as a scaffold for in vivo recellularization: deleterious effects of granulocyte colony-stimulating factor. J Thorac Cardiovasc Surg 2006;131:843—52. https://doi.org/10.1016/j.jtcvs.2005.11.037.

[58] De Visscher G, Mesure L, Meuris B, Ivanova A, Flameng W. Improved endothelialization and reduced thrombosis by coating a synthetic vascular graft with fibronectin and stem cell homing factor SDF-1alpha. Acta Biomater 2012;8:1330—8. https://doi.org/10.1016/j.actbio.2011.09.016.

[59] Hibino N, Yi T, Duncan DR, Rathore A, Dean E, Naito Y, Dardik A, Kyriakides T, Madri J, Pober JS, Shinoka T, Breuer CK. A critical role for macrophages in neovessel formation and the development of stenosis in tissue-engineered vascular grafts. FASEB J 2011;25:4253—63. https://doi.org/10.1096/fj.11-186585.

[60] Roh JD, Sawh-Martinez R, Brennan MP, Jay SM, Devine L, a Rao D, Yi T, Mirensky TL, Nalbandian A, Udelsman B, Hibino N, Shinoka T, Saltzman WM, Snyder E, Kyriakides TR, Pober JS, Breuer CK. Tissue-engineered vascular grafts transform into mature blood vessels via an inflammation-mediated process of vascular remodeling. Proc Natl Acad Sci USA 2010;107: 4669—74. https://doi.org/10.1073/pnas.0911465107.

[61] Talacua H, Smits AIPM, Muylaert DEP, van Rijswijk JW, Vink A, Verhaar MC, Driessen-Mol A, van Herwerden LA, Bouten CVCVC, Kluin J, Baaijens FPT. In situ tissue engineering of functional small-diameter blood vessels by host circulating cells only. Tissue Eng 2015;21:2583—94. https://doi.org/10.1089/ten.TEA.2015.0066.

[62] Shi Q, Wu MH, Hayashida N, Wechezak AR, Clowes AW, Sauvage LR. Proof of fallout endothelialization of impervious Dacron grafts in the aorta and inferior vena cava of the dog. J Vasc Surg 1994;20:546—56. http://www.ncbi.nlm.nih.gov/pubmed/7933256.

[63] Pennel T, Bezuidenhout D, Koehne J, Davies NH, Zilla P. Transmural capillary ingrowth is essential for confluent vascular graft healing. Acta Biomater 2018;65:237—47. https://doi.org/10.1016/j.actbio.2017.10.038.

[64] Capulli AK, Emmert MY, Pasqualini FS, Kehl D, Caliskan E, Lind JU, Sheehy SP, Park SJ, Ahn S, Weber B, Goss JA, Hoerstrup SP, Parker KK. JetValve: rapid manufacturing of biohybrid scaffolds for biomimetic heart valve replacement. Biomaterials 2017;133:229—41. https://doi.org/10.1016/j.biomaterials.2017.04.033.

[65] Dekker S, van Geemen D, Driessen-Mol A, van den Bogaerdt AJ, Aikawa E, Smits AIPM. Sheep-specific immunohistochemical panel for the evaluation of regenerative and inflammatory processes in tissue-engineered heart valves. Front Cardiovasc Med 2018. https://doi.org/10.3389/fcvm.2018.00105.

[66] De Visscher G, Plusquin R, Mesure L, Flameng W. Selection of an immunohistochemical panel for cardiovascular research in sheep. Appl Immunohistochem Mol Morphol 2010;18:382—91. https://doi.org/10.1097/PAI.0b013e3181cd32e7.

[67] Moreira R, Neusser C, Kruse M, Mulderrig S, Wolf F, Spillner J, Schmitz-Rode T, Jockenhoevel S, Mela P. Tissue-engineered fibrin-based heart valve with bio-inspired textile reinforcement. Adv Healthc Mater 2016;5:2113—21. https://doi.org/10.1002/adhm.201600300.

[68] Duan B, Hockaday LA, Kang KH, Butcher JT. 3D Bioprinting of heterogeneous aortic valve conduits with alginate/gelatin hydrogels. J Biomed Mater Res A 2013;101:1255—64. https://doi.org/10.1002/jbm.a.34420.

[69] Chen Q, Bruyneel A, Carr C, Czernuszka J. Bio-mechanical properties of novel Bi-layer collagen-elastin scaffolds for heart valve tissue engineering. Procedia Eng 2013;59:247—54. https://doi.org/10.1016/j.proeng.2013.05.118.

[70] Taylor PM, Sachlos E, Dreger SA, Chester AH, Czernuszka JT, Yacoub MH. Interaction of human valve interstitial cells with collagen matrices manufactured using rapid prototyping. Biomaterials 2006;27:2733–7. https://doi.org/10.1016/j.biomaterials.2005.12.003.

[71] Colazzo F, Sarathchandra P, Smolenski RT, Chester AH, Tseng Y-T, Czernuszka JT, Yacoub MH, Taylor PM. Extracellular matrix production by adipose-derived stem cells: implications for heart valve tissue engineering. Biomaterials 2011;32:119–27. https://doi.org/10.1016/j.biomaterials.2010.09.003.

[72] Dreger SA, Thomas P, Sachlos E, Chester AH, Czernuszka JT, Taylor PM, Yacoub MH. Potential for synthesis and degradation of extracellular matrix proteins by valve interstitial cells seeded onto collagen scaffolds. Tissue Eng 2006;12:2533–40. https://doi.org/10.1089/ten.2006.12.2533.

[73] Neidert MR, Tranquillo RT. Tissue-engineered valves with commissural alignment. Tissue Eng 2006;12:891–903. https://doi.org/10.1089/ten.2006.12.891.

[74] Brougham CM, Levingstone TJ, Shen N, Cooney GM, Jockenhoevel S, Flanagan TC, O'Brien FJ. Freeze-drying as a novel biofabrication method for achieving a controlled microarchitecture within large, complex natural biomaterial scaffolds. Adv Healthc Mater 2017;6. https://doi.org/10.1002/adhm.201700598.

[75] Ye Q, Zünd G, Benedikt P, Jockenhoevel S, Hoerstrup SP, Sakyama S, Hubbell JA, Turina M. Fibrin gel as a three dimensional matrix in cardiovascular tissue engineering. Eur J Cardiothorac Surg 2000;17:587–91. http://www.ncbi.nlm.nih.gov/pubmed/10814924.

[76] Cholewinski E, Dietrich M, Flanagan TC, Schmitz-Rode T, Jockenhoevel S. Tranexamic acid–an alternative to aprotinin in fibrin-based cardiovascular tissue engineering. Tissue Eng 2009;15:3645–53. https://doi.org/10.1089/ten.tea.2009.0235.

[77] Dietrich M, Heselhaus J, Wozniak J, Weinandy S, Mela P, Tschoeke B, Schmitz-Rode T, Jockenhoevel S. Fibrin-based tissue engineering: comparison of different methods of autologous fibrinogen isolation. Tissue Eng C Methods 2013;19:216–26. https://doi.org/10.1089/ten.TEC.2011.0473.

[78] Jockenhoevel S, Chalabi K, Sachweh JS, V Groesdonk H, Demircan L, Grossmann M, Zund G, Messmer BJ. Tissue engineering: complete autologous valve conduit–a new moulding technique. Thorac Cardiovasc Surg 2001;49:287–90. https://doi.org/10.1055/s-2001-17807.

[79] Long JL, Tranquillo RT. Elastic fiber production in cardiovascular tissue-equivalents. Matrix Biol 2003;22:339–50. http://www.ncbi.nlm.nih.gov/pubmed/12935818.

[80] Mol A, van Lieshout MI, Dam-de Veen CG, Neuenschwander S, Hoerstrup SP, Baaijens FPT, Bouten CVC. Fibrin as a cell carrier in cardiovascular tissue engineering applications. Biomaterials 2005;26:3113–21. https://doi.org/10.1016/j.biomaterials.2004.08.007.

[81] Weber M, Gonzalez de Torre I, Moreira R, Frese J, Oedekoven C, Alonso M, Rodriguez Cabello CJ, Jockenhoevel S, Mela P. Multiple-step injection molding for fibrin-based tissue-engineered heart valves. Tissue Eng C Methods 2015;21:832–40. https://doi.org/10.1089/ten.TEC.2014.0396.

[82] Flanagan TC, Cornelissen C, Koch S, Tschoeke B, Sachweh JS, Schmitz-Rode T, Jockenhoevel S. The in vitro development of autologous fibrin-based tissue-engineered heart valves through optimised dynamic conditioning. Biomaterials 2007;28:3388–97. https://doi.org/10.1016/j.biomaterials.2007.04.012.

[83] Robinson PS, Johnson SL, Evans MC, Barocas VH, Tranquillo RT. Functional tissue-engineered valves from cell-remodeled fibrin with commissural alignment of cell-produced collagen. Tissue Eng 2008;14:83–95. https://doi.org/10.1089/ten.a.2007.0148.

[84] Robinson PS, Tranquillo RT. Planar biaxial behavior of fibrin-based tissue-engineered heart valve leaflets. Tissue Eng 2009;15:2763–72. https://doi.org/10.1089/ten.tea.2008.0426.

[85] Weber M, Heta E, Moreira R, Gesche VN, Schermer T, Frese J, Jockenhoevel S, Mela P. Tissue-engineered fibrin-based heart valve with a tubular leaflet design. Tissue Eng C Methods 2014;20:265–75. https://doi.org/10.1089/ten.TEC.2013.0258.

[86] Moreira R, Gesche VN, Hurtado-Aguilar LG, Schmitz-Rode T, Frese J, Jockenhoevel S, Mela P. TexMi: development of tissue-engineered textile-reinforced mitral valve prosthesis. Tissue Eng C Methods 2014;20:741–8. https://doi.org/10.1089/ten.tec.2013.0426.

[87]   Gonzalez de Torre I, Weber M, Quintanilla L, Alonso M, Jockenhoevel S, Rodríguez Cabello JC, Mela P. Hybrid elastin-like recombinamer-fibrin gels: physical characterization and in vitro evaluation for cardiovascular tissue engineering applications. Biomater Sci 2016;4:1361–70. https://doi.org/10.1039/C6BM00300A.

[88]   Vesely I. The role of elastin in aortic valve mechanics. J Biomech 1997;31:115–23. https://doi.org/10.1016/S0021-9290(97)00122-X.

[89]   Scott M, Vesely I. Aortic valve cusp microstructure: the role of elastin. Ann Thorac Surg 1995;60:S391–4. http://www.ncbi.nlm.nih.gov/pubmed/7646194.

[90]   Almine JF, V Bax D, Mithieux SM, Nivison-Smith L, Rnjak J, Waterhouse A, Wise SG, Weiss AS. Elastin-based materials. Chem Soc Rev 2010;39:3371–9. https://doi.org/10.1039/b919452p.

[91]   Hinderer S, Shena N, Ringuette L-J, Hansmann J, Reinhardt DP, Brucker SY, Davis EC, Schenke-Layland K. In vitro elastogenesis: instructing human vascular smooth muscle cells to generate an elastic fiber-containing extracellular matrix scaffold. Biomed Mater 2015;10:034102. https://doi.org/10.1088/1748-6041/10/3/034102.

[92]   Wanjare M, Agarwal N, Gerecht S. Biomechanical strain induces elastin and collagen production in human pluripotent stem cell-derived vascular smooth muscle cells. Am J Physiol Cell Physiol 2015;309:C271–81. https://doi.org/10.1152/ajpcell.00366.2014.

[93]   Gilbert TW, Sellaro TL, Badylak SF. Decellularization of tissues and organs. Biomaterials 2006;27:3675–83. https://doi.org/10.1016/j.biomaterials.2006.02.014.

[94]   Iop L, Gerosa G. Guided tissue regeneration in heart valve replacement: from preclinical research to first-in-human trials. BioMed Res Int 2015;2015:1–13. https://doi.org/10.1155/2015/432901.

[95]   Cicha I, Rüffer A, Cesnjevar R, Glöckler M, Agaimy A, Daniel WG, Garlichs CD, Dittrich S. Early obstruction of decellularized xenogenic valves in pediatric patients: involvement of inflammatory and fibroproliferative processes. Cardiovasc Pathol 2011;20:222–31. https://doi.org/10.1016/j.carpath.2010.04.006.

[96]   Liao J, Joyce EM, Sacks MS. Effects of decellularization on the mechanical and structural properties of the porcine aortic valve leaflet. Biomaterials 2008;29:1065–74. https://doi.org/10.1016/j.biomaterials.2007.11.007.

[97]   Spina M, Ortolani F, El Messlemani A, Gandaglia A, Bujan J, Garcia-Honduvilla N, Vesely I, Gerosa G, Casarotto D, Petrelli L, Marchini M. Isolation of intact aortic valve scaffolds for heart-valve bioprostheses: extracellular matrix structure, prevention from calcification, and cell repopulation features. J Biomed Mater Res A 2003;67:1338–50. https://doi.org/10.1002/jbm.a.20025.

[98]   Cebotari S, Lichtenberg A, Tudorache I, Hilfiker A, Mertsching H, Leyh R, Breymann T, Kallenbach K, Maniuc L, Batrinac A, Repin O, Maliga O, Ciubotaru A, Haverich A. Clinical application of tissue engineered human heart valves using autologous progenitor cells. Circulation 2006;114:I132–7. https://doi.org/10.1161/CIRCULATIONAHA.105.001065.

[99]   Cebotari S, Tudorache I, Ciubotaru A, Boethig D, Sarikouch S, Goerler A, Lichtenberg A, Cheptanaru E, Barnaciuc S, Cazacu A, Maliga O, Repin O, Maniuc L, Breymann T, Haverich A. Use of fresh decellularized allografts for pulmonary valve replacement may reduce the reoperation rate in children and young adults: early report. Circulation 2011;124:115–24. https://doi.org/10.1161/CIRCULATIONAHA.110.012161.

[100] Neumann A, Sarikouch S, Breymann T, Cebotari S, Boethig D, Horke A, Beerbaum P, Westhoff-Bleck M, Bertram H, Ono M, Tudorache I, Haverich A, Beutel G. Early systemic cellular immune response in children and young adults receiving decellularized fresh allografts for pulmonary valve replacement. Tissue Eng 2014;20:1003–11. https://doi.org/10.1089/ten.TEA.2013.0316.

[101] Costa F, Dohmen P, Vieira E, Lopes SV, Colatusso C, Pereira EWL, Matsuda CN, Cauduro S. Ross Operation with decellularized pulmonary allografts: medium-term results. Rev Bras Cir Cardiovasc 2007;22:454. 62, http://www.ncbi.nlm.nih.gov/pubmed/18488113.

[102] da Costa FDA, Dohmen PM, Duarte D, von Glenn C, Lopes SV, Filho HH, da Costa MBA, Konertz W. Immunological and echocardiographic evaluation of decellularized versus cryopreserved allografts during the Ross operation. Eur J Cardiothorac Surg 2005;27:572–8. https://doi.org/10.1016/j.ejcts.2004.12.057.

[103] da Costa FDA, Etnel JRG, Torres R, Balbi Filho EM, Torres R, Calixto A, Mulinari LA. Decellularized allografts for right ventricular outflow tract reconstruction in children. World J Pediatr Congenit Heart Surg 2017;8:605−12. https://doi.org/10.1177/2150135117723916.

[104] da Costa FDA, Etnel JRG, Charitos EI, Sievers H-H, Stierle U, Fornazari D, Takkenberg JJM, Bogers AJJC, Mokhles MM. Decellularized versus standard pulmonary allografts in the Ross procedure: propensity-matched analysis. Ann Thorac Surg 2018;105:1205−13. https://doi.org/10.1016/j.athoracsur.2017.09.057.

[105] Etnel JRG, Suss PH, Schnorr GM, Veloso M, Colatusso DF, Balbi Filho EM, da Costa FDA. Fresh decellularized versus standard cryopreserved pulmonary allografts for right ventricular outflow tract reconstruction during the Ross procedure: a propensity-matched study. Eur J Cardiothorac Surg 2018. https://doi.org/10.1093/ejcts/ezy079.

[106] da Costa FDA, Costa ACBA, Prestes R, Domanski AC, Balbi EM, Ferreira ADA, Lopes SV. The early and midterm function of decellularized aortic valve allografts. Ann Thorac Surg 2010;90:1854−60. https://doi.org/10.1016/j.athoracsur.2010.08.022.

[107] Gallo M, Naso F, Poser H, Rossi A, Franci P, Bianco R, Micciolo M, Zanella F, Cucchini U, Aresu L, Buratto E, Busetto R, Spina M, Gandaglia A, Gerosa G. Physiological performance of a detergent decellularized heart valve implanted for 15 months in Vietnamese pigs: surgical procedure, follow-up, and explant inspection. Artif Organs 2012;36:E138−50. https://doi.org/10.1111/j.1525-1594.2012.01447.x.

[108] Iop L, Bonetti A, Naso F, Rizzo S, Cagnin S, Bianco R, Dal Lin C, Martini P, Poser H, Franci P, Lanfranchi G, Busetto R, Spina M, Basso C, Marchini M, Gandaglia A, Ortolani F, Gerosa G. Decellularized allogeneic heart valves demonstrate self-regeneration potential after a long-term preclinical evaluation. PLoS One 2014;9. https://doi.org/10.1371/journal.pone.0099593.

[109] Simon P, Kasimir MT, Seebacher G, Weigel G, Ullrich R, Salzer-Muhar U, Rieder E, Wolner E. Early failure of the tissue engineered porcine heart valve SYNERGRAFT in pediatric patients. Eur J Cardiothorac Surg 2003;23:1002−6. https://doi.org/10.1016/S1010-7940(03)00094-0. discussion 1006.

[110] Brown JW, Ruzmetov M, Eltayeb O, Rodefeld MD, Turrentine MW. Performance of SynerGraft decellularized pulmonary homograft in patients undergoing a Ross procedure. Ann Thorac Surg 2011;91:416−22. https://doi.org/10.1016/j.athoracsur.2010.10.069. discussion 422−3.

[111] Dohmen PM, Konertz W. Results with decellularized xenografts. Circ Res 2006;99:e10. https://doi.org/10.1161/01.RES.0000242261.29251.49.

[112] Erdbrügger W, Stein-Konertz M. Re: early failure of xenogenous de-cellularised pulmonary valve conduits: a word of caution! Eur J Cardiothorac Surg 2011;39:283−4. https://doi.org/10.1016/j.ejcts.2010.05.004. author reply 284.

[113] Konertz W, Angeli E, Tarusinov G, Christ T, Kroll J, Dohmen PM, Krogmann O, Franzbach B, Pace Napoleone C, Gargiulo G. Right ventricular outflow tract reconstruction with decellularized porcine xenografts in patients with congenital heart disease. J Heart Valve Dis 2011;20:341−7. http://www.ncbi.nlm.nih.gov/pubmed/21714427.

[114] Rüffer A, Purbojo A, Cicha I, Glöckler M, Potapov S, Dittrich S, Cesnjevar RA. Early failure of xenogenous de-cellularised pulmonary valve conduits−a word of caution! Eur J Cardiothorac Surg 2010;38:78−85. https://doi.org/10.1016/j.ejcts.2010.01.044.

[115] Perri G, Polito A, Esposito C, Albanese SB, Francalanci P, Pongiglione G, Carotti A. Early and late failure of tissue-engineered pulmonary valve conduits used for right ventricular outflow tract reconstruction in patients with congenital heart disease. Eur J Cardiothorac Surg 2012;41:1320−5. https://doi.org/10.1093/ejcts/ezr221.

[116] Voges I, Bräsen JH, Entenmann A, Scheid M, Scheewe J, Fischer G, Hart C, Andrade A, Pham HM, Kramer H-H, Rickers C. Adverse results of a decellularized tissue-engineered pulmonary valve in humans assessed with magnetic resonance imaging. Eur J Cardiothorac Surg 2013. https://doi.org/10.1093/ejcts/ezt328.

[117] Stamm C, Khosravi A, Grabow N, Schmohl K, Treckmann N, Drechsel A, Nan M, Schmitz K-P, Haubold A, Steinhoff G. Biomatrix/polymer composite material for heart valve tissue engineering. Ann Thorac Surg 2004;78:2084—92. https://doi.org/10.1016/j.athoracsur.2004.03.106. discussion 2092—3.

[118] Wu S, Liu YL, Cui B, Qu XH, Chen GQ. Study on decellularized porcine aortic valve/poly (3-hydroxybutyrate-co-3- hydroxyhexanoate) hybrid heart valve in sheep model. Artif Organs 2007; 31:689—97. https://doi.org/10.1111/j.1525-1594.2007.00442.x.

[119] Ropcke DM, Ilkjaer C, Tjornild MJ, Skov SN, Ringgaard S, Hjortdal VE, Nielsen SL. Small intestinal submucosa tricuspid valve tube graft shows growth potential, remodelling and physiological valve function in a porcine model. Interact Cardiovasc Thorac Surg 2017;24:918—24. https://doi.org/10.1093/icvts/ivx017.

[120] Ropcke DM, Rasmussen J, Ilkjær C, Skov SN, Tjørnild MJ, Baandrup UT, Christian Danielsen C, Hjortdal VE, Nielsen SL. Mid-term function and remodeling potential of tissue engineered tricuspid valve: histology and biomechanics. J Biomech 2018;71:52—8. https://doi.org/10.1016/j.jbiomech.2018.01.019.

[121] Hofmann M, Schmiady MO, Burkhardt BE, Dave HH, Hübler M, Kretschmar O, Bode PK. Congenital aortic valve repair using CorMatrix®: a histologic evaluation. Xenotransplantation 2017;24:1—9. https://doi.org/10.1111/xen.12341.

[122] Zaidi AH, Nathan M, Emani S, Baird C, Del Nido PJ, Gauvreau K, Harris M, Sanders SP, Padera RF. Preliminary experience with porcine intestinal submucosa (CorMatrix) for valve reconstruction in congenital heart disease: histologic evaluation of explanted valves. J Thorac Cardiovasc Surg 2014;148:2216—2225.e1. https://doi.org/10.1016/j.jtcvs.2014.02.081.

[123] Woo JS, Fishbein MC, Reemtsen B. Histologic examination of decellularized porcine intestinal submucosa extracellular matrix (CorMatrix) in pediatric congenital heart surgery. Cardiovasc Pathol 2016;25:12—7. https://doi.org/10.1016/j.carpath.2015.08.007.

[124] Syedain ZH, Bradee AR, Kren S, Taylor DA, Tranquillo RT. Decellularized tissue-engineered heart valve leaflets with recellularization potential. Tissue Eng 2013;19:759—69. https://doi.org/10.1089/ten.TEA.2012.0365.

[125] Syedain ZH, a Meier L, Reimer JM, Tranquillo RT. Tubular heart valves from decellularized engineered tissue. Ann Biomed Eng 2013. https://doi.org/10.1007/s10439-013-0872-9.

[126] Dijkman PE, Driessen-Mol A, Frese L, Hoerstrup SP, Baaijens FPT. Decellularized homologous tissue-engineered heart valves as off-the-shelf alternatives to xeno- and homografts. Biomaterials 2012;33:4545—54. https://doi.org/10.1016/j.biomaterials.2012.03.015.

[127] Spriestersbach H, Prudlo A, Bartosch M, Sanders B, Radtke T, Baaijens FPT, Hoerstrup SP, Berger F, Schmitt B. First percutaneous implantation of a completely tissue-engineered self-expanding pulmonary heart valve prosthesis using a newly developed delivery system: a feasibility study in sheep. Cardiovasc Interv Ther 2017;32:36—47. https://doi.org/10.1007/s12928-016-0396-y.

[128] Loerakker S, Argento G, Oomens CWJ, Baaijens FPT. Effects of valve geometry and tissue anisotropy on the radial stretch and coaptation area of tissue-engineered heart valves. J Biomech 2013;46: 1792—800. https://doi.org/10.1016/j.jbiomech.2013.05.015.

[129] Emmert MY, Schmitt BA, Loerakker S, Sanders B, Spriestersbach H, Fioretta ES, Bruder L, Brakmann K, Motta SE, Lintas V, Dijkman PE, Frese L, Berger F, Baaijens FPT, Hoerstrup SP. Computational modeling guides tissue-engineered heart valve design for long-term in vivo performance in a translational sheep model. Sci Transl Med 2018;10. https://doi.org/10.1126/scitranslmed.aan4587.

[130] Blum KM, Drews JD, Breuer CK. Tissue-engineered heart valves: a call for mechanistic studies. Tissue Eng B Rev 2018. https://doi.org/10.1089/ten.TEB.2017.0425.

[131] Hinderer S, Layland SL, Schenke-Layland K. ECM and ECM-like materials - biomaterials for applications in regenerative medicine and cancer therapy. Adv Drug Deliv Rev 2016;97:260—9. https://doi.org/10.1016/j.addr.2015.11.019.

[132] Stassen OMJA, Muylaert DEP, Bouten CVC, Hjortnaes J. Current challenges in translating tissue-engineered heart valves. Curr Treat Options Cardiovasc Med 2017;19:71. https://doi.org/10.1007/s11936-017-0566-y.

[133] Hinderer S, Sudrow K, Schneider M, Holeiter M, Layland SL, Seifert M, Schenke-Layland K. Surface functionalization of electrospun scaffolds using recombinant human decorin attracts circulating endothelial progenitor cells. Sci Rep 2018;8:110. https://doi.org/10.1038/s41598-017-18382-y.

[134] Hinderer S, Brauchle E, Schenke-Layland K. Generation and assessment of functional biomaterial scaffolds for applications in cardiovascular tissue engineering and regenerative medicine. Adv Healthc Mater 2015;4:2326–41. https://doi.org/10.1002/adhm.201400762.

[135] Garash R, Bajpai A, Marcinkiewicz BM, Spiller KL. Drug delivery strategies to control macrophages for tissue repair and regeneration. Exp Biol Med 2016;241:1054–63. https://doi.org/10.1177/1535370216649444.

[136] Lee J, Cuddihy MJ, Kotov NA. Three-dimensional cell culture matrices: state of the art. Tissue Eng B Rev 2008;14:61–86. https://doi.org/10.1089/teb.2007.0150.

[137] Milleret V, Hefti T, Hall H, Vogel V, Eberli D. Influence of the fiber diameter and surface roughness of electrospun vascular grafts on blood activation. Acta Biomater 2012;8:4349–56. https://doi.org/10.1016/j.actbio.2012.07.032.

[138] Lamichhane S, Anderson JA, Remund T, Sun H, Larson MK, Kelly P, Mani G. Responses of endothelial cells, smooth muscle cells, and platelets dependent on the surface topography of polytetrafluoroethylene. J Biomed Mater Res A 2016;104:2291–304. https://doi.org/10.1002/jbm.a.35763.

[139] Sodian R, Hoerstrup SP, Sperling JS, Martin DP, Daebritz S, Mayer JE, Vacanti JP. Evaluation of biodegradable, three-dimensional matrices for tissue engineering of heart valves. Am Soc Artif Intern Organs J 2000;46:107–10. https://doi.org/10.1097/00002480-200001000-00025.

[140] Liberski A, Ayad N, Wojciechowska D, Zielińska D, Struszczyk MH, Latif N, Yacoub M. Knitting for heart valve tissue engineering. Glob Cardiol Sci Pract 2016;2016. https://doi.org/10.21542/gcsp.2016.31.

[141] Van Lieshout M, Peters G, Rutten M, Baaijens F, Knitted A. Fibrin-covered polycaprolactone scaffold for tissue engineering of the aortic valve. Tissue Eng 2006;12:481–7. https://doi.org/10.1089/ten.2006.12.481.

[142] van Lieshout MI, Vaz CM, Rutten MCM, Peters GWM, Baaijens FPT. Electrospinning versus knitting: two scaffolds for tissue engineering of the aortic valve. J Biomater Sci Polym Ed 2006;17:77–89. http://www.ncbi.nlm.nih.gov/pubmed/16411600.

[143] Wu S, Duan B, Liu P, Zhang C, Qin X-H, Butcher JT. Fabrication of aligned nanofiber polymer yarn networks for anisotropic soft tissue scaffolds. ACS Appl Mater Interfaces 2016. https://doi.org/10.1021/acsami.6b05199. acsami.6b05199.

[144] Wu S, Duan B, Qin X, Butcher JT. Living nano-micro fibrous woven fabric/hydrogel composite scaffolds for heart valve engineering. Acta Biomater 2017;51:89–100. https://doi.org/10.1016/j.actbio.2017.01.051.

[145] Cheung DY, Duan B, Butcher JT. Current progress in tissue engineering of heart valves: multiscale problems, multiscale solutions. Expert Opin Biol Ther 2015:1–18. https://doi.org/10.1517/14712598.2015.1051527.

[146] Vukicevic M, Puperi DS, Jane Grande-Allen K, Little SH. 3D printed modeling of the mitral valve for catheter-based structural interventions. Ann Biomed Eng 2017;45:508–19. https://doi.org/10.1007/s10439-016-1676-5.

[147] Sodian R, Loebe M, Hein A, Martin DP, Hoerstrup SP, V Potapov E, Hausmann H, Lueth T, Hetzer R. Application of stereolithography for scaffold fabrication for tissue engineered heart valves. Am Soc Artif Intern Organs J 2002;48:12–6. http://www.ncbi.nlm.nih.gov/pubmed/11814091.

[148] Lueders C, Jastram B, Hetzer R, Schwandt H. Rapid manufacturing techniques for the tissue engineering of human heart valves. Eur J Cardiothorac Surg 2014;46:593–601. https://doi.org/10.1093/ejcts/ezt510.

[149]   Hockaday LA, Kang KH, Colangelo NW, Cheung PYC, Duan B, Malone E, Wu J, Girardi LN, Bonassar LJ, Lipson H, Chu CC, Butcher JT. Rapid 3D printing of anatomically accurate and mechanically heterogeneous aortic valve hydrogel scaffolds. Biofabrication 2012;4:035005. https://doi.org/10.1088/1758-5082/4/3/035005.

[150]   Hinderer S, Seifert J, Votteler M, Shen N, Rheinlaender J, Schäffer TE, Schenke-Layland K. Engineering of a bio-functionalized hybrid off-the-shelf heart valve. Biomaterials 2014;35:2130—9. https://doi.org/10.1016/j.biomaterials.2013.10.080.

[151]   Masoumi N, Larson BL, Annabi N, Kharaziha M, Zamanian B, Shapero KS, Cubberley AT, Camci-Unal G, Manning KB, Mayer JE, Khademhosseini A, Electrospun PGS. PCL microfibers align human valvular interstitial cells and provide tunable scaffold anisotropy. Adv Healthc Mater 2014;3: 929—39. https://doi.org/10.1002/adhm.201300505.

[152]   Li Q, Bai Y, Jin T, Wang S, Cui W, Stanciulescu I, Yang R, Nie H, Wang L, Zhang X. Bioinspired engineering of poly(ethylene glycol) hydrogels and natural protein fibers for layered heart valve constructs. ACS Appl Mater Interfaces 2017;9:16524—35. https://doi.org/10.1021/acsami.7b03281.

[153]   Breuer CK, Shin'oka T, Tanel RE, Zund G, Mooney DJ, Ma PX, Miura T, Colan S, Langer R, Mayer JE, Vacanti JP. Tissue engineering lamb heart valve leaflets. Biotechnol Bioeng 1996;50: 562—7. https://doi.org/10.1002/(SICI)1097-0290(19960605)50:5<562::AID-BIT11>3.0.CO;2-L.

[154]   Stock UA, Nagashima M, Khalil PN, Nollert GD, Herden T, Sperling JS, Moran A, Lien J, Martin DP, Schoen FJ, Vacanti JP, Mayer JE. Tissue-engineered valved conduits in the pulmonary circulation. J Thorac Cardiovasc Surg 2000;119:732—40. http://www.ncbi.nlm.nih.gov/pubmed/10733761.

[155]   Alavi SH, Kheradvar A. Metal mesh scaffold for tissue engineering of membranes. Tissue Eng C Methods 2012;18:293—301. https://doi.org/10.1089/ten.tec.2011.0531.

[156]   Alavi SH, Soriano Baliarda M, Bonessio N, Valdevit L, Kheradvar A. A tri-leaflet nitinol mesh scaffold for engineering heart valves. Ann Biomed Eng 2017;45:413—26. https://doi.org/10.1007/s10439-016-1778-0.

[157]   Alavi SH, Liu WF, Kheradvar A. Inflammatory response assessment of a hybrid tissue-engineered heart valve leaflet. Ann Biomed Eng 2013;41:316—26. https://doi.org/10.1007/s10439-012-0664-7.

[158]   Dankers PYW, Harmsen MC, Brouwer LA, Van Luyn MJA, Meijer EW. A modular and supramolecular approach to bioactive scaffolds for tissue engineering. Nat Mater 2005;4:568—74. https://doi.org/10.1038/nmat1418.

[159]   Goor OJGM, Keizer HM, Bruinen AL, Schmitz MGJ, Versteegen RM, Janssen HM, Heeren RMA, Dankers PYW. Efficient functionalization of additives at supramolecular material surfaces. Adv Mater 2017;29. https://doi.org/10.1002/adma.201604652.

[160]   Schilling JA, Shurley HM, Joel W, Richter KM, White BN. Fibrocollagenous tubes structured in vivo. Morphology and biological characteristics. Arch Pathol 1961;71:548—53. http://www.ncbi.nlm.nih.gov/pubmed/13747660.

[161]   Sparks CH. Die-grown reinforced arterial grafts: observations on long-term animal grafts and clinical experience. Ann Surg 1970;172:787—94. http://www.ncbi.nlm.nih.gov/pubmed/5477654.

[162]   Guidoin R, Noël HP, Marois M, Martin L, Laroche F, Béland L, Côté R, Gosselin C, Descotes J, Chignier E, Blais P. Another look at the Sparks-Mandril arterial graft precursor for vascular repair. -pathology by scanning electron microscopy. Biomater Med Devices Artif Organs 1980;8:145—67. http://www.ncbi.nlm.nih.gov/pubmed/6446943.

[163]   Hallin RW, Sweetman WR. The Sparks' mandril graft. A seven year follow-up of mandril grafts placed by Charles H. Sparks and his associates. Am J Surg 1976;132:221—3. http://www.ncbi.nlm.nih.gov/pubmed/133619.

[164]   Mizuno T, Takewa Y, Sumikura H, Ohnuma K, Moriwaki T, Yamanami M, Oie T, Tatsumi E, Uechi M, Nakayama Y. Preparation of an autologous heart valve with a stent (stent-biovalve) using the stent eversion method. J Biomed Mater Res B Appl Biomater 2014;102:1038—45. https://doi.org/10.1002/jbm.b.33086.

[165]   Sakai O, Kanda K, Takamizawa K, Sato T, Yaku H, Nakayama Y. Faster and stronger vascular "Biotube" graft fabrication in vivo using a novel nicotine-containing mold. J Biomed Mater Res B Appl Biomater 2009;90:412—20. https://doi.org/10.1002/jbm.b.31300.

[166] Hayashida K, Kanda K, Yaku H, Ando J, Nakayama Y. Development of an in vivo tissue-engineered, autologous heart valve (the biovalve): preparation of a prototype model. J Thorac Cardiovasc Surg 2007;134:152—9. https://doi.org/10.1016/j.jtcvs.2007.01.087.

[167] Kishimoto S, Takewa Y, Nakayama Y, Date K, Sumikura H, Moriwaki T, Nishimura M, Tatsumi E. Sutureless aortic valve replacement using a novel autologous tissue heart valve with stent (stent biovalve): proof of concept. J Artif Organs 2015;18:185—90. https://doi.org/10.1007/s10047-015-0817-1.

[168] Yamanami M, Yahata Y, Uechi M, Fujiwara M, Ishibashi-Ueda H, Kanda K, Watanabe T, Tajikawa T, Ohba K, Yaku H, Nakayama Y. Development of a completely autologous valved conduit with the sinus of Valsalva using in-body tissue architecture technology: a pilot study in pulmonary valve replacement in a beagle model. Circulation 2010;122:S100—6. https://doi.org/10.1161/CIRCULATIONAHA.109.922211.

[169] L'Heureux N, Stoclet JC, Auger FA, Lagaud GJ, Germain L, Andriantsitohaina R. A human tissue-engineered vascular media: a new model for pharmacological studies of contractile responses. FASEB J 2001;15:515—24. https://doi.org/10.1096/fj.00-0283com.

[170] L'Heureux N, Pâquet S, Labbé R, Germain L, Auger FA. A completely biological tissue-engineered human blood vessel. FASEB J 1998;12:47—56. http://www.ncbi.nlm.nih.gov/pubmed/9438410.

[171] Tremblay C, Ruel J, Bourget J-M, Laterreur V, Vallières K, Tondreau MY, Lacroix D, Germain L, Auger FA. A new construction technique for tissue-engineered heart valves using the self-assembly method. Tissue Eng C Methods 2014;20:905—15. https://doi.org/10.1089/ten.TEC.2013.0698.

[172] Dubé J, Bourget J-M, Gauvin R, Lafrance H, Roberge CJ, Auger FA, Germain L. Progress in developing a living human tissue-engineered tri-leaflet heart valve assembled from tissue produced by the self-assembly approach. Acta Biomater 2014;10:3563—70. https://doi.org/10.1016/j.actbio.2014.04.033.

[173] Picard-Deland M, Ruel J, Galbraith T, Tremblay C, Kawecki F, Germain L, Auger FA. Tissue-engineered tubular heart valves combining a novel precontraction phase with the self-assembly method. Ann Biomed Eng 2017;45:427—38. https://doi.org/10.1007/s10439-016-1708-1.

[174] Engelmayr Jr GC, Sales VL, Mayer Jr JE, Sacks MS. Cyclic flexure and laminar flow synergistically accelerate mesenchymal stem cell-mediated engineered tissue formation: implications for engineered heart valve tissues. Biomaterials 2006;27:6083—95. https://doi.org/10.1016/j.biomaterials.2006.07.045.

[175] Smits AIPM, Driessen-Mol A, Bouten CVC, Baaijens FPT. A mesofluidics-based test platform for systematic development of scaffolds for in situ cardiovascular tissue engineering. Tissue Eng C Methods 2012;18:475—85. https://doi.org/10.1089/ten.TEC.2011.0458.

[176] Balachandran K, Alford PW, Wylie-Sears J, Goss JA, Grosberg A, Bischoff J, Aikawa E, Levine RA, Parker KK. Cyclic strain induces dual-mode endothelial-mesenchymal transformation of the cardiac valve. Proc Natl Acad Sci USA 2011;108:19943. https://doi.org/10.1073/pnas.1106954108. 8.

[177] El-Hamamsy I, Balachandran K, Yacoub MH, Stevens LM, Sarathchandra P, Taylor PM, Yoganathan AP, Chester AH. Endothelium-dependent regulation of the mechanical properties of aortic valve cusps. J Am Coll Cardiol 2009;53:1448—55. https://doi.org/10.1016/j.jacc.2008.11.056.

[178] van Haaften EE, Wissing TB, Rutten MCM, Bulsink JA, Gashi K, van Kelle MAJ, Smits AIPM, Bouten CVC, Kurniawan NA. Decoupling the effect of shear stress and stretch on tissue growth and remodeling in a vascular graft. Tissue Eng C Methods 2018;24:418—29. https://doi.org/10.1089/ten.TEC.2018.0104.

[179] van Kelle MAJ, Oomen PJA, Bulsink JA, Janssen-van den Broek MWJT, Lopata RGP, Rutten M, Loerakker S, Bouten C. A bioreactor to identify the driving mechanical stimuli of tissue growth and remodeling. Tissue Eng C Methods 2017;23. https://doi.org/10.1089/ten.TEC.2017.0141. ten.TEC.2017.0141.

[180] Ryan AJ, Brougham CM, Garciarena CD, Kerrigan SW, O'Brien FJ. Towards 3D in vitro models for the study of cardiovascular tissues and disease. Drug Discov Today 2016;00:1—9. https://doi.org/10.1016/j.drudis.2016.04.014.

[181] Wolf F, Vogt F, Schmitz-Rode T, Jockenhoevel S, Mela P. Bioengineered vascular constructs as living models for in vitro cardiovascular research. Drug Discov Today 2016;00. https://doi.org/10.1016/j.drudis.2016.04.017.

[182] Nejad SP, Blaser MC, Paul Santerre J, Caldarone CA, Simmons CA. Biomechanical conditioning of tissue engineered heart valves: too much of a good thing? Adv Drug Deliv Rev 2015. https://doi.org/10.1016/j.addr.2015.11.003.

[183] Hildebrand DK, Wu ZJ, Mayer JE, Sacks MS. Design and hydrodynamic evaluation of a novel pulsatile bioreactor for biologically active heart valves. Ann Biomed Eng 2004;32:1039—49. http://www.ncbi.nlm.nih.gov/pubmed/15446500.

[184] Vismara R, Soncini M, Talò G, Dainese L, Guarino A, Redaelli A, Fiore GB. A bioreactor with compliance monitoring for heart valve grafts. Ann Biomed Eng 2010;38:100—8. https://doi.org/10.1007/s10439-009-9803-1.

[185] Sodian R, Hoerstrup SP, Sperling JS, Daebritz SH, Martin DP, Schoen FJ, Vacanti JP, Mayer Jr JE. Tissue engineering of heart valves: in vitro experiences. Ann Thorac Surg 2000;70:140—4. https://doi.org/10.1016/S0003-4975%2800%2901255-8.

[186] Hoerstrup SP, Sodian R, Sperling JS, Vacanti JP, Mayer JE. New pulsatile bioreactor for in vitro formation of tissue engineered heart valves. Tissue Eng 2000;6:75—9. https://doi.org/10.1089/107632700320919.

[187] Beelen MJ, Neerincx PE, Van De Molengraft MJG. Control of an air pressure actuated disposable bioreactor for cultivating heart valves. Mechatronics 2011;21:1288—97. https://doi.org/10.1016/j.mechatronics.2011.09.003.

[188] Dumont K, Yperman J, Verbeken E, Segers P, Meuris B, Vandenberghe S, Flameng W, Verdonck PR. Design of a new pulsatile bioreactor for tissue engineered aortic heart valve formation. Artif Organs 2002;26:710. 4, http://www.ncbi.nlm.nih.gov/pubmed/12139499.

[189] Gheewala N, Grande-Allen KJ. Design and mechanical evaluation of a physiological mitral valve organ culture system. Cardiovasc Eng Technol 2010;1:123—31. https://doi.org/10.1007/s13239-010-0012-8.

[190] Kaasi A, Cestari IA, Stolf NAG, Leirner AA, Hassager O, Cestari IN. A new approach to heart valve tissue engineering: mimicking the heart ventricle with a ventricular assist device in a novel bioreactor. J Tissue Eng Regenerat Med 2011;5:292—300. https://doi.org/10.1002/term.315.

[191] König F, Hollweck T, Pfeifer S, Reichart B, Wintermantel E, Hagl C, Akra B. A pulsatile bioreactor for conditioning of tissue-engineered cardiovascular constructs under endoscopic visualization. J Funct Biomater 2012;3:480—96. https://doi.org/10.3390/jfb3030480.

[192] Sierad LN, Simionescu A, Albers C, Chen J, Maivelett J, Tedder ME, Liao J, Simionescu DT. Design and testing of a pulsatile conditioning system for dynamic endothelialization of polyphenol-stabilized tissue engineered heart valves. Cardiovasc Eng Technol 2010;1:138—53. https://doi.org/10.1007/s13239-010-0014-6.

[193] Zeltinger J, Landeen LK, Alexander HG, Kidd ID, Sibanda B. Development and characterization of tissue-engineered aortic valves. Tissue Eng 2001;7:9—22. https://doi.org/10.1089/107632701300003250.

[194] Stock UA, Vacanti JP. Cardiovascular physiology during fetal development and implications for tissue engineering. Tissue Eng 2001;7:1—7. https://doi.org/10.1089/107632701300003241.

[195] Mol A, Driessen NJB, Rutten MCM, Hoerstrup SP, Bouten CVC, Baaijens FPT. Tissue engineering of human heart valve leaflets: a novel bioreactor for a strain-based conditioning approach. Ann Biomed Eng 2005;33:1778—88. https://doi.org/10.1007/s10439-005-8025-4.

[196] Syedain ZH, Tranquillo RT. Controlled cyclic stretch bioreactor for tissue-engineered heart valves. Biomaterials 2009;30:4078—84. https://doi.org/10.1016/j.biomaterials.2009.04.027.

[197] D'Amore A, Luketich SK, Raffa GM, Olia S, Menallo G, Mazzola A, D'Accardi F, Grunberg T, Gu X, Pilato M, V Kameneva M, Badhwar V, Wagner WR. Heart valve scaffold fabrication: bioinspired control of macro-scale morphology, mechanics and micro-structure. Biomaterials 2018;150:25—37. https://doi.org/10.1016/j.biomaterials.2017.10.011.

[198]   Sanders B, Loerakker S, Fioretta ES, Bax DJPP, Driessen-Mol A, Hoerstrup SP, Baaijens FPT. Improved geometry of decellularized tissue engineered heart valves to prevent leaflet retraction. Ann Biomed Eng 2016;44:1061—71. https://doi.org/10.1007/s10439-015-1386-4.

[199]   Kortsmit J, Driessen NJB, Rutten MCM, Baaijens FPT. Nondestructive and noninvasive assessment of mechanical properties in heart valve tissue engineering. Tissue Eng 2009;15:797—806. https://doi.org/10.1089/ten.tea.2008.0197.

[200]   Kortsmit J, Rutten MCM, Wijlaars MW, Baaijens FPT. Deformation-controlled load application in heart valve tissue engineering. Tissue Eng C Methods 2009;15:707—16. https://doi.org/10.1089/ten.TEC.2008.0658.

[201]   Ziegelmueller JA, Zaenkert EK, Schams R, Lackermair S, Schmitz C, Reichart B, Sodian R. Optical monitoring during bioreactor conditioning of tissue-engineered heart valves. Am Soc Artif Intern Organs J 2010;56:228—31. https://doi.org/10.1097/MAT.0b013e3181cf3bdd.

[202]   Hurtado-Aguilar LG, Mulderrig S, Moreira R, Hatam N, Spillner J, Schmitz-Rode T, Jockenhoevel S, Mela P. Ultrasound for *in vitro* , noninvasive real-time monitoring and evaluation of tissue-engineered heart valves. Tissue Eng C Methods 2016;22:974—81. https://doi.org/10.1089/ten.tec.2016.0300.

[203]   Kreitz S, Dohmen G, Hasken S, Schmitz-Rode T, Mela P, Jockenhoevel S. Nondestructive method to evaluate the collagen content of fibrin-based tissue engineered structures via ultrasound. Tissue Eng C Methods 2011;17:1021—6. https://doi.org/10.1089/ten.TEC.2010.0669.

[204]   Oomen PJA, van Kelle MAJ, Oomens CWJ, Bouten CVC, Loerakker S. Nondestructive mechanical characterization of developing biological tissues using inflation testing. J Mech Behav Biomed Mater 2017;74:438—47. https://doi.org/10.1016/j.jmbbm.2017.07.009.

[205]   Ayoub S, Ferrari G, Gorman RC, Gorman JH, Schoen FJ, Sacks MS. Heart valve biomechanics and underlying mechanobiology. Comp Physiol 2016;6:1743—80. https://doi.org/10.1002/cphy.c150048.

[206]   Obbink-Huizer C, Oomens CWJ, Loerakker S, Foolen J, Bouten CVC, Baaijens FPT. Computational model predicts cell orientation in response to a range of mechanical stimuli. Biomechanics Model Mechanobiol 2014;13:227—36. https://doi.org/10.1007/s10237-013-0501-4.

[207]   Loerakker S, Ristori T, Baaijens FPT. A computational analysis of cell-mediated compaction and collagen remodeling in tissue-engineered heart valves. J Mech Behav Biomed Mater 2016;58:173—87. https://doi.org/10.1016/j.jmbbm.2015.10.001.

[208]   Hamid MS, Sabbah HN, Stein PD. Influence of stent height upon stresses on the cusps of closed bioprosthetic valves. J Biomech 1986;19:759—69. http://www.ncbi.nlm.nih.gov/pubmed/3793750.

[209]   Thubrikar MJ. The aortic valve. CRC Press; 1989.

[210]   Amoroso NJ, D'Amore A, Hong Y, Wagner WR, Sacks MS. Elastomeric electrospun polyurethane scaffolds: the interrelationship between fabrication conditions, fiber topology, and mechanical properties. Adv Mater 2011;23:106—11. https://doi.org/10.1002/adma.201003210.

[211]   Argento G, Simonet M, Oomens CWJ, Baaijens FPT. Multi-scale mechanical characterization of scaffolds for heart valve tissue engineering. J Biomech 2012;45:2893—8. https://doi.org/10.1016/j.jbiomech.2012.07.037.

[212]   Fan R, Bayoumi AS, Chen P, Hobson CM, Wagner WR, Mayer JE, Sacks MS. Optimal elastomeric scaffold leaflet shape for pulmonary heart valve leaflet replacement. J Biomech 2013;46:662—9. https://doi.org/10.1016/j.jbiomech.2012.11.046.

[213]   Miller KS, Lee YU, Naito Y, Breuer CK, Humphrey JD. Computational model of the in vivo development of a tissue engineered vein from an implanted polymeric construct. J Biomech 2014;47:2080—7. https://doi.org/10.1016/j.jbiomech.2013.10.009.

[214]   Miller KS, Khosravi R, Breuer CK, Humphrey JD. A hypothesis-driven parametric study of effects of polymeric scaffold properties on tissue engineered neovessel formation. Acta Biomater 2015;11:283—94. https://doi.org/10.1016/j.actbio.2014.09.046.

[215] Khosravi R, Miller KS, Best C a, Shih YC, Lee Y-U, Yi T, Shinoka T, Breuer CK, Humphrey JD. Biomechanical diversity despite mechanobiological stability in tissue engineered vascular grafts two years post-implantation. Tissue Eng 2015;21:1529—38. https://doi.org/10.1089/ten.tea.2014.0524.

[216] Szafron JM, Khosravi R, Reinhardt J, Best CA, Bersi MR, Yi T, Breuer CK, Humphrey JD. Immuno-driven and mechano-mediated neotissue formation in tissue engineered vascular grafts. Ann Biomed Eng 2018. https://doi.org/10.1007/s10439-018-2086-7.

[217] Lutter G, Metzner A, Jahnke T, Bombien R, Boldt J, Iino K, Cremer J, Stock UA. Percutaneous tissue-engineered pulmonary valved stent implantation. Ann Thorac Surg 2010;89:259—63. https://doi.org/10.1016/j.athoracsur.2009.06.048.

[218] Metzner A, Stock UA, Iino K, Fischer G, Huemme T, Boldt J, Braesen JH, Bein B, Renner J, Cremer J, Lutter G. Percutaneous pulmonary valve replacement: autologous tissue-engineered valved stents. Cardiovasc Res 2010;88:453—61. https://doi.org/10.1093/cvr/cvq212.

[219] Stock UA, Degenkolbe I, Attmann T, Schenke-Layland K, Freitag S, Lutter G. Prevention of device-related tissue damage during percutaneous deployment of tissue-engineered heart valves. J Thorac Cardiovasc Surg 2006;131:1323—30. https://doi.org/10.1016/j.jtcvs.2006.01.053.

[220] Emmert MY, Weber B, Behr L, Frauenfelder T, Brokopp CE, Grünenfelder J, Falk V, Hoerstrup SP. Transapical aortic implantation of autologous marrow stromal cell-based tissue-engineered heart valves: first experiences in the systemic circulation. JACC Cardiovasc Interv 2011;4:822—3. https://doi.org/10.1016/j.jcin.2011.02.020.

[221] Emmert MY, Weber B, Behr L, Sammut S, Frauenfelder T, Wolint P, Scherman J, Bettex D, Grünenfelder J, Falk V, Hoerstrup SP. Transcatheter aortic valve implantation using anatomically oriented, marrow stromal cell-based, stented, tissue-engineered heart valves: technical considerations and implications for translational cell-based heart valve concepts. Eur J Cardiothorac Surg 2014;45: 61—8. https://doi.org/10.1093/ejcts/ezt243.

[222] Cabrera MS, Oomens CWJ, Baaijens FPT. Understanding the requirements of self-expandable stents for heart valve replacement: radial force, hoop force and equilibrium. J Mech Behav Biomed Mater 2017;68:252—64. https://doi.org/10.1016/j.jmbbm.2017.02.006.

[223] Cabrera MS, Sanders B, Goor OJGM, Driessen-Mol A, Oomens CW, Baaijens FPT. Computationally designed 3D printed self-expandable polymer stents with biodegradation capacity for minimally invasive heart valve implantation: a proof-of-concept study, 3D print. Addit Manuf 2017;4:19—29. https://doi.org/10.1089/3dp.2016.0052.

[224] Lee TC, Midura RJ, Hascall VC, Vesely I. The effect of elastin damage on the mechanics of the aortic valve. J Biomech 2001;34:203—10. http://www.ncbi.nlm.nih.gov/pubmed/11165284.

[225] Wagenseil JE, Mecham RP. Vascular extracellular matrix and arterial mechanics. Physiol Rev 2009; 89:957—89. https://doi.org/10.1152/physrev.00041.2008.

[226] Wagenseil JE, Mecham RP. New insights into elastic fiber assembly. Birth Defects Res C Embryo Today 2007;81:229—40. https://doi.org/10.1002/bdrc.20111.

[227] Wise SG, Weiss AS. Tropoelastin Int J Biochem Cell Biol 2009;41:494—7. https://doi.org/10.1016/ j.biocel.2008.03.017.

[228] Wise SG, Yeo GC, Hiob MA, Rnjak-Kovacina J, Kaplan DL, Ng MKC, Weiss AS. Tropoelastin: a versatile, bioactive assembly module. Acta Biomater 2014;10:1532—41. https://doi.org/10.1016/ j.actbio.2013.08.003.

[229] Mithieux SM, Wise SG, Weiss AS. Tropoelastin - a multifaceted naturally smart material. Adv Drug Deliv Rev 2013;65:421—8. https://doi.org/10.1016/j.addr.2012.06.009.

[230] Votteler M, Berrio DAC, Horke A, Sabatier L, Reinhardt DP, Nsair A, Aikawa E, Schenke-Layland K. Elastogenesis at the onset of human cardiac valve development. Development 2013; 140:2345—53. https://doi.org/10.1242/dev.093500.

[231] Krettek A, Sukhova GK, Libby P. Elastogenesis in human arterial disease: a role for macrophages in disordered elastin synthesis. Arterioscler Thromb Vasc Biol 2003;23:582—7. https://doi.org/ 10.1161/01.ATV.0000064372.78561.A5.

[232] Crapo PM, Wang Y. Hydrostatic pressure independently increases elastin and collagen co-expression in small-diameter engineered arterial constructs. J Biomed Mater Res A 2011;96 A:673—81. https://doi.org/10.1002/jbm.a.33019.

[233] Huang AH, Balestrini JL, V Udelsman B, Zhou KC, Zhao L, Ferruzzi J, Starcher BC, Levene MJ, Humphrey JD, Niklason LE. Biaxial stretch improves elastic fiber maturation, collagen arrangement, and mechanical properties in engineered arteries. Tissue Eng C Methods 2016;22:524—33. https://doi.org/10.1089/ten.TEC.2015.0309.

[234] Lin S, Sandig M, Mequanint K. Three-dimensional topography of synthetic scaffolds induces elastin synthesis by human coronary artery smooth muscle cells, tissue. Eng Times Part A 2011;17:1561—71. https://doi.org/10.1089/ten.tea.2010.0593.

[235] Battiston KG, Labow RS, Simmons CA, Santerre JP. Immunomodulatory polymeric scaffold enhances extracellular matrix production in cell co-cultures under dynamic mechanical stimulation. Acta Biomater 2015;24:74—86. https://doi.org/10.1016/j.actbio.2015.05.038.

[236] Yeo GC, Aghaei-Ghareh-Bolagh B, Brackenreg EP, Hiob MA, Lee P, Weiss AS. Fabricated elastin. Adv Healthc Mater 2015;4:2530—56. https://doi.org/10.1002/adhm.201400781.

[237] Sugiura T, Agarwal R, Tara S, Yi T, Lee YU, Breuer CK, Weiss AS, Shinoka T. Tropoelastin inhibits intimal hyperplasia of mouse bioresorbable arterial vascular grafts. Acta Biomater 2017;52: 74—80. https://doi.org/10.1016/j.actbio.2016.12.044.

[238] Landau S, Szklanny AA, Yeo GC, Shandalov Y, Kosobrodova E, Weiss AS, Levenberg S. Tropoelastin coated PLLA-PLGA scaffolds promote vascular network formation. Biomaterials 2017;122: 72—82. https://doi.org/10.1016/j.biomaterials.2017.01.015.

[239] Wise SG, Liu H, Yeo GC, Michael PL, Chan AHP, Ngo AKY, Bilek MMM, Bao S, Weiss AS. Blended polyurethane and tropoelastin as a novel class of biologically interactive elastomer. Tissue Eng 2016;22:524—33. https://doi.org/10.1089/ten.TEA.2015.0409.

[240] Wise SG, Byrom MJ, Waterhouse A, Bannon PG, Weiss AS, Ng MKC. A multilayered synthetic human elastin/polycaprolactone hybrid vascular graft with tailored mechanical properties. Acta Biomater 2011;7:295—303. https://doi.org/10.1016/j.actbio.2010.07.022.

[241] Porras AM, van Engeland NCA, Marchbanks E, McCormack A, Bouten CVC, Yacoub MH, Latif N, Masters KS. Robust generation of quiescent porcine valvular interstitial cell cultures. J Am Heart Assoc 2017;6. https://doi.org/10.1161/JAHA.116.005041.

[242] Duan B, Yin Z, Kang LH, Magin RL, Butcher JT. Active tissue stiffness modulation controls valve interstitial cell phenotype and osteogenic potential in 3D culture. Acta Biomater 2016. https://doi.org/10.1016/j.actbio.2016.03.007.

[243] Hjortnaes J, Camci-Unal G, Hutcheson JD, Jung SM, Schoen FJ, Kluin J, Aikawa E, Khademhosseini A. Directing valvular interstitial cell myofibroblast-like differentiation in a hybrid hydrogel platform. Adv Healthc Mater 2015;4:121—30. https://doi.org/10.1002/adhm.201400029.

[244] van der Valk DC, van der Ven CFT, Blaser MC, Grolman JM, Wu P-J, Fenton OS, Lee LH, Tibbitt MW, Andresen JL, Wen JR, Ha AH, Buffolo F, van Mil A, Bouten CVC, Body SC, Mooney DJ, Sluijter JPG, Aikawa M, Hjortnaes J, Langer R, Aikawa E. Engineering a 3D-bioprinted model of human heart valve disease using nanoindentation-based biomechanics. Nanomaterials 2018;8:296. https://doi.org/10.3390/nano8050296.

[245] Mabry KM, Lawrence RL, Anseth KS. Dynamic stiffening of poly(ethylene glycol)-based hydrogels to direct valvular interstitial cell phenotype in a three-dimensional environment. Biomaterials 2015; 49:47—56. https://doi.org/10.1016/j.biomaterials.2015.01.047.

[246] Kirschner CM, Alge DL, Gould ST, Anseth KS. Clickable, photodegradable hydrogels to dynamically modulate valvular interstitial cell phenotype. Adv Healthc Mater 2014;3:649—57. https://doi.org/10.1002/adhm.201300288.

[247] Gould ST, Darling NJ, Anseth KS. Small peptide functionalized thiol-ene hydrogels as culture substrates for understanding valvular interstitial cell activation and de novo tissue deposition. Acta Biomater 2012;8:3201—9. https://doi.org/10.1016/j.actbio.2012.05.009.

[248] Gould ST, Anseth KS. Role of cell-matrix interactions on VIC phenotype and tissue deposition in 3D PEG hydrogels. J Tissue Eng Regenerat Med 2016;10:E443—53. https://doi.org/10.1002/term.1836.

[249] Sridharan R, Cameron AR, Kelly DJ, Kearney CJ, O'Brien FJ. Biomaterial based modulation of macrophage polarization: a review and suggested design principles. Mater Today 2015;18: 313—25. https://doi.org/10.1016/j.mattod.2015.01.019.

[250] Hotaling NA, Tang L, Irvine DJ, Babensee JE. Biomaterial strategies for immunomodulation. Annu Rev Biomed Eng 2015;17. https://doi.org/10.1146/annurev-bioeng-071813-104814. annurev-bioeng-071813-104814.

[251] Matheson LA, Maksym GN, Santerre JP, Labow RS. Cyclic biaxial strain affects U937 macrophage-like morphology and enzymatic activities. J Biomed Mater Res A 2006;76:52—62. https://doi.org/10.1002/jbm.a.30448.

[252] Matheson LA, Maksym GN, Santerre JP, Labow RS. The functional response of U937 macrophage-like cells is modulated by extracellular matrix proteins and mechanical strain. Biochem Cell Biol 2006;84:763—73. https://doi.org/10.1139/o06-093.

[253] Matheson LA, Maksym GN, Santerre JP, Labow RS. Differential effects of uniaxial and biaxial strain on U937 macrophage-like cell morphology: influence of extracellular matrix type proteins. J Biomed Mater Res A 2007;81A:971—81. https://doi.org/10.1002/jbm.a.31117.

[254] Ballotta V, Driessen-Mol A, Bouten CVC, Baaijens FPT. Strain-dependent modulation of macrophage polarization within scaffolds. Biomaterials 2014;35:4919—28. https://doi.org/10.1016/j.biomaterials.2014.03.002.

[255] Battiston KG, Ouyang B, Labow RS, Simmons C a, Santerre JP. Monocyte/macrophage cytokine activity regulates vascular smooth muscle cell function within a degradable polyurethane scaffold. Acta Biomater 2014;10:1146—55. https://doi.org/10.1016/j.actbio.2013.12.022.

[256] Smits AIPM, Ballotta V, Driessen-Mol A, Bouten CVC, Baaijens FPT. Shear flow affects selective monocyte recruitment into MCP-1-loaded scaffolds. J Cell Mol Med 2014;18:2176—88. https://doi.org/10.1111/jcmm.12330.

[257] Książek AA, Mitchell KJ, Cesarovic N, Schwarzwald CC, Hoerstrup SP, Weber B, Ksiazek AA, Mitchell KJ, Cesarovic N, Schwarzwald CC, Hoerstrup SP, Weber B. PGA (polyglycolic acid)-P4HB (poly-4-hydroxybutyrate)-Based bioengineered valves in the rat aortic circulation. J Heart Valve Dis 2016;25:380—8. http://www.ncbi.nlm.nih.gov/pubmed/27989051.

[258] Usprech J, Chen WLK, Simmons CA. Heart valve regeneration: the need for systems approaches. Wiley Interdiscip Rev Syst Biol Med 2016;8:169—82. https://doi.org/10.1002/wsbm.1329.

[259] Moraes C, Likhitpanichkul M, Lam CJ, Beca BM, Sun Y, a Simmons C. Microdevice array-based identification of distinct mechanobiological response profiles in layer-specific valve interstitial cells. Integr Biol (Camb) 2013;5:673—80. https://doi.org/10.1039/c3ib20254b.

[260] Chen MB, Srigunapalan S, Wheeler AR, Simmons CA. A 3D microfluidic platform incorporating methacrylated gelatin hydrogels to study physiological cardiovascular cell-cell interactions. Lab Chip 2013;13:2591—8. https://doi.org/10.1039/c3lc00051f.

[261] Usprech J, Romero DA, Amon CH, Simmons CA. Combinatorial screening of 3D biomaterial properties that promote myofibrogenesis for mesenchymal stromal cell-based heart valve tissue engineering. Acta Biomater 2017;58:34—43. https://doi.org/10.1016/j.actbio.2017.05.044.

[262] Hjortnaes J, Goettsch C, Hutcheson JD, Camci-Unal G, Lax L, Scherer K, Body S, Schoen FJ, Kluin J, Khademhosseini A, Aikawa E. Simulation of early calcific aortic valve disease in a 3D platform: a role for myofibroblast differentiation. J Mol Cell Cardiol 2016;94:13—20. https://doi.org/10.1016/j.yjmcc.2016.03.004.

[263] Porras AM, Westlund JA, Evans AD, Masters KS. Creation of disease-inspired biomaterial environments to mimic pathological events in early calcific aortic valve disease. Proc Natl Acad Sci USA 2017. https://doi.org/10.1073/pnas.1704637115. 201704637.

[264] Boersema GSA, Utomo L, Bayon Y, Kops N, van der Harst E, Lange JF, Bastiaansen-Jenniskens YM. Monocyte subsets in blood correlate with obesity related response of macrophages to biomaterials in vitro. Biomaterials 2016;109:32—9. https://doi.org/10.1016/j.biomaterials.2016.09.009.

[265] Krawiec JT, Weinbaum JS, St Croix CM, Phillippi JA, Watkins SC, Peter Rubin J, Vorp DA. A cautionary tale for autologous vascular tissue engineering: impact of human demographics on the ability of adipose-derived mesenchymal stem cells to recruit and differentiate into smooth muscle cells. Tissue Eng 2015;21:426—37. https://doi.org/10.1089/ten.tea.2014.0208.

[266] Krawiec JT, Weinbaum JS, Liao HT, Ramaswamy AK, Pezzone DJ, Josowitz AD, D'Amore A, Rubin JP, Wagner WR, Vorp DA. In vivo functional evaluation of tissue-engineered vascular grafts fabricated using human adipose-derived stem cells from high cardiovascular risk populations. Tissue Eng 2016;22:765—75. https://doi.org/10.1089/ten.tea.2015.0379.

[267] Wang Z, Zheng W, Wu Y, Wang J, Zhang X, Wang K, Zhao Q, Kong D, Ke T, Li C. Differences in the performance of PCL-based vascular grafts as abdominal aorta substitutes in healthy and diabetic rats. Biomater Sci 2016;4:1485—92. https://doi.org/10.1039/c6bm00178e.

[268] Sicari BM, Johnson SA, Siu BF, Crapo PM, Daly KA, Jiang H, Medberry CJ, Tottey S, Turner NJ, Badylak SF. The effect of source animal age upon the in vivo remodeling characteristics of an extracellular matrix scaffold. Biomaterials 2012;33:5524—33. https://doi.org/10.1016/j.biomaterials.2012.04.017.

CHAPTER 7

Computer modeling and simulation of heart valve function and intervention

Wei Sun[1], Wenbin Mao[1], Boyce E. Griffith[2]

[1]Tissue Mechanics Laboratory, The Wallace H. Coulter Department of Biomedical Engineering, Georgia Institute of Technology and Emory University, Atlanta, GA, United States; [2]Department of Mathematics, Carolina Center for Interdisciplinary Applied Mathematics, Computational Medicine Program, and McAllister Heart Institute, University of North Carolina, Chapel Hill, NC, United States

Contents

7.1 Introduction

Computational modeling and simulation using various approaches, including structural finite element analysis (FEA), computational fluid dynamics (CFD), and fluid—structure interaction (FSI), have been very effective tools for the engineering analysis of heart valve function in health and disease [1]. Numerical simulations of tissue stresses or hemodynamics in native aortic valves [2—8], native mitral valves [9—17], and prosthetic valves [18—31] have greatly enhanced our understanding of native and artificial valve function. *In silico* analyses provide alternative approaches to quantifying the behavior

Principles of Heart Valve Engineering
ISBN 978-0-12-814661-3, https://doi.org/10.1016/B978-0-12-814661-3.00007-1

of both native and prosthetic heart valves that complement ex vivo bench testing, in vivo animal experiments, and in vivo human trials. Such models are particularly useful in prosthetic heart valve design because numerical simulations offer comparatively fast and inexpensive ways to analyze conceptual designs and design optimization [32]. In particular, computer modeling and simulation facilitate parametric testing, in which individual design parameters can be isolated and systematically investigated. Furthermore, modeling and simulation provide detailed data on device kinematics, flow patterns, and stress distributions that are not readily available in experiments. This chapter reviews the fundamental formulation of models of valve function and methods for solving the equations and then describes applications of structural and FSI modeling to native and bioprosthetic heart valves. Challenges related to constructing realistic models from clinical image data are also detailed.

7.2 Governing equations

We begin by briefly stating the fundamental equations of momentum conservation that describe FSI in heart valves. These equations will then be simplified to describe only the structural response of the valves to static or dynamic loading conditions.

In an FSI model, the computational region to be simulated includes both solid and fluid subdomains. Let Ω_t denote the computational region at time t. We identify Ω_t^f as the region occupied by the fluid (e.g., blood in models of in vivo systems, or a blood analogue such as saline or glycerol solution in vitro) at time t, and Ω_t^s as the region occupied by the structure (e.g., the valve leaflets or other cardiac structures). Let $\mathbf{x} \in \Omega_t$ denote current coordinates in the computational domain at time t, let $\mathbf{u}^i(\mathbf{x}, t)$, for $i = $ f, s, denote the velocity of the fluid ($i = $ f) or solid ($i = $ s) at position $\mathbf{x}$ and time t, and let ρ^i, for $i = $ f, s, denote the mass density of the fluid and solid, respectively. For simplicity, we assume that both the fluid and solid have uniform mass densities and are incompressible, so that $\nabla \cdot \mathbf{u}^f(\mathbf{x},t) = 0$ and $\nabla \cdot \mathbf{u}^s(\mathbf{x},t) = 0$, although it is common in engineering practice to use a *nearly* incompressible formulation both in purely structural models and also in FSI models. The fundamental equations of motion for fluid and solid are the equations of momentum balance, which, in Eulerian form, are

$$\rho^f \left(\frac{\partial \mathbf{u}}{\partial t}(\mathbf{x}, t) + \mathbf{u}(\mathbf{x}, t) \cdot \nabla \mathbf{u}(\mathbf{x}, t) \right) = \nabla \cdot \boldsymbol{\sigma}^f(\mathbf{x}, t) \text{ for } \mathbf{x} \in \Omega_t^f, \tag{7.1}$$

$$\rho^s \left(\frac{\partial \mathbf{u}}{\partial t}(\mathbf{x}, t) + \mathbf{u}(\mathbf{x}, t) \cdot \nabla \mathbf{u}(\mathbf{x}, t) \right) = \nabla \cdot \boldsymbol{\sigma}^s(\mathbf{x}, t) \text{ for } \mathbf{x} \in \Omega_t^s, \tag{7.2}$$

in which $\boldsymbol{\sigma}^f$ and $\boldsymbol{\sigma}^s$ are the Cauchy stress tensors associated with the fluid and solid, respectively. Eqns. (7.1) and (7.2) must be supplemented with matching conditions at the fluid—solid interface, $\Gamma_t^{fs} = \overline{\Omega_t^f} \cap \overline{\Omega_t^s}$, typically continuity of displacement (i.e., the no–slip condition) and continuity of traction (i.e., equal and opposite forces), along with suitable loading or tethering conditions along the remainder of the boundaries of the fluid and solid subdomains.

It remains to specify the material response of the fluid and solid. In these equations, the forms of the Cauchy stress tensors associated with the fluid and solid determine these responses. At the scale of the heart, it is common to describe the blood as a viscous incompressible fluid so that $\boldsymbol{\sigma}^f = -p^f(\mathbf{x}, t)\mathbb{I} + \mu^f\left(\boldsymbol{\nabla}\mathbf{u}(\mathbf{x}, t) + \boldsymbol{\nabla}\mathbf{u}(\mathbf{x}, t)^T\right)$, in which p^f is the hydrostatic pressure of the fluid and μ^f is the fluid viscosity. With this definition for the fluid stress along with the incompressibility constraint, the equations of motion for the fluid become the incompressible Navier–Stokes equations,

$$\rho^f\left(\frac{\partial \mathbf{u}}{\partial t}(\mathbf{x}, t) + \mathbf{u}(\mathbf{x}, t) \cdot \boldsymbol{\nabla}\mathbf{u}(\mathbf{x}, t)\right) = -\boldsymbol{\nabla}p^f(\mathbf{x}, t) + \mu^f\nabla^2\mathbf{u}(\mathbf{x}, t), \tag{7.3}$$

$$\boldsymbol{\nabla} \cdot \mathbf{u}(\mathbf{x}, t) = 0, \tag{7.4}$$

for $\mathbf{x} \in \Omega_t^f$. Typical engineering approaches to solving these equations numerically include finite element and finite volume methods, although finite difference methods are also still widely used. See Elman et al. [33], Ferziger and Peric [34], or Versteeg and Malalasekra [35] for further details on numerical methods for CFD.

Because heart valve tissues exhibit a highly nonlinear and anisotropic material response, it is appropriate to describe the biomechanics of heart valves at the macroscopic level using the framework of large deformation nonlinear elasticity. (See, e.g., Holzapfel [36] for further details on nonlinear continuum mechanics.) To do so, let $\mathbf{X} \in \Omega_0^s$ denote Lagrangian material coordinates attached to the reference configuration of the structure, and let $\boldsymbol{\chi}(\mathbf{X}, t) \in \Omega_t^s$ denote the current position of material point $\mathbf{X}$ at time t. The deformation gradient tensor associated with the motion from the reference configuration to the current configuration is $\mathbb{F} = \frac{\partial \boldsymbol{\chi}}{\partial \mathbf{X}}$, and the associated Jacobian determinant is $J = \det(\mathbb{F})$. The Cauchy stress of the solid $\boldsymbol{\sigma}^s(\mathbf{x}, t)$ can be related to the first Piola-Kirchhoff stress, $\mathbb{P}^s(\mathbf{X}, t)$, at corresponding spatial positions via $\boldsymbol{\sigma}^s = J^{-1}\mathbb{P}^s\mathbb{F}^T$. At the macroscopic level, it is common to describe heart valve leaflets as hyperelastic materials, for which $\mathbb{P}^s = \frac{\partial W}{\partial \mathbb{F}}$, in which $W = W(\mathbb{F})$ is a (local) strain energy function. Examples of specific forms of W are provided below.

In Lagrangian coordinates, the momentum equation for the solid becomes

$$\rho^s\frac{\partial^2 \boldsymbol{\chi}}{\partial t^2}(\mathbf{X}, t) = \boldsymbol{\nabla}_{\mathbf{X}} \cdot \mathbb{P}^s(\mathbf{X}, t) \text{ for } \mathbf{X} \in \Omega_0^s, \tag{7.5}$$

in which $\frac{\partial^2 \boldsymbol{\chi}}{\partial t^2}(\mathbf{X}, t) = \mathbf{u}(\boldsymbol{\chi}(\mathbf{X}, t), t)$ is the acceleration of the structure in Lagrangian (material) coordinates. For a purely structural analysis, it can suffice to ignore dynamic effects, and to consider only static or quasi-static conditions. In this case, the solid is assumed to be instantaneously at equilibrium, and the solid momentum equation simplifies to

$$\nabla_\mathbf{X} \cdot \mathbb{P}^s(\mathbf{X}, t) = \mathbf{0} \text{ for } \mathbf{X} \in \Omega_0^s, \tag{7.6}$$

which, along with suitable loading and tethering conditions, is a boundary-value problem for the structural deformation $\chi(\mathbf{X}, t)$ at "pseudo-time" t. Conventional finite element (FE) methods that are widely used in engineering practice for static or quasi-static structural analyses will solve a related weak form of Eq. (7.6),

$$\int_{\Omega_0^s} \mathbb{P}^s(\mathbf{X}, t) : \nabla_\mathbf{X} \delta \mathbf{V}(\mathbf{X}) d\mathbf{X} = 0, \text{ for all test functions } \delta \mathbf{V}(\mathbf{X}), \tag{7.7}$$

along with suitable boundary conditions, which would appear as constraints on the space of solutions or as additional surface forcing terms in Eq. (7.7). A formal method to obtain a weak form like Eq. (7.7) is to integrate Eq. (7.6) against a test function $\delta\mathbf{V}(\mathbf{X})$ and then to integrate by parts, to move the divergence off of the stress tensor $\mathbb{P}^s$ and onto the test function.

In practice, conventional FE methods assume that the solid domain Ω_0^s can be described in terms of a collection of geometrically simple elements, typically triangles and quadrilaterals in two spatial dimensions, or tetrahedra and hexahedra in three dimensions, so that integrals like that appearing in Eq. (7.5) can be treated using simple numerical quadrature rules defined at the level of individual elements. (See, e.g., Belytschko et al. [37] or Bonet and Wood [38] for additional details on FE methods for structural analyses.). Thus, in an FE analysis, it is imperative to construct a high-quality mesh that describes the geometry of the structure. Constructing meshes to describe such complex anatomical geometries is a key challenge in patient-specific modeling of heart valves and is addressed below.

7.3 Structural modeling

Structural FEA can provide a detailed quantitative stress and strain analysis of heart valves. Current imaging modalities, including three-dimensional echocardiography, magnetic resonance imaging (MRI), and, in particular, computed tomography (CT), allow for the accurate reconstruction of patient-specific anatomical structures, which, as described above, is one of the first steps in a realistic computational analysis. However, material properties and boundary conditions also affect the fidelity of computational simulations. Over the last decade, significant improvements have been made in patient-specific computational models of heart valves. In detailing structural FEA of native and bioprosthetic heart valves, we shall review the state-of-the-art techniques, including the modeling of valve geometries, material properties, and loading and boundary conditions.

7.3.1 Geometrical modeling

Three-dimensional cardiac image data are readily available in modern hospitals, but such data have been considerably underutilized in both clinical and engineering analyses of cardiac function. In clinical practice, valvular disease evaluation is routinely performed using static two dimensional images, although most imaging modalities offer time-resolved volumetric data.

Such data enable the reconstruction of the three-dimensional geometry of the heart valves at a patient-specific level. When integrated with computational models, native heart valve biomechanical function can be investigated, and preoperative planning tools can be developed. Furthermore, provided that sufficiently large numbers of such patient-specific computational models are available, patient population-based analyses can be performed to quantify uncertainty in device and treatment factors, and can be used to perform virtual clinical trials to accelerate device development. The concept of integrating three-dimensional anatomical geometries of patient heart valves into computational analysis to study the natural history of valve disease as well as patient specific conditions has been developed over the past 3 decades. However, to date, patient-specific aortic valve (AV) and mitral valve (MV) models have only been developed and reported in a relatively small number of cases. This limited patient-specific modeling can be attributed to several bottlenecks of current technologies:

1. Thin structures, such as the delicate valve leaflets (Fig. 7.1), cannot be automatically segmented using traditional methods provided by most commercial image analysis software, which are mainly based on image intensity algorithms. Thus, these structures must be manually traced, and, to some degree, estimated—even when using high-resolution CT data. The normal thickness of the leaflets is usually very small (~ 1 mm), which is smaller than two pixels in high-resolution (e.g., $0.49 \times 0.49 \times 1.25$ mm) CT images, and the shape of the aortic valve varies across

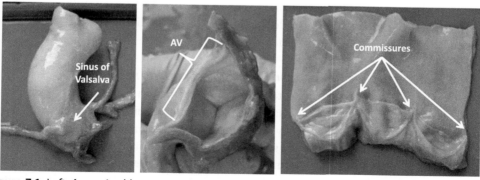

Figure 7.1 Left: An excised human aortic root. Middle: View of the ventricular side of the AV. Right: The splayed open aortic root and aortic valve.

patients. Simple image analysis algorithms (e.g., intensity thresholding) are not suitable for leaflet segmentation, and although level-set based image segmentation algorithms [39,40] can identify the surface of relatively large objects such as the surface of the aortic root, such methods often do not produce acceptable results for small, thin objects such as the valve leaflets. Moreover, the geometry that results from the initial segmentation must then be further processed (surfaces smoothed, holes filled, calcification spots segmented separately) to create a mesh suitable for FE analysis. The process is time-consuming, labor-intensive, and subject to human digitization errors and inconsistencies.

2. Mesh generation for complex patient-specific anatomical geometries, such as the valve leaflets, aortic root, ascending aorta, or left ventricle, takes many hours to perform manually. Furthermore, FE meshes generated in this fashion lack mesh correspondence, i.e., the element and nodal connectivity vary between different patients. This makes direct comparisons between different models more difficult.

3. Simulation can be computationally demanding. For example, a single transcatheter aortic valve replacement (TAVR) FE analysis typically requires days to complete even with modern high-performance computing resources, which further inhibits analysis and rapid feedback for clinical use.

7.3.1.1 Manual reconstruction of aortic valve geometries

Despite the challenges described above, image analysis software can be used to extract the anatomical geometry of the aortic valve and root. Wang et al. [41–43] developed a custom image processing code to extract the aortic and mitral valve geometries. Briefly, for the aortic root reconstruction, a series of two-dimensional cross sectional planes were generated along the longitudinal axis of the aortic root. The three-dimensional coordinates of the aortic root and coronary arteries were manually digitized and output as a point cloud. The point cloud was then imported into HyperMesh (Altair Engineering, Troy, MI), and smooth contour lines of the valve surface were generated, which were subsequently used to generate FE meshes of the aortic root. Fig. 7.2 illustrates three-dimensional valve geometries at different phases of the cardiac cycle that were reconstructed from the CT data shown in the figure.

7.3.1.2 Automatic valve estimation from clinical cardiac images

Most recent studies on automatic cardiac image segmentation and geometry reconstruction [44–50] focus on the cardiac chambers, the ventricular outflow tract, and the aorta. Studies on automatic valve geometry segmentation have been very limited. Zheng et al. [51] developed a method for aorta segmentation and landmark detection from three-dimensional C-Arm CT images, which was targeted to three-dimensional visualization of the calcified aortic valve during transapical aortic valve implantation. Pouch et al. [52] proposed a method based on intensity-based image registration to estimate aortic

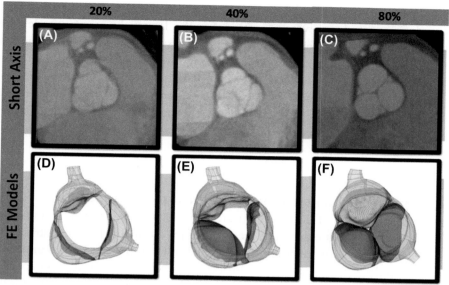

Figure 7.2 CT images of one patient at (A) 20% (fully opened), (B) 40% (half opened), and (C) 80% (closed) of the cardiac cycle from short-views; (D,E,F) Reconstructed 3D valve geometries from the CT scans shown in a,b,c). *(Adapted from Sun W, Martin C, Pham T. Computational modeling of cardiac valve function and intervention. Annu Rev Biomed Eng. 2014;16:53—76.)*

leaflet geometries from three-dimensional echocardiographic images that only requires three landmarks chosen by the user. Ionasec et al. [53] proposed a method for geometry reconstruction of the mitral-aortic complex from three-dimensional CT images. The method used more than 600 sets of aortic valve geometries that were manually segmented from three-dimensional images as the data to train their models. In the study by Liang et al. [54], a dictionary learning based image analysis method was developed to automatically reconstruct the three-dimensional geometries of the aortic valve from patient CT images. The method was evaluated by comparing the reconstructed geometries at mid-diastole from 10 patients to those manually created by human experts, and a mean discrepancy of 0.69 ± 0.13 mm was obtained.

7.3.1.3 Importance of mesh correspondence in valve geometry reconstruction

Manually constructed meshes of the same complex anatomical structures (valve leaflets, aortic root, ascending aorta) from different subjects will generally lack mesh correspondence. Specifically, meshes of the same structures from different patients will typically have different element and nodal connectivity. This is a critical limitation of manual mesh generation, because the lack of mesh correspondence makes the automated analysis of large patient population responses impractical: model set-up (i.e., assignment of material properties, loading, and boundary conditions) and data analysis must be repeated for each independent model. In the study by Liang et al. [54], patient-specific FE meshes with mesh correspondence were generated. For the 10 patient-specific models created, only the nodal coordinates and boundary conditions had to be redefined: the material property and surface definitions (for the pressure loading conditions) remained unchanged. Mesh correspondence also facilitated the specification of patient-specific boundary conditions from middle-to end-diastole, and quickly generated new meshes at different phases, which is particularly useful for model validation. Robust automatic methods for establishing mesh correspondence will streamline the FE analysis of large patient cohorts.

7.3.2 Tissue properties

7.3.2.1 Experimental characterization of valve tissue properties

The material properties of heart valve tissues are now routinely quantified using planar biaxial tensile tests conditions, which are essential for estimating the valve stress distribution. The biaxial properties of porcine and ovine leaflets have been studied under various loading-controlled or displacement-controlled protocols, as well as in layer-specific and age-dependent conditions. Many studies have used animal valves as surrogates for human tissues [56–59]. Human valve tissue material properties have been measured more recently. Martin & Sun [60] reported the biaxial properties of aged human and animal aortic valve (AV) leaflets. May-Newman & Yin [61] and Kunzelman & Cochran [62] pioneered studies on the planar biaxial mechanical response of porcine mitral valve (MV) leaflets. A few studies extracted the material properties of the human MV through ex vivo studies [63,64]. *In vivo* animal MV material properties have been investigated through animal experiments [65–67]. Recent work has also considered the right heart, including the tricuspid valve (TV) and pulmonary valve (PV) [68,69]. For instance, the mechanical properties of four human heart valves were tested and compared by Pham et al. [70]. That study not only provided a baseline quantification of aged human valves, but also offered a better understanding of the age-dependent differences among the four valves. Such experimental studies have demonstrated significant differences between the material properties of animal and aged human tissues. Notably, in all cases, the aged human tissues were much stiffer than the animal tissues. Consequently, using animal tissue properties for simulations of valve repair may reduce the clinical relevancy of the results.

Bioprosthetic heart valves (BHV) have been widely used to treat valvular disease for over 40 years. These devices offer excellent hemodynamic performance and a low risk of thrombosis. Glutaraldehyde-treated bovine pericardium (BP) or porcine pericardium (PP) are used for the leaflets of most current BHVs due to the attractive mechanical properties of these tissues. As a result, the mechanical properties of these tissues have been widely investigated and characterized through uniaxial and biaxial tensile tests [71–76]. Their out-of-plane, flexural behavior has also been studied [77,78].

7.3.2.2 Constitutive models of heart valve tissues

Constitutive models of heart valve tissues must incorporate the nonlinear and aniso-tropic responses of the real tissues. Extensive work has been devoted to developing constitutive models to describe the mechanics of the cardiac and cardiovascular tissues. Modeling approaches range from phenomenological descriptions to unit-cell models that are completely derived from a microscopic cellular structure [79].

i) **Fung-type models.** The nonlinear orthotropic model proposed by Fung [80] remains one of the most commonly used hyperelastic models for characterizing the mechanical response of valve tissues [80,81]. A typical Fung-elastic strain energy function W takes the form:

$$W = \frac{c}{2}\left[e^Q - 1\right], Q = A_1 E_{11}^2 + A_2 E_{22}^2 + 2A_3 E_{11} E_{22} + A_4 E_{12}^2 + 2A_5 E_{11} E_{22}$$
$$+ 2A_6 E_{22} E_{12}, \tag{7.8}$$

in which c and A_i, $i = 1, \dots, 6$, are material constants, and E_{ij}, $i, j = 1,2$, are in-plane components of the Green strain tensor. Although there are several variants of the Green strain-based exponential form of Eq. (7.8), all such models are typically classified as Fung-elastic models. Eq. (7.8) is often used to model planar biaxial mechanical responses of valve tissues and can be easily implemented with plane stress elements, such as shell or membrane elements [82].

ii) **Strain invariant-based fiber-reinforced hyperelastic models.** Although several different invariant-based constitutive models have been developed, a particular formulation is a version of the model introduced by Holzapfel et al. [83,84]:

$$W = C_{10}\left\{\exp[C_{01}(I_1 - 3)] - 1\right\} + \frac{k_1}{2k_2}\sum_{i=1}^{2}\left[\exp\left\{k_2[\kappa I_1 + (1 - 3\kappa)I_{4i} - 1]^2\right\} - 1\right]. \tag{7.9}$$

In this model, briefly, the valve tissues are assumed to be composed of a matrix material with two families of embedded fibers, each of which has a preferred direction. The strain invariant $I_1 = \mathrm{tr}(\mathbb{F}^T\mathbb{F})$, which is defined in terms of the deformation gradient tensor $\mathbb{F}$, describes the isotropic matrix material, and the strain invariant I_{4i}, which is the square of the stretch in the fiber direction, describes the mechanical response of the embedded fibers. C_{10} and C_{01} characterize the matrix material, and k_1 and k_2 characterize the fiber response. In addition, a dispersion parameter κ accounts for fiber angle dispersions. If $\kappa = 0$, the fibers are perfectly aligned (no dispersion). When $\kappa = 0.33$, the fibers are randomly distributed and the material becomes isotropic. Computational simulation studies using this type of model can be found for the AV [85,86] and MV [41,87,88].

7.3.2.3 Loading boundary conditions

The pressure load on the heart valve is the consequence of the blood flow, which leads to a spatial variation of pressure. However, without a full FSI analysis, realistic pressure distributions are usually not available. Therefore, in pure structural analyses, a uniform pressure field is typically applied on the leaflets. Depending on the type of the valves and the disease conditions, physiological or pathological transvalvular pressures should be used. For instance, in a quasi-static simulation of AV closure, a pressure load of 100 mmHg is usually applied on the aortic side of the leaflets. If the dynamics of the leaflet are of interest, time-varying pressure waveforms on the two sides of the valve should be applied over several cardiac cycles to reach periodic steady state. However, because viscous dissipation from the surrounding fluid is absent in purely structural models, the leaflet kinematics may exhibit excessive vibrations, especially during early systole. Therefore, damping models that account for the surrounding blood are often incorporated in the dynamic simulations [89,90]. To account for the leaflet coaptation during the valve closure, contact between leaflets must also be considered.

7.3.3 Computational structural analysis of heart valve function and intervention

In engineering practice, the primary numerical tool used for heart valve stress analysis is the finite element (FE) method. Because of the large amount of computational work on the four cardiac valves in the literature, this review focuses on work related to the aortic valve function and intervention.

7.3.3.1 Modeling native aortic valves

Early studies employing FEA to investigate the native AV were mainly to understand basic valve function. Although these studies can offer a wealth of knowledge, they are somewhat limited by simplified assumptions, such as linear elastic tissue properties [4,6] or idealized geometries [91]. In recent years, several studies have been conducted to investigate the influence of the material properties on the results from computational modeling [90,92,93]. For instance, Koch et al. [92] found that the transverse isotropy and hyperelastic material resulted in better coaptation and more uniform stress and strain distributions. Becker et al. [93] adopted a statistical method to analyze the uncertainties in loading and material properties. Notably, they found that the material properties of the leaflets were less important than the properties of the sinus and aorta in determining the leaflet stress distributions.

With the advance of clinical imaging techniques, patient-specific geometries have been utilized in FE simulations to improve the accuracy of the model predictions [94,95]. Conti et al. [94] found that anatomical differences between leaflet-sinus units could cause differences in stress and strain patterns, especially for the leaflets. Ten patient-specific AV models were reconstructed from echocardiography images and the simulated valve deformation in late diastole matched with the in vivo measurement [95].

In addition to modeling the normal state of the aortic valve, FE models have been employed to investigate the effects of pathological conditions, including dilated aortic root, bicuspid aortic valve (BAV), and calcific aortic valve. For instance, the effect of annular and/or sino-tubular junction dilatation on leaflet stress was evaluated [96]. It was found that annular dilatation is the key factor causing increased stress on aortic leaf-lets, independent from sino-tubular junction dilatation. The presence of annular dilata-tion may greatly decrease the feasibility of a valve-sparing procedure. Congenital BAV is another disease that has been studied using FE modeling [97—99]. Conti et al. [97] reported abnormal leaflet kinematics and stress distribution of a right/left fusion BAV compared to the normal valve. Higher stresses were obtained in the central basal region of the conjoint cusp (+800%). They concluded that the altered aortic wall stress distri-butions could play a role in developing the unique form of aortopathy that frequently develops with BAV. Increased ascending aortic wall stress in patients with BAV was also found in another patient-specific study [99], in which 20 patients with BAV were compared to 20 patients with tricuspid aortic valves and matched aortic diameters. Calcific aortic valve disease is a progressive pathology characterized by calcification within the AV cusps. FE models were also developed to investigate the progression of calcification [100,101]. Arzani & Mofrad [100] developed a method to model the evo-lution of calcification based on the mechanical strain on the aortic side of the valve. They found that the radial expansion of calcification starting from the attachment region was in agreement with the reported clinical data (Fig. 7.3).

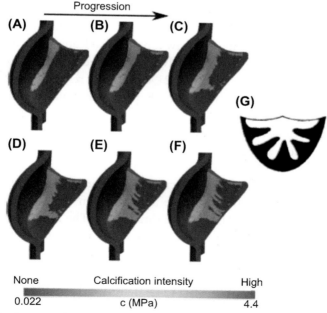

Figure 7.3 Calcification growth at different snapshots (A) — (F) from simulations. The color bar rep-resents the calcification intensity. (G) One of the two calcification growth patterns observed in most of the patients. (Adapted from Arzani A, Mofrad MR. A strain-based finite element model for calcification progression in aortic valves. J Biomech. 2017;65:216—220.)

7.3.3.2 Modeling bioprosthetic heart valves

Because the preferred treatment for many AV diseases is the replacement of the diseased valve by a prosthesis, many computational studies have focused on prosthetic valve function and design. Current prosthetic valves are divided into mechanical heart valves and BHVs. The former is made of highly durable pyrolytic carbon and metal alloys, whereas the latter is fabricated from biological soft tissues. BHVs are often preferred in clinical practice because of their excellent hemodynamic characteristics, which, unlike mechanical valves, do not require the use of sustained anticoagulation therapy. Considering the prevalence and continuing development of BHVs, we only review computational studies relevant to BHVs.

Computational studies have shown that leaflet stress distributions can be modulated by changing BHV design through varying manufacturing techniques [102], leaflet shapes [103–105], and frame mounting methods [20,106–108]. Auricchio et al. [104] simulated a patient-specific stentless aortic valve implantation using FEA. Their results indicated that both the valve size and the anatomical asymmetry of the sinuses of Valsalva affect prosthesis placement. Xiong et al. [105] utilized FEA to investigate the effect of leaflet geometries on the dynamic behavior, leaflet coaptation, and stress distribution on the stentless pericardial aortic valve. Two leaflet designs were considered: a simple tubular valve and a complex three-dimensional molded valve. It was found that the molded leaflet design, which resembled the native valve, was superior to the tubular valve in all aspects.

Aortic valve reconstruction using leaflet grafts made from autologous pericardium is an effective surgical treatment for some forms of aortic regurgitation. The procedure is underutilized because of the difficulty of sizing grafts to effectively seal with the native leaflets. Hammer et al. [109] explored how a pericardial leaflet graft of various sizes interacts with two native leaflets when the valve is closed and loaded. They found that graft width and height must both be increased 21% and 27%, respectively, compared to the native leaflet to achieve proper valve closure. Experimental validation in excised porcine aortas confirmed the results of simulations.

With the advent of tissue engineering and 3D printing techniques, synthetically manufactured valves, such as polymeric heart valve and tissue-engineered heart valve (TEHV), have been extensively studied [110–113] because of their biocompatibility, hemocompatibility, and resistance to calcification. Loerakker et al. [111] investigated the influence of valve geometry and tissue anisotropy on the deformation profile and closed configuration of TEHVs through FE simulations. The results suggested that valve geometry and tissue anisotropy are both important to maximize the radial strains and thereby the coaptation area. Loosdregt et al. [112] used a combined experimental and computational approach to analyze the balance of cell-mediated leaflet shortening and the stress imposed on the leaflets of TEHVs. Serrani et al. [113] studied the optimization of the material microstructure in a polymeric heart valve. They found that the optimal microstructure was characterized by the circumferential orientation of the cylinders within the valve leaflet, similar to the native valve orientation.

Transcatheter aortic valve (TAV) replacement, or TAVR, has emerged as a less-invasive treatment of aortic stenosis in which a bioprosthetic valve is implanted via a catheter within the diseased AV. Since the first-in-human implantation by Cribier and colleagues [114] in 2002, there has been explosive growth in its use throughout the world. This revolutionary therapy, which involves less surgical trauma and affords shorter recovery times, was recently approved by US Food and Drug Administration (FDA) to treat intermediate-risk patients [115].

FE analysis has been used to evaluate the structural performance of TAV leaflets and stents. Li & Sun [116] simulated the deformation of a pericardial bioprosthetic valve under quasi-static loading conditions to examine the effects of different material properties and tissue thickness on the leaflet stress distribution. Their results showed that, as expected, leaflet stresses decrease with increasing tissue thickness, and that under the same loading and boundary conditions, bovine pericardial leaflets have a lower peak stress than porcine pericardial leaflets. In a comparable study, Smuts et al. [117] investigated the influence the leaflet material and orientation on pericardial bioprosthetic valve function. It was concluded that kangaroo pericardium is superior to bovine pericardium for TAV applications because of its smaller thickness and greater extensibility. In both studies [116,117], it was concluded that the tissue fiber orientation should be aligned with the circumferential direction of the valve. Sun et al. [118] used a similar approach to investigate the impact of elliptical TAV deployment due to severe aortic calcification on valve function. They found that the distorted elliptical TAV configuration induced elevated leaflet stresses and central aortic regurgitation (Fig. 7.4), which has since been confirmed by in vitro experiments [119–121]. The impact of incomplete TAV expansion on leaflet stress and strain distributions was also investigated by Abbasi et al. [122]. They found that 2–3 mm incomplete TAV stent expansion induced localized high stress regions within the commissure region, while 4–5 mm incomplete expansion induced localized high stress within the belly of the TAV leaflets. The results from these studies have important implications for TAV design.

Because the TAV leaflets are generally thinner than their surgical counterparts, the durability of TAV devices may be limited. FE simulations have been developed to investigate the device durability [107,123]. Martin & Sun [123] developed a computational framework to assess leaflet fatigue under cyclic loading by incorporating a permanent set term in the material model. They found that the TAV leaflets sustained higher stresses, strains, and fatigue damage compared to the surgical aortic valve (SAV) leaflets (Fig. 7.5). The simulation results suggested that the durability of TAVs is about 7.8 years. Besides the durability of leaflets, the stent-frame fatigue response was studied by Petrini et al. [124]. The specific failure location of Nitinol stent was correctly recognized in the simulations, which is in agreement with experimental tests. They found that the size and stiffness of the surrounding wall had a strong influence on the fatigue response of the stent.

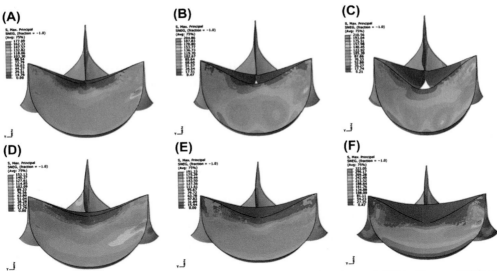

Figure 7.4 Maximum principal stress distribution of fully loaded asymmetric TAV with eccentricity of (A) 0.3, (B) 0.5, and (C) 0.68 in scenario 1 (the major axis of an elliptical TAV was aligned with one of the leaflet coaptation lines), and scenario 2 (the major axis was perpendicular to the coaptation line) with eccentricity of (D) 0.3, (E) 0.5, and (F) 0.68. *(Adapted from Sun W, Li K, Sirois E. Simulated elliptical bioprosthetic valve deformation: implications for asymmetric transcatheter valve deployment. J Biomech. 2010;43(16):3085–3090.)*

Using computational methods, *in vivo* TAV device interactions with human tissues can be predicted. To quantify TAV device-tissue interaction forces, several FE models have been developed. The first approach involves an inverse analysis of the deployed stent shape to determine the radial forces. Gessat et al. [125] developed a spline-based method to reconstruct deformed CoreValve stent geometry from post-TAVR CT images. Local stent deformations were determined by comparing the nominal and deformed stent geometries and were described by a set of displacements vectors. These displacement boundary conditions were then applied to an undeformed CoreValve FE model using an iterative relaxation algorithm until the deformed stent geometry was achieved [126], and the radial components of the stent-tissue interaction force could be extracted. Hopf et al. [126] have also used a similar approach to determine patient-specific contact forces between a CoreValve implant and the surrounding tissue. Although these studies provide meaningful insight for improved stent design, it is difficult to determine the possible mechanisms for altered stent deformation when the deployment process and surrounding tissue are not included in the analysis. Through TAV deployment simulations in idealized LVOT geometries, Tzamtzis et al. [127] have shown that the host tissue geometry and stiffness may impact the radial contact force between the stent and tissue.

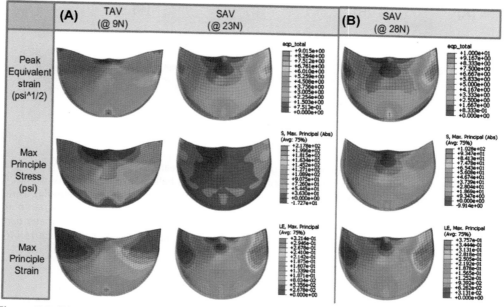

Figure 7.5 (A) Contours of the peak equivalent strain, maximum principal stress, and maximum principal strain of the TAV at the 9N state and the SAV at the 23N state showing similar patterns and peak values. (B) Contours of the peak equivalent strain, maximum principal stress, and maximum principal strain of the SAV at failure (28N state). *(Adapted from Martin C, Sun W. Comparison of transcatheter aortic valve and surgical bioprosthetic valve durability: a fatigue simulation study. J Biomech. 2015;48(12):3026–3034.)*

There has been a recent push in the field to develop computational methods for the patient-specific analysis of TAV deployment to determine the precise effects of the aortic geometry and tissue properties on clinical outcomes. Capelli et al. [128] and Wang et al. [42] were among the first groups to publish patient-specific FE analyses of TAV deployment. These analyses showed that patient-specific geometry induces asymmetric stress distributions on the device and tissue [42,128], and impacts a particular patient's risk of paravalvular leak (PVL), coronary occlusion, and rupture (Fig. 7.6) [42,129]. Auricchio et al. [130], Gunning et al. [131], and Morganti et al. [132] have since shown that TAV leaflets may become distorted when the device is deployed within a realistic rather than an idealized and circular aortic root geometry; thus, patient-specific analyses are important for predicting device function in addition to device-tissue interactions. Device positioning to align the TAV commissures with the native aortic valve commissures may also be important for reducing TAV leaflet stresses and consequently achieving a durable result [131]. More sophisticated FE models, which include the aortic valve calcification along with the aortic root and valve leaflets [85,126,128,132–134], have demonstrated that the extent and location of

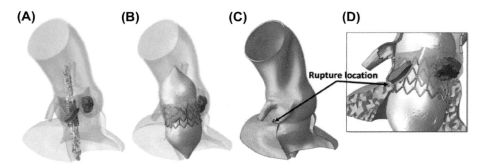

Figure 7.6 (A) Pre- and (B) postdeployment geometries of a 94-year-old patient's aorta, including the entire aortic root with coronary arteries, calcified leaflets, and a balloon-expandable TAV device. (C) Full and (D) local views of the deformed aortic root and balloon deployment indicates annulus tearing under the left coronary ostium due to the dislodgement of calcification into the vulnerable part of the aortic sinus. *(Adapted from Wang Q, Kodali S, Primiano C, Sun W. Simulations of transcatheter aortic valve implantation: implications for aortic root rupture. Biomechanics Model Mechanobiol. 2015;14(1): 29–38.)*

calcium deposits significantly affect the stent expansion process and a patient's risk of aortic root rupture during deployment [134]. Simulation results for an exemplary retrospective case indicated that SAPIEN TAV deployment would induce aortic root rupture below the left main coronary artery due to the stent pushing a large calcium deposit into the left coronary sinus during expansion (Fig. 7.5). Rupture in this location was also observed in the actual clinical procedure for the same patient. Sturla et al. [135] investigated the impact of different aortic valve calcification patterns on the outcome of TAVR. Numerical results highlighted the dependency on the specific calcification pattern of the "dog-bone" of the stent. High stresses acting on calcium deposits may be a risk factor for embolization of calcific material.

Recently, computational models have been applied to new-generation TAV device with the improvement on the modeling accuracy, including precise three-dimensional geometry of TAV through micro CT scanning [136,137], TAV crimping [138,139], delivery apparatus [140], comparison against postprocedural clinical data [141,142], and population-specific response based on multiple patients data [141–143]. Bosi et al. [141] studied the TAVR with SAPIEN XT devices in 14 patients. The implantation site material parameters were adjusted using a design of experiments approach to minimize the error between computational prediction and clinical results. Bailey et al. [140] simulated SAPIEN XT deployment into an patient-specific aortic root model with different orientations. It was found that the preferred orientation is for the prosthetic leaflets to be aligned with the native leaflets. Hopf et al. [143] performed FEA to extract the contact forces between implanted Nitinol stent and the host tissue from 46 patients underwent TAVR procedure with the Medtronic CoreValve. A comparison of systolic and diastolic data revealed slightly higher contact forces during diastole. By analyzing clinical outcomes, they found that an increase in mean force is associated with higher incidents of paravalvular leakage and the need for a permanent pacemaker. Bianchi et al. [144] evaluated the effect of various TAV deployment locations on the procedural outcome by assessing the risk for valve migration in a patient with Edwards SAPIEN valve. They found proximal deployment led to lower contact area, which poses an increased risk for valve migration. Simulations of TAV crimping, with and without models of the bioprosthetic leaflets, were performed, and larger radial deformations were observed in stent without leaflets, but the difference was relatively small. It was also demonstrated by Bailey et al. [139] that including leaflets in the TAVR simulation has only a marginal impact on the results.

7.4 Fluid—structure interaction

It is generally appreciated that many of the difficulties of prosthetic heart valves are directly related to the fluid dynamics of the replacement valve [145,146]. Although structural analyses provide substantial information about stress distributions in valve leaflets, as mentioned above, it is challenging to define realistic pressure distributions in purely structural simulations. Because heart valve leaflets are thin structures that move with the blood but, at the same time, impart forces on the blood that alter its motion, describing either the fluid dynamics of the valve, or the dynamic structural kinematics of the valve leaflets, motivates the use FSI analyses. Two widely used numerical approaches to FSI include arbitrary Lagrangian-Eulerian (ALE) methods [147] and immersed boundary (IB) methods [148,149]. ALE methods directly approximate Eqs. (7.1) and (7.2), which separately describe the fluid and solid subdomains of the overall computational domain. Such methods are well suited to FSI problems involving relatively modest deformations, but mesh regeneration can be challenging for models involving large structural deformations, which are characteristic of simulations of heart valve dynamics. Furthermore, because ALE methods use separate, geometrically conforming meshes for the fluid and solid subdomains, it can be challenging to use them for applications that involve transient contact or near-contact between structures, which is another characteristic of models of heart valves.

The IB method uses a different approach. Instead of constructing and maintaining geometrically conforming meshes for the fluid and solid subdomains, it treats the case in which the solid subdomain is fully immersed in the fluid subdomain. This allows the IB method to employ a relatively simple description of the fluid and, in particular, allows it to avoid the use of bodyconforming meshes for the fluid and solid. Ultimately, IB and ALE methods use the same fundamental equations of momentum conservation, but the IB formulation facilitates implementations that are suited for models with large structural deformations, thin elastic structures, and contact or near-contact between structures.

Peskin introduced the IB method in the 1970s to simulate heart valve dynamics [150–152]. Although this simulation methodology enabled some of the earliest FSI models of the heart and its valves [152–156], a limitation of the original implementations of the IB formulation was that they used relatively simple structural models, in which the immersed structure was described as a system of springs and beams immersed in fluid. This limited the ability of classical IB models to use realistic descriptions of the biomechanics of the immersed structures. Beginning in the 2000s, modern versions of the IB method [157–161] were developed that allowed for the use of structural models based on large deformation nonlinear mechanics. Such schemes use FE descriptions of the structural response, so that structural formulations like those detailed above may be used to model the valve leaflets. More recent extensions of the IB method, such as methods based on isogeometric analysis, have also seen applications to simulating heart valve dynamics [162–164].

To briefly state an IB formulation of the equations of motion that is suitable for using nonlinear continuum mechanics descriptions of the valve leaflets or other cardiac structures, we assume for simplicity that the overall computational domain is independent of time, so that $\Omega = \Omega_t^f \cup \Omega_t^s$, and that the fluid and solid mass densities are equal, $\rho = \rho^f = \rho^s$. Neither of these assumptions is fundamental, and both can be relaxed. The equations of motion become

$$\rho\left(\frac{\partial \mathbf{u}}{\partial t}(\mathbf{x}, t) + \mathbf{u}(\mathbf{x}, t)\cdot\nabla\mathbf{u}(\mathbf{x}, t)\right) = -\nabla p(\mathbf{x}, t) + \mu\nabla^2\mathbf{u}(\mathbf{x}, t) + \mathbf{f}(\mathbf{x}, t) + \mathbf{t}(\mathbf{x}, t),$$

(7.10)

$$\nabla \cdot \mathbf{u}(\mathbf{x}, t) = 0,$$

(7.11)

$$\mathbf{f}(\mathbf{x}, t) = \int_{\Omega_0^s} \nabla\cdot\mathbb{P}^s(\mathbf{X}, t)\delta(\mathbf{x} - \boldsymbol{\chi}(\mathbf{X}, t))\mathrm{d}\mathbf{X},$$

(7.12)

$$\mathbf{t}(\mathbf{x}, t) = \int_{\partial\Omega_0^s} -\mathbb{P}^s(\mathbf{X}, t)\mathbf{N}(\mathbf{X})\delta(\mathbf{x} - \boldsymbol{\chi}(\mathbf{X}, t))\mathrm{d}A,$$

(7.13)

$$\frac{\partial\boldsymbol{\chi}}{\partial t}(\mathbf{X}, t) = \int_{\Omega} \mathbf{u}(\mathbf{x}, t)\delta(\mathbf{x} - \boldsymbol{\chi}(\mathbf{X}, t))\mathrm{d}\mathbf{x} = \mathbf{u}(\boldsymbol{\chi}(\mathbf{X}, t), t),$$

(7.14)

in which Eqns. (7.10)–(7.13) hold for all $\mathbf{x} \in \Omega$, and Eq. (7.14) holds for all $\mathbf{X}\in\Omega_0^s$. Here, $\delta(\mathbf{x})$ is the three-dimensional Dirac delta function and $\mathbf{N}(\mathbf{X})$ is the exterior unit normal to $\partial\Omega_0^s$, the boundary of the reference configuration of the solid subdomain. In this formulation, $\mathbf{u}(\mathbf{x}, t)$ is the velocity of whatever material happens to be at position $\mathbf{x}$ at time t. In particular, it describes the velocity of the fluid for $\mathbf{x}\in\Omega_t^f$, and it describes the velocity of the solid for $\mathbf{x}\in\Omega_t^f$. Because of the presence of viscosity in Eq. (7.10), the velocity field $\mathbf{u}(\mathbf{x}, t)$ is continuous on the computational domain Ω.

Consequently, the IB formulation of FSI automatically satisfies the no-slip condition at along the fluid-solid interface, Γ_t^{fs}. Furthermore, in the formulation (7.5)–(7.14), the resulting total Cauchy stress tensor of the fluid-solid system takes the form

$$\sigma(\mathbf{x}, t) = \sigma^f(\mathbf{x}, t) + \begin{cases} \sigma^s(\mathbf{x}, t) & \text{for } \mathbf{x} \in \Omega_t^s, \\ 0 & \text{otherwise,} \end{cases} \tag{7.15}$$

so that the solid is treated as a viscoelastic body rather than a purely elastic body. In practice, the viscous stresses are relatively small within the structure and are neglected. Traction continuity is maintained at the fluid-solid interface because the traction-like force $\mathbf{t}(\mathbf{x}, t)$ that is applied along Γ_t^{fs} in the momentum equation implies equal-and-opposite traction discontinuities in σ^f and σ^s at the interface [161,165]. Thus, although σ^f and σ^s are discontinuous on the full computational domain Ω, the total stress, σ, defined in Eq. (7.15) satisfies traction continuity.

7.4.1 FSI models of heart valve dynamics

FSI models can predict the three-dimensional flow patterns and leaflet kinematics of native valves [31,166–173] as well as mechanical heart valves [174–182], bioprosthetic valves [162–164,183–185], and polymeric valves [170,186,187]. Despite this substantial body of work, however, most experimentally validated FSI valve models consider MHVs. For instance, Nobili et al. [188] report validation studies with a St. Jude bileaflet MHV using high-speed cinematography, and Dasi et al. [175], Ge at al. [189], Guivier-Curien et al. [190], and Jun et al. [191] report validation studies of bileaflet valves using particle image velocimetry (PIV) [192–194]. Fewer studies have validated flexible valve models. Quaini et al. [195] validated a simplified elastic aperture model of a flexible heart valve using Doppler ultrasound. Recently, Wu et al. [185] compared leaflet kinematics and valve open areas for an in vitro TAVR model, but they do not provide detailed comparisons for hemodynamic parameters (pressures and flow rates) or comparisons to detailed flow patterns. A major challenge for FSI models of bioprosthetic valves remains in their careful validation against experimental data, but such validation studies are crucial for enabling the full impact of these models on the design of devices or clinical interventions.

7.4.2 In vitro models

In vitro experimental models of heart valve dynamics are useful in valve design. They also provide a controlled environment that is ideal for validating the predictions of computational models. Pulse duplicators are experimental systems that can be used to design prosthetic heart valves, validate valve performance, and study valve dysfunction. Typically, these systems model only a portion of the cardiac circulation. Griffith, Scotten, and co-workers developed an IB model of the aortic test section of a ViVitro Systems Inc. (VSI) pulse duplicator (ViVitro Systems Inc., Victoria, BC, Canada) [196], which is closely related to the commercial pulse duplicator system produced by ViVitro Labs Inc. Both ViVitro pulse duplicator systems provide similar experimental models of the left heart, including a rigid left atrium (LA), a flexible left ventricle (LV), and a rigid left ventricular outflow tract (LVOT), aortic root, and ascending aorta. Valves are inserted between the model LA and LV, and in the model LVOT. A programmable piston pump modulates the left ventricular chamber volume to drive cardiac output, and afterload is provided by a system of compliances and resistances that is analogous to a classical Windkessel circulation model. Initial computer simulations of the AV test section considered FSI models of mechanical heart valve prostheses [196]. More recent extensions of this model have simulated the performance of porcine tissue BHVs as well as pericardial BHVs. These simulations use fiber-reinforced hyperelastic constitutive models to describe the biomechanics of the flexible BHV leaflets. Figs. 7.7 and 7.8 shows renderings of results from a representative computational model. In this simulation, flow is driven by experimental pressure waveforms obtained from the VSI pulse duplicator using a glutaraldehyde-fixed porcine BHV, and downstream loading is provided by a three-element Windkessel (RCR circuit-type) model [197] fit to experimental pressure and flow recordings from the VSI system. The fiber-reinforced constitutive model of Driessen et al. [198] is used to describe the flexible leaflets of the porcine BHV. This model uses a neo-Hookean description of the leaflet matrix along with a fiber-aligned stress with an exponential length-tension relationship for the collagen fibers distributed within the bioprosthetic valve leaflet. The constitutive model parameters, including fiber angle distributions, are based on experimental

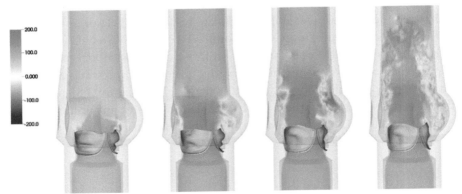

Figure 7.7 Three-dimensional fluid—structure interaction (FSI) simulation of the opening dynamics of a porcine bioprosthetic heart valve in the aortic test section of a ViVitro Systems Inc. (VSI) pulse duplicator using an immersed boundary (IB) method with a nonlinear finite element (FE) model of the leaflet biomechanics. The flow direction is from bottom to top, with red indicating forward flow and blue reverse flow. Flow is driven by experimental pressure waveforms obtained from the ViVitro testing device using a glutaraldehyde-fixed porcine bioprosthetic heart valve. Downstream loading is provided by a three-element Windkessel model [197] fit to experimental data obtained from the pulse duplicator.

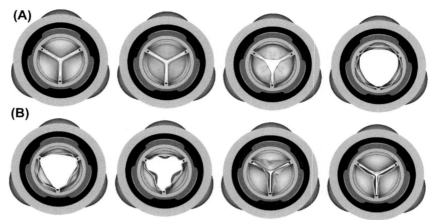

Figure 7.8 Leaflet kinematics, as viewed from the distal end of the aortic test section, during (A) opening and (B) closure for the simulation shown in Fig. 7.7.

studies of Billiar and Sacks [59,199] and correspond to porcine heart valve tissue that has been chemically fixed using glutaraldehyde. The detailed leaflet fiber structure is not available, and so it must be modeled. These simulations use a model valve fiber architecture generated by an approach [200] based on Poisson interpolation (also known as harmonic interpolation) [201–203]. Fig. 7.9 compares experimental and simulated pressure and flow data, which are in excellent agreement. The flow rates and leaflet kinematics are not prescribed in the computational model. Nonetheless, stroke volumes agree to 1%. Furthermore, the closed valve opens against a realistic pressure load, and closes with a small but realistic amount of regurgitant flow. To date, however, only the bulk flow properties

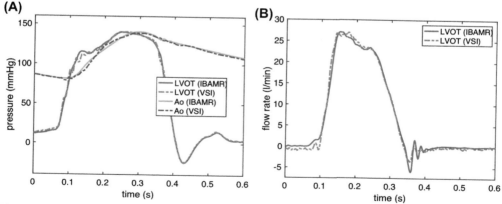

Figure 7.9 Comparisons between simulated (*IBAMR*, solid lines) and experimental (*VSI*, dashed lines) (A) pressures and (B) flow rates for the porcine bioprosthetic valve model in a ViVitro Systems Inc. (VSI) pulse duplicator. Panel (A) shows that the closed valve supports a realistic pressure load, and panel (B) shows that it does so without leakage. Stroke volumes agree to 1%. See also Figs. 7.7 and 7.8.

have been compared between the experimental and computational models. Current research aims to perform more detailed comparisons of the experimental and computational flow patterns using PIV to obtain quantitative flow mappings.

7.4.3 Subject-specific models

As in structural mechanics models, numerical simulations of FSI can predict device performance in the in vivo environment. All of the challenges of patient-specific structural models are also present in patient-specific FSI models. These include specifying the anatomy, constructing meshes, assigning material properties, and determining suitable driving, loading, and tethering conditions. Fig. 7.10 shows results from a subject-specific simulation of FSI in the aortic root and ascending aorta [200]. As in the in vitro model described above, in this simulation, flow is driven by an upstream pressure waveform, and a Windkessel model provides downstream loading [197]. In this case, however, the models that provide upstream and downstream driving and loading conditions were derived from clinical data from healthy patients [204]. Figs. 7.11–7.13 provide details of leaflet kinematics, fiber strains, and von Mises stresses using this FSI model with biomechanical models of native and bioprosthetic aortic valves. In these simulations, the native leaflets experience the largest deformations in diastole, whereas the glutaraldehyde-fixed porcine leaflets with 4 mmHg fixation pressure deform the least. The corresponding fiber strains are in agreement with these findings: the native leaflets experience the largest fiber stretches and the glutaraldehyde-fixed porcine leaflets with 4 mmHg fixation pressure; see the smallest fiber stretches. However, the differences between the different types of valves become substantially less pronounced as the valves open. The native valve leaflets also experience slightly larger stresses than the glutaraldehyde-fixed leaflets, and the BHV leaflets at 0 mmHg fixation pressure experience slightly larger stresses that the BHV leaflets at 4 mmHg fixation pressure. These results are in good qualitative agreement with purely structural simulations performed

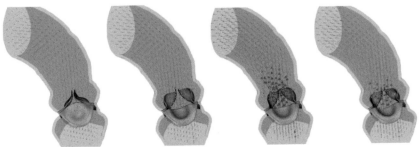

Figure 7.10 Aortic valve opening dynamics in an immersed boundary (IB) simulation of fluid–structure interaction (FSI) in a subject-specific model of the aortic root and ascending aorta. The panels show the leaflet kinematics and fluid dynamics. Peak flow velocities during early systole, highlighted in red, are approximately 2 m/s. (*Adapted from Hasan A, Kolahdouz EM, Enquobahrie A, Caranasos TG, Vavalle JP, Griffith BE. Image-based immersed boundary model of the aortic root. Med Eng Phys 2017;47:72–84.*)

Figure 7.11 Leaflet displacements (cm) obtained during late diastole and early systole, using the subject-specific FSI model shown in Fig. 7.10 along with aortic valve leaflet constitutive parameters for (A) glutaraldehyde-fixed porcine leaflets with 4 mmHg fixation pressure, (B) glutaraldehyde-fixed porcine leaflets with 0 mmHg fixation pressure, and (C) fresh porcine leaflets. The glutaraldehyde-fixed porcine leaflets with 4 mmHg fixation pressure deform the least in diastole, and that the fresh leaflets experience the largest deformations. (*Adapted from Hasan A, Kolahdouz EM, Enquobahrie A, Caranasos TG, Vavalle JP, Griffith BE. Image-based immersed boundary model of the aortic root. Med Eng Phys 2017;47:72—84.*)

Figure 7.12 Similar to Fig. 7.11, but here showing the fiber stretch for the glutaraldehyde-fixed porcine leaflets with 4 mmHg fixation pressure. (*Adapted from Hasan A, Kolahdouz EM, Enquobahrie A, Caranasos TG, Vavalle JP, Griffith BE. Image-based immersed boundary model of the aortic root. Med Eng Phys 2017;47:72—84.*)

Figure 7.13 Similar to Figs. 7.11 and 7.12, but here showing the von Mises stresses (kPa) for the glutaraldehyde-fixed porcine leaflets with 4 mmHg fixation pressure. (*Adapted from Hasan A, Kolahdouz EM, Enquobahrie A, Caranasos TG, Vavalle JP, Griffith BE. Image-based immersed boundary model of the aortic root. Med Eng Phys 2017;47:72—84.*)

by Driessen et al. [198] using the same biomechanics models, but the FSI simulations provide additional information about leaflet kinematics during opening and closing that are not available in quasi-static models.

7.5 Conclusions and future outlook

Computer modeling and simulation are already widely used in the design of medical devices, and advanced modeling techniques, such as FSI analyses, are providing increasingly complete descriptions of heart valve dynamics. Key challenges for modeling and simulation remain in the rigorous validation of the model results, especially for dynamic FSI models of heart valves, and the acceptance of results from such models by regulatory agencies such as the US Food and Drug Administration (FDA). The FDA has a long-standing interest in understanding how to interpret the results from computational fluid dynamics and FSI analyses [205–208], and we believe that certain predictions from these types of models will ultimately serve as evidence in regulatory submissions. Models also can help to make clinical treatment decisions but likewise require very rigorous experimental validation before such applications become a reality.

Designing the next generation of devices for transcatheter mitral valve (TMV) replacement [209] also motivates the use of more comprehensive computational models. For instance, because of the complex and dynamic anatomical geometry of the mitral apparatus, useful modeling platforms for TMV device design may require coupled descriptions of the structural mechanics and fluid dynamics of at least the left heart. For these same reasons, computer models may play a larger role in TMV device selection than they do in current TAV clinical practice. The development of comprehensive whole-heart FSI models that include realistic descriptions of the biomechanics of the heart and its valves is an active area of research [210].

Finally, it is well-known that BHV durability remains a major challenge, and an important emerging computational tool for heart valve design that was not described in detail in this chapter is the description of leaflet degradation and failure. Models of leaflet damage accumulation are being actively developed [123], but computer models cannot perform direct numerical simulations of the full durable lifetime of a BHV, nor can current models realistically predict the material fracture processes that lead to most BHV failure events. Because of the immense computational demands of such models, mathematical techniques for bridging the fast time scales of the cardiac cycle and the slow time scales of damage acquisition are clearly needed to make such simulations tractable. Emerging heart valve models incorporate descriptions of biological mechanisms at both the macroscale and the microscale levels [211] and promise to provide a more mechanistic model of leaflet function, especially dysfunction, and may ultimately play a key role in the design of the engineered heart valves of the future.

Acknowledgments

W.S. acknowledges support from NIH awards HL104080, HL127570, and HL142036. B.E.G. acknowledges support from NIH award HL117063 and NSF awards OAC 1450327 and OAC 1652541. B.E.G. also acknowledges assistance from Ebrahim Kolahdouz, Lawrence Scotten, and Benjamin Vadala-Roth in developing the geometrical models for the VSI pulse duplicator.

References

[1] Humphrey JD. Cardiovascular solid mechanics : cells, tissues, and organs. New York: Springer; 2002. xvi, 757 pp.

[2] Grande KJ, Cochran RP, Reinhall PG, Kunzelman KS. Stress variations in the human aortic root and valve: the role of anatomic asymmetry. Ann Biomed Eng 1998;26(4):534−45. PubMed PMID: 9662146.

[3] Grande KJ, Cochran RP, Reinhall PG, Kunzelman KS. Mechanisms of aortic valve incompetence in aging: a finite element model. J Heart Valve Dis 1999;8(2):149−56. PubMed PMID: 10224573.

[4] Grande KJ, Cochran RP, Reinhall PG, Kunzelman KS. Mechanisms of aortic valve incompetence: finite element modeling of aortic root dilatation. Ann Thorac Surg 2000;69(6):1851−7. PubMed PMID: 10892936.

[5] Grande-Allen KJ, Cochran RP, Reinhall PG, Kunzelman KS. Re-creation of sinuses is important for sparing the aortic valve: a finite element study. J Thorac Cardiovasc Surg 2000;119(4 Pt 1):753−63. PubMed PMID: 10733765.

[6] Grande-Allen KJ, Cochran RP, Reinhall PG, Kunzelman KS. Mechanisms of aortic valve incompetence: finite-element modeling of Marfan syndrome. J Thorac Cardiovasc Surg 2001;122(5):946−54. PubMed PMID: 11689800.

[7] Grande-Allen KJ, Cochran RP, Reinhall PG, Kunzelman KS. Finite-element analysis of aortic valve-sparing: influence of graft shape and stiffness. IEEE Trans Biomed Eng 2001;48(6):647−59. PubMed PMID: 11396595.

[8] Gould PL, Cataloglu A, Chattopadyhay A, Dhatt G, Clark RE. Stress analysis of the human aortic valve. Comput Struct 1973;3:377.

[9] Kunzelman KS, Cochran RP, Chuong C, Ring WS, Verrier ED, Eberhart RD. Finite element analysis of the mitral valve. J Heart Valve Dis 1993;2(3):326−40. PubMed PMID: 8269128.

[10] Reimink MS, Kunzelman KS, Verrier ED, Cochran RP. The effect of anterior chordal replacement on mitral valve function and stresses. A finite element study. Am Soc Artif Intern Organs J 1995; 41(3).

[11] Kunzelman KS, Reimink MS, Cochran RP. Annular dilatation increases stress in the mitral valve and delays coaptation: a finite element computer model. Cardiovasc Surg 1997;5(4):427−34.

[12] Kunzelman KS, Quick DW, Cochran RP. Altered collagen concentration in mitral valve leaflets: biochemical and finite element analysis. Ann Thorac Surg 1998;66(6 Suppl. l):S198−205. PubMed PMID: 9930448.

[13] Kunzelman KS, Reimink MS, Cochran RP. Flexible versus rigid ring annuloplasty for mitral valve annular dilatation: a finite element model. J Heart Valve Dis 1998;7(1):108−16.

[14] Kunzelman KS, Cochran RP. Hemi-homograft replacement of the mitral valve: a finite element model. Stentless Bioprosthesis 1999:31−8.

[15] Einstein DR, Kunzelman KS, Reinhall PG, Nicosia MA, Cochran RP. Haemodynamic determinants of the mitral valve closure sound: a finite element study. Med Biol Eng Comput 2004;42(6): 832−46.

[16] Einstein DR, Kunzelman KS, Reinhall PG, Nicosia MA, Cochran RP. Non-linear fluid-coupled computational model of the mitral valve. J Heart Valve Dis 2005;14(3):376−85.

[17] Kunzelman KS, Einstein DR, Cochran RP. Fluid-structure interaction models of the mitral valve: function in normal and pathological states. Phil Trans Biol Sci 2007;1484(362):1393−406.

[18] Black MM, Howard IC, Huang XC, Patterson EA. A three-dimensional analysis of a bioprosthetic heart valve. J Biomech 1991;24:793–801.

[19] Burriesci G, Howard IC, Patterson EA. Influence of anisotropy on the mechanical behaviour of bioprosthetic heart valves. J Med Eng Technol 1999;23(6):203–15.

[20] Cacciola G, Peters GW, Baaijens FP. A synthetic fiber-reinforced stentless heart valve. J Biomech 2000;33(6):653–8. PubMed PMID: 10807985.

[21] Cataloglu A, Clark RE, Gould PL. Stress analysis of aortic valve leaflets with smoothed geometrical data. J Biomech 1977;10:153–8.

[22] De Hart J, Cacciola G, Schreurs PJ, Peters GW. A three-dimensional analysis of a fibre-reinforced aortic valve prosthesis. J Biomech 1998;31(7):629–38. PubMed PMID: 9796685.

[23] Hamid M, Sabbah H, Stein P. Finite element evaluation of stresses on closed leaflets of bioprosthetic heart valves with flexible stents. Finite Elem Anal Des 1985;1:213–25.

[24] Hamid MS, Sabbah HN, Stein PD. Influence of stent height upon stresses on the cusps of closed bioprosthetic valves. J Biomech 1986;19:759–69.

[25] Hamid MS, Sabbah HN, Stein PD. Some computational aspects of nonlinear stress analysis of bioprosthetic heart valves. In: Spilker R, Simon B, editors. Computational methods in bioengineering: ASME; 1988. p. 335–46.

[26] Huang X, Black MM, Howard IC, Patterson EA. A two dimensional finite element analysis of a bioprosthetic heart valve. J Biomech 1990;23:753–62.

[27] Li J, Luo XY, Kuang ZB. A nonlinear anisotropic model for porcine aortic heart valves. J Biomech 2001;34(10):1279–89. PubMed PMID: 11522307.

[28] Patterson EA, Howard IC, Thornton MA. A comparative study of linear and nonlinear simulations of the leaflets in a bioprosthetic heart valve during the cardiac cycle. J Med Eng Technol 1996;20(3): 95–108. PubMed PMID: 0008877750.

[29] Rousseau E, van Steenhoven A, Janssen J. A mechanical analysis of the closed Hancock heart valve prosthesis. J Biomech 1988;21(7):545–62.

[30] Thornton MA, Howard IC, Patterson EA. Three-dimensional stress analysis of polypropylene leaflets for prosthetic heart valves. Med Eng Phys 1997;19(6):588–97.

[31] De Hart J, Baaijens FP, Peters GW, Schreurs PJ. A computational fluid-structure interaction analysis of a fiber-reinforced stentless aortic valve. J Biomech 2003;36(5):699–712. PubMed PMID: 12695000.

[32] Sacks MS, Mirnajafi A, Sun W, Schmidt P. Bioprosthetic heart valve heterograft biomaterials: structure, mechanical behavior and computational simulation. Expert Rev Med Devices 2006; 3(6):817–34. PubMed PMID: 17280546.

[33] Elman H, Silvester D, Wathen A. Finite elements and fast iterative solvers with applications in incompressible fluid dynamics. 2nd ed. Oxford University Press; 2014.

[34] Ferziger JH, Peric M. Computational methods for fluid dynamics. 3rd ed. Springer; 2001.

[35] Versteeg H, Malalasekera W. An introduction to computational fluid dynamics: the finite volume method. 2nd ed. Pearson; 2007.

[36] Holzapfel GA. Nonlinear solid mechanics: a continuum approach for engineering. Hoboken, NJ, USA: John Wiley & Sons; 2000.

[37] Belytschko T, Liu WK, Moran B. Nonlinear finite elements for continua and structures. Hoboken, NJ, USA: John Wiley & Sons; 2000.

[38] Bonet J, Wood RD. Nonlinear continuum mechanics for finite element analysis. 2nd ed. New York, NY, USA: Cambridge University Press; 2008.

[39] Yang J, Staib L, Duncan JS. Neighbor-constrained segmentation with level set based 3-D deformable models. IEEE Trans Med Imaging 2004;23(8):940–8.

[40] Yushkevich PA, Piven J, Hazlett HC, Smith RG, Ho S, Gee JC, Gerigb G. User-guided 3D active contour segmentation of anatomical structures: significantly improved efficiency and reliability. Neuroimage 2006;31(3):1116–28.

[41] Wang Q, Sun W. Finite element modeling of mitral valve dynamic deformation using patientspecific multi-slices computed tomography scans. Ann Biomed Eng 2013;41(1):142–53. https://doi.org/10.1007/s10439-012-0620-6. PubMed PMID: 22805982.

[42]   Wang Q, Sirois E, Sun W. Patient-specific modeling of biomechanical interaction in transcatheter aortic valve deployment. J Biomech 2012;45(11):1965–71. https://doi.org/10.1016/j.jbiomech.2012.05.008. Epub 2012/06/16. PubMed PMID: 22698832.

[43]   Wang Q, Book G, Contreras Ortiz SH, Primiano C, McKay R, Kodali S, Sun W. Dimensional analysis of aortic root geometry during diastole using 3D models reconstructed from clinical 64-slice computed tomography images. Cardiovas Eng Technol 2011;2(4):324–33. https://doi.org/10.1007/s13239-011-0052-8.

[44]   Huang X, Dione DP, Compas C, Papademetris X, Lin B, Bregasi A, Sinusas AJ, Staib L, Duncan JS. Contour tracking in echocardiographic sequences via sparse representation and dictionary learning. Med Image Anal 2014;18(2):253–71.

[45]   Zhu Y, Papademetris X, Sinusas AJ, Duncan JS. Segmentation of the left ventricle from cardiac MR images using a subject-specific dynamical model. IEEE Trans Med Imaging 2010;29(3):669–87.

[46]   Ecabert O, Peters J, Weese J, editors. Modeling shape variability for full heart segmentation in cardiac computed-tomography images. SPIE Medical Imaging; 2006.

[47]   Zheng Y, Barbu A, Georgescu B, Scheuering M, Comaniciu D. Four-chamber heart modeling and automatic segmentation for 3-D cardiac CT volumes using marginal space learning and steerable features. IEEE Trans Med Imaging 2008;27(11):1668–81.

[48]   Lorenz C, Berg J. A comprehensive shape model of the heart. Med Image Anal 2006;10(4):657–70.

[49]   Lin N, Papademetris X, Sinusas AJ, Duncan JS, editors. Analysis of left ventricular motion using a general robust point matching algorithm. Medical image computing and computer assisted intervention; 2003.

[50]   Schneider RJ, Perrin DP, Vasilyev NV, Marx GR, PJd N, Howe RD. Itral annulus segmentation from four-dimensional ultrasound using a valve state predictor and constrained optical flow. Med Image Anal 2012;16(2):497–504.

[51]   Zheng Y, John M, Liao R, Nöttling A, Boese J, Kempfert J, Walther T, Brockmann G, Comaniciu D. Automatic aorta segmentation and valve landmark detection in C-arm CT for transcatheter aortic valve implantation. IEEE Trans Med Imaging 2012;31(12):2307–21.

[52]   Pouch AM, Wang H, Takabe M, Jackson BM, Sehgal CM, Gorman III JH, Gorman RC, Yushkevich PA. Automated segmentation and geometrical modeling of the tricuspid aortic valve in 3D echocardiographic images. Med Image Comput Comput Assist Interv 2013;1:485–92.

[53]   Ionasec RI, Voigt I, Georgescu B, Wang Y, Houle H, Vega-Higuera F, Navab N, Comaniciu D. Patient-specific modeling and quantification of the aortic and mitral valves from 4-D cardiac CT and TEE. IEEE Trans Med Imaging 2010;29(9):1636–51.

[54]   Liang L, Kong F, Martin C, Pham T, Wang Q, Duncan J, Sun W. Machine learning-based 3-D geometry reconstruction and modeling of aortic valve deformation using 3-D computed tomography images. Int J Numer Method Biomed Eng 2017;33(5). https://doi.org/10.1002/cnm.2827. Epub 2016 Aug 24. PubMed PMID: 27557429; PMCID: PMC5325825.

[55]   Sun W, Martin C, Pham T. Computational modeling of cardiac valve function and intervention. Annu Rev Biomed Eng 2014;16:53–76.

[56]   Stella JA, Sacks MS. On the biaxial mechanical properties of the layers of the aortic valve leaflet. J Biomech Eng 2007;129(5):757–66.

[57]   Gundiah N, Matthews PB, Karimi R, Azadani A, Guccione J, Guy TS, Saloner D, Tseng EE. Significant material property differences between the porcine ascending aorta and aortic sinuses. J Heart Valve Dis 2008;17(6):606.

[58]   Stephens EH, Grande-Allen KJ. Age-related changes in collagen synthesis and turnover in porcine heart valves. J Heart Valve Dis 2007;16(6):672–82.

[59]   Billiar KL, Sacks MS. Biaxial mechanical properties of the natural and glutaraldehyde treated aortic valve cusp—part I: experimental results. J Biomech Eng 2000;122(1):23–30.

[60]   Martin C, Sun W. Biomechanical characterization of aortic valve tissue in humans and common animal models. J Biomed Mater Res A 2012;100(6):1591–9.

[61]   May-Newman K, Yin F. Biaxial mechanical behavior of excised porcine mitral valve leaflets. Am J Physiol Heart Circ Physiol 1995;269(4):H1319–27.

[62] Kunzelman KS, Cochran RP. Stress/strain characteristics of porcine mitral valve tissue: parallel versus perpendicular collagen orientation. J Card Surg 1992;7(1):71–8.

[63] Pham T, Sun W. Material properties of aged human mitral valve leaflets. J Biomed Mater Res A 2014;102(8):2692–703.

[64] Prot V, Skallerud B, Sommer G, Holzapfel GA. On modelling and analysis of healthy and pathological human mitral valves: two case studies. J Mech Behav Biomed Mater 2010;3(2):167–77.

[65] Sacks MS, Enomoto Y, Graybill JR, Merryman WD, Zeeshan A, Yoganathan AP, Levy RJ, Gorman RC, Gorman JH. In-vivo dynamic deformation of the mitral valve anterior leaflet. Ann Thorac Surg 2006;82(4):1369–77.

[66] Krishnamurthy G, Itoh A, Bothe W, Swanson JC, Kuhl E, Karlsson M, Miller DC, Ingels NB. Stress–strain behavior of mitral valve leaflets in the beating ovine heart. J Biomech 2009;42(12): 1909–16.

[67] Rausch MK, Bothe W, Kvitting J-PE, Göktepe S, Miller DC, Kuhl E. In vivo dynamic strains of the ovine anterior mitral valve leaflet. J Biomech 2011;44(6):1149–57.

[68] Stradins P, Lacis R, Ozolanta I, Purina B, Ose V, Feldmane L, Kasyanov V. Comparison of biomechanical and structural properties between human aortic and pulmonary valve. Eur J Cardiothorac Surg 2004;26(3):634–9.

[69] Brazile B, Wang B, Wang G, Bertucci R, Prabhu R, Patnaik SS, Butler JR, Claude A, Brinkman-Ferguson E, Williams LN. On the bending properties of porcine mitral, tricuspid, aortic, and pulmonary valve leaflets. J Long Term Eff Med Implant 2015;25(1–2).

[70] Pham T, Sulejmani F, Shin E, Wang D, Sun W. Quantification and comparison of the mechanical properties of four human cardiac valves. Acta Biomaterialia 2017;54:345–55.

[71] Caballero A, Sulejmani F, Martin C, Pham T, Sun W. Evaluation of transcatheter heart valve biomaterials: biomechanical characterization of bovine and porcine pericardium. J Mech Behav Biomed Mater 2017;75:486–94.

[72] Páez JG, Jorge E, Rocha A, Castillo-Olivares J, Millan I, Carrera A, Cordon A, Tellez G, Burgos R. Mechanical effects of increases in the load applied in uniaxial and biaxial tensile testing. Part II. Porcine pericardium. J Mater Sci Mater Med 2002;13(5):477–83.

[73] Aguiari P, Fiorese M, Iop L, Gerosa G, Bagno A. Mechanical testing of pericardium for manufacturing prosthetic heart valves. Interact Cardiovasc Thorac Surg 2015;22(1):72–84.

[74] Hülsmann J, Grün K, El Amouri S, Barth M, Hornung K, Holzfuß C, Lichtenberg A, Akhyari P. Transplantation material bovine pericardium: biomechanical and immunogenic characteristics after decellularization vs. glutaraldehyde-fixing. Xenotransplantation 2012;19(5):286–97.

[75] Gauvin R, Marinov G, Mehri Y, Klein J, Li B, Larouche D, Guzman R, Zhang Z, Germain L, Guidoin R. A comparative study of bovine and porcine pericardium to highlight their potential advantages to manufacture percutaneous cardiovascular implants. J Biomater Appl 2013;28(4):552–65.

[76] Oswal D, Korossis S, Mirsadraee S, Wilcox H, Watterson K, Fisher J, Ingham E. Biomechanical characterization of decellularized and cross-linked bovine pericardium. J Heart Valve Dis 2007;16(2):165.

[77] Murdock K, Martin C, Sun W. Characterization of mechanical properties of pericardium tissue using planar biaxial tension and flexural deformation. J Mech Behav Biomed Mater 2018;77:148–56.

[78] Mirnajafi A, Raymer J, Scott MJ, Sacks MS. The effects of collagen fiber orientation on the flexural properties of pericardial heterograft biomaterials. Biomaterials 2005;26(7):795–804.

[79] Kheradvar A, Groves EM, Falahatpisheh A, Mofrad MK, Alavi SH, Tranquillo R, Dasi LP, Simmons CA, Grande-Allen KJ, Goergen CJ. Emerging trends in heart valve engineering: Part IV. Computational modeling and experimental studies. Ann Biomed Eng 2015;43(10):2314–33.

[80] Fung YC. Biomechanics: mechanical properties of living tissues. 2nd ed. New York: Springer Verlag; 1993. 568 pp.

[81] Humphery JD. Cardiovascular solid mechanics. Springer Verlag; 2002.

[82] Sun W. Biomechanical simulation of heart valve biomaterials. Pittsburgh: University of Pittsburgh; 2003. PhD Dissertation.

[83] Holzapfel GA, Gasser TC, Ogden RW. A new constitutive framework for arterial wall mechanics and a comparative study of material models. J Elast 2000;61:1–48.

[84]   Gasser TC, Ogden RW, Holzapfel GA. Hyperelastic modelling of arterial layers with distributed collagen fibre orientations. J R Soc Interface 2006;3(6):15—35.

[85]   Wang Q, Primiano C, McKay R, Kodali S, Sun W. CT image-based engineering analysis of transcatheter aortic valve replacement. JACC Cardiovasc Imaging 2014;7(5):526—8. https://doi.org/10.1016/j.jcmg.2014.03.006. PubMed PMID: 24831213; PMCID: PMC4034127.

[86]   Wang Q, Kodali S, Sun W. Analysis of aortic root rupture during transcatheter aortic valve implantation using computational simulations. In: 12th US national congress on computational mechanics (USNCCM12); July 22—25, 2013; Raleigh, North Carolina, USA; 2013.

[87]   Wang Q, Primiano C, Sun W. Can isolated annular dilatation cause significant ischemic mitral regurgitation? Another look at the causative mechanisms. J Biomech 2014;47(8):1792—9. https://doi.org/10.1016/j.jbiomech.2014.03.033. PubMed PMID: 24767703.

[88]   Kalra K, Wang Q, McIver BV, Shi W, Guyton RA, Sun W, Sarin EL, Thourani VH, Padala M. Temporal changes in interpapillary muscle dynamics as an active indicator of mitral valve and left ventricular interaction in ischemic mitral regurgitation. J Am Coll Cardiol 2014;64(18):1867—79. https://doi.org/10.1016/j.jacc.2014.07.988. PubMed PMID: 25444139.

[89]   Kim H, Lu J, Sacks MS, Chandran KB. Dynamic simulation of bioprosthetic heart valves using a stress resultant shell model. Ann Biomed Eng 2008;36(2):262—75.

[90]   Saleeb A, Kumar A, Thomas V. The important roles of tissue anisotropy and tissue-to-tissue contact on the dynamical behavior of a symmetric tri-leaflet valve during multiple cardiac pressure cycles. Med Eng Phys 2013;35(1):23—35.

[91]   Katayama S, Umetani N, Hisada T, Sugiura S. Bicuspid aortic valves undergo excessive strain during opening: a simulation study. J Thorac Cardiovasc Surg 2013;145(6):1570—6.

[92]   Koch T, Reddy B, Zilla P, Franz T. Aortic valve leaflet mechanical properties facilitate diastolic valve function. Comput Methods Biomech Biomed Eng 2010;13(2):225—34.

[93]   Becker W, Rowson J, Oakley J, Yoxall A, Manson G, Worden K. Bayesian sensitivity analysis of a model of the aortic valve. J Biomech 2011;44(8):1499—506.

[94]   Conti CA, Votta E, Della Corte A, Del Viscovo L, Bancone C, Cotrufo M, Redaelli A. Dynamic finite element analysis of the aortic root from MRI-derived parameters. Med Eng Phys 2010;32(2):212—21.

[95]   Labrosse MR, Beller CJ, Boodhwani M, Hudson C, Sohmer B. Subject-specific finite-element modeling of normal aortic valve biomechanics from 3D+ t TEE images. Med Image Anal 2015; 20(1):162—72.

[96]   Weltert L, de Tullio MD, Afferrante L, Salica A, Scaffa R, Maselli D, Verzicco R, De Paulis R. Annular dilatation and loss of sino-tubular junction in aneurysmatic aorta: implications on leaflet quality at the time of surgery. A finite element study. Interact Cardiovasc Thorac Surg 2013;17(1):8—12.

[97]   Conti CA, Della Corte A, Votta E, Del Viscovo L, Bancone C, De Santo LS, Redaelli A. Biomechanical implications of the congenital bicuspid aortic valve: a finite element study of aortic root function from in vivo data. J Thorac Cardiovasc Surg 2010;140(4):890—6. e2.

[98]   Jermihov PN, Jia L, Sacks MS, Gorman RC, Gorman JH, Chandran KB. Effect of geometry on the leaflet stresses in simulated models of congenital bicuspid aortic valves. Cardiovasc Eng Technol 2011; 2(1):48—56.

[99]   Nathan DP, Xu C, Plappert T, Desjardins B, Gorman JH, Bavaria JE, Gorman RC, Chandran KB, Jackson BM. Increased ascending aortic wall stress in patients with bicuspid aortic valves. Ann Thorac Surg 2011;92(4):1384—9.

[100] Arzani A, Mofrad MR. A strain-based finite element model for calcification progression in aortic valves. J Biomech 2017;65:216—20.

[101] Weinberg EJ, Schoen FJ, Mofrad MR. A computational model of aging and calcification in the aortic heart valve. PLoS One 2009;4(6):e5960.

[102] Leat M, Fisher J. The influence of manufacturing methods on the function and performance of a synthetic leaflet heart valve. Proc IME H J Eng Med 1995;209(1):65—9.

[103] Leat M, Fisher J. A synthetic leaflet heart valve with improved opening characteristics. Med Eng Phys 1994;16(6):470—6.

[104] Auricchio F, Conti M, Morganti S, Totaro P. A computational tool to support pre-operative planning of stentless aortic valve implant. Med Eng Phys 2011;33(10):1183—92.

[105] Xiong FL, Goetz WA, Chong CK, Chua YL, Pfeifer S, Wintermantel E, Yeo JH. Finite element investigation of stentless pericardial aortic valves: relevance of leaflet geometry. Ann Biomed Eng 2010;38(5):1908–18.

[106] Cacciola G, Peters G, Schreurs P. A three-dimensional mechanical analysis of a stentless fibrereinforced aortic valve prosthesis. J Biomech 2000;33(5):521–30.

[107] Martin C, Sun W. Simulation of long-term fatigue damage in bioprosthetic heart valves: effects of leaflet and stent elastic properties. Biomechanics Model Mechanobiol 2014;13(4):759–70.

[108] Avanzini A, Battini D. Structural analysis of a stented pericardial heart valve with leaflets mounted externally. Proc IME H J Eng Med 2014;228(10):985–95.

[109] Hammer PE, Chen PC, Pedro J, Howe RD. Computational model of aortic valve surgical repair using grafted pericardium. J Biomech 2012;45(7):1199–204.

[110] Mohammadi H, Boughner D, Millon L, Wan W. Design and simulation of a poly (vinyl alcohol)—bacterial cellulose nanocomposite mechanical aortic heart valve prosthesis. Proc IME H J Eng Med 2009;223(6):697–711.

[111] Loerakker S, Argento G, Oomens CW, Baaijens FP. Effects of valve geometry and tissue anisotropy on the radial stretch and coaptation area of tissue-engineered heart valves. J Biomech 2013;46(11): 1792–800.

[112] van Loosdregt IA, Argento G, Driessen-Mol A, Oomens CW, Baaijens FP. Cell-mediated retraction versus hemodynamic loading—A delicate balance in tissue-engineered heart valves. J Biomech 2014; 47(9):2064–9.

[113] Serrani M, Brubert J, Stasiak J, De Gaetano F, Zaffora A, Costantino ML, Moggridge G. A computational tool for the microstructure optimization of a polymeric heart valve prosthesis. J Biomech Eng 2016;138(6):061001.

[114] Cribier A, Eltchaninoff H, Bash A, Borenstein N, Tron C, Bauer F, Derumeaux G, Anselme F, Laborde F, Leon MB. Percutaneous transcatheter implantation of an aortic valve prosthesis for calcific aortic stenosis: first human case description. Circulation 2002;106(24):3006–8.

[115] Cribier A. The development of transcatheter aortic valve replacement (TAVR). Global Cardiology Science and Practice 2017;2016(4).

[116] Li K, Sun W. Simulated thin pericardial bioprosthetic valve leaflet deformation under static pressure-only loading conditions: implications for percutaneous valves. Epub March 25, 2010 Ann Biomed Eng 2010;38(8):2690–701. https://doi.org/10.1007/s10439-010-0009-3. PubMed PMID: 20336372.

[117] Smuts AN, Blaine DC, Scheffer C, Weich H, Doubell AF, Dellimore KH. Application of finite element analysis to the design of tissue leaflets for a percutaneous aortic valve. J Mech Behav Biomed Mater 2010;4(1):85–98. S1751-6161(10)00132-3 [pii] 10.1016/j.jmbbm.2010.09.009. Epub 2010/ 11/26. PubMed PMID: 21094482.

[118] Sun W, Li K, Sirois E. Simulated elliptical bioprosthetic valve deformation: implications for asymmetric transcatheter valve deployment. J Biomech 2010;43(16):3085–90.

[119] Gunning PS, Saikrishnan N, McNamara LM, Yoganathan AP. An in vitro evaluation of the impact of eccentric deployment on transcatheter aortic valve hemodynamics. Ann Biomed Eng 2014;42(6): 1195–206. https://doi.org/10.1007/s10439-014-1008-6. PubMed PMID: 24719050.

[120] Kuetting M, Sedaghat A, Utzenrath M, Sinning JM, Schmitz C, Roggenkamp J, Werner N, Schmitz- Rode T, Steinseifer U. In vitro assessment of the influence of aortic annulus ovality on the hydrodynamic performance of self-expanding transcatheter heart valve prostheses. J Biomech 2014;47(5):957–65. https://doi.org/10.1016/j.jbiomech.2014.01.024. PubMed PMID: 24495752.

[121] Scharfschwerdt M, Meyer-Saraei R, Schmidtke C, Sievers HH. Hemodynamics of the Edwards Sapien XT transcatheter heart valve in noncircular aortic annuli. J Thorac Cardiovasc Surg 2014; 148(1):126–32. https://doi.org/10.1016/j.jtcvs.2013.07.057. Epub 2013/09/28. PubMed PMID: 24071472.

[122] Abbasi M, Azadani AN. Leaflet stress and strain distributions following incomplete transcatheter aortic valve expansion. J Biomech 2015;48(13):3672–80.

[123] Martin C, Sun W. Comparison of transcatheter aortic valve and surgical bioprosthetic valve durability: a fatigue simulation study. J Biomech 2015;48(12):3026—34.

[124] Petrini L, Dordoni E, Allegretti D, Pott D, Kütting M, Migliavacca F, Pennati G. Simplified multi-stage computational approach to assess the fatigue behavior of a niti transcatheter aortic valve during in vitro tests: a proof-of-concept study. J Med Dev 2017;11(2):021009.

[125] Cubic hermite bezier spline based reconstruction of implanted aortic valve stents from CT images. In: Gessat M, Altwegg L, Frauenfelder T, Plass A, Falk V, editors. Engineering in medicine and biology society, EMBC, 2011 annual international conference of the IEEE August 30 2011-September 3 2011; 2011.

[126] Gessat M, Hopf R, Pollok T, Russ C, Frauenfelder T, Sundermann SH, Hirsch S, Mazza E, Szekely G, Falk V. Image-based mechanical analysis of stent deformation: concept and exemplary implementation for aortic valve stents. IEEE Trans Biomed Eng 2014;61(1):4—15. https://doi.org/10.1109/TBME.2013.2273496.

[127] Tzamtzis S, Viquerat J, Yap J, Mullen MJ, Burriesci G. Numerical analysis of the radial force produced by the Medtronic-CoreValve and Edwards-SAPIEN after transcatheter aortic valve implantation (TAVI). Med Eng Phys 2013;35(1):125—30. https://doi.org/10.1016/j.medengphy.2012.04.009. PubMed PMID: 22640661.

[128] Capelli C, Bosi GM, Cerri E, Nordmeyer J, Odenwald T, Bonhoeffer P, Migliavacca F, Taylor AM, Schievano S. Patient-specific simulations of transcatheter aortic valve stent implantation. Med Biol Eng Comput 2012;50(2):183—92.

[129] Wang Q, Kodali S, Primiano C, Sun W. Simulations of transcatheter aortic valve implantation: implications for aortic root rupture. Biomechanics Model Mechanobiol 2015;14(1):29—38.

[130] Auricchio F, Conti M, Morganti S, Reali A. Simulation of transcatheter aortic valve implantation: a patient-specific finite element approach. Epub 2013/02/14 Comput Methods Biomech Biomed Eng 2013. https://doi.org/10.1080/10255842.2012.746676. PubMed PMID: 23402555.

[131] Gunning PS, Vaughan TJ, McNamara LM. Simulation of self expanding transcatheter aortic valve in a realistic aortic root: implications of deployment geometry on leaflet deformation. Ann Biomed Eng 2014. https://doi.org/10.1007/s10439-014-1051-3. PubMed PMID: 24912765.

[132] Morganti S, Conti M, Aiello M, Valentini A, Mazzola A, Reali A, Auricchio F. Simulation of transcatheter aortic valve implantation through patient-specific finite element analysis: two clinical cases. J Biomech 2014;47(11):2547—55. https://doi.org/10.1016/j.jbiomech.2014.06.007.

[133] Grbic S, Mansi T, Ionasec R, Voigt I, Houle H, John M, Schoebinger M, Navab N, Comaniciu D. Image-based computational models for TAVI planning: from CT images to implant deployment. Med Image Comput Comput Assist Interv 2013;16(Pt 2):395—402. PubMed PMID: 24579165.

[134] Wang Q, Kodali S, Primiano C, Sun W. Simulations of transcatheter aortic valve implantation: implications for aortic root rupture. Biomechanics Model Mechanobiol 2014:1—10. https://doi.org/10.1007/s10237-014-0583-7.

[135] Sturla F, Ronzoni M, Vitali M, Dimasi A, Vismara R, Preston-Maher G, Burriesci G, Votta E, Redaelli A. Impact of different aortic valve calcification patterns on the outcome of transcatheter aortic valve implantation: a finite element study. J Biomech 2016;49(12):2520—30.

[136] Xuan Y, Krishnan K, Ye J, Dvir D, Guccione JM, Ge L, Tseng EE. Stent and leaflet stresses in a 26-mm first-generation balloon-expandable transcatheter aortic valve. J Thorac Cardiovasc Surg 2017;153(5):1065—73.

[137] Xuan Y, Krishnan K, Ye J, Dvir D, Guccione JM, Ge L, Tseng EE. Stent and leaflet stresses in 29- mm second-generation balloon-expandable transcatheter aortic valve. Ann Thorac Surg 2017;104(3):773—81.

[138] Morganti S, Brambilla N, Petronio A, Reali A, Bedogni F, Auricchio F. Prediction of patientspecific post-operative outcomes of TAVI procedure: the impact of the positioning strategy on valve performance. J Biomech 2016;49(12):2513—9.

[139] Bailey J, Curzen N, Bressloff NW. Assessing the impact of including leaflets in the simulation of TAVI deployment into a patient-specific aortic root. Comput Methods Biomech Biomed Eng 2016;19(7):733—44.

[140] Bailey J, Curzen N, Bressloff NW. The impact of imperfect frame deployment and rotational orientation on stress within the prosthetic leaflets during transcatheter aortic valve implantation. J Biomech 2017;53:22–8.

[141] Bosi GM, Capelli C, Cheang MH, Delahunty N, Mullen M, Taylor AM, Schievano S. Populations-pecific material properties of the implantation site for transcatheter aortic valve replacement finite element simulations. J Biomech 2018;71:236–44.

[142] Schultz C, Rodriguez-Olivares R, Bosmans J, Lefevre T, De Santis G, Bruining N, Collas V, Dezutter T, Bosmans B, Rahhab Z, El Faquir N, Watanabe Y, Segers P, Verhegghe B, Chevalier B, van Mieghem N, De Beule M, Mortier P, de Jaegere P. Patient-specific image-based computer simulation for theprediction of valve morphology and calcium displacement after TAVI with the Medtronic CoreValve and the Edwards SAPIEN valve. EuroIntervention 2016;11(9): 1044–52. https://doi.org/10.4244/EIJV11I9A212. PubMed PMID: 26788707.

[143] Hopf R, Sündermann SH, Born S, Ruiz CE, Van Mieghem NM, de Jaegere PP, Maisano F, Falk V, Mazza E. Postoperative analysis of the mechanical interaction between stent and host tissue in patients after transcatheter aortic valve implantation. J Biomech 2017;53:15–21.

[144] Bianchi M, Marom G, Ghosh RP, Fernandez HA, Taylor JR, Slepian MJ, Bluestein D. Effect of balloon-expandable transcatheter aortic valve replacement positioning: a patient-specific numerical model. Artif Organs 2016;40(12).

[145] Yoganathan AP, He ZM, Jones SC. Fluid mechanics of heart valves. Annu Rev Biomed Eng 2004;6: 331–62.

[146] Dasi LP, Simon HA, Sucosky P, Yoganathan AP. Fluid mechanics of artificial heart valves. Clin Exp Pharmacol Physiol 2009;36(2):225–37.

[147] Donea J, Giuliani S, Halleux JP. An arbitrary Lagrangian-Eulerian finite element method for transient dynamic fluid-structure interactions. Comput Methods Appl Mech Eng 1982;33:689–723.

[148] Peskin CS. The immersed boundary method. Acta Numer 2002;11:479–517.

[149] Mittal R, Iaccarino G. Immersed boundary methods. Annu Rev Fluid Mech 2005;37:239–61.

[150] Peskin CS. Flow patterns around heart valves: a digital computer method for solving the equations of motion. Albert Einstein College of Medicine; 1972.

[151] Peskin CS. Flow patterns around heart valves: a numerical method. J Comput Phys 1972;10(2):252–71.

[152] Peskin CS. Numerical analysis of blood flow in the heart. J Comput Phys 1977;25(3):220–52.

[153] Peskin CS, McQueen DM. Modeling prosthetic heart valves for numerical analysis of blood flow in the heart. J Comput Phys 1980;37(1):113–32.

[154] McQueen DM, Peskin CS. Computer-aided design of pivoting-disc prosthetic mitral valves. J Thorac Cardiovasc Surg 1983;86(1):126–35.

[155] McQueen DM, Peskin CS. Computer-assisted design of butterfly-bileaflet valves for the mitral position. Scand J Thorac Cardiovasc Surg 1985;19(2):139–48.

[156] Meisner JS, McQueen DM, Ishida Y, Vetter HO, Bortolotti U, Strom JA, Frater RWM, Peskin CS, Yellin EL. Effects of timing of atrial systole on ventricular filling and mitral valve closure: computer and dog studies. Am J Physiol Heart Circ Physiol 1985;249(3):H604–19.

[157] Zhang L, Gerstenberger A, Wang X, Liu WK. Immersed finite element method. Comput Methods Appl Mech Eng 2004;193(21–22):2051–67.

[158] Liu WK, Liu Y, Farrell D, Zhang L, Wang XS, Fukui Y, Patankar N, Zhang Y, Bajaj C, Lee J, Hong J, Chen X, Hsu H. Immersed finite element method and its applications to biological systems. Comput Methods Appl Mech Eng 2006;195(13–16):1722–49.

[159] Boffi D, Gastaldi L, Heltai L, Peskin CS. On the hyper-elastic formulation of the immersed boundary method. Comput Methods Appl Mech Eng 2008;197(25–28):2210–31.

[160] Devendran D, Peksin CS. An energy-based immersed boundary method for incompressible viscoelasticity. J Comput Phys 2012;231(14):4613–42.

[161] Griffith BE, Luo XY. Hybrid finite difference/finite element version of the immersed boundary method. Int J Numer Meth Biomed Eng 2017;33(11):e2888. 31.

[162] Hsu MC, Kamensky D, Bazilevs Y, Sacks MS, Hughes TJR. Fluid–structure interaction analysis of bioprosthetic heart valves: significance of arterial wall deformation. Comput Mech 2014;54(4): 1055–71.

[163] Hsu MC, Kamensky D, Xu F, Kiendl J, Wang CL, Wu MCH, Mineroff J, Reali A, Bazilevs Y, Sacks MS. Dynamic and fluid-structure interaction simulations of bioprosthetic heart valves using parametric design with T-splines and Fung-type material models. Comput Mech 2015;55(6): 1211–25.

[164] Hsu MC, Sacks MS, Kamensky D, Schillinger D, Evans JA, Aggarwal A, Bazilevs Y, Hughes TJR. An immersogeometric variational framework for fluid-structure interaction: application to bioprosthetic heart valves. Comput Methods Appl Mech Eng 2015;284:1005–53.

[165] Gao H, Wang HM, Berry C, Luo XY, Griffith BE. Quasi-static image-based immersed boundary-finite element model of left ventricle under diastolic loading. Int J Numer Meth Biomed Eng 2014; 30(11):1199–222.

[166] de Hart J, Peters GWM, Schreurs PJG, Baaijens FPT. A three-dimensional computational analysis of fluid-structure interaction in the aortic valve. J Biomech 2003;36(1):102–12.

[167] de Hart J, Peters GWM, Schreurs PJG, Baaijens FPT. Collagen fibers reduce stresses and stabilize motion of aortic valve leaflets during systole. J Biomech 2004;37(3):303–11.

[168] Carmody CJ, Gurriesci G, Howard IC, Patterson EA. An approach to the simulation of fluidstructure interaction in the aortic valve. J Biomech 2006;39:158–69.

[169] Marom G, Peleg M, Halevi R, Rosenfeld M, Raanani E, Hamdan A, Haj-Ali R. Fluid-structure interaction model of aortic valve with porcine-specific collagen fiber alignment in the cusps. J Biomech Eng 2013;135(10):101001–6.

[170] Gilmanov A, Le TB, Sotiropoulos F. A numerical approach for simulating fluid structure interaction of flexible thin shells undergoing arbitrarily large deformations in complex domains. J Comput Phys 2015;300:814–43.

[171] Gilmanov A, Sotiropoulos F. Comparative hemodynamics in an aorta with bicuspid and trileaflet valves. Theor Comput Fluid Dynam 2016;30(1–2):67–85.

[172] Laadhari A, Quarteroni A. Numerical modeling of heart valves using resistive Eulerian surfaces. Int J Numer Meth Biomed Eng 2016;32(5):e02743.

[173] Mega M, Marom G, Halevi R, Hamdan A, Bluestein D, Haj-Ali R. Imaging analysis of collagen fiber networks in cusps of porcine aortic valves: effect of their local distribution and alignment on valve functionality. Comput Methods Biomech Biomed Eng 2016;19(9):1002–8.

[174] Yang JM, Balaras E. An embedded-boundary formulation for large-eddy simulation of turbulent flows interacting with moving boundaries. J Comput Phys 2006;215(1):12–40.

[175] Dasi LP, Ge L, Simon HA, Sotiropoulos F, Yoganathan AP. Vorticity dynamics of a bileaflet mechanical heart valve in an axisymmetric aorta. Phys Fluid 2007;19(6):067105.

[176] Borazjani I, Ge L, Sotiropoulos F. Curvilinear immersed boundary method for simulating fluid structure interaction with complex 3D rigid bodies. J Comput Phys 2008;227(16):7587–620.

[177] de Tullio MD, De Palma P, Iaccarino G, Pascazio G, Napolitano M. An immersed boundary method for compressible flows using local grid refinement. J Comput Phys 2007;225(2):2098–117.

[178] Borazjani I, Ge L, Sotiropoulos F. High-resolution fluid-structure interaction simulations of flow through a Bi-leaflet mechanical heart valve in an anatomic aorta. Ann Biomed Eng 2010;38(2): 326–44.

[179] Simon HA, Ge L, Sotiropoulos F, Yoganathan AP. Simulation of the three-dimensional hinge flow fields of a bileaflet mechanical heart valve under aortic conditions. Ann Biomed Eng 2010;38(3): 841–53.

[180] de Tullio MD, Nam J, Pascazio G, Balaras E, Verzicco R. Computational prediction of mechanical hemolysis in aortic valved prostheses. Eur J Mech B Fluid 2012;35:47–53.

[181] Le TB, Sotiropoulos F. Fluid-structure interaction of an aortic heart valve prosthesis driven by an animated anatomic left ventricle. J Comput Phys 2013;244:41–62.

[182] Annerel S, Claessens T, Degroote J, Segers P, Vierendeels J. Validation of a numerical FSI simulation of an aortic BMHV by in vitro PIV experiments. Med Eng Phys 2014;36(8):1014–23.

[183] Borazjani I. Fluid-structure interaction, immersed boundary-finite element method simulations of bio-prosthetic heart valves. Comput Methods Appl Mech Eng 2013;257:103–16.

[184] Kemp I, Dellimore K, Rodriguez R, Scheffer C, Blaine D, Weich H, Doubell A. Experimental validation of the fluid-structure interaction simulation of a bioprosthetic aortic heart valve. Australas Phys Eng Sci Med 2013;36(3):363—73.

[185] Wu W, Pott D, Mazza B, Sironi T, Dordoni E, Chiastra C, Petrini L, Pennati G, Dubini G, Steinseifer U, Sonntag S, Kuetting M, Migliavacca F. Fluid-structure interaction model of a percutaneous aortic valve: comparison with an in vitro test and feasibility study in a patient-specific case. Ann Biomed Eng 2016;44(2):590—603.

[186] Gharaie SH, Morsi Y. A novel design of a polymeric aortic valve. Int J Artif Organs 2015;38(5): 259—70.

[187] Piatti F, Sturla F, Marom G, Sheriff J, Claiborne TE, Slepian MJ, Redaelli A, Bluestein D. Hemodynamic and thrombogenic analysis of a trileaflet polymeric valve using a fluid-structure interaction approach. J Biomech 2015;48(13):3641—9.

[188] Nobili M, Morbiducci U, Ponzini R, Del Gaudio C, Balducci A, Grigioni M, Montevecchi FM, Redaelli A. Numerical simulation of the dynamics of a bileaflet prosthetic heart valve using a fluid-structure interaction approach. J Biomech 2008;41(11):2539—50.

[189] Ge L, Dasi LP, Sotiropoulos F, Yoganathan AP. Characterization of hemodynamic forces induced by mechanical heart valves: Reynolds vs. viscous stresses. Ann Biomed Eng 2008;36(2):276—97.

[190] Guivier-Curien C, Deplano V, Bertrand E. Validation of a numerical 3-D fluid-structure interaction model for a prosthetic valve based on experimental PIV measurements. Med Eng Phys 2009;31(8): 986—93.

[191] Jun BH, Saikrishnan N, Arjunon S, Yun BM, Yoganathan AP. Effect of hinge gap width of a St. Jude medical bileaflet mechanical heart valve on blood damage potential-an in vitro micro particle image velocimetry study. J Biomech Eng 2014;136(9):091008.

[192] Willert CE, Gharib M. Digital particle image velocimetry. Exp Fluid 1991;10(4):181—93.

[193] Raffel M, Willert C, Kompenhans J. Particle image velocimetry: a practical guide. New York, NY, USA: Springer-Verlag; 1998.

[194] Prasad AK. Particle image velocimetry. Curr Sci 2000;79(1):51—60.

[195] Quaini A, Canic S, Glowinski R, Igo S, Hartley CJ, Zoghbi W, Little S. Validation of a 3D computational fluid-structure interaction model simulating flow through an elastic aperture. J Biomech 2012;45(2):310—8.

[196] Griffith BE, Flamini V, DeAnda A, Scotten L. Simulating the dynamics of an aortic valve prosthesis in a pulse duplicator: numerical methods and initial experience. J Med Dev 2013;7(4):0409121—2.

[197] Stergiopulos N, Westerhof BE, Westerhof N. Total arterial inertance as the fourth element of the windkessel model. Am J Physiol Heart Circ Physiol 1999;276(1):H81—8.

[198] Driessen NJ, Bouten CV, Baaijens FP. A structural constitutive model for collagenous cardiovascular tissues incorporating the angular fiber distribution. J Biomech Eng 2005;127(3):494—503.

[199] Billiar KL, Sacks MS. Biaxial mechanical properties of the native and glutaraldehyde-treated aortic valve cusp: Part II: a structural constitutive model. J Biomech Eng 2000;122(4):327—35.

[200] Hasan A, Kolahdouz EM, Enquobahrie A, Caranasos TG, Vavalle JP, Griffith BE. Image-based immersed boundary model of the aortic root. Med Eng Phys 2017;47:72—84.

[201] Bayer JD, Blake RC, Plank G, Trayanova NA. A novel rule-based algorithm for assigning myocardial fiber orientation to computational heart models. Ann Biomed Eng 2012;40(10):2243—54.

[202] Rossi S, Lassila T, Ruiz-Baier R, Sequeira A, Quarteroni A. Thermodynamically consistent orthotropic activation model capturing ventricular systolic wall thickening in cardiac electromechanics. Eur J Mech A Solid 2014;48:129—42.

[203] Wong J, Kuhl E. Generating fibre orientation maps in human heart models using Poisson interpolation. Comput Methods Biomech Biomed Eng 2014;11(17):1217—26.

[204] Murgo JP, Westerhof N, Giolma JP, Altobelli SA. Aortic input impedance in normal man: relationship to pressure wave forms. Circulation 1980;62(1):105—16.

[205] Harihan P, Giarra M, Reddy V, Day SW, Manning KB, Deutsch S, Stewart SF, Myers MR, Berman MR, Burgreen GW, Paterson EG, Malinauskas RA. Multilaboratory particle image velocimetry analysis of the FDA benchmark nozzle model to support validation of computational fluid dynamics simulations. J Biomech Eng 2011;133(4):041002.

[206] Stewart SFC, Paterson EG, Burgreen GW, Harihan P, Giarra M, Reddy VY, Day SW, Manning KB, Deutsch S, Berman MR, Myers MR, Malinauskas RA. Assessment of CFD performance in simulations of an idealized medical device: results of FDA's first computational interlaboratory study. Cardiovasc Eng Technol 2012;3(2):139–60.

[207] Stewart SFC, Harihan P, Paterson EG, Burgreen WW, Reddy V, Day SW, Giarra M, Manning KB, Deutsch S, Berman MR, Myers MR, Malinauskas RA. Results of FDA's first interlaboratory computational study of a nozzle with a sudden contraction and conical diffuser. Cardiovasc Eng Technol 2013;4:374–91. 4.

[208] Malinauskas RA, Harihan P, Day SW, Herbertson LH, Buesen M, Steinseifer U, Aycock KI, Good BC, Deutsch S, Manning KB, Craven BA. FDA benchmark medical device flow models for CFD validation. Am Soc Artif Intern Organs J 2017;63(2):150–60.

[209] De Backer O, Piazza N, Banai S, Lutter G, Maisano F, Herrmann HC, Franzen OW, Søndergaard L. Percutaneous transcatheter mitral valve replacement: an overview of devices in preclinical and early clinical evaluation. Circ Cardiovasc Interv 2014;7(3):400–9.

[210] Baillargeon B, Rebelo N, Fox DD, Taylor RL, Kuhl E. The Living Heart Project: a robust and integrative simulator for human heart function. Eur J Mech A Solid 2014;48:38–47.

[211] Ayoub S, Ferrari G, Gorman RC, Gorman JH, Schoen FJ, Sacks MS. Heart valve biomechanics and underlying mechanobiology. Comp Physiol 2016;6(4):1743–80.

CHAPTER 8

In vitro experimental methods for assessment of prosthetic heart valves

Ali N. Azadani

Department of Mechanical & Materials Engineering, Ritchie School of Engineering and Computer Science, University of Denver, Denver, CO, United States

Contents

In vitro experimental testing is an integral part of design and verification of implantable medical devices, including prosthetic heart valves. Preclinical in vitro evaluation of prosthetic heart valves is required to ensure safety and improve structural and hemodynamic performance of the devices in a well-controlled environment. In addition, in vitro bench testing reduces reliance and use of animals in safety and risk assessment of medical devices. It is generally accepted that preclinical animal models are time-consuming, expensive, and raise ethical concerns [1,2]. Moreover, in recent years, the role of computational modeling and simulation is becoming increasingly important and evident in design and development of medical devices [3]. Computational simulations reduce the need to perform expensive preclinical tests to optimize design and reduce the associated risk factors. In addition, regulatory agencies such as the US Food and Drug Administration (FDA) and EU Medical Device Regulatory (MDR) currently accept validated computational modeling and simulation as a scientific evidence in regulatory submissions [4]. As a result, in vitro tests play an important role in validation and verification of the computational simulations.

The International Organization for Standardization (ISO) provides guidance for in vitro testing of prosthetic heart valves. The ISO-5840: 2015 is the current active standard that covers heart valve substitutes. The standard is structured in three parts: (part-1) general requirements, (part-2) surgically implanted heart valve substitutes,

Principles of Heart Valve Engineering
ISBN 978-0-12-814661-3, https://doi.org/10.1016/B978-0-12-814661-3.00008-3

and (part-3) heart valve substitutes implanted by transcatheter techniques. In addition to the ISO standard, the FDA provided detailed data requirements for in vitro, animal, and clinical testing necessary in its draft guidance to support the approval of new prosthetic heart valves. The FDA draft guidance, "Heart Valves-Investigational Device Exemption (IDE) and Premarket Approval (PMA) Applications," was widely used for preclinical assessment of prosthetic heart valves. However, the draft guidance was withdrawn in April 2015 by the FDA, and the agency relies on the recommendations set forth by the ISO standard. As a result, ISO-5840 is currently the primary guidance for preclinical testing of prosthetic heart valves. The purpose of this chapter is to undertake a review of the currently available and commonly used in vitro tests in preclinical assessment of prosthetic heart valves. The chapter is divided into two major sections. In the first section, hydrodynamic evaluation of prosthetic heart valves is reviewed. Then hydrodynamic evaluation is followed by a discussion of the structural assessment of prosthetic heart valves using in vitro tests. The chapter concludes with a discussion of the implications of the in vitro tests and future directions.

8.1 Hydrodynamic evaluation

Hydrodynamic evaluation of prosthetic heart valves provides information on the fluid mechanical performance of the substitutes. According to ISO-5840, heart valve substitutes should be tested under the expected physiological environment of the intended patient population [5]. Therefore, the loading environment varies as a function of implant site, patient blood pressure conditions, and patient cardiac output. Table 8.1, adopted from ISO-5840, defines the physiological parameters for normal and pathological conditions in adult population. The physiological parameters for pediatric population can be found in the ISO standard [5]. To verify hydrodynamic performance of heart valve substitutes, the valves should be tested under both normal and pathological conditions. Prosthetic heart valves being tested in in vitro systems must be of clinical grade to ensure the prosthetic valves represent the valves that will be implanted in patients. ISO-5840 recommends hydrodynamic testing under both steady and pulsatile flow conditions. A summary of the test conditions and practical implications is provided below in this section.

8.1.1 Steady flow testing

While heart valve substitutes are exposed to unsteady (pulsatile) flow under physiological condition, steady forward and backward flow testing of prosthetic heart valve are helpful in verifying the accuracy of in vitro pulsatile flow tests, which will be discussed later in this chapter. Multiple studies in the scientific literature have reported the use of steady flow loops to examine hemodynamic performance of heart valve substitutes [6–14]. For steady flow testing of prosthetic heart valves, a straight tube with an internal diameter

Table 8.1 Heart valve substitute operational environment for left and right sides of heart in adult population [5].

Table 1.		Normotensive	Hypotensive	Mild hypertensive	Moderate hypertensive	Severe hypertensive	Very severe hypertensive
Arterial peak systolic pressure (mmHg)		120	60	140–159	160–179	180–209	≥210
Arterial end diastolic pressure (mmHg)		80	40	90–99	100–109	110–119	≥120
Right ventricle peak systolic pressure (mmHg)		18–35	15	40–49	50–59	60–84	≥85
Pulmonary artery end diastolic pressure (mmHg)		8–15	5	15–19	20–24	25–34	≥35
Peak differential pressure across closed valve (mmHg)	Aortic	100	50	115–129	130–144	145–164	≥165
	Mitral	120	60	140–159	160–179	180–209	≥210
	Pulmonary	13–25	10	28–34	35–42	43–59	≥60
	Tricuspid	18–35	15	40–49	50–59	60–84	≥85

General conditions: heart rate: 30–200 bpm, cardiac output: 3–15 L/min, forward flow volume: 25–100 mL, temperature: 34–42°C.

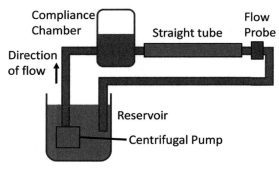

Figure 8.1 Schematic of flow loop for testing heart valve substitute under steady flow conditions. *(Reprinted and modified with permission from Yap CH, Saikrishnan N, Tamilselvan G, Yoganathan AP. Experimental technique of measuring dynamic fluid shear stress on the aortic surface of the aortic valve leaflet. J Biomech Eng 2011;133(6):061007.)*

of 35 mm is recommended by ISO-5840 (Fig. 8.1). Normal saline solution with density of 1005 kg/m^3 and viscosity of 1.0 cP is recommended to be used in the experiments. In steady forward flow testing, pressure difference across the valve has to be measured over a flow rate range of 5—30 L/min in 5 L/min increments. Pressure taps should be located one tube diameter upstream and three tube diameters downstream of the prosthetic valve. In steady backflow testing, on the other hand, static leakage across the heart valve substitute should be conducted at five back pressures over a range of 40 —200 mmHg. In addition to heart valve substitutes, standard nozzles with details listed in ISO-5840 should be tested to characterize the accuracy of pressure, forward flow, and leakage measurements [15]. The hydrodynamic results should be included in the regulatory submission packages.

8.1.2 Pulsatile flow systems

Pulsatile flow systems can closely mimic blood flow characteristics within the heart and vasculature system. For testing prosthetic heart valves, a pulse duplicator system can be utilized to replicate ventricular and arterial function. Pulse duplicator is an in vitro closed flow loop system that is designed to generate a pulsatile flow and simulate hemodynamic condition in the left- or right side of heart. In the past few decades, several homemade [16—29] and commercial [30—33] pulse duplicator systems have been developed and used to examine prosthetic heart valves (Fig. 8.2). A reciprocating hydraulic or pneumatic-driven silicone diaphragm is often used to create the pulsatile flow. Physiological condition can be simulated by controlling flow resistance and local compliance within the pulse duplicator system. ISO-5840 lists the required test configurations for qualitative and quantitative assessments of prosthetic heart valves in pulsatile flow experiments. For instance, transvalvular pressure gradient should be

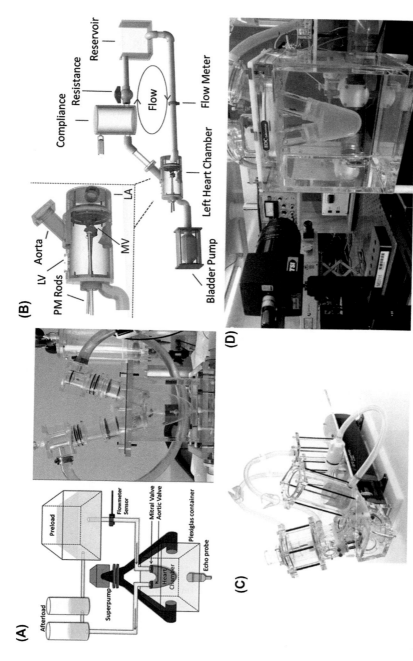

Figure 8.2 (A) Homemade pulse duplicator system for testing aortic and mitral valves. (B) Homemade pulse duplicator system for testing mitral valves. (C) Vivitro commercial pulse duplicator system. (D) BDC custom-built commercial pulse duplicator system. ((A) Reprinted with permission from Morisawa D, Falahatpisheh A, Avenatti E, Little SH, Kheradvar A. Intraventricular vortex interaction between transmitral flow and paravalvular leak. Sci Rep. 2018;8(1):15657. (B) Reprinted with permission from Bloodworth CH, Pierce EL, Easley TF, Drach A, Khalighi AH, Toma M, et al. Ex vivo methods for informing computational models of the mitral valve. Ann Biomed Eng. 2017;45(2):496—507. (D) Reprinted with permission from Vahidkhah K, Barakat M, Abbasi M, Javani S, Azadani PN, Tandar A, et al. Valve thrombosis following transcatheter aortic valve replacement: significance of blood stasis on the leaflets. Eur J Cardiothorac Surg. 2017;51(5):927—935.)

measured at a constant heart rate (e.g., 70 beats per minute) and variable flow rate (e.g., 2, 3.5, 5, and 7 L/min). In addition, regurgitation volumes should be measured at three different pressure conditions representative of hypotensive, normotensive, and severe hypertensive conditions, as described previously in Table 8.1. The obtained pressure and flow waveforms should closely resemble physiological conditions for the intended prosthetic valve application.

To closely mimic blood flow characteristics in a pulse duplicator system, kinematic similarity should be properly adopted. Kinematic similarity is generally met when both geometric and dynamic similarity conditions are satisfied. Geometric similarity involves the similarity of shape, which means the geometry of pulse duplicator should resemble human heart geometry. Therefore, to test aortic heart valves, realistic glass, acrylic, silicone, and cadaveric aortic roots have been previously utilized in pulse duplicator systems (Fig. 8.3) [30−32,36,38−40]. On the other hand, to satisfy dynamic similarity, viscosity and density of the working fluid in a pulse duplicator has to match both blood viscosity (3−4 cP) and blood density (1060 kg/m^3). Under the above condition, both Reynolds and Womersley numbers for the working fluid and blood will be equivalent. Therefore, aqueous glycerin solutions have been widely used for in vitro testing of heart valve substitutes [41]. At room temperature (22°C), a recirculating fluid of 37% by volume glycerin solution (99% purity) in normal saline solution, and at 37°C, a recirculating fluid of 45% by volume glycerol (99% purity) in normal saline solution can be used as a blood analog fluid with viscosity of 3.5 cP. In case refractive index of the working fluid has to be matched with solid boundaries for optical velocimetry measurements, other blood analog solutions such as sodium iodide solutions have been utilized in the past in the in vitro tests [11,42,43].

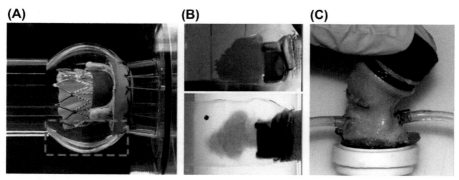

Figure 8.3 (A) Acrylic aortic root chamber designed on the basis of anatomic average values of human aortic root. (B) Aortic root phantoms created in transparent and compliant silicone. Red dye was used as contrast agent. (C) Human cadaveric aortic root was implanted in a pulse duplicator system [36]. *((A) Reprinted with permission from Midha PA, Raghav V, Condado JF, Arjunon S, Uceda DE, Lerakis S, et al. How can we help a patient with a small failing bioprosthesis?: an in vitro case study. JACC Cardiovasc Interv 2015;8(15):2026−2033. (B) Reprinted with permission from Jahren SE, Heinisch PP, Hasler D, Winkler BM, Stortecky S, Pilgrim T, et al. Can bioprosthetic valve thrombosis be promoted by aortic root morphology? An in vitro study. Interact Cardiovasc Thorac Surg 2018. (C) Used with permission from Vivitro labs.)*

Accuracy of quantitative flow measurements depends on calibration and type of measuring instruments. In the pulsatile flow systems, disposable blood pressure transducers have been commonly used for pressure measurements. Pressure transducers should have a differential measurement accuracy of at least ± 2 mmHg and an upper frequency limit of at least 30 Hz. Besides, ultrasonic and electromagnetic flow meters have been utilized to measure flow rate within the systems. Flow meters should have an upper frequency of at least 30 Hz. It is important to note that for accurate timing of flow measurements, there should be no flexible connection (e.g., flexible pipe) between the flowmeter and the prosthetic heart valve. Calibration of pressure transducers and flow meters is essential for accurate quantitative evaluation of prosthetic heart valves. Data acquisition should be run over at least 10 cardiac cycles, either consecutive or randomly selected. Quantitative hemodynamic endpoints such as mean transvalvular pressure gradient, effective orifice area, and regurgitation volume and regurgitation fraction should be calculated for the heart valve substitutes. In addition, transvalvular energy loss can be calculated to determine the potential impact of valve performance on the ventricular function [44,45]. Moreover, high-speed cameras can be utilized in the in vitro tests to determine leaflet kinematic of prosthetic valves during different phases of a cardiac cycle and detect events such as formation of cavitation bubbles in mechanical heart valves, as shown below in Fig. 8.4 [24,46,47].

8.2 Particle image velocimetry

Flow field characterization of prosthetic heart valves plays a critical role in assessment of the hemolytic and thrombogenic potential of various prosthetic valve designs. The hemodynamic results will manifest themselves in clinical outcome of patients who receive the heart valve prostheses. Laser–imaging techniques such as particle image

(A) **(B)** **(C)** **(D)**

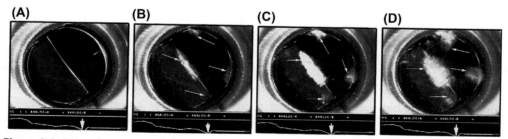

Figure 8.4 A timed sequence of photographs of gas-filled microbubbles formation in a 29 mm mitral St. Jude Medical bileaflet mechanical valve at closure. (A) End-diastolic period (<1 ms prior to complete closure), (B) the burst at the instant of closure and formation of clouds of microbubbles (t_0), (C) dissipation of clouds at about t_0+5 ms, and (D) pressure recovery and growth of persisting gas-filled bubbles at approximately t_0+8 ms. *(Adapted from Rambod E, Beizaie M, Shusser M, Milo S, Gharib M. A physical model describing the mechanism for formation of gas microbubbles in patients with mitral mechanical heart valves. Ann Biomed Eng. 1999;27(6):774–792, reproduced with permission.)*

velocimetry (PIV) and laser Doppler anemometry are widely available and used by scientific community to measure kinematic properties of fluid flow [48,49]. The techniques are noninvasive and require optical access to the flow field. Detailed quantitative measurements of flow velocity, viscous shear stress, Reynolds stress, and vorticity in the vicinity of prosthetic heart valves are feasible by means of the imaging techniques. The measurements can be highly accurate under appropriate settings, and the experimental data can be used for design modification and flow optimization of prosthetic heart valves. Moreover, the data can be analyzed for verification and validation of computational simulations.

PIV is a well-established and powerful method for flow velocity measurement. Two-dimensional (2D), three-dimensional (3D), and time-resolved three-dimensional (4D) imaging are feasible and would enable one to obtain a comprehensive characterization of flow structures. PIV enables full field flow measurements, which is a significant advantage as compared with single-point velocity measurement techniques such as LDA. In PIV measurements, flow field is seeded with small particles. The particles should be small enough to follow the flow and large enough to generate a strong scattering signal. The seeded flow is illuminated by a light source, typically a double-pulse laser. Light optics, commonly a combination of cylindrical and spherical lenses, is required to illuminate the region of interest. Furthermore, charge-coupled device cameras and high-speed cameras have been used frequently to record the particle image field. Good particle image focus is essential to collect high-quality raw data in PIV measurements. A control system, known as a synchronizer, is used as a timing device to synchronize the camera with the laser pulses. Two images are taken with a short known time delay between each capture. The two images are compared to determine particle displacement field. Particle displacement is then translated into instantaneous velocity measurements by dividing the displacement vector field by the known time delay.

For prosthetic heart valves, PIV measurements have been carried out by multiple research teams in steady [9,11,50,51] and pulsatile [17,19,23,33,38,39,52—68] flow systems (Fig. 8.5). Refractive index matching between the working fluid and solid nonflat boundaries is essential for accurate flow velocity measurements. For simple nondeforming solid boundaries, such as straight pipes, calibration functions can be applied to the PIV measurements to remove the optical distortion caused by the refractive index mismatch between the working fluid and the valve-mounting chamber. For prosthetic heart valves, hollow glass spheres with diameter of 8—12 μm are commonly used for seeding the flow. The particles provide a strong scattering signal and follow the flow structures. A concentration of 15—20 particles/mm^3 has been commonly used for 2D PIV measurements. Lower concentration of seeding particles should be considered in 3D PIV measurements to reduce the number of ghost particles in postprocessing. Moreover, calibration of the PIV system in situ just before or just after data acquisition is essential for achieving accurate and reliable measurements.

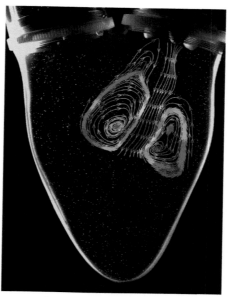

Figure 8.5 Streamlines of transmitral jet downstream a 25 mm St. Jude Mitral Biocor bioprosthesis [66]. *(Figure courtesy of Dr. Arash Kheradvar, University of California, Irvine.)*

In a pulse duplicator system, flow velocity passing through prosthetic heart valves changes during each cardiac cycle. Therefore, the time delay in PIV measurements has to be adjusted accordingly to obtain high-quality raw data. As a result, phase-locked PIV measurements have been commonly used to determine instantaneous pulsatile flow velocity field of heart valve substitutes. In phase-locked PIV measurements, the synchronizer can be triggered by a pulse generator that drives the piston pump. Triggering would allow data acquisition over multiple instances regularly spaced over the entire cardiac cycle. The time delay of the image acquisition can be adjusted to optimize particle image displacements. A sample of phase-locked PIV velocity measurement downstream of a surgical Carpentier-Edwards PERIMOUNT Magna aortic heart valve (Edwards Lifesciences, Irvine, CA) is presented in Fig. 8.6. The instantaneous flow velocity field can be used to determine instantaneous viscous shear stress downstream of the bioprosthetic heart valve. It is well-known that both shear stress magnitude and exposure time are important in thrombotic and hemolytic complications associated with prosthetic heart valves [70,71]. The higher the exposure time, the lower would be the threshold level of shear stress for blood cell impairment [72–75]. In addition to flow-induced blood damage downstream of prosthetic valves, micro-PIV measurements have been carried out to quantify velocity magnitude and shear stress in the hinge of mechanical heart valves to identify potential for blood cell damage and platelet activation [76,77].

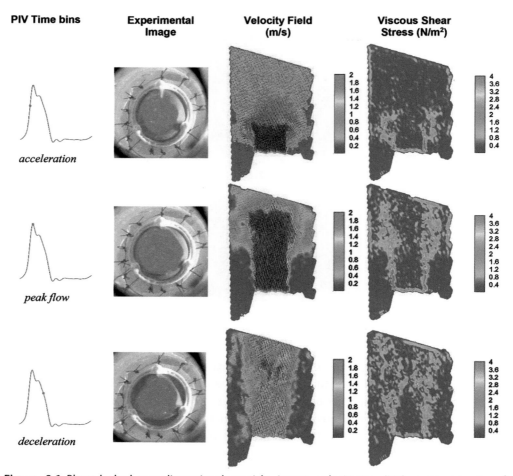

| PIV Time bins | Experimental Image | Velocity Field (m/s) | Viscous Shear Stress (N/m²) |

Figure 8.6 Phase-locked two-dimensional particle image velocimetry (PIV) measurements of Carpentier-Edwards PERIMOUNT Magna aortic bioprosthesis during (A) acceleration, (B) peak flow, and (C) deceleration phases of systole. *(Reprinted with permission from Barakat M, Dvir D, Azadani AN. Fluid dynamic characterization of transcatheter aortic valves using particle image velocimetry. Artif Organs 2018;42(11):E357—E368.)*

2D PIV measurements have widely been utilized to characterize flow field downstream of mechanical, bioprosthetic, and polymeric heart valves [17,71]. However, flow passing through prosthetic heart valves is 3D and spatially inhomogeneous. Recent advances in PIV techniques such as stereo-PIV and tomographic PIV have enabled instantaneous measurement of all three components of velocity and vorticity vectors [78]. Furthermore, components of velocity gradient tensor can be determined by time-resolved tomographic PIV. In addition to conventional PIV measurements, echocardiographic PIV using a combination of ultrasound imaging and PIV algorithms has been used in the in vitro systems to determine blood velocity field [79—82]. Echocardiographic PIV can be used to obtain a better understanding of hemodynamic response of prosthetic heart valves in in vivo setting. For further information regarding PIV data collection, data processing, and uncertainty analysis, the readers are referred to references such as [48,83—86] for more details.

8.3 Accelerated wear testing

Accelerated wear testing is a commonly used in vitro testing method to simulate wear and fatigue in prosthetic heart valves. In the past few decades, several homemade [87—89] and commercial [90—94] systems have been used to examine durability of prosthetic heart valves (Fig. 8.7). Heart valve durability testers normally operate at significantly higher frequency than the normal physiological heart rate (i.e., 30—200 bpm). The operating frequency range varies among different systems, but a typical frequency range of operation is 5—50 Hz (i.e., 300—3000 bpm). The ISO-5840 prescribes guidelines for durability testing of prosthetic heart valves. The standard mandates durability testing of bioprosthetic heart valves for a minimum of 200 million cycles (equivalent to 5 years in vivo). For mechanical heart valves, it is recommended by the regulatory agencies to test the valves for 400 million cycles (equivalent to 10 years in vivo). In the durability

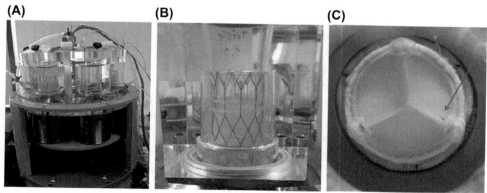

Figure 8.7 Accelerated durability testing. (A) Dynatek Labs M6 Heart Valve Tester. (B) An example of a transcatheter heart valve in the durability tester (FoldaValve [95], FOLDA, LLC, Rancho Santa Margarita, CA). (C) 23-mm Carpentier-Edwards PERIMOUNT Magna Ease Bioprosthesis after approximately 500 million cycles (equivalent to approximately 12.5 years). A pin hole was observed in the leaflet caused by a loose cloth. *((B) Reprinted with permission from Kheradvar A, Groves EM, Falahatpisheh A, Mofrad MK, Alavi SH, Tranquillo R, et al. Emerging trends in heart valve engineering: Part IV. Computational modeling and experimental studies. Ann Biomed Eng. 2015;43(10):2314—2333. (C) Reprinted with permission from Raghav V, Okafor I, Quach M, Dang L, Marquez S, Yoganathan AP. Long-term durability of Carpentier-Edwards Magna Ease valve: a one billion cycle in vitro study. Ann Thorac Surg. 2016;101(5):1759—1765.)*

tests, it is important to adjust the settings to attain complete valve opening and closing in each cycle. In addition, the peak differential closing pressure should be greater than 95 mmHg. Moreover, the transvalvular pressure difference must be maintained over 95% of the cycles. The temperature of the working fluid (e.g., 0.2% glutaraldehyde/water solution) should be kept at 37°C throughout the tests. Additionally, it is recommended to consider evaluation of transvalvular pressure every day, macroscopic inspection every 25 million cycles, and a microscopic inspection and hydrodynamic evaluation every 75 million cycles. Qualitative valve failure rating can be based on the size and number of pin holes and tears on the leaflets as well as the degree of fraying, incomplete cooptation, delamination, and excessive deformation of the leaflets (Fig. 8.7) [93]. The evaluation criteria and failure rating should be adjusted based on the individual heart valve design.

Although durability assessment of prosthetic heart valves is considered to be an essential component of preclinical testing, the applied load on the prosthetic heart valves in the accelerated durability testers is not physiologic. As a result, the prosthetic valves are often damaged under the excessive loading condition that exists in the accelerated durability testers. Valve damage is more pronounced in bioprosthetic heart valves. Bioprosthetic heart valve leaflets are made from fixed animal tissue, and soft tissue is well known to be viscoelastic in nature. Leaflet deformation is highly dependent on the rate of loading applied to the specimens. Consequently, varying degrees of correlation have been observed between in vitro accelerated durability test results and clinical data [87,88,97]. In addition, it has been shown that different test setups and methods may produce totally different results and outcomes [98]. As a result, in addition to accelerated wear testing, wear testing at physiological heart rate is recommended by ISO-5840. The results of the real-time wear testing can be used for validation of results obtained from accelerated durability tests.

Bioprosthetic heart valves degenerate through two distinct but potentially synergistic mechanisms: (1) mechanical wear and (2) calcification [99,100]. As a result, in vitro calcification solutions have been also used in the durability testers to assess durability in presence of calcification and study the factors that promote bioprosthetic valve calcification (Fig. 8.8) [103]. A number of different calcification solutions have been used in the in vitro durability testers [101,102,104,105]. In the in vitro tests, depletion of calcium and calcium uptake by the prosthetic heart valves can be measured to determine the rate and progression of calcification. Multiple different imaging modalities, such as X-ray, micro-CT, and scanning electron microscopy (SEM), can be used to quantify the degree of calcification and characterize morphology and chemical composition. Although in vitro calcification assessment is a promising approach, further studies are needed to confirm the validity and accuracy of the assessment compared to clinical results.

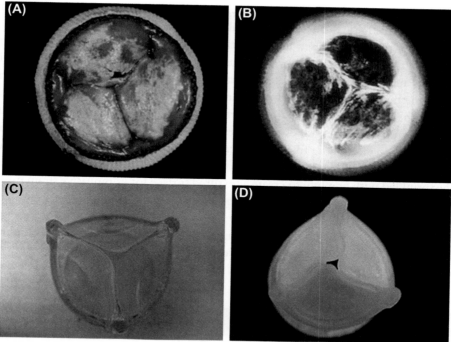

Figure 8.8 (A) A St. Jude Medical porcine bioprosthetic valve examined in vitro with a pulsatile accelerated calcification testing device. (B) X-ray image of the bioprosthesis shows massive mineralization after 19 million cycles. Angioflex polymeric valve (C) before and (D) after being subjected to cyclic loading in a calcification solution for over 50 million cycles. ((B) Reprinted with permission from Pettenazzo E, Deiwick M, Thiene G, Molin G, Glasmacher B, Martignago F, et al. Dynamic in vitro calcification of bioprosthetic porcine valves: evidence of apatite crystallization. J Thorac Cardiovasc Surg. 2001;121(3):500−509. (D) Reprinted with permission from Zadeh PB, Corbett SC, Nayeb-Hashemi H. In-vitro calcification study of polyurethane heart valves. Mater Sci Eng C. 2014;35:335−340.)

8.4 Structural assessment

The role of computational modeling and simulation is becoming increasingly important in design and development of prosthetic heart valves. Computational simulations reduce the need (and associated cost) to perform in vitro experimental tests and in vivo preclinical animal studies. Prosthetic heart valves feature materials such as metals (e.g., stainless steel, cobalt alloys, and Nitinol), ceramics (e.g., polycarbonate), polymers (e.g., Dacron, polyester, and polytetrafluoroethylene), and biological tissue (e.g., fixed bovine and porcine pericardium). In computational simulations, considering accurate constitutive models for the materials is essential to properly simulate mechanical response of prosthetic heart valves under physiological loading conditions. Constitutive models essentially estimate physical behavior of materials under a specific condition of interest [106−108].

As a result, in vitro experimental tests have been carried out to determine mechanical properties of prosthetic heart valve materials.

A number of materials used in bioprosthetic heart valves such as the fixed biological tissue exhibit nonlinear anisotropic viscoelastic mechanical behavior. As a result, uniaxial [109—116] and biaxial [117—124] tensile tests have been widely carried out to determine the in-plane mechanical properties of the leaflets. The tensile tests provide useful insight regarding the in-plane stress—strain behavior of the samples. Uniaxial tests are limited because the applied loading does not represent in vivo physiological loading condition. The planar mechanics of the fixed biological tissue can be more appropriately characterized by biaxial tests. In biaxial tensile tests, samples are mounted on four linear arms and can be sprinkled with graphite to create a textured surface for noninvasive strain measurements (Fig. 8.9). A normal saline bath heated to 37°C can provide a physiological environment for the tissue. Force can be measured by two load cells located on the two orthogonal arms. Real-time displacement of the graphite markers can be obtained using a camera placed over the top surface of the samples. Soft tissue preconditioning should be considered before the tests. In the preconditioning phase, the soft tissue is

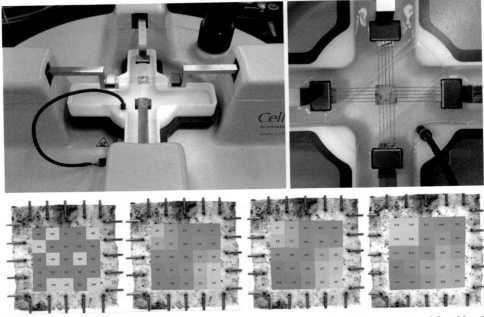

Figure 8.9 Planar biaxial stretching system used to determine mechanical properties of fixed bovine pericardium specimen (*top row*). Strain maps of the bovine pericardium sample (*bottom row*). Strain increases from left to right during equibiaxial stretching. (*Reprinted with permission from Abbasi M, Azadani AN. Leaflet stress and strain distributions following incomplete transcatheter aortic valve expansion. J Biomech. 2015;48(13):3663—3671.*)

subjected to a cycling load for a number of times (e.g., 10 cycles) from the unloaded state to a loaded state, which typically is a percentage of the load that is being tested. Soft tissue preconditioning will allow the tissue to align itself with the load before the actual measurement. After preconditioning, either force- or displacement–controlled cycles can be applied to the specimen to measure its stress—strain characteristics. Subsequently, image tracking software can be used to obtain strain maps of the samples (Fig. 8.9, bottom row).

To incorporate detailed structural features of soft tissue in computational models (e.g., orientation and concentration of collagen fibers), in vitro techniques such as small-angle light scattering technique [124—126] and second-harmonic generation (SHG) microscopy [122] have been utilized previously (Fig. 8.10). Small-angle light scattering technique is a reliable tool that can be used to map the gross collagen fiber orientation of thin biological tissue. For example, Hiester and Sacks [127,128] quantified the collagen fiber architecture of bovine pericardium sac. The fiber architecture of the sac demonstrated considerable intra- and intersac variability, and therefore there were no anatomical regions within the sac from which structurally consistent tissues could be obtained for bioprosthetic heart valve leaflet fabrication [128]. Recently, Alavi et al. used SHG microscopy to characterize the 3D collagen fiber architecture of fixed bovine pericardium leaflets in response to a variety of different loading conditions [122]. SHG microscopy is a powerful technique to characterize the organization of collagen fibers in biological tissue. The real-time visualization method showed that the collagen fibers do not necessarily all align with the load in each layer throughout the depth of tissue [122].

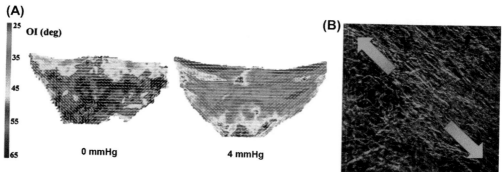

Figure 8.10 (A) Small-angle light scattering data of two porcine aortic valve leaflets fixed at 0 and 4 mmHg. (B) The architecture of the collagen fibers in the leaflet of a porcine mitral valve. The second-harmonic generation microscopy shows the arrangement of collagen fibers in their relaxed state. ((A) Reprinted with permission from Sacks MS, Smith DB, Hiester ED. A small angle light scattering device for planar connective tissue microstructural analysis. Ann Biomed Eng. 1997;25(4):678—689. (B) Reprinted with permission from Kheradvar A, Groves EM, Falahatpisheh A, Mofrad MK, Alavi SH, Tranquillo R, et al. Emerging trends in heart valve engineering: Part IV. Computational modeling and experimental studies. Ann Biomed Eng. 2015;43(10):2314—2333.)

Although uniaxial and biaxial tensile tests have been widely used to characterize mechanical properties of bioprosthetic valve leaflets, the planer tests cannot determine the out-of-plane mechanical properties of the tissue [129]. To determine the 3D mechanical properties of soft biological tissue, triaxial mechanical testing has been developed and used in the past few years [130–132]. In triaxial testing techniques, deformation modes in three dimensions can be applied on small cubic specimens. Although triaxial mechanical testing device is a powerful tool, the device is not applicable to thin specimens such as bioprosthetic heart valve leaflets [130,131]. In addition, significant interspecimen and intraspecimen exist in mechanical properties of biological soft tissue [127,128]. As a result, mechanical properties of the tissue examined by tensile tests may be significantly different from the actual materials used in prosthetic heart valves. As a result, inverse finite element analysis is a valuable method that has been used to estimate 3D mechanical properties of bioprosthetic heart valve leaflets [133–136]. In addition, inverse finite element methods would allow one to determine mechanical properties of the specific tissue of interest used in bioprosthetic heart valves under intact condition. For inverse finite element simulations of bioprosthetic heart valves, experimental optical measurements such as digital image correlation (DIC) technique can be used to determine leaflet deformation noninvasively under physiological loading condition (Fig. 8.11). 3D DIC measurement systems are becoming more readily available and have been widely used in experimental mechanics [138–143]. The DIC technique is particularly suited well to determine leaflet displacement and strain fields of bioprosthetic heart valves because of its high temporal and special resolution.

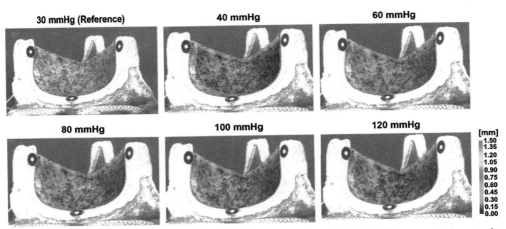

Figure 8.11 Displacement contour plots of Carpentier-Edwards PERIMOUNT Magna aortic heart valve (Edwards Lifesciences, CA) at back pressures of 30–120 mmHg. *(Reprinted with permission from Abbasi M, Qiu D, Behnam Y, Dvir D, Clary C, Azadani AN. High resolution three-dimensional strain mapping of bioprosthetic heart valves using digital image correlation. J Biomech. 2018.)*

The DIC system consists of two high-speed cameras to obtain a stereometric view of the object. The surface of the object has to be covered with a random speckle pattern for data acquisition (Fig. 8.11). Camera calibration is essential to achieve accurate and reliable measurements. The experimental displacement field can be used in inverse finite element simulations to determine 3D anisotropic mechanical properties of leaflets on a valve-specific basis.

Few in vitro studies have investigated fatigue damage of soft biological tissue as applied to bioprosthetic heart valves. Fatigue experiments are often time-consuming, complex, and not always reflect the true in vivo mechanical and chemical environment. Therefore, the existing challenges to some extent have hindered the development of a reliable fatigue damage model for bioprosthetic heart valves. Because of the simplicity of the approach, uniaxial fatigue loading experiments under either load or displacement control settings were commonly adopted to determine progression of fatigue in isolated tissue [124,144–148]. Uniaxial fatigue tests are limited because the applied loading does not represent in vivo physiological loading condition. Therefore, biaxial fatigue experiments may potentially be utilized to determine tissue fatigue properties under more physiologically relevant loading conditions. Other experimental techniques, such as cyclic bending, have also been developed and utilized to determine fatigue properties of soft tissue in conditions closer to physiological reality [149–157]. For further information regarding fatigue damage of collagenous tissues, the readers are referred to Ref. [158].

8.5 Structural component fatigue assessment

Structural component fatigue assessment of prosthetic heart valve substitutes is one of the key elements of preclinical studies mandated by the regulatory agencies. Therefore, stress/strain analysis of the prosthetic heart valve components should be performed under in vivo loading conditions using finite element analysis. The analysis should account for all of the steps from initial fabrication to actual placement of the prosthetic heart valve in the heart. In addition, all of the loading conditions that may be applied to the prosthetic heart valve in in vivo clinical applications should be considered in the simulations (e.g., normal, hypertensive, and hypotensive conditions). Validation of the stress/strain analysis is required to demonstrate assurance in the predicted results. Following stress/strain analysis, fatigue characterization of the prosthetic heart valve materials should be conducted to determine the material fatigue strength. Fatigue characterization generally falls into the following three categories: (1) stress/life (S/N) characterization for materials such as engineering alloys and polymers, (2) strain/life (ε/N) characterization for materials such as Nitinol, and (3) fatigue crack growth for use in damage tolerance analysis characterization for materials such as pyrolytic carbon. The stress/strain analysis results are then compared with the material fatigue characterization test results to assess fatigue lifetime and determine the fatigue safety factor of the component. Finally, device-level fatigue tests should be performed in addition to structural component fatigue assessment to confirm structural reliability predictions. For further details and references, the readers are referred to ISO-5840 [5].

8.6 Corrosion assessment

Corrosion resistance is an indispensable property of implantable devices with metallic parts [159,160]. Corrosion may occur simultaneously through several pathways, e.g., pitting corrosion, crevice corrosion, and galvanic corrosion. Corrosion can lead to mechanical failure of the implantable device components. Furthermore, corrosion by-products could cause adverse biological responses such as local immune response and inflammatory and allergic reactions [161,162]. In transcatheter heart valves, the existence of a metal stent obligates use of in vitro tests to assess corrosion resistance of the stent during preclinical studies. Commonly used standard methods include ASTM F2129 and ASTM F746. In addition, nondestructive test techniques, such as electrochemical impedance spectroscopy (ASTM G106) and electrochemical noise measurements (ASTM G199), might be beneficial for monitoring corrosion properties. Moreover, surface morphology of the stent can be assessed by SEM. In addition, surface chemistry of the stent can be determined by X-ray photoelectron spectroscopy and Auger electron spectroscopy. It is recommended to determine the corrosion resistance of heart valve substitutes in a physiological environment both prefatigue and postfatigue testing. Although corrosion assessment is mandated by the regulatory agencies, there are no universal acceptance criteria for preclinical corrosion testing, which is mainly because of difficulties in correlating the in vitro results with in vivo response [160]. As a result, using proper controls (e.g., another marketed device) is recommended to be considered in the in vitro tests. In addition, the manufacturers should provide rationale for the test methods and conditions, number of samples considered in the in vitro tests, and the adopted final acceptance criteria.

8.7 Summary

In conclusion, the primary purpose of this chapter was to review the commonly used in vitro methods to characterize hemodynamic and structural performance of prosthetic heart valves. We also reviewed the most relevant studies regarding in vitro assessment of prosthetic heart valve. There are additional in vitro tests and standards beyond those described in this chapter (e.g., MRI compatibility and biocompatibility). In vitro experimental testing has been an integral part of verification and assessment of heart valve substitutes. The advancements in experimental techniques reduce the use of animal models in safety and risk assessment of prosthetic heart valves. Furthermore, experimental methods play an important role in validation of computational models and simulations. A synergic combination of experimental techniques and computer-aided design tools would allow us to design and develop the future generation of heart valve substitutes. As Chaux and associates state [163], the elusive "Holy Grail" for heart valve replacement continues is a prosthesis that has excellent hemodynamic performance, does not require anticoagulation, and is durable throughout the projected lifetime of all possible recipients. Considering the enhanced capabilities of modern scientific testing and measurement, vigorous efforts are under way to achieve this longstanding goal.

References

[1]   Salerno C, Droel J, Bianco R. Current state of in vivo preclinical heart valve evaluation. J Heart Valve Dis 1998;7(2):158—62.
[2]   Bianco RW, Gallegos RP, Rivard AL, Voight J, Dalmasso AP. Animal models for cardiac research. In: Handbook of cardiac anatomy, physiology, and devices. Springer; 2009. p. 393—410.
[3]   Oberkampf WL, Trucano TG, Hirsch C. Verification, validation, and predictive capability in computational engineering and physics. Appl Mech Rev 2004;57(5):345—84.
[4]   Food, Administration D. Reporting of computational modeling studies in medical device submissions—draft guidance for industry and food and Drug administration staff only. Rockville, MD: Food and Drug Administration; 2014.
[5]   ISO 5840-1:2015. Cardiovascular implants – Cardiac valve prostheses – Part 1: general requirements. 2015.
[6]   Yap CH, Saikrishnan N, Tamilselvan G, Yoganathan AP. Experimental technique of measuring dynamic fluid shear stress on the aortic surface of the aortic valve leaflet. J Biomech Eng 2011; 133(6):061007.
[7]   Seaman C, Akingba AG, Sucosky P. Steady flow hemodynamic and energy loss measurements in normal and simulated calcified tricuspid and bicuspid aortic valves. J Biomech Eng 2014;136(4): 041001.
[8]   Steady flow characterization of prosthetic heart valves using computational fluid dynamics techniques. In: Ranjith G, Rufus R, Arun Kumar N, Sajin Raj A, Muraleedharan C, editors. International conference on system dynamics and control-ICSDC; 2010.
[9]   Lim W, Chew Y, Chew T, Low H. Steady flow dynamics of prosthetic aortic heart valves: a comparative evaluation with PIV techniques. J Biomech 1998;31(5):411—21.
[10] Smadi O, Fenech M, Hassan I, Kadem L. Flow through a defective mechanical heart valve: a steady flow analysis. Med Eng Phys 2009;31(3):295—305.
[11] Browne P, Ramuzat A, Saxena R, Yoganathan AP. Experimental investigation of the steady flow downstream of the St. Jude bileaflet heart valve: a comparison between laser Doppler velocimetry and particle image velocimetry techniques. Ann Biomed Eng 2000;28(1):39—47.
[12] Ge L, Leo H-L, Sotiropoulos F, Yoganathan AP. Flow in a mechanical bileaflet heart valve at laminar and near-peak systole flow rates: CFD simulations and experiments. J Biomech Eng 2005;127(5): 782—97.
[13] Yoganathan AP, Corcoran WH, Harrison EC. Pressure drops across prosthetic aortic heart valves under steady and pulsatile flow—in vitro measurements. J Biomech 1979;12(2):153—64.
[14] Garcia D, Pibarot P, Dumesnil JG, Sakr F, Durand L-G. Assessment of aortic valve stenosis severity: a new index based on the energy loss concept. Circulation 2000;101(7):765—71.
[15] ISO 5840-1:2015. Cardiovascular implants – Cardiac valve prostheses – Part 2: surgically implanted heart valve substitutes. 2015.
[16] Kheradvar A, Gharib M. On mitral valve dynamics and its connection to early diastolic flow. Ann Biomed Eng 2009;37(1):1—13.
[17] Falahatpisheh A, Kheradvar A. High-speed particle image velocimetry to assess cardiac fluid dynamics in vitro: from performance to validation. Eur J Mech B Fluid 2012;35:2—8.

[18]   Midha PA, Raghav V, Condado JF, Arjunon S, Uceda DE, Lerakis S, et al. How can we help a patient with a small failing bioprosthesis?: an in vitro case study. JACC Cardiovasc Interv 2015; 8(15):2026—33.

[19]   Hatoum H, Moore BL, Maureira P, Dollery J, Crestanello JA, Dasi LP. Aortic sinus flow stasis likely in valve-in-valve transcatheter aortic valve implantation. J Thorac Cardiovasc Surg 2017;154(1): 32—43. e1.

[20]   McNally A, Madan A, Sucosky P. Morphotype-dependent flow characteristics in bicuspid aortic valve ascending aortas: a benchtop particle image velocimetry study. Front Physiol 2017;8:44.

[21]   Rabbah J-P, Saikrishnan N, Yoganathan AP. A novel left heart simulator for the multi-modality characterization of native mitral valve geometry and fluid mechanics. Ann Biomed Eng 2013; 41(2):305—15.

[22]   Siefert AW, Rabbah JPM, Koomalsingh KJ, Touchton SA, Saikrishnan N, McGarvey JR, et al. In vitro mitral valve simulator mimics systolic valvular function of chronic ischemic mitral regurgitation ovine model. Ann Thorac Surg 2013;95(3):825—30.

[23]   Groves EM, Falahatpisheh A, Su JL, Kheradvar A. The effects of positioning of transcatheter aortic valve on fluid dynamics of the aortic root. Am Soc Artif Intern Organs J 2014;60(5):545.

[24]   Azadani AN, Reardon M, Simonato M, Aldea G, Nickenig G, Kornowski R, et al. Effect of transcatheter aortic valve size and position on valve-in-valve hemodynamics: an in vitro study. J Thorac Cardiovasc Surg 2017;153(6):1303—15.

[25]   Shandas R, Kwon J, Valdes-Cruz L. A method for determining the reference effective flow areas for mechanical heart valve prostheses: in vitro validation studies. Circulation 2000;101(16):1953—9.

[26]   Robicsek F, Thubrikar MJ. Role of sinus wall compliance in aortic leaflet function. Am J Cardiol 1999;84(8):944—6.

[27]   De Paulis R, Schmitz C, Scaffa R, Nardi P, Chiariello L, Reul H. In vitro evaluation of aortic valve prosthesis in a novel valved conduit with pseudosinuses of Valsalva. J Thorac Cardiovasc Surg 2005; 130(4):1016—21.

[28]   Sedaghat A, Sinning J-M, Utzenrath M, Ghalati PF, Schmitz C, Werner N, et al. Hydrodynamic performance of the Medtronic CoreValve and the Edwards SAPIEN XT transcatheter heart valve in surgical bioprostheses: an in vitro valve-in-valve model. Ann Thorac Surg 2016;101(1):118—24.

[29]   Meschini V, De Tullio M, Querzoli G, Verzicco R. Flow structure in healthy and pathological left ventricles with natural and prosthetic mitral valves. J Fluid Mech 2018;834:271—307.

[30]   Azadani AN, Jaussaud N, Matthews PB, Chuter TA, Ge L, Guy TS, et al. Aortic valve-in-valve implantation: impact of transcatheter- bioprosthesis size mismatch. J Heart Valve Dis 2009;18(4): 367—73.

[31]   Azadani AN, Jaussaud N, Matthews PB, Ge L, Chuter TA, Tseng EE. Transcatheter aortic valves inadequately relieve stenosis in small degenerated bioprostheses. Interact Cardiovasc Thorac Surg 2010;11(1):70—7.

[32]   Vahidkhah K, Barakat M, Abbasi M, Javani S, Azadani PN, Tandar A, et al. Valve thrombosis following transcatheter aortic valve replacement: significance of blood stasis on the leaflets. Eur J Cardiothorac Surg 2017;51(5):927—35.

[33]   Ducci A, Tzamtzis S, Mullen MJ, Burriesci G. Hemodynamics in the Valsalva sinuses after transcatheter aortic valve implantation (TAVI). J Heart Valve Dis 2013;22(5):688—96.

[34]   Morisawa D, Falahatpisheh A, Avenatti E, Little SH, Kheradvar A. Intraventricular vortex interaction between transmitral flow and paravalvular leak. Sci Rep 2018;8(1):15657.

[35]   Bloodworth CH, Pierce EL, Easley TF, Drach A, Khalighi AH, Toma M, et al. Ex vivo methods for informing computational models of the mitral valve. Ann Biomed Eng 2017;45(2):496—507.

[36]   Azadani AN, Jaussaud N, Ge L, Chitsaz S, Chuter TA, Tseng EE. Valve-in-valve hemodynamics of 20-mm transcatheter aortic valves in small bioprostheses. Ann Thorac Surg 2011;92(2):548—55.

[37]   Jahren SE, Heinisch PP, Hasler D, Winkler BM, Stortecky S, Pilgrim T, et al. Can bioprosthetic valve thrombosis be promoted by aortic root morphology? An in vitro study. Interact Cardiovasc Thorac Surg 2018;27:108—11.

[38]   Kheradvar A, Kasalko J, Johnson D, Gharib M. An in vitro study of changing profile heights in mitral bioprostheses and their influence on flow. Am Soc Artif Intern Organs J 2006;52(1):34—8.

[39] Saikrishnan N, Yap C-H, Milligan NC, Vasilyev NV, Yoganathan AP. In vitro characterization of bicuspid aortic valve hemodynamics using particle image velocimetry. Ann Biomed Eng 2012; 40(8):1760—75.

[40] Bark DL, Vahabi H, Bui H, Movafaghi S, Moore B, Kota AK, et al. Hemodynamic performance and thrombogenic properties of a superhydrophobic Bileaflet mechanical heart valve. Ann Biomed Eng 2017;45(2):452—63.

[41] Association GP. Physical properties of glycerine and its solutions. Glycerine Producers' Association; 1963.

[42] Budwig R. Refractive index matching methods for liquid flow investigations. Exp Fluid 1994;17(5): 350—5.

[43] Bai K, Katz J. On the refractive index of sodium iodide solutions for index matching in PIV. Exp Fluid 2014;55(4):1704.

[44] Akins CW, Travis B, Yoganathan AP. Energy loss for evaluating heart valve performance. J Thorac Cardiovasc Surg 2008;136(4):820—33.

[45] Azadani AN, Jaussaud N, Matthews PB, Ge L, Guy TS, Chuter TA, et al. Energy loss due to paravalvular leak with transcatheter aortic valve implantation. Ann Thorac Surg 2009;88(6):1857—63.

[46] Rambod E, Beizaie M, Shusser M, Milo S, Gharib M. A physical model describing the mechanism for formation of gas microbubbles in patients with mitral mechanical heart valves. Ann Biomed Eng 1999;27(6):774—92.

[47] Brennen CE. Cavitation in medicine. Interface focus 2015;5(5):20150022.

[48] Adrian RJ, Westerweel J. Particle image velocimetry. Cambridge University Press; 2011.

[49] Principles of laser Doppler anemometers. In: Durst F, editor. Von karman inst of fluid dyn meas and predictions of complex turbulent flows, vol 1 11; 1980 (SEE N81-15263 06-34).

[50] Shandas R, Kwon J. Digital particle image velocimetry (DPIV) measurements of the velocity profiles through bileaflet mechanical valves: in vitro steady. Biomed Sci Instrum 1996;32:161—7.

[51] Lim WL, Chew YT, Chew TC, Low HT. Steady flow velocity field and turbulent stress mappings downstream of a porcine bioprosthetic aortic valveIn Vitro. Ann Biomed Eng 1997;25(1):86—95.

[52] Manning KB, Kini V, Fontaine AA, Deutsch S, Tarbell JM. Regurgitant flow field characteristics of the St. Jude bileaflet mechanical heart valve under physiologic pulsatile flow using particle image velocimetry. Artif Organs 2003;27(9):840—6.

[53] Leo HL, Dasi LP, Carberry J, Simon HA, Yoganathan AP. Fluid dynamic assessment of three polymeric heart valves using particle image velocimetry. Ann Biomed Eng 2006;34(6):936—52.

[54] Brücker C, Steinseifer U, Schröder W, Reul H. Unsteady flow through a new mechanical heart valve prosthesis analysed by digital particle image velocimetry. Meas Sci Technol 2002;13(7):1043.

[55] Subramanian A, Mu H, Kadambi J, Wernet M, Brendzel A, Harasaki H. Particle image velocimetry investigation of intravalvular flow fields of a bileaflet mechanical heart valve in a pulsatile flow. J Heart Valve Dis 2000;9(5):721—31.

[56] Lim W, Chew Y, Chew T, Low H. Particle image velocimetry in the investigation of flow past artificial heart valves. Ann Biomed Eng 1994;22(3):307—18.

[57] Kini V, Bachmann C, Fontaine A, Deutsch S, Tarbell J. Integrating particle image velocimetry and laser Doppler velocimetry measurements of the regurgitant flow field past mechanical heart valves. Artif Organs 2001;25(2):136—45.

[58] Marassi M, Castellini P, Pinotti M, Scalise L. Cardiac valve prosthesis flow performances measured by 2D and 3D-stereo particle image velocimetry. Exp Fluid 2004;36(1):176—86.

[59] Kaminsky R, Kallweit S, Weber HJ, Claessens T, Jozwik K, Verdonck P. Flow visualization through two types of aortic prosthetic heart valves using stereoscopic high-speed particle image velocimetry. Artif Organs 2007;31(12):869—79.

[60] Lim W, Chew Y, Chew T, Low H. Pulsatile flow studies of a porcine bioprosthetic aortic valve in vitro: PIV measurements and shear-induced blood damage. J Biomech 2001;34(11):1417—27.

[61] Akutsu T, Saito J. Dynamic particle image velocimetry flow analysis of the flow field immediately downstream of bileaflet mechanical mitral prostheses. J Artif Organs 2006;9(3):165—78.

[62] Shi Y, Yeo TJH, Zhao Y, Hwang NH. Particle image velocimetry study of pulsatile flow in bi-leaflet mechanical heart valves with image compensation method. J Biol Phys 2006;32(6):531—51.

[63] Akutsu T, Fukuda T. Time-resolved particle image velocimetry and laser Doppler anemometry study of the turbulent flow field of bileaflet mechanical mitral prostheses. J Artif Organs 2005;8(3):171—83.

[64] Balducci A, Grigioni M, Querzoli G, Romano G, Daniele C, D'Avenio G, et al. Investigation of the flow field downstream of an artificial heart valve by means of PIV and PTV. Exp Fluid 2004;36(1): 204—13.

[65] Lu P-C, Liu J-S, Huang R-H, Lo C-W, Lai H-C, Hwang NH. The closing behavior of mechanical aortic heart valve prostheses. Am Soc Artif Intern Organs J 2004;50(4):294—300.

[66] Kheradvar A, Falahatpisheh A. The effects of dynamic saddle annulus and leaflet length on transmittal flow pattern and leaflet stress of a bileaflet bioprosthetic mitral valve. J Heart Valve Dis 2012;21(2):225.

[67] Stühle S, Wendt D, Hou G, Wendt H, Schlamann M, Thielmann M, et al. In-vitro investigation of the hemodynamics of the Edwards Sapien™ transcatheter heart valve. J Heart Valve Dis 2011;20(1):53.

[68] Vahidkhah K, Cordasco D, Abbasi M, Ge L, Tseng E, Bagchi P, et al. Flow-induced damage to blood cells in aortic valve stenosis. Ann Biomed Eng 2016:1—13.

[69] Barakat M, Dvir D, Azadani AN. Fluid dynamic characterization of transcatheter aortic valves using particle image velocimetry. Artif Organs 2018;42(11):E357—68.

[70] Yoganathan AP, He Z, Casey Jones S. Fluid mechanics of heart valves. Annu Rev Biomed Eng 2004; 6:331—62.

[71] Dasi LP, Simon HA, Sucosky P, Yoganathan AP. Fluid mechanics of artificial heart valves. Clin Exp Pharmacol Physiol 2009;36(2):225—37.

[72] Leverett L, Hellums J, Alfrey C, Lynch E. Red blood cell damage by shear stress. Biophys J 1972; 12(3):257—73.

[73] Hung TC, Hochmuth RM, Joist JH, Sutera SP. Shear-induced aggregation and lysis of platelets. T Am Soc Art Int Org 1976;22:285—91.

[74] Williams AR. Release of serotonin from human platelets by acoustic microstreaming. J Acoust Soc Am 1974;56(5):1640—9.

[75] Ramstack JM, Zuckerman L, Mockros LF. Shear-induced activation of platelets. J Biomech 1979; 12(2):113—25.

[76] Jun BH, Saikrishnan N, Yoganathan AP. Micro particle image velocimetry measurements of steady diastolic leakage flow in the hinge of a St. Jude Medical® regent™ mechanical heart valve. Ann Biomed Eng 2014;42(3):526—40.

[77] Jun BH, Saikrishnan N, Arjunon S, Yun BM, Yoganathan AP. Effect of hinge gap width of a St. Jude medical bileaflet mechanical heart valve on blood damage potential—an in vitro micro particle image velocimetry study. J Biomech Eng 2014;136(9):091008.

[78] Hinsch KD, Hinrichs H. Three-dimensional particle velocimetry. In: Three-dimensional velocity and vorticity measuring and image analysis techniques. Springer; 1996. p. 129—52.

[79] Kim H, Hertzberg J, Shandas R. Development and validation of echo PIV. Exp Fluid 2004;36(3): 455—62.

[80] Kheradvar A, Houle H, Pedrizzetti G, Tonti G, Belcik T, Ashraf M, et al. Echocardiographic particle image velocimetry: a novel technique for quantification of left ventricular blood vorticity pattern. J Am Soc Echocardiogr 2010;23(1):86—94.

[81] Voorneveld J, Muralidharan A, Hope T, Vos HJ, Kruizinga P, van der Steen AF, et al. High frame rate ultrasound particle image velocimetry for estimating high velocity flow patterns in the left ventricle. IEEE Trans Ultrason Ferroelectr Freq Control Dec. 2018;65(12):2222—32.

[82] Hong G-R, Pedrizzetti G, Tonti G, Li P, Wei Z, Kim JK, et al. Characterization and quantification of vortex flow in the human left ventricle by contrast echocardiography using vector particle image velocimetry. J Am Coll Cardiol 2008;1(6):705—17.

[83] Raffel M, Willert CE, Wereley ST, Kompenhans J. Particle image velocimetry: a practical guide. Springer; 2013.

[84] Raghav V, Sastry S, Saikrishnan N. Experimental assessment of flow fields associated with heart valve prostheses using particle image velocimetry (PIV): recommendations for best practices. Cardiovasc Eng Technol 2018:1—15.

[85] Schroeder A, Willert CE. Particle image velocimetry: new developments and recent applications. Springer Science & Business Media; 2008.

[86] Coleman HW, Steele WG. Experimentation, validation, and uncertainty analysis for engineers. John Wiley & Sons; 2009.

[87] Iwasaki K, Umezu M, Iijima K, Imachi K. Implications for the establishment of accelerated fatigue test protocols for prosthetic heart valves. Artif Organs 2002;26(5):420—9.

[88] Clark RE, Swanson W, Kardos JL, Hagen RW, Beauchamp RA. Durability of prosthetic heart valves. Ann Thorac Surg 1978;26(4):323—35.

[89] Grigioni M, Daniele C, D'avenio G, Calcagnini G, Barbaro V. Wear patterns from fatigue life testing of a prosthetic heart valve. Wear 1999;225:743—8.

[90] Cox JL, Ad N, Myers K, Gharib M, Quijano R. Tubular heart valves: a new tissue prosthesis design—preclinical evaluation of the 3F aortic bioprosthesis. J Thorac Cardiovasc Surg 2005; 130(2):520—7.

[91] Alavi SH, Baliarda MS, Bonessio N, Valdevit L, Kheradvar A. A tri-leaflet nitinol mesh scaffold for engineering heart valves. Ann Biomed Eng 2017;45(2):413—26.

[92] Raghav V, Okafor I, Quach M, Dang L, Marquez S, Yoganathan AP. Long-term durability of Carpentier-Edwards Magna Ease valve: a one billion cycle in vitro study. Ann Thorac Surg 2016; 101(5):1759—65.

[93] Gallocher SL. Durability assessment of polymer trileaflet heart valves. Florida International University; 2007.

[94] Butterfield M, Fisher J. Fatigue analysis of clinical bioprosthetic heart valves manufactured using photooxidized bovine pericardium. J Heart Valve Dis 2000;9(1):161—6. discussion 7.

[95] Kheradvar A, Groves EM, Tseng E. Proof of concept of FOLDAVALVE, a novel 14 Fr totally repositionable and retrievable transcatheter aortic valve. 2015.

[96] Kheradvar A, Groves EM, Falahatpisheh A, Mofrad MK, Alavi SH, Tranquillo R, et al. Emerging trends in heart valve engineering: Part IV. Computational modeling and experimental studies. Ann Biomed Eng 2015;43(10):2314—33.

[97] Vesely I, Boughner DR, Leeson-Dietrich J. Bioprosthetic valve tissue viscoelasticity: implications on accelerated pulse duplicator testing. Ann Thorac Surg 1995;60:S379—83.

[98] Lu P-C, Liu J-s, Xi B, Li S, Wu J, Hwang NH. On accelerated fatigue testing of prosthetic heart valves. In: Frontiers in biomedical engineering. Springer; 2003. p. 185—96.

[99] Schoen FJ. Mechanisms of function and disease of natural and replacement heart valves. Annu Rev Pathol 2012;7:161—83.

[100] Schoen F, Levy R, Nelson A, Bernhard W, Nashef A, Hawley M. Onset and progression of experimental bioprosthetic heart valve calcification. Lab Invest 1985;52(5):523—32.

[101] Pettenazzo E, Deiwick M, Thiene G, Molin G, Glasmacher B, Martignago F, et al. Dynamic in vitro calcification of bioprosthetic porcine valves: evidence of apatite crystallization. J Thorac Cardiovasc Surg 2001;121(3):500—9.

[102] Zadeh PB, Corbett SC, Nayeb-Hashemi H. In-vitro calcification study of polyurethane heart valves. Mater Sci Eng C 2014;35:335—40.

[103] Bernacca G, Fisher A, Mackay T, Wheatley D. A dynamic in vitro method for studying bioprosthetic heart valve calcification. J Mater Sci Mater Med 1992;3(4):293—8.

[104] Golomb G, Wagner D. Development of a new in vitro model for studying implantable polyurethane calcification. Biomaterials 1991;12(4):397—405.

[105] Starcher BC, Urry DW. Elastin coacervate as a matrix for calcification. Biochem Biophys Res Commun 1973;53(1):210—6.

[106] Holzapfel G. Nonlinear solid mechanics: a continuum approach for engineering. West Sussex, England: John Wiley & Sons, Ltd; 2000.

[107] Lai WM, Rubin DH, Krempl E, Rubin D. Introduction to continuum mechanics. Butterworth-Heinemann; 2009.

[108] Humphrey JD. Cardiovascular solid mechanics: cells, tissues, and organs. Springer Science & Business Media; 2013.

[109]   Trowbridge E, Black M, Daniel C. The mechanical response of glutaraldehyde-fixed bovine pericardium to uniaxial load. J Mater Sci 1985;20(1):114—40.

[110]   Lee JM, Haberer SA, Boughner DR. The bovine pericardial xenograft: I. Effect of fixation in aldehydes without constraint on the tensile viscoelastic properties of bovine pericardium. J Biomed Mater Res A 1989;23(5):457—75.

[111]   Crofts C, Trowbridge E. The tensile strength of natural and chemically modified bovine pericardium. J Biomed Mater Res A 1988;22(2):89—98.

[112]   Zioupos P, Barbenel J. Mechanics of native bovine pericardium: II. A structure based model for the anisotropic mechanical behaviour of the tissue. Biomaterials 1994;15(5):374—82.

[113]   JGa P, Herrero EJ, Sanmartın AC, Millan I, Cordon A, Maestro MM, et al. Comparison of the mechanical behaviors of biological tissues subjected to uniaxial tensile testing: pig, calf and ostrich pericardium sutured with Gore-Tex. Biomaterials 2003;24(9):1671—9.

[114]   Zioupos P, Barbenel J, Fisher J. Anisotropic elasticity and strength of glutaraldehyde fixed bovine pericardium for use in pericardial bioprosthetic valves. J Biomed Mater Res A 1994;28(1): 49—57.

[115]   Sánchez-Arévalo F, Farfán M, Covarrubias D, Zenit R, Pulos G. The micromechanical behavior of lyophilized glutaraldehyde-treated bovine pericardium under uniaxial tension. J Mech Behav Biomed Mater 2010;3(8):640—6.

[116]   Oswal D, Korossis S, Mirsadraee S, Wilcox H, Watterson K, Fisher J, et al. Biomechanical characterization of decellularized and cross-linked bovine pericardium. J Heart Valve Dis 2007; 16(2):165.

[117]   Sun W, Sacks MS, Sellaro TL, Slaughter WS, Scott MJ. Biaxial mechanical response of bioprosthetic heart valve biomaterials to high in-plane shear. J Biomech Eng 2003;125(3):372—80.

[118]   Billiar KL, Sacks MS. Biaxial mechanical properties of the natural and glutaraldehyde treated aortic valve cusp-Part I: experimental results. J Biomech Eng 2000;122(1):23—30.

[119]   Abbasi M, Azadani AN. Leaflet stress and strain distributions following incomplete transcatheter aortic valve expansion. J Biomech 2015;48(13):3663—71.

[120]   Lee M, LeWinter MM, Freeman G, Shabetai R, Fung Y. Biaxial mechanical properties of the pericardium in normal and volume overload dogs. Am J Physiol Heart Circ Physiol 1985;249(2): H222—30.

[121]   Páez JG, Jorge E, Rocha A, Castillo-Olivares J, Millan I, Carrera A, et al. Mechanical effects of increases in the load applied in uniaxial and biaxial tensile testing. Part II. Porcine pericardium. J Mater Sci Mater Med 2002;13(5):477—83.

[122]   Alavi SH, Ruiz V, Krasieva T, Botvinick EL, Kheradvar A. Characterizing the collagen fiber orientation in pericardial leaflets under mechanical loading conditions. Ann Biomed Eng 2013;41(3): 547—61.

[123]   Chew PH, Yin FC, Zeger SL. Biaxial stress-strain properties of canine pericardium. J Mol Cell Cardiol 1986;18(6):567—78.

[124]   Wells SM, Sacks MS. Effects of fixation pressure on the biaxial mechanical behavior of porcine bioprosthetic heart valves with long-term cyclic loading. Biomaterials 2002;23(11):2389—99.

[125]   Sacks MS, Smith DB, Hiester ED. A small angle light scattering device for planar connective tissue microstructural analysis. Ann Biomed Eng 1997;25(4):678—89.

[126]   Billiar K, Sacks M. A method to quantify the fiber kinematics of planar tissues under biaxial stretch. J Biomech 1997;30(7):753—6.

[127]   Hiester ED, Sacks MS. Optimal bovine pericardial tissue selection sites. I. Fiber architecture and tissue thickness measurements. J Biomed Mater Res 1998;39(2):207—14.

[128]   Hiester ED, Sacks MS. Optimal bovine pericardial tissue selection sites. II. Cartographic analysis. J Biomed Mater Res 1998;39(2):215—21.

[129]   Holzapfel GA, Ogden RW. On planar biaxial tests for anisotropic nonlinearly elastic solids. A continuum mechanical framework. Math Mech Solids 2009;14(5):474—89.

[130]   Sommer G, Schriefl AJ, Andrä M, Sacherer M, Viertler C, Wolinski H, et al. Biomechanical properties and microstructure of human ventricular myocardium. Acta Biomaterialia 2015;24: 172—92.

[131] Avazmohammadi R, Li DS, Leahy T, Shih E, Soares JS, Gorman JH, et al. An integrated inverse model-experimental approach to determine soft tissue three-dimensional constitutive parameters: application to post-infarcted myocardium. Biomechanics Model Mechanobiol 2017:1—23.

[132] Dokos S, LeGrice IJ, Smaill BH, Kar J, Young AA. A triaxial-measurement shear-test device for soft biological tissues. J Biomech Eng 2000;122(5):471—8.

[133] Abbasi M, Barakat MS, Vahidkhah K, Azadani AN. Characterization of three-dimensional anisotropic heart valve tissue mechanical properties using inverse finite element analysis. J Mech Behav Biomed Mater 2016;62:33—44.

[134] Aggarwal A, Sacks MS. An inverse modeling approach for semilunar heart valve leaflet mechanics: exploitation of tissue structure. Biomechanics Model Mechanobiol 2016;15(4):909—32.

[135] Murdock K, Martin C, Sun W. Characterization of mechanical properties of pericardium tissue using planar biaxial tension and flexural deformation. J Mech Behav Biomed Mater 2018;77:148—56.

[136] Abbasi M, Barakat MS, Dvir D, Azadani AN. A non-invasive material characterization framework for bioprosthetic heart valves. Ann Biomed Eng 2019;47(1):97—112.

[137] Abbasi M, Qiu D, Behnam Y, Dvir D, Clary C, Azadani AN. High resolution three-dimensional strain mapping of bioprosthetic heart valves using digital image correlation. J Biomech 2018 Jul 25;76:27—34.

[138] Chu T, Ranson W, Sutton MA. Applications of digital-image-correlation techniques to experimental mechanics. Exp Mech 1985;25(3):232—44.

[139] Palanca M, Brugo TM, Cristofolini L. Use of digital image correlation to investigate the biomechanics of the vertebra. J Mech Med Biol 2015;15(02):1540004.

[140] Validation of digital image correlation techniques for strain measurement in biomechanical test models. In: Rogge RD, Small SR, Archer DB, Berend ME, Ritter MA, editors. Proceedings of the ASME summer bioengineering conference; 2013.

[141] Luyckx T, Verstraete M, De Roo K, De Waele W, Bellemans J, Victor J. Digital image correlation as a tool for three-dimensional strain analysis in human tendon tissue. J Exp Orthop 2014;1(1):7.

[142] Sun W, Abad A, Sacks MS. Simulated bioprosthetic heart valve deformation under quasi-static loading. J Biomech Eng 2005;127(6):905—14.

[143] Heide-Jørgensen S, Krishna SK, Taborsky J, Bechsgaard T, Zegdi R, Johansen P. A novel method for optical high spatiotemporal strain analysis for transcatheter aortic valves in vitro. J Biomech Eng 2016; 138(3):034504.

[144] Broom N. The stress/strain and fatigue behaviour of glutaraldehyde preserved heart-valve tissue. J Biomech 1977;10(11—12):707—24.

[145] Broom N. Fatigue-induced damage in glutaraldehyde-preserved heart valve tissue. J Thorac Cardiovasc Surg 1978;76(2):202—11.

[146] Broom ND. An 'in vitro'study of mechanical fatigue in glutaraldehyde-treated porcine aortic valve tissue. Biomaterials 1980;1(1):3—8.

[147] Sun W, Sacks M, Fulchiero G, Lovekamp J, Vyavahare N, Scott M. Response of heterograft heart valve biomaterials to moderate cyclic loading. J Biomed Mater Res A 2004;69(4):658—69.

[148] Sellaro TL, Hildebrand D, Lu Q, Vyavahare N, Scott M, Sacks MS. Effects of collagen fiber orientation on the response of biologically derived soft tissue biomaterials to cyclic loading. J Biomed Mater Res A 2007;80(1):194—205.

[149] Garcia Paez J, Claramunt R, Jorge Herrero E, Millan I, Tolmos J, Alvarez L, et al. Energy consumption as a predictor test of the durability of a biological tissue employed in cardiac bioprosthesis. J Biomed Mater Res A 2009;89(2):336—44.

[150] Paez J, Sanmartín AC, Herrero EJ, Millan I, Cordon A, Rocha A, et al. Durability of a cardiac valve leaflet made of calf pericardium: fatigue and energy consumption. J Biomed Mater Res A 2006;77(4): 839—49.

[151] Mirnajafi A, Zubiate B, Sacks MS. Effects of cyclic flexural fatigue on porcine bioprosthetic heart valve heterograft biomaterials. J Biomed Mater Res A 2010;94(1):205—13.

[152] Raghavan D, Starcher BC, Vyavahare NR. Neomycin binding preserves extracellular matrix in bioprosthetic heart valves during in vitro cyclic fatigue and storage. Acta Biomaterialia 2009;5(4): 983—92.

[153] Billiar KL, Sacks MS. Long-term mechanical fatigue response of porcine bioprosthetic heart valves. ASME-Publications-Bed 1998;39:343–4.

[154] Smith DB, Sacks MS, Pattany PM, Schroeder R. Fatigue-induced changes in bioprosthetic heart valve three-dimensional geometry and the relation to tissue damage. J Heart Valve Dis 1999;8(1): 25–33.

[155] Sacks MS, Smith DB. Effects of accelerated testing on porcine bioprosthetic heart valve fiber architecture. Biomaterials 1998;19(11–12):1027–36.

[156] Sacks MS. The biomechanical effects of fatigue on the porcine bioprosthetic heart valve. J Long Term Eff Med Implant 2001;11(3&4).

[157] Bernacca G, Mackay T, Wilkinson R, Wheatley D. Polyurethane heart valves: fatigue failure, calcification, and polyurethane structure. J Biomed Mater Res 1997;34(3):371–9.

[158] Martin C, Sun W. Fatigue damage of collagenous tissues: experiment, modeling and simulation studies. J Long Term Eff Med Implant 2015;25(1–2).

[159] Manivasagam G, Dhinasekaran D, Rajamanickam A. Biomedical implants: corrosion and its prevention-A review. Recent Pat Corros Sci 2010;2:40–54.

[160] Nagaraja S, Di Prima M, Saylor D, Takai E. Current practices in corrosion, surface characterization, and nickel leach testing of cardiovascular metallic implants. J Biomed Mater Res B Appl Biomater 2017;105(6):1330–41.

[161] Santin M, Mikhalovska L, Lloyd AW, Mikhalovsky S, Sigfrid L, Denyer SP, et al. In vitro host response assessment of biomaterials for cardiovascular stent manufacture. J Mater Sci Mater Med 2004;15(4):473–7.

[162] Wataha JC, O'Dell NL, Singh BB, Ghazi M, Whitford GM, Lockwood PE. Relating nickel-induced tissue inflammation to nickel release in vivo. J Biomed Mater Res A 2001;58(5):537–44.

[163] Chaux A, Gray RJ, Stupka JC, Emken MR, Scotten LN, Siegel R. Anticoagulant independent mechanical heart valves: viable now or still a distant holy grail. Ann Transl Med 2016;4(24).

CHAPTER 9

Transvalvular flow

Daisuke Morisawa[1,3], Arash Kheradvar[1,2,3]
[1]Department of Biomedical Engineering, University of California Irvine, Irvine, CA, United States; [2]Department of Medicine, Division of Cardiology, University of California Irvine, CA, United States; [3]The Edwards Lifesciences Center for Advanced Cardiovascular Technology, University of California Irvine, CA, United States

Contents

9.1 Fluid dynamics of transmitral flow

Diastolic function of the left ventricle (LV) mainly depends on transmitral flow. During early diastole, transmitral flow (TMF) is the integral outcome of ventricular relaxation, compliance, and filling pressures. Diastolic suction or transvalvular pressure drop is considered to actively contribute to early ventricular filling [1].

9.1.1 Transvalvular pressure drop

Pressure drop in a fluid is generally defined as the difference in total pressure between the two points in the fluid, where the total pressure in the first point is higher than the second point. This can be mathematically described by the general Bernoulli's equation, if the flow's inertial forces dominate the viscous forces (Fig. 9.1A):

$$P_1 - P_2 = \frac{1}{2}\rho\left(v_1^2 - v_2^2\right) + \rho \int_1^2 \frac{dv}{dt} \cdot ds \qquad (9.1)$$

where P_1 and P_2 are downstream and upstream pressure, ρ is density of fluid, v_2 and v_1 are downstream and upstream flow velocity, dv, dt, and ds are differential element related to

Principles of Heart Valve Engineering
ISBN 978-0-12-814661-3, https://doi.org/10.1016/B978-0-12-814661-3.00009-5

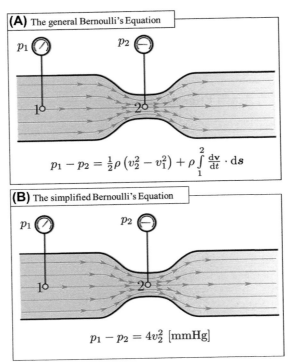

(A) The general Bernoulli's Equation

p_1

p_2

$$p_1 - p_2 = \tfrac{1}{2}\rho\left(v_2^2 - v_1^2\right) + \rho\int_1^2 \tfrac{dv}{dt}\cdot d\boldsymbol{s}$$

(B) The simplified Bernoulli's Equation

p_1

p_2

$$p_1 - p_2 = 4v_2^2\ [\mathrm{mmHg}]$$

Figure 9.1 *The general and simplified Bernoulli's equation.* (A) General Bernoulli's equation and (B) simplified Bernoulli's equation, which is modified based on the premise that v_1^2 is smaller enough than v_2^2 and $\rho\int_1^2\frac{dv}{dt}\,ds$ can be neglected. *(Modified from Falahatpisheh A, Rickers C, Gabbert D, Heng EL, Stalder A, Kramer HH, Kilner PJ, Kheradvar A. Simplified Bernoulli's method significantly underestimates pulmonary transvalvular pressure drop. J Magn Reson Imag 2016;43:1313–19.)*

velocity, time, and path, respectively. The last term of the equation accounts for the fluid inertia.

In human heart, pressure drop often occurs due to sudden changes in a chamber volume, which leads to opening of a heart valve and transvalvular blood stream. In some situations due to heart valve stenosis, the heart chamber upstream of the stenotic valve develops a pressure overload that alters the natural pressure drop phenomenon and significantly affects the blood flow stream. Thus, precise assessment of the stenosis severity is essential to establish the strategy to approach the stenotic valve disease. In clinical practice, transmitral pressure drop is essential to determine the severity of the diastolic dysfunction. In the case of mitral insufficiency with valve regurgitation, left atrium (LA) develops volume overload as characterized by left atrial volume index [2]. In many patients with left heart failure, mitral regurgitation is often present, and the severity of the regurgitation changes according to the fluid accumulation, indicating systolic dysfunction and worsening of the heart failure. Tricuspid regurgitant flow is commonly used for the estimation of pulmonary systolic pressure indicating the presence of pulmonary edema, elevation of left atrial pressure, and/or left ventricular end–diastolic pressure.

Although cardiac catheterization is currently the gold standard for measurement of transvalvular pressure drop, it is not always practical for routine follow-up due to its invasive nature. In 1978, Hetle et al., for the first time, introduced the method for non-invasive assessment of pressure drop using Doppler echocardiography [3]. In their study, they applied the simplified Bernoulli's equation for clinical use. They considered that when inertial effect and the v_1 are small enough, they can be neglected to simplify the general Bernoulli's equation (Eq. 9.1) as follows:

$$P_1 - P_2 = \frac{1}{2}\rho v_2^2 \tag{9.2}$$

This method is often used to calculate the pressure gradient from the blood flow velocity, and if the velocity and pressure are expressed in terms of [m/s] and [mmHg], Eq. (9.2) can be reduced to

$$\Delta P = 4v_2^2 \tag{9.3}$$

where ΔP is the transvalvular pressure drop (Fig. 9.1B). After the introduction of the simplified Bernoulli's equation to clinical practice, it has been widely used worldwide.

Although the simplified Bernoulli's method is very popular in clinical practice, some limitations have been reported in the previous studies. Falahatpisheh et al. compared the values of pressure drop calculated from the simplified Bernoulli's equation with those from the general Bernoulli's in patients with valve stenosis and reported that the simplified Bernoulli's method systematically underestimates the pressure drop [4]. More recently, Ha et al. described a method using 4D flow MRI and showed that turbulence-based pressure drop estimation can complement the simplified Bernoulli method for better assessment of valve diseases [5].

More information about the Bernoulli's equation's advantages and limitations are provided in the Appendix section.

9.1.2 Transmitral vortex formation

In fluid dynamics, vortex has been described as a swirling motion in the flow. Different types of vortices have been studied and characterized in the past 50 years. In human left heart, vortex formation phenomenon is observed during LV early filling, and the formed vortex undergoes interaction and translation during the isovolumic and ejection phases of the cardiac cycle [6,7]. Intraventricular vortex formation has been reported to be associated with the efficiency of blood ejection. Therefore, precise visualization and quantification of intraventricular vortex formation can provide additional insights into the pathophysiology of heart disease. Cardiac flow is highly three-dimensional, and presently, vortex formation can be visualized by two methods, namely phase-contrast cardiac magnetic resonance image (PCMRI) or 4D flow MRI [8,9] and echocardiographic particle image velocimetry (echo-PIV). While 4D flow MRI can provide the 3D flow field and its reliability has been validated, 4D flow cardiac MRI is expensive, requires extensive postprocessing, and is not accessible in most clinics. Further, 4D flow cardiac MRI does not provide real-time information, which limits its application to capturing only relatively large-scale flow structures, as effective phase–averaging over several cardiac cycles prevents imaging of local flow instabilities. Nominal 2D echo-PIV has been performed in B-mode and can be performed in real time at the bedside even for patients with unstable conditions in the intensive care unit. Thus, echo-PIV is expected to be widely used in real clinical setting in the near future.

In a healthy heart, TMF rapidly moves from LA to LV during the early diastole because of intraventricular pressure gradient and develops asymmetric shear layers at the trailing edges of the mitral valve [10,11]. Compared with the mitral valve's posterior leaflet, a larger shear layer emerges at the anterior leaflet, not only because of the asymmetric shape of mitral leaflet whose anterior leaflet is larger than posterior (Fig. 9.2A) but also because of the proximity of the leaflet tip to the ventricular wall [11]. This shear layer moves toward the LV apex and rolls up into a vortex (Fig. 9.2B). During the vortex formation process, vortex ring eventually pinches off from the transmitral jet. Some previous studies have reported that isovolumic relaxation and contraction phases also significantly contribute to the LV vortex dynamics. As isovolumic contraction period begins after mitral valve closure, transmitral vortex moves toward the vicinity of the anterior mitral leaflet (Fig. 9.3G,H), turning from the LV apex to outflow tract. Immediately following the aortic valve opening, the vortex dissipates and blood flow is ejected through the systolic phase with the constant flow direction from apex toward aortic valve (Fig. 9.3I) [6,12]. In isovolumic relaxation period, flow direction reverses from the LV base to the apex

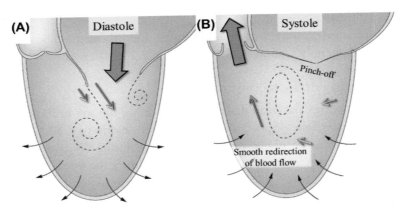

Figure 9.2 *Schematic layout of left ventricular fluid dynamics.* (A) Transmitral flow (TMF) during diastole: The difference in the length of the anterior and posterior leaflets leads to asymmetric shear layers at the edges of the mitral valve. The TMF rolls up into a nonaxisymmetric vortex ring. (B) Intraventricular vortex flow during systole: The pinched off vortex from the TMF changes its direction toward the outflow tract. *(Modified from Pedrizzetti G, La Canna G, Alfieri O, Tonti G. The vortex—an early predictor of cardiovascular outcome? Nat Rev Cardiol 2014;11:545—53.)*

Normal Heart

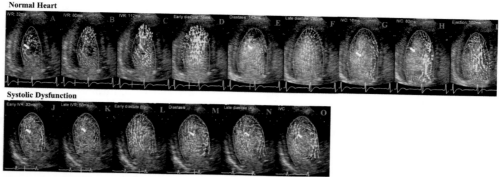

Systolic Dysfunction

Figure 9.3 *Intraventricular flow visualization by echocardiographic particle image velocimetry.* (upper panel) Intraventricular vortex formation in a normal heart. (lower panel) Intraventricular vortex formation in a patient with systolic dysfunction. During isovolumic contraction, transmitral vortex moves toward the outflow tract. During isovolumic relaxation period, flow direction reverses from the left ventricular base toward the apex despite the absence of any inflow into the left ventricle. *(Modified from Hong G-R, Pedrizzetti G, Tonti G, Li P, Wei Z, Kim JK, Baweja A, Liu S, Chung N, Houle H, Narula J, Vannan MA. Characterization and quantification of vortex flow in the human left ventricle by contrast echocardiography using vector particle image velocimetry. J Am Coll Cardiol Imag 2008;1:705–17.)*

despite the absence of any inflow into the LV (Fig. 9.3A–C) [6]. It is suggested that the appropriate transmitral vortex formation helps efficient diastolic filling, minimizes kinetic energy dissipation, and prevents intraventricular thrombus formation[13].

The 3D flow field acquired by 4D flow MRI in healthy subjects shows torus–shaped vortex ring core being formed from TMF throughout the diastole (Fig. 9.4). The vortex ring continues to develop toward the apex during E-wave filling phase and reaches its full development at the peak E-wave. During E-wave deceleration phase, the vortex core deforms, and its orientation aligns with the LV long axis. At the beginning of the A-wave, a similar vortex core is formed and persists at the LV base, pending the end of diastole [9]. This sequence of transmitral vortex formation can be abnormally deviated in diseased hearts. Hong et al. have reported that the vortex shape in patients with systolic dysfunction is spherical, which differs from healthy subjects' whose vortex shape is elliptical [6]. Other studies have reported that LV dilatation or dyskinesia generates an inefficient vortex leading to local flow stagnation and reducing mixing flow, which possibly trigger thrombus formation [14,15].

Compared with aortic, pulmonary and tricuspid valves that are symmetrically trileaf-let, mitral valve is asymmetrically bileaflet with a large anterior and small posterior leaflet. This asymmetric design produces a D–shaped orifice, which can contribute to develop-ment of a non-axisymmetric vortex ring downstream of the mitral valve. Falahatpisheh et al. have examined the three-dimensional vortex formation in the downstream of the D–shaped orifice in vitro using digital PIV and reported that D–shaped orifice and large

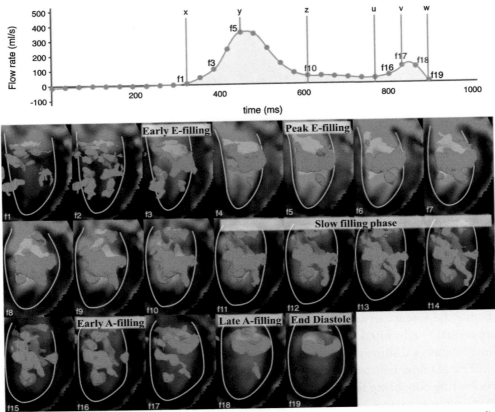

Figure 9.4 *Three-dimensional visualization of intraventricular flow using phase-contrast cardiac magnetic resonance image.* The transmitral vortex ring continues to develop toward the apex during E-wave filling phase and reaches to its full development at the peak E-wave. During E-wave deceleration phase, the vortex core deforms, and its orientation aligns with the left ventricular long axis. *(Modified from Elbaz MSM, Calkoen EE, Westenberg JJM, Lelieveldt BPF, Roest AAW, van der Geest RJ. Vortex flow during early and late left ventricular filling in normal subjects: quantitative characterization using retrospectively-gated 4d flow cardiovascular magnetic resonance and three-dimensional vortex core analysis. J Cardiovasc Magn Reson 2014;16:78.)*

anterior leaflet lead to the dynamic interaction of the shear layer rolling up from the anterior leaflet and reduce the nonuniformity in vorticity generation [10,16].

Several in vitro and in vivo studies on LV fluid dynamics have consistently shown the intraventricular vortex formation process in healthy subjects using echo–PIV and 4D flow cardiac MRI [6,9,17]. Transmitral vortex formation modulates kinetic energy and momentum transfer and contributes to efficient blood transport from LA toward the aorta [108]. Furthermore, many clinical studies have reported the characteristics of transmitral vortex formation in heart diseases.

9.1.3 Vortex formation time index

Vortex formation time (VFT) index is a nondimensional measure to characterize the formation of the vortex ring, which has been introduced by Gharib et al. [18] According to the experiments using a piston–cylinder setup (Fig. 9.5), they showed that the ratio of the piston stroke to the diameter of the cylinder nozzle conveys important information about the flow propulsion. They described VFT index as

$$\text{VFT} = \frac{L}{D} = \frac{\overline{U}t}{D}$$

where L is the piston stroke, D is the diameter of the cylinder nozzle, $\overline{U}$ is the piston's average velocity, and t is the duration of the pulse. According to Gharib et al., a formed vortex ring continues to grow until VFT reaches to ~ 4, and then the vortex ring growth terminates as VFT exceeds 4. In that case, the additional ejected fluid in the pulse would follow the leading vortex ring as a trailing jet (Fig. 9.6). This study further explains that the vortex formation completes at a VFT of ~ 4 because of energetic constraints. Therefore, a VFT of ~ 4 is considered an optimal fluid dynamics status considering efficient fluid transport through a pulsed jet. Later on, VFT index was further expended to characterize TMF in the human heart [19].

$$\text{VFT} = \frac{4(1 - \beta)}{\pi} \alpha^3 \times EF \tag{9.4}$$

where α is a nondimensional LV volumetric parameter and β is the fraction of the stroke volume contributed from the atrial contraction (A-wave) component of LV filling. Considering that the total blood volume contributed to the LV during E- and A-waves should be equal to the systolic stroke volume, ε can be described as

$$\varepsilon = \frac{V_E}{SV} = \frac{VTI_E}{VTI_E + VTI_A} = 1 - \beta \tag{9.5}$$

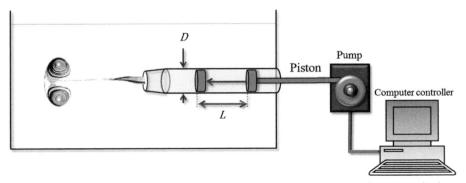

Figure 9.5 *General schematic of a vortex ring generator.* The experimental piston/cylinder setup used by Gharib M to establish the concept of vortex formation time. *(The vortex picture is extracted and modified from Gharib M, Rambod E, Kheradvar A, Sahn DJ, Dabiri JO. Optimal vortex formation as an index of cardiac health. Proc Natl Acad Sci USA 2006;103:6305–08.)*

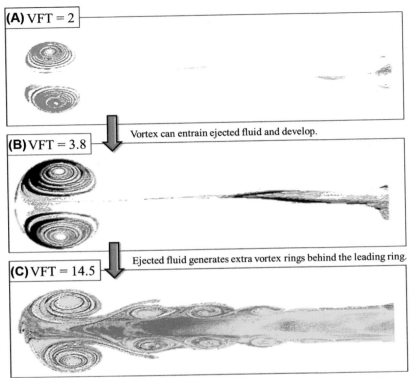

Figure 9.6 *Visualization of vortex rings at different stroke ratios.* (A) and (B) A formed vortex ring continues to grow until vortex formation time (VFT) reaches to ~4, and then the vortex ring growth terminates as VFT exceeds 4. (C) The additional ejected fluid in the pulse follows the leading vortex ring as a trailing jet when VFT exceeds ~4. *(Modified from Gharib M, Rambod, E., and Shariff, KA. universal time scale for vortex ring. Formation. J Fluid Mech 1998;360:121—40.)*

where V_A is the flow volume that enters the LV during atrial contraction, EDV is the left ventricular end-diastolic volume, VTI_E and VTI_A are the velocity time integral of E- and A-wave, respectively, and D_E is the effective diameter of mitral valve geometric orifice area (GOA):

$$D_E = 2\sqrt{\frac{GOA}{\pi}} \tag{9.6}$$

Nondimensional LV volumetric parameter, α, is obtained as

$$\alpha = \left(\frac{EDV}{D_E^3}\right)^{\frac{1}{3}} \tag{9.7}$$

Figure 9.7 *Vortex formation time (VFT) in the left ventricle.* VFT can be interpreted as an index to quantify the performance of an invisible piston transferring blood from the left atrium to left ventricle. *(Heart illustration is adopted and modified from Flachskampf et al. J Am Coll Cardiol Img 2015;8(9): 1071–93. Vortex images are extracted and modified from Gharib M, Rambod E, Kheradvar A, Sahn DJ, Dabiri JO. Optimal vortex formation as an index of cardiac health. Proc Natl Acad Sci USA 2006;103: 6305–08.)*

VFT can also be obtained based on conventional echocardiographic parameters by rewriting Eq. (9.4) as

$$\text{VFT} = \frac{4}{\pi} \times \frac{VTI_E}{VTI_E + VTI_A} \times \frac{EDV}{D_E^3} \times EF \tag{9.8}$$

$$\text{VFT} = \frac{4}{\pi} \times \frac{VTI_E}{VTI_E + VTI_A} \times \frac{SV}{D_E^3} \tag{9.9}$$

For the human heart, VFT can be interpreted as an index to quantify the performance of an invisible piston transferring blood from the LA to LV. Alternatively, VFT indicates the ability of LV diastolic suction (Fig. 9.7). More recent studies report that transmitral vortex pinches off from the transmitral jet in a VFT range of 3.3–5.5 rather than 4 [1,20,21].

9.1.4 Diastolic dysfunction and transmitral flow

Heart failure with preserved ejection fraction (HFpEF) affects over 50% of patients with heart failure symptoms [22]. Thus, assessment of LV diastolic function is crucial to make therapeutic decisions for this group of patients with diastolic dysfunction. The term "diastolic dysfunction" refers to abnormality in several key components of cardiac function, which leads to relaxation impairment, reduction in restoring force, and worsening of the chamber compliance. These functional impairments would eventually lead to elevation of LV filling and LA pressures, respectively. As the LV diastolic function deteriorates, transmitral blood flow velocity (E- and A-waves), mitral annular velocity (e′), and their respective ratios would change. The severity of diastolic dysfunction is generally classified as mild, moderate, and severe dysfunction, according to echocardiographic evaluation.

The ratio of transmitral E- to A-wave (E/A), the peak velocity of mitral annulus (e′), and E/e′ are established indices to assess the diastolic function. However, these noninvasive echocardiographic-based indices come with limitations in accuracy as they cannot directly measure LV filling pressure. E/A ratio is one of the most common indices to differentiate among the grades of diastolic dysfunction (i.e., impaired relaxation, pseudonormal and restrictive patterns). However, E/A cannot differentiate pseudonormal pattern from normal diastolic function because abnormal E/A ratio in mild diastolic dysfunction falsely changes toward normal pattern as the grade of diastolic dysfunction evolves. Presently, there is no single noninvasive imaging-based index that can directly and efficiently quantify LV filling pressure and chamber's relaxation performance to ultimately assess the severity of diastolic dysfunction. Therefore, clinicians integrate and process the information from the patient's clinical history, physical examinations, and multiple imaging modalities to accurately evaluate the diastolic function.

Recent advancements in cardiac imaging technology (e.g., echo-PIV or 4D flow MRI) have provided the means to visualize and assess the intraventricular fluid dynamics, particularly vortex formation. Although there is currently no well-established knowledge about the intraventricular fluid dynamics in patients with HFpEF, previous studies have reported the relationship between diastolic dysfunction and transmitral VFT. For example, Kheradvar et al. have reported that a statistically significant difference in transmitral VFT index exists between healthy subjects and patients with different grades of diastolic dysfunction (Fig. 9.8) [23]. They also found that most subjects with diastolic dysfunction whose e′ < 8 cm/s present with a lower range for VFT (<3.5) (Figs. 9.8 and 9.9). These results and studies by others [24,25] suggest that transmitral VFT can be used as an index to distinguish among grades of diastolic dysfunction.

9.1.5 Mitral annulus recoil

Dynamics of mitral valve annulus has been reported to be an important consideration to evaluate LV systolic and diastolic function. Carlhäll et al. showed that mitral annular excursion volume calculated by temporally integrated product of the annulus area and its incremental excursion represents $19 \pm 3\%$ of the stroke volume (Fig. 9.10) [26]. During diastole, mitral annulus moves toward the LA, regarded as mitral annulus recoil [107]. During diastole's rapid filling phase when transmitral jet flows from LA toward the

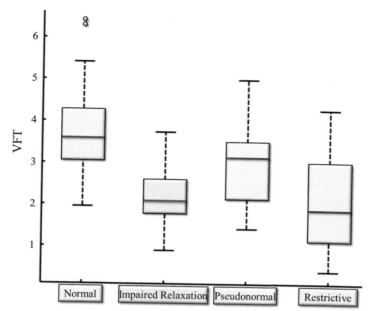

Figure 9.8 *Transmitral vortex formation time (VFT) index in normal subjects and patients with various grade of diastolic dysfunction.* Transmitral VFT index can be differentiated between healthy subjects and patients with different grades of diastolic dysfunction. *(Modified from Kheradvar A, Assadi R, Falahatpisheh A, Sengupta PP. Assessment of transmitral vortex formation in patients with diastolic dysfunction. J Am Soc Echocardiogr 2012;25:220—27.)*

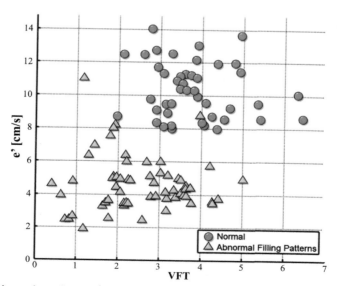

Figure 9.9 *Mitral annulus velocity (e′) versus vortex formation time (VFT) in normal subject compared with patients with abnormal diastolic filling patterns.* Patients with diastolic dysfunction whose e′ < 8 cm/s present with lower range of transmitral VFT. *(Modified from Kheradvar A, Assadi R, Falahatpisheh A, Sengupta PP. Assessment of transmitral vortex formation in patients with diastolic dysfunction. J Am Soc Echocardiogr 2012;25:220—27.)*

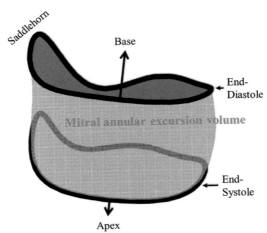

Figure 9.10 *Mitral annular excursion volume.* Mitral annular excursion volume is calculated by temporally integrated product of the annulus area and its incremental excursion, and it represents $19 \pm 3\%$ of the stroke volume. *(Modified from Carlhall C, Wigstrom L, Heiberg E, Karlsson M, Bolger AF, Nylander E. Contribution of mitral annular excursion and shape dynamics to total left ventricular volume change. Am J Physiol Heart Circ Physiol 2004;287:H1836—41.)*

LV apex, mitral valve annulus behaves according to the Newton's second and third laws. Thrust of the TMF due to LV suction derives the recoil force that leads to mitral annular movement away from the apex, in an opposite direction to the TMF. This mitral annulus movement defines the longitudinal LV deformation, which is strongly correlated to LV diastolic function. Accordingly, several studies have suggested that the mitral annulus motion and recoil force provide additional information to evaluate LV diastolic function [1,21,27,28]. In clinical practice, mitral annulus velocity is commonly measured by tissue Doppler imaging, and its peak velocity (e') during early diastole has been established as a representation of mitral valve recoil, a useful index to assess LV diastolic function. It is now common knowledge that e' diminishes in patients with impaired diastolic function [29]. Kheradvar et al. showed that mitral valve recoil is strongly associated with the transmitral vortex formation, and the maximal mitral valve recoil occurs when the transmitral vortex is pinched off from the transmitral jet, regardless of the valve size or the characteristics of ventricular pressure drop [20,21]. As previously mentioned, VFT is an index describing vortex formation, and the vortex's pinch-off phenomenon occurs in the VFT range of 3.5—5.5, denoting optimal momentum transfer for TMF. Kheradvar et al. also clinically confirmed the correlation between VFT and e' and found that diastolic dysfunction patients with reduced mitral recoil ($e' > 8$) usually exhibit lower than optimal VFT range (<3.5) [23]. Further clinical studies are needed to establish VFT index for grading diastolic dysfunction.

9.1.6 Consequences of mitral valve dysfunction

Mitral valve separates the LV from LA and has a unique structure characterized by its bileaflet and dynamic saddle-shaped annulus, evidently different from the other three heart valves. Mitral valve's unique structure may be attributed to the fact that it should endure the harsh LV pressure during systole. Mitral valve complex is a sophisticated structure that comprises of mitral annulus and two dissimilar leaflets that are connected to the papillary muscles and ventricular wall via the chordae tendineae. To maintain the appropriate hemodynamics, mitral valve complex is required to perform two major tasks namely, ensuing the smooth TMF during diastole and rigorous leaflet coaptation to avoid flow regurgitation during systole. Thus, malfunction of the mitral valve complex can lead to conditions ranging from stenosis to insufficiency. Although mitral stenosis is more common in developing countries due to prevalence of rheumatic heart disease, mitral insufficiency is currently more prevailing in developed countries.

Mitral valve uniquely possesses a nonplanar, saddle-shaped annulus. Some clinical studies have indicated that the geometry of mitral annulus is an important factor in establishing functional mitral regurgitation, as seen in mitral valve prolapse and acute ischemic mitral regurgitation [30—33]. Furthermore, the dynamic saddle-shaped annulus of mitral valve has been reported to bear the load and reduce the stress exerted on the mitral valve's leaflets [11,34]. It is suggested that inappropriate geometry of mitral valve due to various heart conditions can lead to the elevation of the mechanical stress on the leaflets, which eventually leads to mitral regurgitation during systole and calcification and stenosis, that ultimately affect diastolic TMF.

Functional mitral regurgitation—defined as the mitral regurgitation due to the adverse LV remodeling and/or abnormally shaped mitral annulus with structurally normal leaflet—is commonly seen in cardiology clinics. It has been reported that 20%—35% of the cases of functional mitral regurgitation occur because of ischemic heart disease [35—37]. In a healthy heart, each component of mitral complex coordinately works during both systole and diastole. Principally, the precise coordination is required in systole to prevent regurgitation when exposed to high systolic pressure. In patients with functional mitral regurgitation, lack of coordination among the mitral complex can be observed due to the deformation of annulus and LV and/or displacement of papillary muscles. As mentioned earlier, normal mitral annulus appears nonplanar saddle-shaped. However, in functional mitral regurgitation, annulus shape is usually dilated and flattened with reduction in circumferential contraction, which leads to suboptimal coaptation (Fig. 9.11) [38,39]. The LV chamber dilatation commonly observed in heart failure patients can occasionally lead to papillary muscle displacement and mitral leaflet "tethering" via the chordae tendineae (Fig. 9.12) [40,41]. Furthermore, reduced LV systolic function diminishes the closing force of mitral leaflets such that eventually results in functional mitral regurgitation in patients with ischemic heart disease, dilated

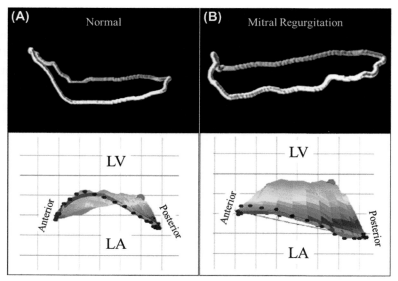

Figure 9.11 *Shape of the mitral annulus in normal subject and patients with mitral regurgitation.* (A) Normal mitral annulus appears nonplanar saddle-shaped. (B) In functional mitral regurgitation, the annulus shape is usually dilated and flattened with reduction in circumferential contraction, which leads to suboptimal coaptation. *(Modified from Watanabe N, Maltais S, Nishino S, O'Donoghue TA, Hung J. Functional mitral regurgitation: imaging insights, clinical outcomes and surgical principles. Prog Cardiovasc Dis 2017;60:351—60.)*

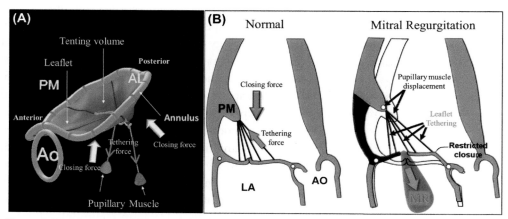

Figure 9.12 Schematic of (A) mitral complex and (B) mechanism of functional mitral regurgitation. In heart failure patients with LV dilatation, the papillary muscles are displaced, which leads to the mitral leaflet "tethering," which eventually results in functional mitral regurgitation. *(Modified from Watanabe N, Maltais S, Nishino S, O'Donoghue TA, Hung J. Functional mitral regurgitation: imaging insights, clinical outcomes and surgical principles. Prog Cardiovasc Dis 2017;60:351—60 and Liel-Cohen et al. Circulation 2000;101:2756—63.)*

cardiomyopathy, and hypertensive heart disease (in dilated phase). Surgical treatment such as mitral valve replacement, mitral valve repair, and annuloplasty are conventionally performed to mitigate symptomatic mitral valve dysfunction. Nowadays, thanks to the advancement of medical device technology, percutaneous approaches such as edge-to-edge repair, MitraClip (Abbott Laboratories, Chicago, IL), Edwards PASCAL Transcatheter Valve Repair System (Edwards Lifesciences, Irvine, CA) and coronary sinus cinching devices are also available. Nevertheless, the success of both surgical and percutaneous treatments depends on precise information of mitral valve geometry, and therefore, 3D-echocardiography seems to play a more important role in the future.

9.1.7 Flow through the mechanical valves

To replace a diseased valve, patients currently have only two choices: mechanical heart valve (MHV) or bioprosthetic heart valve (BHV), each comes with advantages and drawbacks. MHVs tend to last longer than BHVs, but they carry a greater long-term risk for thromboembolism that may lead to stroke and arterial thrombosis. Besides, high shear environment due to rigid leaflets can lead to hemolysis and anemia. Compared with BHVs, some clinical studies have reported lower long-term mortality in patients with MHV in their mitral position. In particular, MHVs are considered superior among patients younger than 70 years old [42,43]. Although BHV implantation is currently growing, more MHVs will potentially be implanted in the future, according to recent clinical studies [42,43]. Therefore, better understanding of MHVs' fluid dynamics is crucial to develop valves with optimal performance. As monoleaflet MHVs are currently less preferred for mitral position, in this section, we mainly focus on the fluid dynamics of bileaflet mechanical valves to replace a native mitral valve.

Flow through a bileaflet MHV implanted at mitral position is strongly affected by the valve orientation [44,45]. In 1996, Van Rijk-Zwikker et al. reported that the flow through an MHV placed at the anatomical orientation (i.e., the MHV's hinge line is parallel to a line through the commissures of the native valve) is redirected at the apex and does not smoothly stream toward the outflow tract (Fig. 9.13A). Otherwise, in anti-anatomical orientation (i.e., the MHV's hinge line is oriented toward the LV outflow tract), the redirected flow smoothly streams toward the LV outflow tract (Fig. 9.13B) [44]. Although some previous studies have reported that antianatomical orientation can lead to a favorable LV flow pattern compared with anatomical orientation, optimal orientation has not yet been fully defined [44–47].

Fluid dynamics analysis of the MHVs, in vitro and in vivo, has thus far been comprehensively studied. Faludi et al. have shown that the pattern of flow and vortex formation vary between healthy subjects and patients with an MHV [48]. In patients with bileaflet MHV implanted at the anatomical orientation, the valve splits the TMF into two jets toward the LV septum and the lateral wall, respectively. Subsequently, the jet component

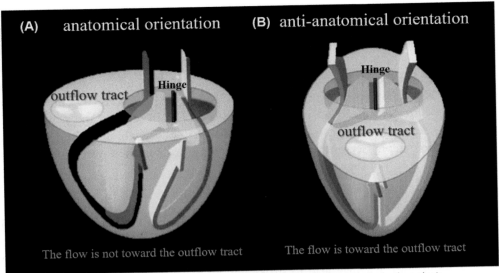

(A) anatomical orientation   **(B) anti-anatomical orientation**

Hinge

outflow tract   Hinge

outflow tract

The flow is not toward the outflow tract   The flow is toward the outflow tract

0° orientation (anatomical orientation)   90° orientation (anti-anatomical orientation)
The hinge line is parallel to the commissure line.   The hinge line is perpendicular to the commissure line.

Figure 9.13 *Intraventricular flow pattern due to different valve orientation.* (A) In case of anatomical orientation (the mechanical heart valve's [MHV's] hinge line is parallel to a line through the commissures of the native valve), the left ventricular (LV) inflow is redirected at the apex and does not smoothly stream toward the outflow tract. (B) In case of antianatomical orientation (the MHV's hinge line is oriented toward the LV outflow tract), the LV inflow redirects and smoothly streams toward the LV outflow tract. *(Modified from Van Rijk-Zwikker GL, Delemarre BJ, Huysmans HA. The orientation of the bi-leaflet carbomedics valve in the mitral position determines left ventricular spatial flow patterns. Eur J Cardiothorac Surg 1996;10:513—20.)*

directing toward the septum forms a large counterclockwise, rotating vortex with an opposite direction compared with those of the healthy subjects' (Fig. 9.14). Moreover, energy dissipation can be higher in presence of an MHV compared with healthy native valve. Su et al. have performed a computational study to investigate the intraventricular flow through bileaflet MHVs (Fig. 9.15) [49]. In that study, the maximal opening angle of the leaflet reached to 85 degrees, which makes the direction of the inflow comparatively straight, and the valve with perpendicularly opened leaflet generated a triple–jet pattern with anterior, central, and posterior side jets, while the anterior jet forming a clockwise vortex in LV. The results of this in vitro study suggested otherwise than the in vivo studies showing counterclockwise vortex in patients with bileaflet MHVs. According to these studies, it can be inferred that the MHV's leaflet opening angle is a considerably important factor for intraventricular flow pattern, and slight deviation of this angle may change the flow direction and vortex formation.

In 2010, Querzoli et al. performed an in vitro study to understand the intraventricular flow pattern, shear stress, and turbulence using two MHVs, a monoleaflet tilting disk and a bileaflet valve [50]. In that study, both mechanical valves that were placed at the mitral

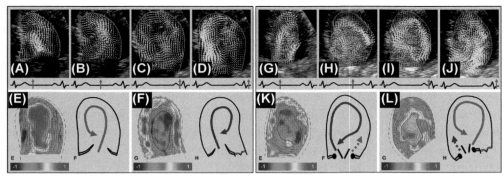

Figure 9.14 *Intraventricular flow pattern in patients with bileaflet mechanical valve versus healthy subject.* The pattern of flow and vortex formation vary between healthy subjects and patients with an mechanical heart valve (MHV). Intraventricular flow visualized by echocardiographic particle image velocimetry (echo-PIV) in healthy subjects (A) to (D) and in patients with bileaflet MHV implanted at the anatomical orientation (G) to (J). The valve splits the transmitral flow into two jets toward the LV septum and the lateral wall, respectively. Subsequently, the jet component directing toward the septum forms a large counterclockwise, rotating vortex with an opposite direction compared with those of the healthy subjects' (E, F, K, L). *(Modified from Faludi R, Szulik M, D'Hooge J, Herijgers P, Rademakers F, Pedrizzetti G, Voigt JU. Left ventricular flow patterns in healthy subjects and patients with prosthetic mitral valves: an in vivo study using echocardiographic particle image velocimetry. J Thorac Cardiovasc Surg 2010;139:1501—10.)*

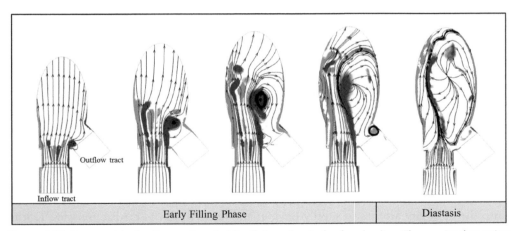

Figure 9.15 *Intraventricular flow pattern in bileaflet mechanical valves in vitro.* The maximal opening angle of the leaflet reached to 85 degrees, which makes the direction of the inflow relatively straight, and the valve with perpendicularly open leaflet generated a triple-jet pattern with anterior, central, and posterior side jets. The anterior jet forms a clockwise vortex in the left ventricle. *(Modified from Su B, Kabinejadian F, Phang HQ, Kumar GP, Cui F, Kim S, Tan RS, Hon JK, Allen JC, Leo HL, Zhong L. Numerical modeling of intraventricular flow during diastole after implantation of BMHV. PLoS One 2015; 10:e0126315.)*

position with anatomical orientation showed higher Reynolds shear stress, which may lead to blood trauma and greater turbulent kinetic energy. More specifically, the monoleaflet MHV showed less-favorable flow behavior (Fig. 9.16).

Unlike BHVs, microbubbles are commonly formed in vicinity of the MHVs at the instant of closure [51]. Microbubble formation has been reported to contribute to platelet activation, hypercoagulability, damage of endothelial lining, and cavitation, which can lead to valve erosion and dysfunction [52–54]. The cavitation risk in bileaflet MHVs is relatively less common than in the monoleaflet MHVs [55,56]. The microbubbles leading to the cavitation are observed in the vicinity of the tip of the leaflets, and the cavitation intensity has been reported to correlate with the tip closing velocity [55,56]. The maximal leaflet tip closing velocity in monoleaflet MHVs (up to 5 m/sec) is reported higher than in the bileaflet MHVs (2.4–3.2 m/sec) [57].

As mentioned earlier, the design and orientation of MHVs strongly affect the intraventricular flow pattern, momentum transfer, blood damage, thrombus formation, and valve durability. The complications related to MHVs implanted at the mitral position still remain unresolved. Thus, improvement in MHV design is still considered an unmet clinical need. As well, further studies are required to clarify which valve orientation is optimal for each clinical condition.

9.2 Fluid dynamics of the aortic valve

9.2.1 Vortex formation in aortic sinus

Aortic valve separates the LV from aorta. It opens during systole to facilitate ejection of the blood from LV and closes during diastole to prevent flow regurgitation from aorta into the LV. Aortic valve consists of three semilunar-shaped leaflets and the sinus of Valsalva, which is a dilated space at the aortic root. The two inlets for the right and the left coronary arteries are located behind two of the three valve cusps, i.e., the right and the left coronary cusps. The third cusp with no inlet is called noncoronary cusp. For the first time, in the early 16th century, the vortex formation within the Valsalva sinus was discovered by Leonardo da Vinci [109]. After mid-20th century, as the prosthetic heart valves developed and advanced, many cardiovascular researchers began studying the Valsalva sinus' fluid dynamics [58,59].

The sinus of Valsalva plays an important role during diastole in which blood flows into the coronary arteries to supply oxygen and nutrient to the cardiac muscles. During late systole, flow recirculation occurs in the root of aorta. Subsequently, the reverse of pressure gradient between the LV and the aorta leads to a small amount of backward flow and generates the closing force of aortic valve (Fig. 9.18A). During diastole, the backward flow bumps the aortic valve's leaflets, and vortices are formed in the sinus of Valsalva, which facilitate smooth transfer of blood into the coronary arteries. Cao et al. have reported that the presence of coronary flow suppresses the vortex formation in

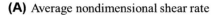

(A) Average nondimensional shear rate

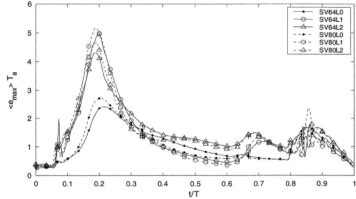

(B) Average nondimensional turbulence kinetic energy

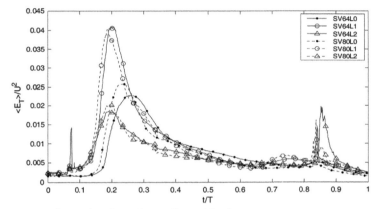

(C) Nondimensional maximum Reynolds shear stresses

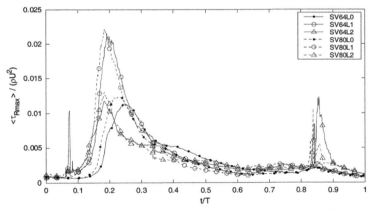

Figure 9.16 *Fluid dynamics comparison between different types of mechanical heart valves at mitral position.* (A) Comparison of the average nondimensional shear rate between a monoleaflet tilting disk ($\bigcirc$) and a bileaflet valve ($\triangle$) indicates higher shear rate compared with control inlet ($\bullet$). (B) Average nondimensional turbulence kinetic energy indicates that monoleaflet tilting disk leads to higher turbulence kinetic energy. (C) Nondimensional maximum Reynolds shear stress is the highest in monoleaflet tilting disk valve. *(Modified from Querzoli G, Fortini S, Cenedese A. Effect of the prosthetic mitral valve on vortex dynamics and turbulence of the left ventricular flow. Phys Fluids 2010;22.)*

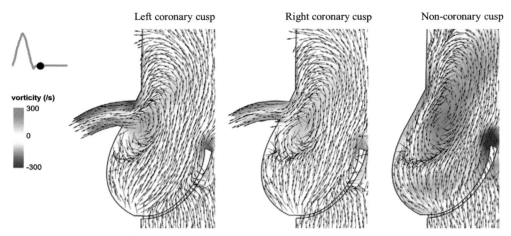

Figure 9.17 *Velocity fields in the aortic sinus and proximal portion of coronary artery during diastole.* Vortex formation in the sinus of Valsalva is affected by the presence or absence of coronary inlet. Noncoronary cusp generates higher temporal shear magnitude and oscillatory shear index compared with cusps with inlet of coronary artery. *(Modified from Cao K, Sucosky P. Aortic valve leaflet wall shear stress characterization revisited: impact of coronary flow. Comput Methods Biomech Biomed Eng 2017;20:468—70.)*

the right and left coronary cusp (Fig. 9.17) [60]. This study suggests that vortex formation in the sinus of Valsalva can be affected based on the presence or absence of coronary inlet. As well, the location of the coronary inlet within the cusp may affect vortex formation. However, to date, no experimental or clinical study has delineated the relationship between vortex formation in the sinus of Valsalva and anatomical characteristics of coronary arteries.

Sinus of Valsalva affects the closing force of the aortic valve, and in particular, the direction of the closing force during late systole is optimized due to the presence of sinus of Valsalva (Fig. 9.18A). The absence of sinus of Valsalva has been reported to increase the stress over the valve leaflets in vitro (Fig. 9.18B) [61]. Alternatively, it has been reported that the presence of pseudosinus of Valsalva after a Bentall procedure does not influence coronary flow reserve [62]. Using 4D flow MRI, Markl et al. compared vortex formation among healthy volunteers, patients underwent David operation with cylindrical tube graft, and patients underwent David operation with neosinus recreation [63]. They showed that the blood vorticity in the coronary cusp is preserved in patients who underwent David operation regardless of the recreation of neosinus. Fukui et al. studied vortex formation in the sinus of Valsalva from the anatomical perspective. In their paper, vortex formation and wall shear stress in the sinus of Valsalva were reported to be strongly influenced by the longitudinal length of the sinus (Fig. 9.19).

Considering the above mentioned studies, it is yet unclear how and to what extent the sinus of Valsalva affects the fluid dynamics around aortic valve and coronary arteries. Furthermore, no clinical study has yet reported the clinical significance of the sinus of

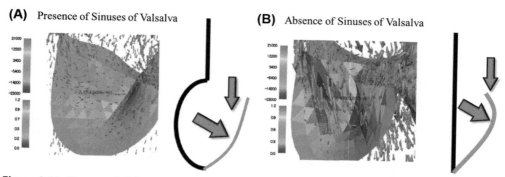

(A) Presence of Sinuses of Valsalva   **(B)** Absence of Sinuses of Valsalva

Figure 9.18 *Stress and deformation in the leaflets.* (A) In presence of sinuses of Valsalva, the reverse of pressure gradient between the LV and aorta leads to a small amount of backward flow, which generates the closing force for aortic valve. (B) In absence of sinuses of Valsalva, the stress over the valve leaflets increases, but closing force would be less compared with presence of sinuses of Valsalva. *(Modified from Katayama S, Umetani N, Sugiura S, Hisada T. The sinus of valsalva relieves abnormal stress on aortic valve leaflets by facilitating smooth closure. J Thorac Cardiovasc Surg 2008;136:1528—35, 1535 e1521.)*

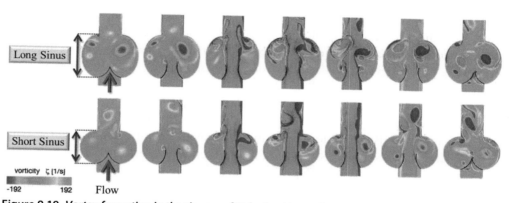

Figure 9.19 *Vortex formation in the sinuses of Valsalva.* Vortex formation and wall shear stress in the sinus of Valsalva are reported to be strongly influenced by the longitudinal length of the sinus. *(More vortices develop in a long sinus compared to a short one. Modified from Fukui et al. J Life Sci Med Res 2013; 3(3):94—102.)*

Valsalva morphology. Further studies are required to provide conclusive answers on clinical significance of Valsalva sinus' fluid dynamics.

9.2.2 Bicuspid aortic valve disease

Bicuspid aortic valve (BAV) morphology is characterized as either the fusion of two of the three leaflets or true presence of two leaflets (Fig. 9.20). BAV is the most common congenital cardiac anomaly whose prevalence is between 0.5% and 2% in adults [64—66]. BAV has various phenotypes, which results in different pathological and clinical outcomes [64]. Accordingly, BAV can lead to symptomatic and/or asymptomatic

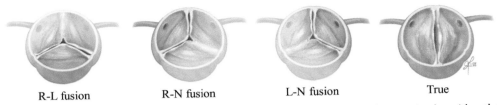

R-L fusion   R-N fusion   L-N fusion   True

Figure 9.20 *Phenotypes of bicuspid aortic valve (BAV).* BAV morphology is characterized as either the fusion of two of the three leaflets (right-left [R-L] fusion, right-noncoronary [R—N] fusion, and left-noncoronary [L-N] fusion) or true presence of two leaflets. *(Modified from Masri A, Kalahasti V, Alkharabsheh S, Svensson LG, Sabik JF, Roselli EE, Hammer D, Johnston DR, Collier P, Rodriguez LL, Griffin BP, Desai MY. Characteristics and long-term outcomes of contemporary patients with bicuspid aortic valves. J Thorac Cardiovasc Surg 2016;151:1650—59 e1651.)*

complications such as aortic stenosis (AS) (12%—37% prevalence in BAV patients) [64,67—69], aortic regurgitation (47%—64%) [68,70—72], infective endocarditis (2%—5%) [67,68,71—73], aortic dilatation, or bicuspid aortopathy (20%—42%) [74,75] whose pathogeneses are complex. In bicuspid aortopathy, vulnerability of the aortic wall due to the presence of abnormal connective tissue and cells are widely recognized [74].

Besides the molecular and cellular mechanisms involved in BAV development, many clinical and experimental studies have reported that fluid dynamics plays an important role in the physiology of the ascending aorta of these patients. The abnormal flow features can be observed downstream of even a normally functioning BAV. Using 4D flow cardiac MRI, Barker et al. studied the relationship between BAV phenotypes (i.e., fusion of the right-left coronary cusp [R-L fusion] with the fusion the right and noncoronary cusp [R—N fusion]) and the fluid dynamics of ascending aorta. They reported that in subjects with R-L fusion, flow jet tends to direct toward the right anterior aortic wall, which leads to a right-handed helical flow direction with an increase in wall shear stress at the right anterior side of aorta (Fig. 9.21 upper panel). Alternatively, in subjects with R—N fusion, flow jet direction tends toward the posterior aorta with an increase in wall shear stress at the right posterior side of the aorta (Fig. 9.21 lower panel) [76]. In BAV, wall shear stress at the aortic wall is also affected by the severity of AS. Using a numerical simulation, Vergara et al. suggested that the asymmetric helical flow and higher wall shear stress depend on the severity of AS (Fig. 9.22) [77]. In another in vitro study using PIV, Saikrishnan et al. reported that the location of BAV orifice significantly affects the fluid dynamics of the aortic sinus as well as the ascending aorta [78]. They also reported that the eccentric location of aortic valve can lead to eccentric systolic jet that forms a vortex with larger turbulent kinetic energy when compared with a control trileaflet valve model (Fig. 9.23). Interestingly, the highest turbulent kinetic energy and shear stress were observed in central orifice BAV model when compared with eccentric orifice BAV model [78]. As an additional factor related to wall shear stress, Kimura et al. reported larger diameter of the midtubular ascending aorta and the higher

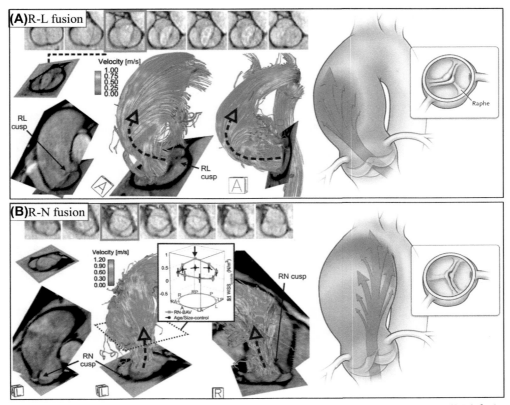

Figure 9.21 *Blood flow direction and helicity in right-left (R-L) and right-noncoronary (R-N) fusion types of bicuspid aortic valve.* Upper panel: In R-L fusion, flow tends to stream toward the right anterior aortic wall leading to a right-handed helical flow with an increase in wall shear stress at the right anterior side of aorta. Lower panel: In R—N fusion, flow direction tends toward the posterior aorta with an increase in wall shear stress at the right posterior side of the aorta. *(Modified from Barker AJ, Markl M, Burk J, Lorenz R, Bock J, Bauer S, Schulz-Menger J, von Knobelsdorff-Brenkenhoff F. Bicuspid aortic valve is associated with altered wall shear stress in the ascending aorta. Circ Cardiovasc Imaging. 2012;5:457—466 and Verma S, Siu SC. Aortic dilatation in patients with bicuspid aortic valve. N Engl J Med 2014;370:1920—29.)*

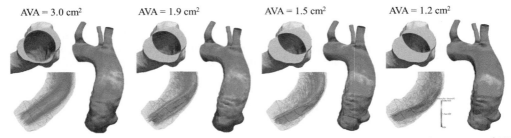

Figure 9.22 *Wall shear stress and blood flow helicity according to aortic valve area.* The severity of AS affects the asymmetric helical flow and higher wall shear stress. As AS progresses, more asymmetric helical flow occurs and wall shear stress increases. Aortic valve area (AVA) is 3.0 cm², 1.9 cm², 1.5 cm², and 1.2 cm² from left to right. *(Modified from Vergara et al. Ann Biomed Eng 2012;40:1760—75.)*

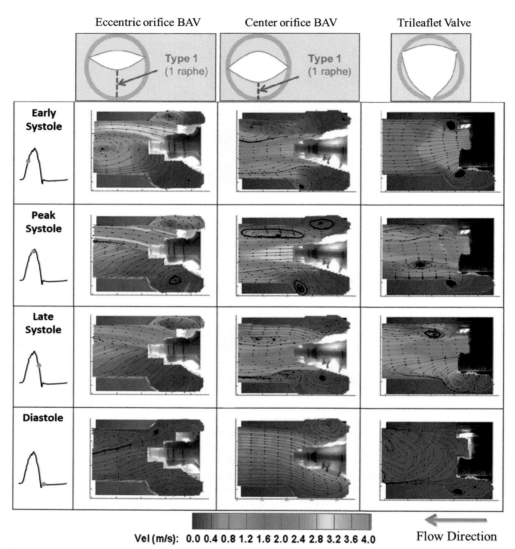

Figure 9.23 *Vortex formation downstream of bicuspid aortic valve.* The eccentric location of aortic valve orifice (left panel) can lead to eccentric systolic jet that forms a vortex with larger turbulent kinetic energy when compared with a center orifice bicuspid aortic valve (BAV) (center panel) and a control trileaflet valve model (right panel). *(Modified from Saikrishnan N, Yap CH, Milligan NC, Vasilyev NV, Yoganathan AP. In vitro characterization of bicuspid aortic valve hemodynamics using particle image velocimetry. Ann Biomed Eng. 2012;40:1760–1775.)*

wall shear stress in BAV patients, compared with control subjects with trileaflet valve. They also found that wall shear stress negatively correlates with the diameter of the mid-tubular ascending aorta in BAV patients [79], which may be attributed to the fact that the jet flow travels a long distance from the aortic valve orifice to reach the aortic wall.

In summary, the current state of the art suggests that the BAV phenotype (e.g., leaflet fusion type), orifice location (i.e., eccentric or central), severity of AS, and size of aortic dilatation are the major factors affecting the fluid dynamics and wall shear stress in the ascending aorta of subjects with BAV. Although many different aspects of the BAV fluid dynamics have been studied so far, no large population clinical trial denoting long-term clinical effects of the aberrant fluid dynamics in these patients has yet been performed. The long-term clinical studies should elucidate how BAV fluid dynamics affects the morphological complications and patients' survival.

9.2.3 Fluid dynamics of paravalvular leak

In developed countries, as the aging society is advancing, the prevalence of AS is gradually expanding. AS is a progressive disease characterized by the valvular calcification that leads to restricting the valve opening. To date, no medical therapy has been reported to reverse the calcification process, and the only effective treatment of AS is via surgical or transcatheter replacement of the aortic valve. Transcatheter aortic valve replacement (TAVR) is a clinically approved therapy for patients with AS, as originally performed by Alain Cribier in 2002 in France [80]. Because TAVR is a significantly less invasive procedure compared with surgical aortic valve replacement (SAVR), the number of TAVR procedures has been rapidly growing as SAVR substitution [81,82]. Although TAVR has many clinical advantages over SAVR, TAVR procedure is suture-less, which sometimes can lead to paravalvular leak (PVL) due to incomplete sealing between the diseased native valve and the implanted stented valve. Despite the TAVR's remarkable technological advancements, post-TAVR PVL still remains an important procedural adverse effect and leads to postoperative morbidity and mortality [83,84].

Unlike more common central aortic regurgitation, PVL jets are often multiple, eccentric, and not straightforward as they are affected by the LV morphology and PVL orifice location (Fig. 9.24) [85]. Further, posteriorly directed PVL jet often interferes with TMF (Fig. 9.25). The variety of PVL jets sometimes precludes its precise assessment. In a set of in vitro experiments, Okafor et al. reported that aortic central regurgitation jet that passes across the TMF interferes with the development of trans-mitral vortex formation and results in increment of turbulence and energy dissipation (Fig. 9.26) [86]. Morisawa et al. performed a set of in vitro experiments using a heart flow simulator to evaluate the intraventricular fluid dynamics in presence of two types of PVL situation, namely anterior and posterior PVL jets, using echo-PIV [87]. In this study, clear interactions between PVL jet and TMF were observed, with anterior

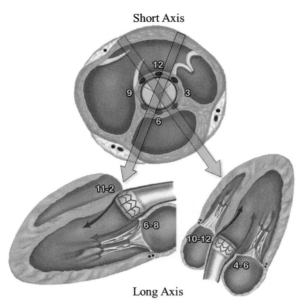

Figure 9.24 *Location of the paravalvular leak jets and its direction.* Paravalvular leak (PVL) jets are often multiple, eccentric, and not straightforward. They are affected by the left ventricular morphology and PVL orifice location. *(Modified from Pibarot P, Hahn RT, Weissman NJ, Monaghan MJ. Assessment of paravalvular regurgitation following TAVR: a proposal of unifying grading scheme. JACC Cardiovasc Imaging 2015;8:340–360.)*

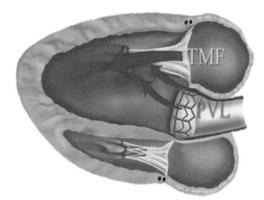

Figure 9.25 *Interaction between paravalvular leak (PVL) and transmitral flow (TMF).* Posterior PVL jet often interferes with TMF. *(Modified fromPibarot P, Hahn RT, Weissman NJ, Monaghan MJ. Assessment of paravalvular regurgitation following TAVR: a proposal of unifying grading scheme. JACC Cardiovasc Imaging. 2015;8:340–360.)*

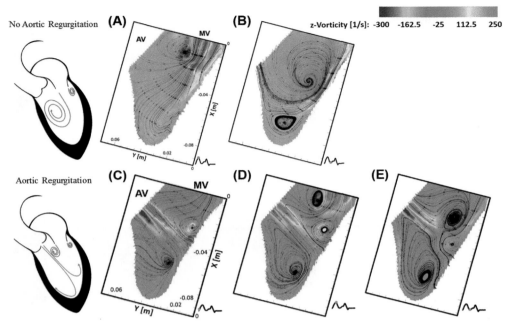

Figure 9.26 *Aortic regurgitation jet inhibiting transmitral flow (TMF) from forming vortex.* In absence of aortic regurgitation (AR), TMF smoothly advances into the left ventricle (LV) during early diastole (A) and forms a large clockwise vortex at the center of the LV (B). In case of aortic regurgitation, the AR jet swiftly flows into the LV (C), followed by TMF. However, advancement of TMF and vortex formation are disturbed by the AR jet (D, E). *(Modified from Okafor I, Raghav V, Condado JF, Midha PA, Kumar G, Yoganathan AP. Aortic regurgitation generates a kinematic obstruction which hinders left ventricular filling. Ann Biomed Eng. 2017;45:1305—14.)*

and posterior PVL situations leading to different fluid dynamics' behaviors (Fig. 9.27). Anterior PVL jet streams into the LV alongside the anterior wall and forms a counterclockwise vortex at the apex (Fig. 9.27B). Subsequently, TMF advances into the LV and collides with PVL jet at the LV center (Fig. 9.27A,C). Alternatively, posterior PVL jet advances toward the posterior wall across the TMF (Fig. 9.27E) and collides with PVL jet after TMF advances into the LV (Fig. 9.27D,F). In addition, it has been reported that greater kinetic energy, circulation, and impulse are required to maintain the cardiac output in PVL situation compared with no PVL situation and particularly in presence of posterior PVL. This study suggests that posterior PVL potentially has more negative impact on the intraventricular fluid dynamics compared with anterior PVL (Figs. 9.26D and 9.27F). To date, the clinical consequences of the aberrant intraventricular fluid dynamics derived from PVL have yet remained unknown. Further clinical and outcome studies are required to establish better management strategies to mitigate PVL.

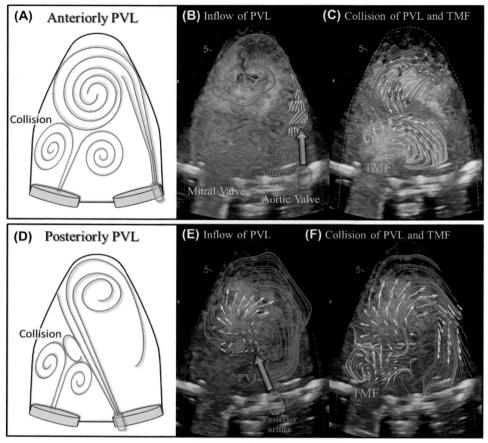

Figure 9.27 *Paravaluvular leaks from anterior and posterior orifice interfering with transmitral flow (TMF).* (A, D) Schematics of intraventricular vortex interaction between paravalvular leak (PVL) jet and TMF. (B) Inflow of anterior PVL jet. (C) Collision of anterior PVL jet and TMF. (E) Inflow of posterior PVL jet. (F) Collision of posterior PVL jet and TMF. *(Modified from Morisawa D, Falahatpisheh A, Avenatti E, Little SH, Kheradvar A. Intraventricular vortex interaction between transmitral flow and paravalvular leak. Sci Rep 2018;8:15657.)*

9.2.4 Flow through the aortic protheses

Prosthetic aortic valves are structurally categorized as mechanical (Fig. 9.28A) and bio-prosthetic (Fig. 9.28B—D). Each valve type has advantages and disadvantages. Although MHVs have robust durability, they need life-long anticoagulation therapy to prevent thrombus formation and embolization, which increases the risk of bleeding. Alternatively, BHVs' durability is suboptimal compared with MHVs', but those valves are less likely to generate flow disturbances compared with MHVs [88,89]. Unlike prosthetic valves implanted at mitral position, those implanted at the aortic valve position are not

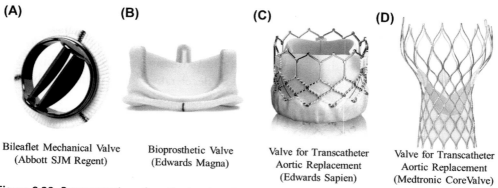

(A)

(B)

(C)

(D)

Bileaflet Mechanical Valve
(Abbott SJM Regent)

Bioprosthetic Valve
(Edwards Magna)

Valve for Transcatheter
Aortic Replacement
(Edwards Sapien)

Valve for Transcatheter
Aortic Replacement
(Medtronic CoreValve)

Figure 9.28 *Representatives of prosthetic valves.* (A) Bileaflet mechanical valve; (B) Bioprosthetic valve; (C) Transcatheter aortic valve (Edwards SAPIEN); (D) Transcatheter aortic valve (Medtronic CoreValve).

exposed to the mechanical stress due to high blood pressure. In fact, recent clinical studies have reported that survival rate is no different between the elderly patients undergoing mechanical or bioprosthetic aortic valve replacement[42].

Although the BHVs are commonly used for aortic valve replacement, bileaflet mechanical heart valves (BMHVs) are still popular [42]. In BMHVs, ejected flow is composed of three separate jets, two lateral orifice jets, and a central orifice jet. Dasi et al. have performed 2D PIV and numerical simulation and precisely described the vorticity dynamics downstream of BMHV [90]. In these valves, the ejected flow is laminar, with the valve housing shear layer rolls up into the ring-like structure expanding to the sinus wall, during the first half of the flow acceleration phase (Fig. 9.29B,C). In the second half of flow acceleration phase, a vortex ring forms at the sinus of Valsalva

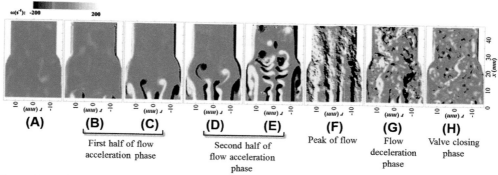

(A)

(B)   **(C)**

First half of flow
acceleration phase

(D)   **(E)**

Second half of
flow acceleration
phase

(F)

Peak of flow

(G)

Flow
deceleration
phase

(H)

Valve closing
phase

Figure 9.29 *Flow field through bileaflet mechanical heart valve.* In mechanical valve, ejected flow is laminar, and the valve housing shear layer rolls up into the ring during the first half of the flow acceleration phase (B, C). In the second half of flow acceleration phase, a vortex ring forms at the sinus of Valsalva and advances further downstream toward the end of the sinus region. The shear layers at the leaflets turn out to be unstable and break down into two von Karman vortices (D, E). At the peak flow, leaflet shear layers rapidly evolve into a chaotic and highly disorganized state (F). This chaotic small-scale vorticity field persists through the flow deceleration phase until the valve closing phase (H). *(Modified from Dasi LP, Ge L, Simon HA, Sotiropoulos F, Yoganathan AP. Vorticity dynamics of a bileaflet mechanical heart valve in an axisymmetric aorta. Phys Fluids 2007;19.)*

advances further downstream toward the end of the sinus region. The shear layers at the leaflets turn out to be unstable and break down into two von Karman vortices (Fig. 9.29D,E). At the peak flow, leaflet shear layers rapidly evolve into a chaotic and highly disorganized state with multiple small vortical structures (Fig. 9.29F). This chaotic small-scale vorticity field persists through the flow deceleration phase until the valve closing phase (Fig. 9.29,H).

Unlike trileaflet BHVs, which are laterally and vertically symmetric, BMHVs are asymmetric. Therefore, valve orientation affects the fluid dynamics. Borazjani et al. have reported that the leaflets motion of BMHVs is affected by the relationship between the valve orientation and plane of the ascending aorta curvature. They concluded that the optimal symmetric motion of the leaflets can be achieved when the BMHV orientation is symmetrical relative to the plane of ascending aorta curvature[91].

One of the important complications related to flow through the MHVs is blood damage due to its configuration and hard leaflets' surface. During diastole, although the BMHVs leaflets are in closed position, their sealing is not perfect, and usually some degrees of transvalvular leakage are present [89]. This transvalvular leak can lead to blood damage due to hemolysis, RBC rapture, or platelet activation, as the first step toward the initiation of the coagulation cascade (Fig. 9.30) [89,92]. The second complication that usually occurs in the long term is unfavorable biological reaction such as pannus formation, which can restrict leaflet motion and eventually leads to valve obstruction. In this situation, surgical reoperation is usually required to mitigate the stenosis [88]. Pannus formation and restriction of leaflet motion can lead to flow disturbance and may even increase the peak pressure gradient (Fig. 9.31B,C).

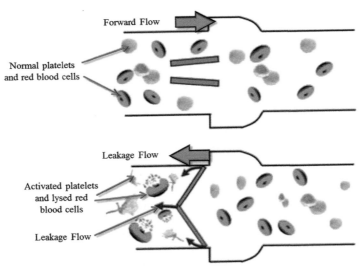

Figure 9.30 Blood damage through bileaflet mechanical heart valve. Transvalvular leak can lead to blood damage due to hemolysis, red blood cells rapture, or platelet activation, as the first step toward the initiation of the coagulation cascade. *(Modified from Dasi LP, Simon HA, Sucosky P, Yoganathan AP. Fluid mechanics of artificial heart valves. Clin Exp Pharmacol Physiol 2009;36:225−37.)*

BHVs have three soft tissue leaflets forming a centrally-opening orifice to eject the blood. BHV's flow behavior downstream the valve is similar to the native aortic valve's when compared with BMHVs'. In BHVs, the forward jet is surrounded by the counter-rotating recirculation regions, where higher turbulent shear stress at the boundary layers of the ejected jet exists [89]. This high shear stress situation may adversely affect the durability of BHVs. Transcatheter aortic valves (TAVs) are also bioprostheses, and two valves, the Edwards' SAPIEN family (Edwards Lifesciences Inc., Irvine, CA) and the Medtronic

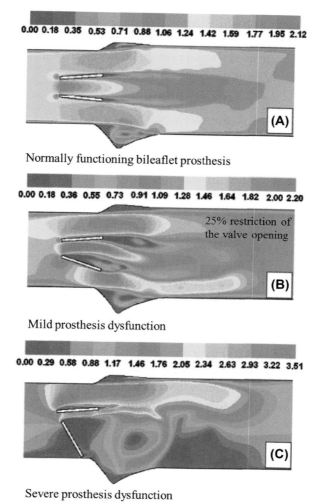

Figure 9.31 *Flow field downstream of prosthesis with the restriction of valve opening.* (A) A normally functioning bileaflet prosthesis. The velocity of flow through the central orifice is higher than that through the lateral orifice. (B) Mild dysfunction of the prosthesis. 25% opening restriction of one leaflet leads to the increase in pressure gradient. (C) Severe prosthesis dysfunction. One leaflet stuck in the closed position shows higher pressure gradient compared with that of mild dysfunction. *(Modified from Pibarot et al. provided by Smadi O et al., Concordia University, Montreal, Quebec, Canada Circulation. 2009;119:1034—48.)*

CoreValve family (Medtronic, Fridley, MN), are widely used in patients (Fig. 9.28C,D). In 2015, Kopanidis et al. numerically compared the fluid dynamics of the Edwards' SAPIEN XT and CoreValve [93]. They reported that the Edwards' SAPIEN XT forms a vortex in the upper ascending aorta, which is more persistent with lower wall shear stress, when compared with the Medtronic's CoreValve (Fig. 9.32). The variations in the fluid dynamics of the valves can be due to the differences in the valve configurations (e.g., Edwards' SAPIEN has an overall shorter valve profile compared to Medtronic's CoreValve). Transvalvular fluid dynamics among other factors such as balloon- or self-expandable deployment and the stent radial force affect patients' clinical outcome. Considering the difference between the two commonly-used TAVR systems, no clinical trial has yet reported a statistically significant difference in mortality, quality of life, heart failure improvement, or major cardiovascular events [94].

9.3 Fluid dynamics of the valves of the right heart

Compared with the studies on left heart's fluid dynamics, not much information is available on the right heart. This is partly due to the difficulties in precise visualization and measurement of the right ventricle (RV). RV has a complex geometry with an asymmetric lunar shape. Furthermore, access to visualize RV by echocardiography is only possible through a narrow acoustic window due to the RV location within the chest. More recently, 4D flow MRI provides better information about the RV function (within its limits) that is used for fluid dynamics characterization through direct measurements and computational modeling [95−98].

In the right heart, blood flows into the right atrium (RA) from the superior vena cava (SVC) and inferior vena cava (IVC) and then transfers to the RV through the tricuspid valve. During systole, the blood is ejected by RV into the pulmonary artery (PA) passing through the pulmonary valve. In healthy subjects, the flow from SVC and IVC merge together to form a vortex sheet in the RA (Fig. 9.33A), and the transtricuspid flow forms a complex vortical flow pattern in the RV during early diastole (Fig. 9.33) where one of its flow components directs toward the RV outflow tract, which leads to a smooth ejection from RV to PA, and the other one directs toward the inferior wall of RV, which does not directly contribute to the blood transfer into the PA (Fig. 9.33B). In the PA, uniform and laminar flow pattern is observed (Fig. 9.33A,B) [97]. Arheden the et al. have studied the kinetic energy of LV and RV in healthy subjects and reported that three peak of kinetic energy are observed during systole, early diastole, and late diastole for both LV and RV [98]. In the RV, the systolic peak is larger than the other two diastolic peaks. Alternatively, the early diastolic peak is stronger than the systolic peak in the LV, which indicates that RV blood filling is less dependent on ventricular suction compared with the LV.

Tetralogy of Fallot (TOF) is one of the most common congenital heart defects, and patients with TOF usually undergo corrective surgery in their first year of life. In these

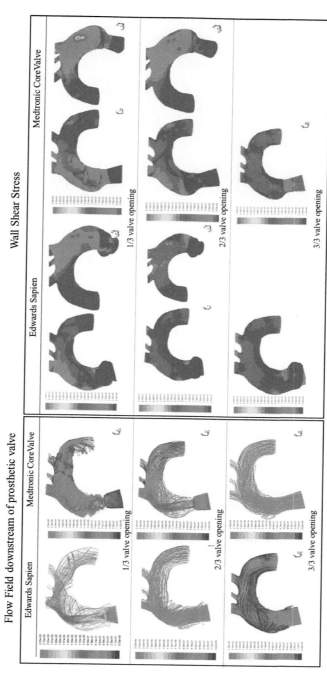

Figure 9.32 Flow field downstream of a transcatheter aortic valve. The Edwards' SAPIEN XT (left side) forms a vortex in the upper ascending aorta, which is more persistent with lower wall shear stress, when compared with the CoreValve (right side). *(Modified from Kopanidis A, Pantos I, Alexopoulos N, Theodorakakos A, Efstathopoulos E, Katritsis D. Aortic flow patterns after simulated implantation of transcatheter aortic valves. Hellenic J Cardiol 2015;56:418–28.)*

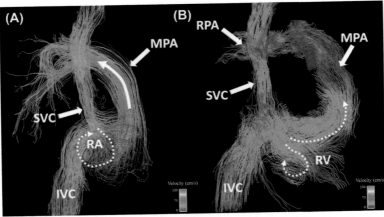

Figure 9.33 *Flow field in the healthy human right heart.* (A) Vortical flow in the right atrium. (B) Vortical flow in the right ventricle. SVC, superior vena cava; IVC, inferior vena cava; RA, right atrium; RV, right ventricle; MPA, main pulmonary artery; RPA, right pulmonary artery. *(Modified by François CJ, Srinivasan S, Schiebler ML, Reeder SB, Niespodzany E, Landgraf BR, Wieben O, Frydrychowicz A. 4d cardiovascular magnetic resonance velocity mapping of alterations of right heart flow patterns and main pulmonary artery hemodynamics in tetralogy of fallot. J Cardiovasc Magn Reson 2012;14:16.)*

patients after repair, given the long-term survival rate (almost 90% at 30 years) [99], pulmonary valve insufficiency is common, which leads to RV volume overload. RV flow in these patients has been visualized and studied using 4D flow MRI [97,100,101]. In repaired TOF patients, an extra vortical component of the flow, curling in the opposite direction to the vortex observed in the healthy subjects, can be observed in the RA (Fig. 9.34A dashed arrow), with the transtricuspid flow directing toward the RV apex (Fig. 9.34B) [97,100]. In these patients, during systole, helical flow and vortex formation are observed in the PA (Fig. 9.34C), which are less likely to be visualized in normal subjects [97]. Comparing with normal subjects, higher peak vorticity in the RA and wall shear stress in the PA have been observed [97,100].

Pulmonary artery hypertension (PAH) is a progressive and life-threatening disease characterized by elevation of pulmonary vascular resistance and PA pressure. PAH causes RV pressure overload, which ultimately leads to RV failure. Previous studies using 4D flow MRI have described RV fluid dynamics in patients with PAH (Fig. 9.35) [102–104]. In patients with PAH, higher helical flow and greater vorticity are observed in the PA compared with healthy subjects [105]. The helicity and vorticity have been reported to be associated with PA stiffness and reduced RV function [103]. In addition, lower pulmonary blood flow velocity, delayed vortex flow in the PA, and lower wall shear stress in the PA are observed in patients with severe PAH and RV dysfunction [102,104].

Although much attention has been recently paid to RV fluid dynamics, there are yet many unknowns about the RV fluid dynamics in case of RV infarction, cardiomyopathy, tricuspid regurgitation, RV pacing, etc. Given that the 4D flow MRI is a useful and

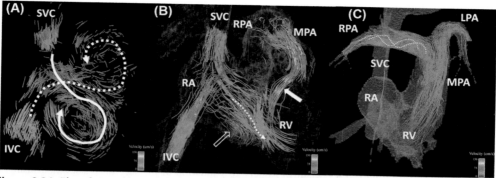

Figure 9.34 *Flow field in the right heart of a patient with tetralogy of Fallot.* (A) Vortical flow in right atrium. (B) Vortical flow in right ventricle. (C) Vortical and helical flow. SVC, superior vena cava; IVC, inferior vena cava; RA, right atrium; RV, right ventricle; MPA, main pulmonary artery; RPA, right pulmonary artery; LPA, left pulmonary artery. *(Modified from François CJ, Srinivasan S, Schiebler ML, Reeder SB, Niespodzany E, Landgraf BR, Wieben O, Frydrychowicz A. 4d cardiovascular magnetic resonance velocity mapping of alterations of right heart flow patterns and main pulmonary artery hemodynamics in tetralogy of fallot. J Cardiovasc Magn Reson 2012;14:16.)*

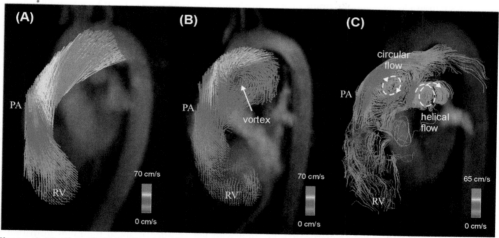

Figure 9.35 *Flow field in the right heart of a patient with pulmonary artery hypertension.* (A) Uniformly directed forward flow in the PA of a healthy subject. (B, C) Vortical and circular flow in the PA of a patient with PAH. RV, right ventricle; PA, pulmonary artery; PAH, pulmonary artery hypertension. *(Modified from Reiter U, Reiter G, Fuchsjager M. Mr phase-contrast imaging in pulmonary hypertension. Br J Radiol 2016;89:20150995.)*

validated modality to visualize the flow in the right heart, it comes with some limitations, including long scan time (10—20 min), low temporal and spatial resolution (2.0—2.5 mm^3), and complex data postprocessing [106]. Further innovations in imaging technologies are needed to improve our understanding on fluid dynamics of the right heart.

References

[1] Kheradvar A, Gharib M. On mitral valve dynamics and its connection to early diastolic flow. Ann Biomed Eng 2009;37:1—13.

[2] Ristow B, Ali S, Whooley MA, Schiller NB. Usefulness of left atrial volume index to predict heart failure hospitalization and mortality in ambulatory patients with coronary heart disease and comparison to left ventricular ejection fraction (from the heart and soul study). Am J Cardiol 2008;102:70—6.

[3] Hatle L, Brubakk A, Tromsdal A, Angelsen B. Noninvasive assessment of pressure drop in mitral stenosis by Doppler ultrasound. Br Heart J 1978;40:131—40.

[4] Falahatpisheh A, Rickers C, Gabbert D, Heng EL, Stalder A, Kramer HH, Kilner PJ, Kheradvar A. Simplified Bernoulli's method significantly underestimates pulmonary transvalvular pressure drop. J Magn Reson Imag 2016;43:1313—9.

[5] Ha H, Kvitting J-PE, Dyverfeldt P, Ebbers T. Validation of pressure drop assessment using 4d flow MRI-based turbulence production in various shapes of aortic stenoses. Magn Reson Med 2018;0.

[6] Hong G-R, Pedrizzetti G, Tonti G, Li P, Wei Z, Kim JK, Baweja A, Liu S, Chung N, Houle H, Narula J, Vannan MA. Characterization and quantification of vortex flow in the human left ventricle by contrast echocardiography using vector particle image velocimetry. J Am Coll Cardiol Img 2008; 1:705—17.

[7] Sengupta PP, Khandheria BK, Korinek J, Jahangir A, Yoshifuku S, Milosevc I, Belohlavek M. Left ventricular isovolumic flow sequence during sinus and paced rhythms: new insights from use of high-resolution Doppler and ultrasonic digital particle imaging velocimetry. J Am Coll Cardiol 2007;49: 899—908.

[8] Markl M, Kilner PJ, Ebbers T. Comprehensive 4d velocity mapping of the heart and great vessels by cardiovascular magnetic resonance. J Cardiovasc Magn Reson 2011;13:7.

[9] Elbaz MSM, Calkoen EE, Westenberg JJM, Lelieveldt BPF, Roest AAW, van der Geest RJ. Vortex flow during early and late left ventricular filling in normal subjects: quantitative characterization using retrospectively-gated 4d flow cardiovascular magnetic resonance and three-dimensional vortex core analysis. J Cardiovasc Magn Reson 2014;16:78.

[10] Falahatpisheh A, Pahlevan N, Kheradvar A. Effect of the mitral valve's anterior leaflet on axisymmetry of transmitral vortex ring. Ann Biomed Eng 2015;43:2349—60.

[11] Kheradvar A, Falahatpisheh A. The effects of dynamic saddle annulus and leaflet length on transmitral flow pattern and leaflet stress of a bileaflet bioprosthetic mitral valve. J Heart Valve Dis 2012;21: 225—33.

[12] Pedrizzetti G, La Canna G, Alfieri O, Tonti G. The vortex—an early predictor of cardiovascular outcome? Nat Rev Cardiol 2014;11:545—53.

[13] Kilner PJ, Yang GZ, Wilkes AJ, Mohiaddin RH, Firmin DN, Yacoub MH. Asymmetric redirection of flow through the heart. Nature 2000;404:759—61.

[14] Son J-W, Park W-J, Choi J-H, Houle H, Vannan MA, Hong G-R, Chung N. Abnormal left ventricular vortex flow patterns in association with left ventricular apical thrombus formation in patients with anterior myocardial infarction: a quantitative analysis by contrast echocardiography. Circulation 2012;76:2640—6.

[15] Carlhall CJ, Bolger A. Passing strange: flow in the failing ventricle. Circ Heart Fail 2010;3.

[16] Falahapisheh A, Kheradvar A. High-speed particle image velocimetry to assess cardiac fluid dynamics in vitro: from performance to validation. Eur J Mech B Fluid 2012;35:2—8.

[17] Kheradvar A, Houle HH, Pedrizzetti G, Tonti G, Belcik T, Ashraf M, Linder JR, Gharib M, Sahn DJ. Echocardiographic particle image velocimetry: a novel technique for quantification of left ventricular blood vorticity pattern. J Am Soc Echocardiogr 2010;23:86—94.

[18] Gharib M, Rambod E, Shariff K. A universal time scale for vortex ring formation. J Fluid Mech 1998; 360:121—40.

[19] Gharib M, Rambod E, Kheradvar A, Sahn DJ, Dabiri JO. Optimal vortex formation as an index of cardiac health. Proc Natl Acad Sci USA 2006;103:6305—8.

[20] Kheradvar A, Milano M, Gharib M. Correlation between vortex ring formation and mitral annulus dynamics during ventricular rapid filling. Am Soc Artif Intern Organs J 2007;53:8—16.

[21] Kheradvar A, Gharib M. Influence of ventricular pressure drop on mitral annulus dynamics through the process of vortex ring formation. Ann Biomed Eng 2007;35:2050—64.

[22] Del Buono MG, Buckley L, Abbate A. Primary and secondary diastolic dysfunction in heart failure with preserved ejection fraction. Am J Cardiol 2018;122:1578—87.

[23] Kheradvar A, Assadi R, Falahatpisheh A, Sengupta PP. Assessment of transmitral vortex formation in patients with diastolic dysfunction. J Am Soc Echocardiogr 2012;25:220—7.

[24] Poh KK, Lee LC, Shen L, Chong E, Tan YL, Chai P, Yeo TC, Wood MJ. Left ventricular fluid dynamics in heart failure: echocardiographic measurement and utilities of vortex formation time. European Heart Journal-Cardiovascular Imaging 2012;13:385—93.

[25] Jiamsripong P, Alharthi MS, Calleja AM, McMahon EM, Katayama M, Westerdale J, Milano M, Heys JJ, Mookadam F, Belohlavek M. Impact of pericardial adhesions on diastolic function as assessed by vortex formation time, a parameter of transmitral flow efficiency. Cardiovasc Ultrasound 2010;8.

[26] Carlhall C, Wigstrom L, Heiberg E, Karlsson M, Bolger AF, Nylander E. Contribution of mitral annular excursion and shape dynamics to total left ventricular volume change. Am J Physiol Heart Circ Physiol 2004;287:H1836—41.

[27] Nagueh SF, Middleton KJ, Kopelen HA, Zoghbi WA, Quinones MA. Doppler tissue imaging: a noninvasive technique for evaluation of left ventricular relaxation and estimation of filling pressures. J Am Coll Cardiol 1997;30:1527—33.

[28] Garcia MJ, Thomas JD, Klein AL. New Doppler echocardiographic applications for the study of diastolic function. J Am Coll Cardiol 1998;32:865—75.

[29] Hasegawa H, Little WC, Ohno M, Brucks S, Morimoto A, Cheng HJ, Cheng CP. Diastolic mitral annular velocity during the development of heart failure. J Am Coll Cardiol 2003;41:1590—7.

[30] Levine RA, Hung J. Ischemic mitral regurgitation, the dynamic lesion: clues to the cure. J Am Coll Cardiol 2003;42:1929—32.

[31] Aikawa K, Sheehan FH, Otto CM, Coady K, Bashein G, Bolson EL. The severity of functional mitral regurgitation depends on the shape of the mitral apparatus: a three-dimensional echo analysis. J Heart Valve Dis 2002;11:627—36.

[32] Gorman 3rd JH, Gorman RC, Jackson BM, Hiramatsu Y, Gikakis N, Kelley ST, Sutton MG, Plappert T, Edmunds Jr LH. Distortions of the mitral valve in acute ischemic mitral regurgitation. Ann Thorac Surg 1997;64:1026—31.

[33] Glasson JR, Komeda M, Daughters 2nd GT, Bolger AF, MacIsaac A, Oesterle SN, Ingels Jr NB, Miller DC. Three-dimensional dynamics of the canine mitral annulus during ischemic mitral regurgitation. Ann Thorac Surg 1996;62:1059—67. discussion 1067—1058.

[34] Salgo IS, Gorman 3rd JH, Gorman RC, Jackson BM, Bowen FW, Plappert T, St John Sutton MG, Edmunds Jr LH. Effect of annular shape on leaflet curvature in reducing mitral leaflet stress. Circulation 2002;106:711—7.

[35] Enriquez-Sarano M, Akins CW, Vahanian A. Mitral regurgitation. Lancet 2009;373:1382—94.

[36] Watanabe S, Bikou O, Hajjar RJ, Ishikawa K. Swine model of mitral regurgitation induced heart failure. In: Ishikawa K, editor. Experimental models of cardiovascular diseases: methods and protocols. New York, NY: Springer New York; 2018. p. 327—35.

[37] Watanabe N, Maltais S, Nishino S, O'Donoghue TA, Hung J. Functional mitral regurgitation: imaging insights, clinical outcomes and surgical principles. Prog Cardiovasc Dis 2017;60:351—60.

[38] Silbiger JJ. Anatomy, mechanics, and pathophysiology of the mitral annulus. Am Heart J 2012;164: 163—76.

[39] Watanabe N, Ogasawara Y, Yamaura Y, Kawamoto T, Akasaka T, Yoshida K. Geometric deformity of the mitral annulus in patients with ischemic mitral regurgitation: a real-time three-dimensional echocardiographic study. J Heart Valve Dis 2005;14:447—52.

[40] Punnoose L, Burkhoff D, Cunningham L, Horn EM. Functional mitral regurgitation: therapeutic strategies for a ventricular disease. J Card Fail 2014;20:252—67.

[41] Levine RA, Schwammenthal E. Ischemic mitral regurgitation on the threshold of a solution: from paradoxes to unifying concepts. Circulation 2005;112:745—58.

[42] Goldstone AB, Chiu P, Baiocchi M, Lingala B, Patrick WL, Fischbein MP, Woo YJ. Mechanical or biologic prostheses for aortic-valve and mitral-valve replacement. N Engl J Med 2017;377:1847—57.

[43] Schnittman SR, Itagaki S, Toyoda N, Adams DH, Egorova NN, Chikwe J. Survival and long-term outcomes after mitral valve replacement in patients aged 18 to 50 years. J Thorac Cardiovasc Surg 2018;155:96—102 e111.

[44] Van Rijk-Zwikker GL, Delemarre BJ, Huysmans HA. The orientation of the bi-leaflet carbomedics valve in the mitral position determines left ventricular spatial flow patterns. Eur J Cardiothorac Surg 1996;10:513—20.

[45] Westerdale JC, Adrian R, Squires K, Chaliki H, Belohlavek M. Effects of bileaflet mechanical mitral valve rotational orientation on left ventricular flow conditions. Open Cardiovasc Med J 2015;9:62—8.

[46] Schoephoerster RT, Chandran KB. Velocity and turbulence measurements past mitral valve prostheses in a model left ventricle. J Biomech 1991;24:549—62.

[47] Machler H, Perthel M, Reiter G, Reiter U, Zink M, Bergmann P, Waltensdorfer A, Laas J. Influence of bileaflet prosthetic mitral valve orientation on left ventricular flow—an experimental in vivo magnetic resonance imaging study. Eur J Cardiothorac Surg 2004;26:747—53.

[48] Faludi R, Szulik M, D'Hooge J, Herijgers P, Rademakers F, Pedrizzetti G, Voigt JU. Left ventricular flow patterns in healthy subjects and patients with prosthetic mitral valves: an in vivo study using echocardiographic particle image velocimetry. J Thorac Cardiovasc Surg 2010;139:1501—10.

[49] Su B, Kabinejadian F, Phang HQ, Kumar GP, Cui F, Kim S, Tan RS, Hon JK, Allen JC, Leo HL, Zhong L. Numerical modeling of intraventricular flow during diastole after implantation of BMHV. PLoS One 2015;10:e0126315.

[50] Querzoli G, Fortini S, Cenedese A. Effect of the prosthetic mitral valve on vortex dynamics and turbulence of the left ventricular flow. Phys Fluids 2010;22.

[51] Milo S, Rambod E, Gutfinger C, Gharib M. Mitral mechanical heart valves: in vitro studies of their closure, vortex and microbubble formation with possible medical implications. Eur J Cardiothorac Surg 2003;24:364—70.

[52] Eichler MJ, Reul HM. Mechanical heart valve cavitation: valve specific parameters. Int J Artif Organs 2004;27:855—67.

[53] Lee H, Homma A, Tatsumi E, Taenaka Y. Observation of cavitation pits on mechanical heart valve surfaces in an artificial heart used in in vitro testing. J Artif Organs 2010;13:17—23.

[54] Brujan EA. Cardiovascular cavitation. Med Eng Phys 2009;31:742—51.

[55] Lee H, Homma A, Taenaka Y. Hydrodynamic characteristics of bileaflet mechanical heart valves in an artificial heart: cavitation and closing velocity. Artif Organs 2007;31:532—7.

[56] Herbertson LH, Deutsch S, Manning KB. Modifying a tilting disk mechanical heart valve design to improve closing dynamics. J Biomech Eng 2008;130:054503.

[57] Mohammadi H, Ahmadian MT, Wan WK. Time-dependent analysis of leaflets in mechanical aortic bileaflet heart valves in closing phase using the finite strip method. Med Eng Phys 2006;28:122—33.

[58] Bellhouse BJ, Bellhouse FH. Mechanism of closure of the aortic valve. Nature 1968;217:86—7.

[59] Bellhouse BJ, Reid KG. Fluid mechanics of the aortic valve. Br Heart J 1969;31:391.

[60] Cao K, Sucosky P. Aortic valve leaflet wall shear stress characterization revisited: impact of coronary flow. Comput Methods Biomech Biomed Eng 2017;20:468—70.

[61] Katayama S, Umetani N, Sugiura S, Hisada T. The sinus of valsalva relieves abnormal stress on aortic valve leaflets by facilitating smooth closure. J Thorac Cardiovasc Surg 2008;136:1528—35. 1535 e1521.

[62] De Paulis R, Bassano C, Scaffa R, Nardi P, Bertoldo F, Chiariello L. Bentall procedures with a novel valved conduit incorporating "sinuses of Valsalva". Surg Technol Int 2004;12:195—200.

[63] Markl M, Draney MT, Miller DC, Levin JM, Williamson EE, Pelc NJ, Liang DH, Herfkens RJ. Time-resolved three-dimensional magnetic resonance velocity mapping of aortic flow in healthy volunteers and patients after valve-sparing aortic root replacement. J Thorac Cardiovasc Surg 2005;130:456—63.

[64] Masri A, Svensson LG, Griffin BP, Desai MY. Contemporary natural history of bicuspid aortic valve disease: a systematic review. Heart 2017;103:1323—30.

[65] Martin M, Lorca R, Rozado J, Alvarez-Cabo R, Calvo J, Pascual I, Cigarran H, Rodriguez I, Moris C. Bicuspid aortic valve syndrome: a multidisciplinary approach for a complex entity. J Thorac Dis 2017;9:S454—64.

[66] Longobardo L, Jain R, Carerj S, Zito C, Khandheria BK. Bicuspid aortic valve: unlocking the morphogenetic puzzle. Am J Med 2016;129:796—805.

[67] Tzemos N, Therrien J, Yip J, Thanassoulis G, Tremblay S, Jamorski MT, Webb GD, Siu SC. Outcomes in adults with bicuspid aortic valves. J Am Med Assoc 2008;300:1317—25.

[68] Kong WK, Delgado V, Poh KK, Regeer MV, Ng AC, McCormack L, Yeo TC, Shanks M, Parent S, Enache R, Popescu BA, Liang M, Yip JW, Ma LC, Kamperidis V, van Rosendael PJ, van der Velde ET, Ajmone Marsan N, Bax JJ. Prognostic implications of raphe in bicuspid aortic valve anatomy. JAMA cardiology 2017;2:285—92.

[69] Michelena HI, Suri RM, Katan O, Eleid MF, Clavel MA, Maurer MJ, Pellikka PA, Mahoney D, Enriquez-Sarano M. Sex differences and survival in adults with bicuspid aortic valves: verification in 3 contemporary echocardiographic cohorts. J Am Heart Assoc 2016;5.

[70] Michelena HI, Khanna AD, Mahoney D, Margaryan E, Topilsky Y, Suri RM, Eidem B, Edwards WD, Sundt 3rd TM, Enriquez-Sarano M. Incidence of aortic complications in patients with bicuspid aortic valves. J Am Med Assoc 2011;306:1104—12.

[71] Michelena HI, Desjardins VA, Avierinos JF, Russo A, Nkomo VT, Sundt TM, Pellikka PA, Tajik AJ, Enriquez-Sarano M. Natural history of asymptomatic patients with normally functioning or minimally dysfunctional bicuspid aortic valve in the community. Circulation 2008;117:2776—84.

[72] Masri A, Kalahasti V, Alkharabsheh S, Svensson LG, Sabik JF, Roselli EE, Hammer D, Johnston DR, Collier P, Rodriguez LL, Griffin BP, Desai MY. Characteristics and long-term outcomes of contemporary patients with bicuspid aortic valves. J Thorac Cardiovasc Surg 2016;151:1650—9. e1651.

[73] Rodrigues I, Agapito AF, de Sousa L, Oliveira JA, Branco LM, Galrinho A, Abreu J, Timoteo AT, Rosa SA, Ferreira RC. Bicuspid aortic valve outcomes. Cardiol Young 2017;27:518—29.

[74] Verma S, Siu SC. Aortic dilatation in patients with bicuspid aortic valve. N Engl J Med 2014;370:1920—9.

[75] Tadros TM, Klein MD, Shapira OM. Ascending aortic dilatation associated with bicuspid aortic valve: pathophysiology, molecular biology, and clinical implications. Circulation 2009;119:880—90.

[76] Barker AJ, Markl M, Burk J, Lorenz R, Bock J, Bauer S, Schulz-Menger J, von Knobelsdorff-Brenkenhoff F. Bicuspid aortic valve is associated with altered wall shear stress in the ascending aorta. Circ Cardiovasc Imaging 2012;5:457—66.

[77] Vergara C, Viscardi F, Antiga L, Luciani GB. Influence of bicuspid valve geometry on ascending aortic fluid dynamics: a parametric study. Artif Organs 2012;36:368—78.

[78] Saikrishnan N, Yap CH, Milligan NC, Vasilyev NV, Yoganathan AP. In vitro characterization of bicuspid aortic valve hemodynamics using particle image velocimetry. Ann Biomed Eng 2012;40:1760—75.

[79] Kimura N, Nakamura M, Komiya K, Nishi S, Yamaguchi A, Tanaka O, Misawa Y, Adachi H, Kawahito K. Patient-specific assessment of hemodynamics by computational fluid dynamics in patients with bicuspid aortopathy. J Thorac Cardiovasc Surg 2017;153:S52—62 e53.

[80] Cribier A, Eltchaninoff H, Bash A, Borenstein N, Tron C, Bauer F, Derumeaux G, Anselme F, Laborde F, Leon MB. Percutaneous transcatheter implantation of an aortic valve prosthesis for calcific aortic stenosis: first human case description. Circulation 2002;106:3006—8.

[81] Hamm CW, Arsalan M, Mack MJ. The future of transcatheter aortic valve implantation. Eur Heart J 2016;37:803—10.

[82] Horne Jr A, Reineck EA, Hasan RK, Resar JR, Chacko M. Transcatheter aortic valve replacement: historical perspectives, current evidence, and future directions. Am Heart J 2014;168:414—23.

[83] Lerakis S, Hayek SS, Douglas PS. Paravalvular aortic leak after transcatheter aortic valve replacement: current knowledge. Circulation 2013;127:397—407.

[84] Reidy C, Sophocles A, Ramakrishna H, Ghadimi K, Patel PA, Augoustides JG. Challenges after the first decade of transcatheter aortic valve replacement: focus on vascular complications, stroke, and paravalvular leak. J Cardiothorac Vasc Anesth 2013;27:184—9.

[85] Pibarot P, Hahn RT, Weissman NJ, Monaghan MJ. Assessment of paravalvular regurgitation following TAVR: a proposal of unifying grading scheme. JACC Cardiovasc Imag 2015;8:340−60.

[86] Okafor I, Raghav V, Condado JF, Midha PA, Kumar G, Yoganathan AP. Aortic regurgitation generates a kinematic obstruction which hinders left ventricular filling. Ann Biomed Eng 2017;45: 1305−14.

[87] Morisawa D, Falahatpisheh A, Avenatti E, Little SH, Kheradvar A. Intraventricular vortex interaction between transmitral flow and paravalvular leak. Sci Rep 2018;8:15657.

[88] Pibarot P, dumesnil JG. Prosthetic heart valves: selection of the optimal prosthesis and long-term management. Circulation 2009;119:1034−48.

[89] Dasi LP, Simon HA, Sucosky P, Yoganathan AP. Fluid mechanics of artificial heart valves. Clin Exp Pharmacol Physiol 2009;36:225−37.

[90] Dasi LP, Ge L, Simon HA, Sotiropoulos F, Yoganathan AP. Vorticity dynamics of a bileaflet mechanical heart valve in an axisymmetric aorta. Phys Fluids 2007;19.

[91] Borazjani I, Sotiropoulos F. The effect of implantation orientation of a bileaflet mechanical heart valve on kinematics and hemodynamics in an anatomic aorta. J Biomech Eng 2010;132:111005.

[92] Min Yun B, Aidun CK, Yoganathan AP. Blood damage through a bileaflet mechanical heart valve: a quantitative computational study using a multiscale suspension flow solver. J Biomech Eng 2014;136: 101009.

[93] Kopanidis A, Pantos I, Alexopoulos N, Theodorakakos A, Efstathopoulos E, Katritsis D. Aortic flow patterns after simulated implantation of transcatheter aortic valves. Hellenic J Cardiol 2015;56: 418−28.

[94] Abdel-Wahab M, Mehilli J, Frerker C, Neumann FJ, Kurz T, Tolg R, Zachow D, Guerra E, Massberg S, Schafer U, El-Mawardy M, Richardt G. Comparison of balloon-expandable vs self-expandable valves in patients undergoing transcatheter aortic valve replacement: the choice randomized clinical trial. J Am Med Assoc 2014;311:1503−14.

[95] Fredriksson AG, Zajac J, Eriksson J, Dyverfeldt P, Bolger AF, Ebbers T, Carlhäll C-J. 4-d blood flow in the human right ventricle. Am J Physiol Heart Circ Physiol 2011;301:H2344−50.

[96] Browning J, Hertzberg J, Schroeder J, Fenster B. 4d flow assessment of vorticity in right ventricular diastolic dysfunction. Bioengineering 2017;4:30.

[97] François CJ, Srinivasan S, Schiebler ML, Reeder SB, Niespodzany E, Landgraf BR, Wieben O, Frydrychowicz A. 4d cardiovascular magnetic resonance velocity mapping of alterations of right heart flow patterns and main pulmonary artery hemodynamics in tetralogy of fallot. J Cardiovasc Magn Reson 2012;14:16.

[98] Carlsson M, Heiberg E, Toger J, Arheden H. Quantification of left and right ventricular kinetic energy using four-dimensional intracardiac magnetic resonance imaging flow measurements. Am J Physiol Heart Circ Physiol 2012;302:H893−900.

[99] Apitz C, Webb GD, Redington AN. Tetralogy of fallot. Lancet 2009;374:1462−71.

[100] Hirtler D, Garcia J, Barker AJ, Geiger J. Assessment of intracardiac flow and vorticity in the right heart of patients after repair of tetralogy of fallot by flow-sensitive 4D MRI. Eur Radiol 2016;26: 3598−607.

[101] Geiger J, Markl M, Jung B, Grohmann J, Stiller B, Langer M, Arnold R. 4d-mr flow analysis in patients after repair for tetralogy of fallot. Eur Radiol 2011;21:1651−7.

[102] Reiter U, Reiter G, Fuchsjager M. Mr phase-contrast imaging in pulmonary hypertension. Br J Radiol 2016;89:20150995.

[103] Schafer M, Barker AJ, Kheyfets V, Stenmark KR, Crapo J, Yeager ME, Truong U, Buckner JK, Fenster BE, Hunter KS. Helicity and vorticity of pulmonary arterial flow in patients with pulmonary hypertension: quantitative analysis of flow formations. J Am Heart Assoc 2017;6.

[104] Odagiri K, Inui N, Hakamata A, Inoue Y, Suda T, Takehara Y, Sakahara H, Sugiyama M, Alley MT, Wakayama T, Watanabe H. Non-invasive evaluation of pulmonary arterial blood flow and wall shear stress in pulmonary arterial hypertension with 3d phase contrast magnetic resonance imaging. SpringerPlus 2016;5:1071.

[105]   Reiter G, Reiter U, Kovacs G, Kainz B, Schmidt K, Maier R, Olschewski H, Rienmueller R. Magnetic resonance-derived 3-dimensional blood flow patterns in the main pulmonary artery as a marker of pulmonary hypertension and a measure of elevated mean pulmonary arterial pressure. Circulation Cardiovascular imaging 2008;1:23—30.

[106]   Lee N, Taylor MD, Banerjee RK. Right ventricle-pulmonary circulation dysfunction: a review of energy-based approach. Biomed Eng Online 2015;14(Suppl. 1):S8.

[107]   Kheradvar A, Gharib M. On mitral valve dynamics and its connection to early diastolic flow. Ann Biomed Eng. 2009 Jan;37(1):1—13.

[108]   Kheradvar A, Rickers C, Morisawa D, Kim M, Hong GR, Pedrizzetti G. Diagnostic and prognostic significance of cardiovascular vortex formation. J Cardiol. 2019 Jun 26, pii: S0914-5087(19)30143-1.

[109]   https://www.nature.com/articles/d41586-019-02144-z.

CHAPTER 10

Heart valve leaflet preparation

Ivan Vesely[1,2]
[1]Class III Medical Device Consulting, Gaithersburg, MD, United States; [2]CroiValve Limited, Dublin, Ireland

Contents

The earliest most versions of prosthetic heart valves were taken from cadavers and implanted as aortic valve allografts (homografts). The first allograft valves were implanted in 1962 by Sir Donald Ross [1]. Homografts proved to be clinically successful from their early days and are still used today [2]. They are limited primarily by their availability and by the technical skill requirements of the surgeon. In the early days, they were preserved by storage in glycerol, but today the most widely used method is cryopreservation.

The harvesting of cadaveric human heart valves for transplantation was forbidden in France in the early days of this field, i.e., the mid 1960s. So when Alain Carpentier first started using animal tissue valves (pig aortic valves), he recognized the need for preservation and implemented first the use of buffered mercurial solutions and formaldehyde [3]. But formaldehyde-preserved leaflets were found to be insufficiently stable when implanted, and that approach was soon replaced by glutaraldehyde, a reagent with double bonds that can truly "crosslink" adjacent amino acid moieties, and thus physically and chemically stabilize the pig leaflet tissue against the proteolytic reactions to which they are subjected when implanted into the recipient. With a readily available source of animal tissues and a chemical process of preserving and sterilizing the material, the early success of the home-made, frame-mounted porcine xenografts [4] introduced by Alain Carpentier (Fig. 10.1) brought about the birth of the entire heart valve industry, situated primarily in the aerospace technology hub of California, Santa Ana.

Principles of Heart Valve Engineering
ISBN 978-0-12-814661-3, https://doi.org/10.1016/B978-0-12-814661-3.00010-1

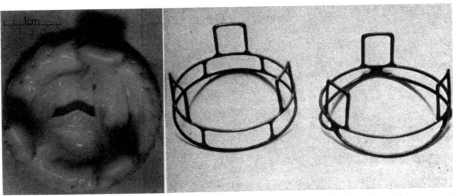

Figure 10.1 Photo of the first xenograft fabricated by Alain Carpentier (left) and the frames on which the porcine tissue was mounted. *(Left image From Carpentier, A., From valvular xenograft to valvular bioprosthesis: 1965-1970. Ann Thorac Surg, 1989. 48(3 Suppl): p. S73-4., right image From Carpentier, A. and Dubost, C. (1972). From xenograft to bioprosthesis: evolution of concepts and techniques of valvular xenografts. In Biological Tissue in Heart Valve Replacement (edited by M. I. Ionescu, D. N. Ross, and G. H. Wooler) p. 515. Butterworths, London.)*

Although clinically very successful, bioprostheses quickly demonstrated their primary weakness—the propensity to fail mechanically and to calcify becomes stenotic just like the original native valve. The heart valve leaflet tissues, with their propensity to fail mechanically and calcify, have become the single greatest weakness of bioprosthetic heart valves [5], something that simply does not occur with mechanical valves [6].

Just like the invention of the bioprosthetic heart valve started an entire heart valve industry, the eventual structural failure of the bioprosthetic valve leaflet tissue started an entire new field of academic research, with now three generations of scientists working on finding a solution to the inevitability of structural valve degeneration.

10.1 Alternative fixation chemistries

Today, five decades after Warren Hancock and Llowel Edwards founded the two main competing technologies in this field and commercialized their first bioprosthesis, all tissue valve leaflets still calcify and mechanically wear out and tear (Fig. 10.2). As noted in the previous chapters, the design of the valve frame, the leaflet shape, and leaflet anchoring are all critical to minimizing leaflet stresses. But in the early days of bioprosthetic valve development, calcification was believed to be the primary mechanism of valve failure and mitigation of calcification was one of the main areas of research, both in academia and in the valve industry. The observation that human allograft valves generally do not calcify suggested that the glutaraldehyde fixative itself was responsible for the calcification and that line of thinking was supported by some experimental evidence [7]. As a result, a large number of studies came about trying to identify alternative fixation chemistries that would replace glutaraldehyde and thus eliminate the calcification problem. Different compounds were tested, such as dye-mediated photooxidation (photofix) [8–10], the use of polyepoxy compounds [11], and carbodiimide [12,13]. Naturally

Leaflet slips downward and prolapses as tear opens up

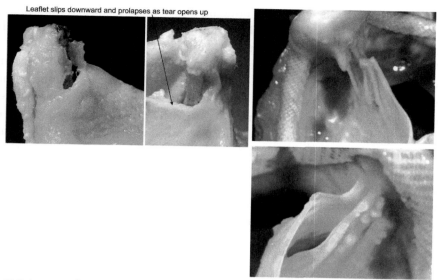

Figure 10.2 Images of commissural tears in bovine pericardial valves (top) resulting from stresses and abrasions against the underlying stent. Image of commissural tear in porcine xenograft valve (upper right) and the nodular calcification (lower right) that often accompanies low-grade collagen disruption before outright material failure.

occurring crosslinking agents, such as genipin [14], have also been tried. The photofix process was perhaps the most interesting, as it did not bind any new chemical agents to the tissue, but rather made use of existing covalent binding sites on the collagen molecules that can be induced to crosslink naturally, in a process similar to that of dehydration of leather or the formation of light-induced cataracts. An excellent review article [15] provides a summary of all the various approaches to crosslinking biological tissues in greater detail than can be reviewed here. Although all of these alternative approaches showed promise in the laboratory and often in animal models, photofix was the rare exception that made it as far as clinical use.

The mode of failure of the photofixed bovine pericardial valves was attributed to design issues, not to the chemistry of photooxidation. It was reported that the greater compliance and softness of this tissue was not sufficiently taken into account in the design of the valve frame and in the shape of the leaflets, and when implanted into patients, the belly of the leaflets abraded against the supporting stent [8]. Apparently, this mode of failure was not noted during animal implants because of the rapid endothelialization of the cloth covering that occurred in the juvenile sheep model in which the valve was tested, which protected the delicate tissue from abrasion against the cloth. It is unclear why this type of abrasion was not noticed during in vitro accelerated wear testing. Regardless, this clinical experience gave photofix a negative reputation and no further improvements to the valve design were attempted and the entire program was abandoned by the manufacturer. An excellent review paper on the chemistry of photofix is provided by

Dr. Moore, one of the key players in bringing this technology through the commercial development process [16]. Photofix has been resurrected a few years ago and is now applied to surgical patches of bovine pericardium [17].

Unfortunately as is the nature of scientific publications, there is no clear postmortem on specifically why and how each of the many alternative fixation chemistries failed to make it out of the laboratory and how they likely failed at the various stages of preclinical testing or if there were other confounding factors why none of these innovative chemistries never made it to market. It thus remains generally unclear specifically what deficiencies over glutaraldehyde they harbor. Scientists and inventors thus continue to tinker with other ways of improving the longevity of bioprosthetic tissues to slow down the apparent inevitability of calcification-related failure and the limitations of glutaraldehyde fixation. In spite of these limitations, glutaraldehyde has withstood the test of time and remains the only crosslinking technique in widespread use for bioprosthetic heart valves.

10.2 Anticalcification strategies

Glutaraldehyde fixation is a far from perfect solution. Glutaraldehyde is a toxic chemical that continues to leach from fixed tissues in small amounts over time. Some say that if glutaraldehyde fixation were to be introduced today as an agent for preserving implantable tissue-based valves, it would have never passed safety testing. But given that over five decades of bioprosthetic valve clinical usage have demonstrated that glutaraldehyde appears to be the only realistic approach to the preservation of animal tissues intended for bioprosthetic valves, it is clearly here to stay—whether it leads to calcification or not.

The degree to which glutaraldehyde fixation is responsible for calcification is reviewed in greater detail in Chapter 12. For the purpose of this chapter, let us simply state that historically, glutaraldehyde fixation was believed to be the main culprit, and that belief started the search for alternative fixation chemistry. An alternative path to mitigating calcification, whether or not it was caused by glutaraldehyde fixation, was to add additional chemistry to the glutaraldehyde-fixed tissues that is specific to the mineralization process or specific to the glutaraldehyde crosslinks.

Chemicals known as diphosphonates [18] inhibit the mineralization process of biological tissues when delivered systemically and have a profoundly negative effect on bone growth and calcium metabolism. Local controlled delivery of the drug was thus tried in animal models with good outcomes [19], but the therapy failed to be translated into a successful tissue treatment process.

An alternative technique that did make it into commercial use is the binding of amino acids to whatever free binding sites may exist in the glutaraldehyde-fixed tissue. This is the α-aminooleic acid approach of Medtronic [20,21], which is still used in both their porcine xenograft valves and in their new bovine pericardial valve. Edwards Lifesciences has an anticalcification approach referred to as Thermafix. There is little information published on the technique, but according to Edwards[1] the approach involves deactivating the free aldehyde (likely by heating [22]) and also extraction of the majority of lipids that naturally exist in the biological tissue because of its cellular makeup. Extraction of lipids using detergents and ethanol [23,24] has been known to reduce the calcification potential of glutaraldehyde-fixed valve tissues since the late 90s.

Edwards has recently introduced a new anticalcification approach called Edwards Integrity-Preservation (EIP), which has shown promise in preclinical studies [25,26]. This approach combines the Thermafix technology with "capping" of free aldehydes and then storage in a glycerol solution after ethylene oxide sterilization. The latter two steps are a major departure from the traditional means of sterilizing and storing glutaraldehyde tissue in the aqueous fixative, its principal advantages being the elimination of the preimplantation rinsing that all other glutaraldehyde-fixed and liquid-stored bioprosthetic valves require. Elimination of the liquid sterilant simplifies the preoperative handling of the valve, simplifies the manufacturing and packaging requirements, and distinguishes the product from its competition. This valve is now on the market in both Europe[2] and the United States[3] and is referred to as the Inspiris Resilia valve.

The third major player in the heart valve space—St.Jude—also has its own anticalcification treatment called Linx technology, which also makes use of ethanol soaking to extract cell membrane—derived lipids, and applies it to both its porcine tissue [27] and bovine pericardium-based heart valves [28]. In the author's experience, ethanol extraction of the cell membrane—derived lipids is perhaps the simplest and most effective first line of defense against the initiation of the mineralization process of glutaraldehyde-fixed bovine pericardial tissue and most likely porcine xenograft tissue as well.

An interesting postscript to the issues of glutaraldehyde and its role in calcification is the observation that while tissues not fixed in glutaraldehyde calcify less (i.e., photofix) [29], tissues fixed in greater concentrations of glutaraldehyde also appear to be calcification-resistant [30]. Clearly, there are multiple competing mechanisms of bioprosthetic tissue calcification.

10.3 No fixation

Glutaraldehyde fixation has been thought to be effective because it does two things: (1) crosslinks and thus masks the chemical markers that would cause a hyperacute reaction in the recipient in response to xenogenic antigens and (2) stabilizes the tissue against the

[1] https://www.edwards.com/eu/Products/HeartValves/Pages/ThermaFixAnimation.aspx.
[2] https://www.edwards.com/ns20160929.
[3] https://www.edwards.com/ns20170705.

proteolytic/enzymatic degradation that happens once implanted because of the foreign body response that is induced in the recipient. The notion that fixed tissue remains antigenic has been known since the early days of this field [31] but has been accepted only relatively recently [32]. To keep tissue soft and pliable, the amount of glutaraldehyde used in the fixative is typically in the range of 0.5%—0.65%. This amount of glutaraldehyde is not sufficient to completely mask the cell membrane—bound antigens or the related glycoproteins. The fixed tissue therefore continues to elicit cellular and humoral immune responses [33]. Inflammatory cells have been found on the surface and within the body of explanted, failed bioprostheses [34—36] demonstrating that this tissue is not as inert as was believed.

With these findings in mind, it was thought that the removal of cells and the associated cell-based antigens would be an appropriate solution. The initial attempts at fabricating nonfixed implants as early as 1984 involved arteries [37] and later valve tissues 10 year later [38], using extraction with sodium dodecyl sulfate. The techniques showed promise and were later combined with various other agents, such as protease inhibitors that protected the matrix from the autolytic enzymes that were released from the dead interstitial cells, and with specific enzymes, such as DNAse to more effective breakdown of the cellular remnants. The addition of these more specific enzymes was based on the observation that often cell-extracted tissues were more antigenic than nontreated tissues, when implanted subcutaneously in the rat model, as cell lysis and subsequent attempts to remove the cell debris with detergents and various hyper- and hypotonic solutions smeared these cellular antigens throughout the tissues. Even when a majority of these cell-based antigens are removed, the decellularized extracellular matrix still has the potential to attract inflammatory cells and induce platelet activation and thrombosis [39].

Perhaps the first measurable success with cellular extraction was by Mayer [40] in 2005, where they showed via rat implants that unfixed, cell-extracted tissues showed less of a cell-mediated antigenic response than untreated tissues. Many other variations on these extraction methods were attempted by many others, and the above summary is not intended to be exhaustive. For a good, detailed chronological review of the various cell extraction approaches, and the chemistries utilized, please see an excellent review article on this topic by Naso and Gandaglia [41].

But as is the nature of product development and preclinical testing, animal models can provide only limited information and can both underestimate and overestimate the severity of the reaction to the implant elicited by the human recipient. There comes a time when promising technologies need to be tried on patients. The boldest of these early attempts was by CryoLife, when in 2001 it decided to implant decellularized porcine aortic valves into a handful of patients. This initial clinical usage was the culmination of nearly 10 years of product development and preclinical testing, which apparently showed promise in sheep implants [42]. The valves were implanted into children that suffered from various congenital valvular defects and required reconstruction of the right ventricular outflow tract.

These valves were thus implanted as part of the Ross procedure [43]. As mechanical valves and anticoagulation therapies are typically not used in children, and conventional bio-prosthetic valves calcify extremely fast in these young patients, children with congenital valve defects have few options and face a future of repeated open chest surgeries and redo procedures for prosthetic valve failure. The implantation of the CryoLife "Synergraft" was thus approved under a compassionate use exemption.

It did not go well. Of the four patients (aged 2.5—9 years) implanted with these valves, one patient died on the seventh postoperative day because of graft rupture, two patients died 6 weeks later, and the final patient died 1 year later because of sudden cardiac death [39,44]. Pathological findings showed inflammation on day 2 of the implant, and all four grafts were encapsulated in a thick fibrous sheath. Histological exam demonstrated a severe foreign body reaction rich with neutrophils, granulocytes, and macrophages [45,46]. CryoLife has subsequently abandoned this product line, but others have pressed onward.

A German company (AutoTissue GmbH) implanted 16 adult patients with its decellularized porcine pulmonary valve in 2007—08 [47]. Within 15 months, 38% of patients required reoperation because of graft failure. As with the CryoLife Synergraft, histological exam of these explanted tissues revealed massive fibrotic response and cellular infiltration.

10.4 Alpha-gal removal

Implantation of nonfixed animal tissues is thus clearly not ready for human use. Interestingly, human allogeneic aortic valves do well in patients, even though there are no attempts to match tissue type, and no immunosuppressant medication is used. This experience is far different from that of solid organs, such as the liver and heart, likely because of the immune-privileged location of the aortic valve. But nonhuman tissues express the notorious α-gal epitope. As noted in an excellent review by Naso and Gandaglia [41], the α-gal antigen family is expressed in mammalian tissues but not in humans and higher primates. The α-gal antigen is expressed by human intestinal bacteria; our immune system is thus continually challenged by this antigen. Indeed, anti−α-gal antibodies account for 1%−3% of circulating immunoglobulins [48]. Once the α-gal antigen on implanted xenogenic tissue is recognized, the complement cascade is activated leading to thrombosis. This is clearly important in protecting the human body from intestinal flora-based infections. Regrettably, the same α-gal antigens persist as membrane debris within the tissue matrix that has been subjected to even rigorous enzymatic and detergent-based extraction protocols. Specific biochemical technique targeting the removal or masking of the remnant α-gal antigens in xenogenic tissues [49] is thus an exciting new approach which may be applicable to both porcine xenograft tissues and bovine pericardium-based valves.

10.5 Different types of tissues

Bovine pericardial heart valves evolved essentially in parallel with porcine xenograft valves. Given that the first implants were human cadaveric aortic valves, and the French clergy put a stop to that, the most logical next approach was the harvesting of aortic valves from slaughtered pigs. Except for a slight asymmetry to the valve because of a "muscular shelf" at the left coronary cusp, the pig valves were nearly identical to human valves. Indeed, it was the early communication between the French surgeon Alain Carpentier and the American entrepreneur Warren Hancock that led to the first commercially available Hancock valve, which was eventually incorporated into the Medtronic product portfolio.

The advantage of using the aortic valve from a pig as a bioprosthesis is self-evident—it is already an anatomically correct, well-functioning biological valve. All that was required was proper fixation and selection for size and symmetry and then trimming of the extraneous tissue to enable mounting on the supporting stent. Heart valves are usually available in sizes of 2 mm increments, typically ranging from 19 mm in diameter for small aortic roots to the mid 30s for the mitral position. For reasons that remain somewhat unclear, almost all valves are sized and numerically labeled in the odd number sizes, rather than even (19—27 mm instead of 20—28 mm). With than in mind, manufacturers fabricating porcine xenograft valves had to sort through a large number of raw aortic valves obtained from slaughter to find a subset that was sufficiently "close" to the sizes of the supporting stent in 2 mm increments. Also, because aortic valves do not have the perfect 120 degree symmetry, selection was also based on avoiding highly asymmetric looking valves because the supporting valve frames were indeed manufactured to be symmetrical. Some in the valve industry say that the rejection rate of raw pig valve tissues was as high as 90% to get the subset of symmetrical, properly dimensioned valves suitable for stent mounting.

Besides the two established lines of bioprosthetic valves—bovine pericardium and porcine xenograft—the early days of this field also witnessed devices constructed from human cadaveric dura matter [50] and fascia lata [51]. These two approaches quickly turned out to be noncompetitive to the bovine pericardium— and pig aortic valve—based approaches.

Because glutaraldehyde fixation was thought to be the culprit behind calcification, nonfixed autologous human pericardium was used to fabricate intraoperatively [52]. The Autogenics "kit" consisted of plastic frame components and cutting templates that would enable a physician to quickly fabricate a valve from the patient's own pericardium (Fig. 10.3). During initial attempts, the raw pericardium quickly failed because of contraction of the tissue, and the valve soon became regurgitant. Only when the pericardial interstitial cells were killed with a quick dip in glutaraldehyde, the valve survived more than a few weeks. But eventually, these valves also experience structural failure much sooner than the commercial leaders, and the approach was eventually abandoned clinically.

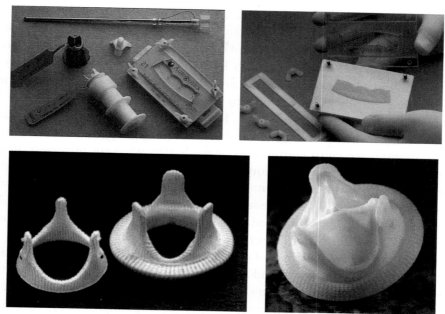

Figure 10.3 Various images of the Autogenics autologous pericardial valve kit components and the final intraoperatively fabricated valve.

Throughout the history of this field, many other tissues were tried, such as kangaroo valves [53] and pericardium from other species (i.e., equine [54], porcine [55], and ostrich [56] pericardium) for reasons of convenience, thickness, or a belief that more long-lived animals will yield more long-lived bioprosthetic valves. Few of these approaches yielded anything with commercial success. Although equine pericardium is used in the Medtronic 3F valve [57] and in the Medtronic CoreValve transcatheter implantable valve [58] and porcine pericardium is used as a wrap in the St. Jude Trifecta [28], the greater majority of the valves sold on the international stage are fabricated from either bovine pericardium or pig aortic valves.

10.6 Physical treatments

As any surgeon can confirm, handling of raw tissues is exceedingly difficult. It is in part because of this challenge that the autologous pericardium of the Autogenics approach required a brief dip in glutaraldehyde to stiffen the tissue and make it amenable to handling and proper positioning on the supporting stent.

For similar reasons, the first-generation Hancock porcine valves were cross-linked with the valve closed and under a pressure difference of 80 mmHg [59]. This created nicely closed valves with well-shaped leaflets. However, it was found through pathological analyses of explants [60] that stiff leaflets of these high-pressure fixed valves had all the crimp of the collagen fibers straightened out during the fixation process. This valve tissue was incapable of responding to cyclic stresses in the appropriate way and was thought to fail sooner than it ought to have. The industry then switched to low-pressure fixation (just several mmHg) so as to preserve the collagen crimp in the main circumferentially oriented collagen fiber bundles. This made the tissue more compliant and apparently more durable. But again, it was found that while much of the crimp of the larger collagen fiber bundles could be preserved through low-pressure fixation, the radial corrugations of the leaflets were not preserved. Radially, the leaflets were stretched out and could thus not respond to valve opening and closing actions in the way that the natural, nonfixed valves did [61]. Only when the tissue was cross-linked at no transvalvular pressure that these radial corrugations in the fibrosal layer of the valve could be maintained. This "zero-pressure fixed" valve was introduced commercially as the Medtronic Intact valve [62].

Paradoxically, what was supposed to be a fairly straightforward process to manufacture porcine xenografts—harvest and fix a pig valve and mount it on a stent—turned out to be hugely challenging from an engineering perspective. Besides the issues of selection for size and symmetry, there were issues of fixation pressure and its impact on the coaptation of the leaflets. Zero-pressure fixation was initially unworkable because of the elastic nature of the aortic root. The aorta and the aortic root that supports the valve leaflets never sees zero pressure when it functions. The root is always cycling between a diastolic pressure of 80 mmHg and a systolic pressure of 120 mmHg, or thereabouts. When the animal is killed and pressure drops to zero, the preloaded elastic lamellae pull the tissue together and the root shrinks by almost 30% [63]. This means that the leaflets inside the aortic root are pulled together away from their nature shape and are profoundly redundant. Valves fixed at this low pressure have excessive leaflet tissue that cannot coapt properly. Therefore, to restore the aortic root to its natural anatomic diameter and the leaflets to their natural shape, these valves need to be mounted on a fixturing system that subjects the inside of the valve to a pressure near 80 mmHg, yet with the leaflets still open and without any pressure difference. This is very challenging. The outflow aspect of the valve can be fitted with a simple tube, but the inflow aspect has only a thin rim of muscle tissue and mitral leaflet. Medtronic had to develop complex fixturing hardware just so that these porcine aortic roots could be fixed at the correct internal pressure, yet still have the leaflets open and fixed at "zero pressure."

By contrast, developing a valve de novo from bovine pericardium seemed far simpler—fix a sheet of bovine pericardium and wrap it around a stent and sew it in place. But that is where good pericardial valve designs are separated from bad designs. The early bovine pericardial valves, such as the Ionescu-Shiley valve, suffered early mechanical failure because of stress concentrations at the commissural stitch [64]. This design flaw is not repeated in the current commercially successful pericardial valves. Today, the two leading valves on the market make use of leaflet clamping (Edwards Magna) or wrapping of the pericardium around the outside of the supporting stent (St. Jude Trifecta). But the design of a valve and the shape of its leaflets cannot be divorced from the mechanical and physical aspects of the leaflet tissue.

As noted in a summary thesis on what features lead to a long-lived, durable bioprosthetic heart valve [65], the so-called "Three Tenets of Good Valve Design" are (1) flexible stent posts, (2) precise control of the central gap, and (3) absolute 120 degrees leaflet symmetry. These are the design features that have enabled the gold standard surgical valves to show zero failure at 20 years in patients older than 70 [66–68]. Symmetry is simply the nature of an engineered valve, such as the bovine pericardial valve design, as long as it is delivered without any deformation. But the precise control of the central gap and the degree of flexure of the stent posts must be related to the physical properties of the tissue that is used. An illustrative example of how valve design related to tissue mechanical properties is evident when the leaflets of porcine xenografts are compared to those of the bovine pericardial valves (see Fig. 10.4).

In the native aortic valve, as in the porcine xenograft valves, the angle of the leaflets is considerably greater than it is in the bovine pericardial valves. This is because of the anisotropy of the porcine tissue. The native aortic valve leaflets are highly extensible in the radial direction (about 60% strain) and come together and coapt largely through their radial elongation [61,69,70]. Circumferentially, however, they extend by less

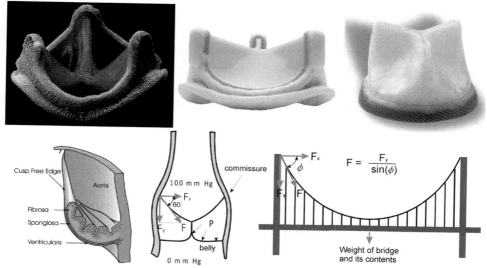

Figure 10.4 Photos of the Hancock porcine valve (top left), the Edwards Magna (top center), and the Sorin Mitroflow (top right), showing the different leaflet free edge angles. Bottom diagrams represent the microstructure of the aortic valve (bottom left) and the geometrical basis for free edge stresses in relation to free edge angles (bottom center and right). *(From I.Vesely, The evolution of bioprosthetic heart valve design and its impact on durability, Cardiovascular Pathology, Volume 12, Issue 5, September–October 2003, Pages 277-286)*

than 10%. During valve opening, the aortic valve leaflets retract and move out of the way radially while at the same time folding back because of the central blood flow. The physical advantage of having large leaflet angles is the far lower tension at the leaflet free edge. By comparison, the free edge stresses of pericardial valves are far greater, simply by virtue of the leaflet angle and the suspension bridge analogy. This is a simplification, however, for illustrative purposes only. In reality, the free edge leaflet stress of the bovine pericardial valve is reduced from the theoretical because the leaflets butt up and lean against each other in the center and thus transfer the free edge stress lower down into the body of the leaflet. But this is exactly why there needs to be precise control of the central gap. If the central gap is too large, there is insufficient leaning of the leaflets against each other and stresses go up. A central gap too small leads to wrinkling, buckling [71], and pinwheeling of the leaflets at the central triple point. The size of the central gap, therefore, is controlled by a combination of three factors: (1) leaflet shape, (2) stent post flexibility, and (3) leaflet tissue compliance. All three have to work together in concert so that the leaflets coapt precisely at maximal loading.

This is where the physical handling of the bovine pericardium during the leaflet preparation process comes in. First of all, bovine pericardial leaflets need to be of the same thickness—thinner leaflets experience greater stress when subjected to tension. They will also stretch more and thus upset the leaflet coaptation balance. Secondly, they need to have the same physical properties, meaning they need to have the same extensibility, again for the same reasons of maintaining proper central coaptation.

Several decades ago, much effort was placed into understanding the fiber structure of bovine pericardium and its impact of the mechanical properties of the tissue [72,73]. The intent of this work was to find optimal harvesting sites of patches of tissue where tissue anisotropy was the least and the most predictable. Small-angle light scattering was a convenient approach to studying the orientation angles of the collagen fibers, and models were developed for representing the dispersion distributions. But the light scattering models were always confounded by the uncertainty associated with angular dispersions resulting from the fibers themselves versus the angles associated with the collagen crimp within each fiber family. Although Edwards Lifesciences reportedly invested considerable amounts of money into studying this phenomenon and attempted to come up with a quantifiable approach that relates fiber distribution to tissue mechanics, to optimal tissue harvesting sites, personal communications with ex Edwards employees suggest that ultimately an empirical, experience-based, "touch-and-feel" approach, combined with some mechanical property measurements, was found to be more reliable that any fiber distribution measurements.

Considerable effort was also expended into modifying tissue anisotropy via preload during glutaraldehyde fixation. The idea was that if one could induce anisotropy into bovine pericardium by tethering and preload, one could "engineer" a leaflet tissue with the same anisotropy as the native aortic valve tissue [74–76]. An anisotropic tissue could then be used in the same way as in the native aortic valves, having greater compliance in the radial direction and less compliance circumferentially. These efforts also did not lead anywhere, as there is no way that anything close to the 50% radial strain of the native aortic valve tissue could be mimicked by selective tethering of bovine pericardium during fixation. All that could be achieved was a stiffer, less compliant pericardium, albeit with some fixation-induced anisotropy. Today, bovine pericardial valves are fabricated from tissue subjected to "some" tensile preload, largely to eliminate the wrinkles that are created when a tissue that is naturally spherical is flattened out. The amount of preload is governed primarily by the empirical experience on the part of the valve industry veterans who are involved in setting up the manufacturing process. There is no unique formula for the "best" way to fix bovine pericardium under the "best" amount of tension. All that can be achieved is to adapt a particular fixation protocol to the compliance of the valve stent, and vice versa, and to the shape of the valve leaflet. Designing a well-functioning, properly coapting bovine pericardial valve is an iterative process, driven largely by design targets, proceeding with the experience of the engineers involved. Tissue compliance therefore subscribes to the Goldilocks doctrine and should therefore be "just right." Not too soft and not too stiff. That is the unfortunate reality of working with a material with natural biological variability.

This natural variability also manifests in the challenges of tissue harvesting and leaflet selection. It is hard enough to find (or fabricate) three leaflets that are mechanically equivalent, when putting together an Edwards Magna valve or a Medtronic Avalus [77]. It is that much more complicated and challenging to find a strip of tissue that is even more uniform so that when this tissue is wrapped around a frame, to form the three leaflets of a St. Jude Trifecta valve, these three leaflets have similar mechanical properties. It is precisely because of the spatially varying distribution of collagen fibers that was studied in detail by Sacks [73], which makes finding a long strip of tissue that is uniform in its compliance that much more difficult. Some of the early work done in the author's lab (published only in conference proceedings) shows that the compliance of strips of tissue varies spatially across the length of the pericardial tissue strip. It is thus highly likely that once the strip of tissue is wrapped around the frame, the two leaflets at opposite ends of the strip will have mechanical properties that differ from each other more than the leaflets adjacent to the central leaflet (see Fig. 10.5). Perhaps the tissue that is used for the manufacture of externally mounted, single-wrap pericardium valves such as the St. Jude Trifecta is fixed at greater tensile preload than that of the individual leaflet, internally mounted pericardial tissue of the Edwards Magna/Medtronic Avalus. Greater tensile preload reduces the natural spatial variability of tissue compliance and makes this type of tissue more uniform and thus easier to manage. Perhaps this is why the free edge angles of the single wrap valves are so shallow, when compared with those of the individual leaflet valves. The lower the leaflet compliance, the shallower the leaflet angles must be to arrive at a properly coapting, well-functioning prosthetic valve. Tissue handling,

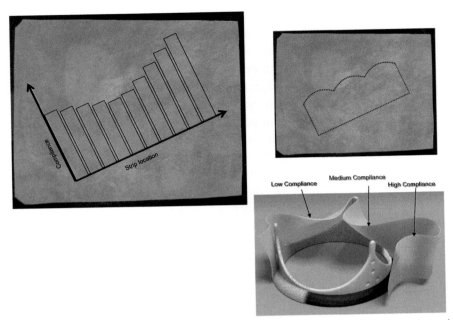

Figure 10.5 Photos (above) of a sheet of bovine pericardium visualized under polarized light, showing the variation in fiber density and fiber orientation. Superimposed over that is a bar graph showing how the compliance of adjacent strips of tissue varies across the sheet of pericardium. If a continuous strip of pericardium is cut from this material (top right), the three leaflets will have different compliance in accordance with that of the region of pericardium, when wrapped around a three postframe (bottom right), as done for the Sorin Mitroflow valve and the St. Jude Trifecta.

tissue selection, material compliance, and valve design thus cannot be managed independently. They all influence each other, as they all interact when the valve opens and closes, and they all affect the ultimate longevity and durability of the device.

10.7 Testing the efficacy of a tissue and its chemical treatments

Once a tissue preparation scheme is finalized, its efficacy needs to be evaluated and verified. The leaflet tissue must be chemically inert and stable, nonantigenic, and not calcify. Tests for these qualities are not only necessary to establish confidence that the valve product will work well once implanted into the patient but also have to incorporate into regulatory standards. Most are essentially mandatory.

Chemical stability can be assessed through digestion with proline following standard protocols. If only a small amount of protein can be digested out of the tissue following incubation with the enzyme, the tissue is considered to be sufficiently well cross-linked. Glutaraldehyde crosslinking also affects the thermal denaturation temperature of the collagen molecule. Using a hydrothermal shrinkage test, one can suspend a strip of tissue between grips, at some preload detected with a load cell, heat up the tissue until the collagen molecules denature and shrink, and then detect the temperature at which this denaturation is detected by a spike in the tension induced in the tissue. Raw, uncrosslinked collagen tissue shrinks at 57—60 degrees, and "properly fixed" tissue shrinks at 80—85 degrees depending on the publications being sourced. The greater the fixation, the greater the hydrothermal shrinkage temperature measured. Interestingly, the photofix treated tissue showed minimal change in its shrinkage temperature compared to unfixed tissue [78], stimulating some concerns as to exactly how the crosslinks manifest and what was the real reason these valves failed clinically.

Most chemical fixation, of course, introduces toxic reagents into a material which itself is supposed to be biocompatible. Numerous other tests have thus been developed, which are intended to detect whether the tissue induces any cytotoxic effects on the recipient. Many tests have been developed and over time, many have been incorporated into regulatory standards or simply good practice when qualifying a valve for human implant use. These tests are as follows:

- Genotoxicity
- Carcinogenicity
- Reproductive toxicity
- Systemic toxicity
- Cytotoxicity
- Sensitization, irritation
- Hemolysis
- Partial thromboplastin time
- Pyrogenicity
- Complement activation
- Mutagenesis
- Chromosomal aberration
- Muscle implantation
- Subcutaneous implantation
- Orthotopic implantation

Perhaps the most relevant test for assessing how well the valve will do is the orthotopic implantation of the valve into an animal model, in most cases, the juvenilely sheep [79]. Valves are implanted into a growing sheep and explanted in about 5 months. If these valves do not calcify, or calcify very little, there is some comfort that they are unlikely to calcify excessively when implanted into patients. Implantation into sheep is not proof that a new candidate valve will do well in patients, but offers some comfort that they are unlikely to fail early due to some physical or chemical features that lead to early calcification.

Preparation of the tissue is relevant to the types of tests chosen to validate the new design. The further one departs in tissue preparation from what has been the norm over the past five decades, the greater is the challenge for the manufacturer to demonstrate that the valve is safe. For example, one cannot test a valve fabricated from autologous human pericardium in a sheep model because in the sheep, the valve would not be autologous. One could use sheep pericardium, but the mechanical properties of the sheep pericardium are different from those of human pericardium, and arguably the design of the leaflets—their shape—is likely to be different. Similarly, if one is to commercialize pig valves that are not glutaraldehyde fixed, but instead decellularized, or treated to remove the α-gal epitopes, should they be tested in sheep or in pigs? These are largely regulatory challenges, but they do impact the logistics of which tissue treatments are feasible for commercialization and are discussed in greater detail in Chapter 15.

10.8 Stentless valves

The pinnacle of porcine xenograft valve evolution was the introduction of the stentless valve. These valves were introduced in the 80s as an alternative to conventional, stent-mounted porcine and bovine tissue valves for the simple reason that removing the sewing cuff enables a slightly larger valve to be implanted into the patient. The cuff and the frame arguably take up space in the patient's aortic root, and if the cuff is absent, a size larger valve can be implanted. Back in those days, valve companies competed with each other with regard to who can get a valve to market with the lowest transvalvular gradient. Stentless valves gave them that competitive edge. Medtronic developed the most successful stentless valve product—the Freestyle valve [80]—with Edwards [81] and St. Jude [82] making less successful variants of the same. Indeed, many companies tried their hand a developing stentless valves to get some market share. All of these valves consisted of a glutaraldehyde–fixed intact porcine aorta with the native leaflets contained and anatomically positioned within. This device was not much different from the human allograft/homograft valve, except that it was from a pig rather than from a cadaver. It required the same, far more complicated implant technique [83] that Carpentier dispensed with when he created the porcine xenograft—a huge step backward—but it offered better gradients and a few other theoretical benefits that are peripheral to this story, and it caught on with surgeons who were now quite expert at complex aortic valve implant procedures.

But in developing these products, it became clear very quickly that placing glutaraldehyde-fixed xenogenic tissue in direct contact with the living host tissue was problematic—there was a severe local tissue response. The reports on exactly what happened are scant because they were experienced largely during development in the valve company laboratories, but the solution to the host response was to cover portions of the xenogenic tissue with Dacron cloth. In the case of the Medtronic Freestyle valve, the so-called "muscular shelf" that intrudes into the base of the right coronary leaflet was covered by cloth. In the Edwards and the St. Jude valve, the entire adventitial surface of the porcine aortic root was covered by cloth.

In cases where these valves have been explanted for early failure, pathological exam has provided insight into the likely modes of failure [84,85]. Being a porcine tissue valve, rather than bovine pericardial, the calcific degeneration was more extrinsic and often nodular on the surface of the porcine leaflets. In bovine valves, the mineralization is more intrinsic with stiffening of the tissue without visual "florets" of mineral on the valve leaflet surface. All valves explanted because of late failure have some level of pannus overgrowth, mainly over the aortic wall component of the graft, particularly at the anastomosis to the native aorta (see Fig. 10.6). The absence of a sewing cuff in these valves, therefore, did not lead to massive tissue overgrowth onto the leaflets. Indeed, in most cases, the leaflets remained free of pannus, even though the adjacent vessel wall and muscular tissue were heavily infiltrated.

The experience with this clinically used, glutaraldehyde-fixed, xenogenic aortic root graft indicates that while in most cases, the classical approach to implanting a prosthetic

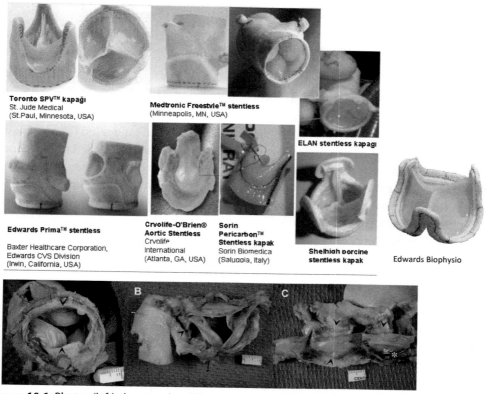

Figure 10.6 Photos (left) showing the different types of stentless valves introduced by all the major heart valve companies. The lack of stent, however, led to early degeneration because of the proximity of the tissue to the host and the resultant inflammatory reaction (bottom). *(From Jagdish Butany et al. Inflammation and infection in nine surgically explanted Medtronic Freestyle® stentless aortic valves, Cardiovascular Pathology, Volume 16, Issue 5, September—October 2007, Pages 258-267)*

valve works, there are infrequent but significant cases of excessive inflammatory response, particularly at the junction of the native and graft aorta. The pig aorta is heavily infiltrated by host lymphocytes and macrophages, and the graft tissue is partially replaced by host fibrotic tissue. In some cases, the cloth covering the muscular shelf of the right coronary porcine cusp is an insufficient barrier and this muscle is completely infiltrated by host cells. Because the implant duration varied from 4 to 108 months, it is unlikely related to normal wear and tear on the implanted device. Most likely, it is not only associated with host factors that vary from patient to patient but also related to the age. Unlike most bioprosthetic valves that are implanted into 65-year-old patients or older, these Freestyle valves were implanted into patients as young as 38, and there appears to be a general correlation between young age and time to failure. The complexity of the stentless valves eventually led to their demise—they were difficult to put in and the inherent residual antigenicity of the material led to early structural failure. Regrettably, this was something that did not come out via the extensive preclinical testing that these valves were subject to before their introduction to the market.

10.9 Surgeon factors

Tissue handling and preparation is not limited to not only manufacture but also to the implant procedure itself. This was embodied best in the example of the Edwards Biophysio valve—a stentless bovine pericardial valve (see Fig. 10.6) that was to be implanted using the assistance of a template and other surgical guides [86,87]. Again, discussions with ex Edwards employees indicated that "apparently only Dr. Carpentier could implant the valve right," making this approach not commercially viable. Even today, aortic allograft valves require tremendous skill on the part of the surgeon, and long-term durability of these valves is highly dependent on the seniority of the surgeon and thus his lifelong experience implanting these valves. Intraoperative handling and preparation during implant is thus particularly important in stentless valves, and it is thus exceedingly difficult to mimic during product development and preclinical testing.

The well-known surgeon Dr. Tirone David once commented during a conference that he did a longitudinal study attempting to correlate the various design, material, and other features of a valve with its longevity in the patient and freedom from various complications. He said that he found that the single greatest factor that contributed to valve longevity was the surgeon. And this was for all valves, not just the stentless valves or the allografts. The implication here is that surgical skill during implantation is a profoundly important variable. How does a manufacturer test for that, and how does the regulatory body ensure that only competent individuals use these products?

Luckily, the complexity of transcatheter valves (reviewed in Chapter 5) has necessitated a rethinking of the way that valves are implanted and valve patients are managed. The new era of transcatheter aortic valve replacement (TAVR) has brought about the concept of Heart Teams [88,89]. Just like most other fields of technology, maturity leads to a reflection of how the field began, where the mistakes were made, and how the mistakes of the past could be avoided in the future.

10.10 Unmet needs and opportunities

Innovation is not a linear process. In my 30-year experience in this field, tissue preparation has become mature in the context of glutaraldehyde fixation and calcification mitigation. Future advances in these areas are most likely to be incremental. What then are the unmet needs and the opportunities for innovation?

The introduction of TAVR is an important lesson here. Before TAVR, those of us who worked in the "golden age of surgical valves" were preoccupied with improvements in tissue durability via tissue treatments and anticalcification strategies so that the inevitability of bioprosthetic valve failure could be postponed and the patient would not have to be subjected to an open heart valve redo. These "traditional" requirements for a good bioprosthesis have been all but forgotten once noninvasive TAVR devices were introduced. Concerns regarding the early failure of TAVR devices often fall on deaf ears.

A case in point is the report that came to light in the summer of 2016 via a presentation at the EuroPCR meeting in Paris, where Dr. Danny Dvir followed 704 patients (mean age of 82 years) who underwent TAVR.[4] Patients were then monitored using repeat echocardiographic exams for up to 10 years. Of the 100 patients who survived at least 5 years after their procedure, 35 showed signs of valve degeneration. The mean time to the onset of degeneration was 61 months (~5 years). The conventional way to study long-term data such as these is via Kaplan—Meier analyses [90]. The TAVR curve of freedom from degeneration drops from 94% at 4 years to 82% at 6 years and to 50% at 8 years among surviving patients. It is well-known that once tissue valves begin to fail, they continue to fail faster. The linearized failure curves are not straight, but instead fall off rapidly (see Fig. 10.7). The current data suggest that the majority of TAVR valves will have failed at 10 years. This is in stark contrast to surgically implanted tissue valves that have durability well beyond 15 years [33], and far, far worse than what one can expect from the gold standard Edwards Perimount Magna class of surgical valves [66] that are increasingly being implanted into younger patients.

The Edwards Perimount Magna class of heart valves represents the pinnacle of surgical valve evolution via good design principles and constant improvements in manufacturing quality control. In patients older than 70 years of age at the time of implant, this valve can last over 20 years [66] (see Fig. 10.7 left panel). Accordingly, valves

---

[4] https://www.mdedge.com/ecardiologynews/article/109297/interventional-cardiology-surgery/tavr-degeneration-estimated-50.

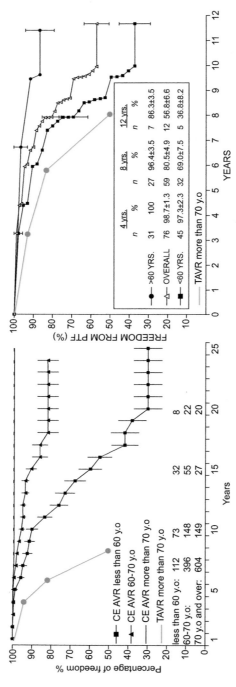

Figure 10.7 Actuarial *curves* of the Edwards valve (left) showing that when implanted into patients older then 70, it can last up to 25 years without failure. Similar *curves* of the Sorin Pericarbon valve (right) that is no longer used because of its comparably poor performance relative to the Edwards valve are shown. Superimposed over both of these *curves* in red are the actuarial *curves* presented by Dvir on the failure rates of transcatheter valves, showing that even in patients older than 70, the transcatheter aortic valve replacement (TAVR) valves fail faster than the no longer used Sorin valves did in patients younger than 60.

with inferior clinical performance, such as the Sorin Pericarbon [91] (Fig. 10.7 right panel) or the Sorin Mitroflow [92], are no longer used in mainstream clinical centers. The same actuarial curves plotted for the inferior valves show the same rapid progression of valve degeneration and failure, but they simply start sooner and decay faster. With that in mind, the early reports of TAVR valve degeneration should raise considerable concern. The early data suggest that when TAVR valves are implanted into patients in their 70s and 80s—the same patients in whom the Edwards valves do not fail at all—TAVR devices fail at a rate that is even faster than that of the worst surgical valve in recent history—the Sorin Mitroflow—and when that valve is implanted into patients younger than 60 (see red lines in Fig. 10.7). If this is true, the implications are profound. TAVR valves will fail faster in the elderly than the worst surgical valves did in the younger patients. Dissemination of TAVR into younger patients clearly cannot take place until these concerns regarding longevity are addressed. It has been 3 years since Dvir brought these concerns to light, but there have been just a scant few publications reporting on TAVR durability issues [93,94]. The reality is that we do not really know how long transcatheter valves will last. Early implants of TAVR devices were done in only the very frail, elderly patients with a limited life expectancy. Indeed, almost all of the recipients of the first-generation Sapien valve that participated in the PARTNER trial were dead after 5 years, preventing any evaluation of transcatheter valve longevity [95].

So even in the age of TAVR, where noninvasive approaches are far more exciting than long-term durability, the need for durability remains because the risk of early valve failure remains. The focus on improving TAVR durability, however, will most likely take place once clear, unequivocal data come to light. In the meantime, the untapped, completely wide open area of advance is most likely in the use of synthetic materials, be they of organic or nonorganic origin, the methods of their preparation and treatment, and their use in niche applications such as the newborn and adolescent patient who simply cannot and should not withstand multiple surgical interventions to replace failed bioprosthetic valve tissues. Perhaps once these new technologies prove themselves in the niche area, not unlike TAVR, then they will move into the mainstream world of the 60+-year-old patient with isolated aortic valve stenosis. The other untapped need is in the developing world, where the demand for prosthetic valves dwarfs that of the developed world [96], and where the $25K+ price tag for TAVR is simply unworkable. In my view, what the world needs is a $250 valve implant solution with a durability of well over a decade. Now that would be a goal worth pursuing!

References

[1] Ross DN. Homograft replacement of the aortic valve. Lancet 1962;2(7254):487.
[2] Lund O, et al. Primary aortic valve replacement with allografts over twenty-five years: valve-related and procedure-related determinants of outcome. J Thorac Cardiovasc Surg 1999;117(1):77—90. discussion 90-1.

[3]   Carpentier A. From valvular xenograft to valvular bioprosthesis: 1965-1970. Ann Thorac Surg 1989; 48(3 Suppl. l):S73—4.

[4]   Carpentier A. D.C., from xenograft to bioprosthesis: evolution of concepts and techniques of valvular xenografts. In: R.a.W. Ionescu, editor. Biological tissue in heart valve replacement. London: Billing & Sons Limited; 1972. p. 515—42.

[5]   Rahimtoola SH. Choice of prosthetic heart valve for adult patients. J Am Coll Cardiol 2003;41(6): 893—904.

[6]   Brown ML, et al. Aortic valve replacement in patients aged 50 to 70 years: improved outcome with mechanical versus biologic prostheses. J Thorac Cardiovasc Surg 2008;135(4):878—84. discussion 884.

[7]   Golomb G, et al. The role of glutaraldehyde-induced cross-links in calcification of bovine pericardium used in cardiac valve bioprostheses. Am J Pathol 1987;127(1):122—30.

[8]   Schoen FJ. Pathologic findings in explanted clinical biprosthetic valves fabricated from photooxidized bovine pericardium. J Heart Valve Dis 1998;7(2):174—9.

[9]   Bianco RW, et al. Feasibility evaluation of a new pericardial bioprosthesis with dye mediated photo-oxidized bovine pericardial tissue. J Heart Valve Dis 1996;5(3):317—22.

[10] Meuris B, et al. Porcine stentless bioprostheses: prevention of aortic wall calcification by dye-mediated photo-oxidation. Artif Organs 2003;27(6):537—43.

[11] Sung HW, et al. Cross-linking characteristics of biological tissues fixed with monofunctional or multi-functional epoxy compounds. Biomaterials 1996;17(14):1405—10.

[12] Girardot JM, Girardot MN. Amide cross-linking: an alternative to glutaraldehyde fixation. J Heart Valve Dis 1996;5(5):518—25.

[13] van Wachem PB, et al. In vivo behavior of epoxy-crosslinked porcine heart valve cusps and walls. J Biomed Mater Res 2000;53(1):18—27.

[14] Sung HW, et al. Mechanical properties of a porcine aortic valve fixed with a naturally occurring cross-linking agent. Biomaterials 1999;20(19):1759—72.

[15] Paul RG, Bailey AJ. Chemical stabilisation of collagen as a biomimetic. ScientificWorldJournal 2003; 3:138—55.

[16] Moore MA. PhotoFix: unraveling the mystery. J Long Term Eff Med Implant 2001;11(3—4): 185—97.

[17] Majeed A, et al. Histology of pericardial tissue substitutes used in congenital heart surgery. Pediatr Dev Pathol 2016;19(5):383—8.

[18] Johnston TP, Schoen FJ, Levy RJ. Prevention of calcification of bioprosthetic heart valve leaflets by Ca^{2+} diphosphonate pretreatment. J Pharm Sci 1988;77(9):740—4.

[19] Levy RJ, et al. Inhibition of calcification of bioprosthetic heart valves by local controlled-release diphosphonate. Science 1985;228(4696):190—2.

[20] Riess FC, et al. Clinical results of the Medtronic mosaic porcine bioprosthesis up to 13 years. Eur J Cardiothorac Surg 2010;37(1):145—53.

[21] Girardot MN, Torrianni M, Girardot JM. Effect of AOA on glutaraldehyde-fixed bioprosthetic heart valve cusps and walls: binding and calcification studies. Int J Artif Organs 1994;17(2):76—82.

[22] Carpentier SM, et al. Biochemical properties of heat-treated valvular bioprostheses. Ann Thorac Surg 2001;71(5 Suppl. l):S410—2.

[23] Vyavahare N, et al. Prevention of biprosthetic heart valve calcification by ethanol preincubation. Efficacy and mechanisms. Circulation 1997;95(2):479—88.

[24] Lee CH, et al. Inhibition of aortic wall calcification in bioprosthetic heart valves by ethanol pretreatment: biochemical and biophysical mechanisms. J Biomed Mater Res 1998;42(1):30—7.

[25] De La Fuente AB, et al. Advanced integrity preservation technology reduces bioprosthesis calcification while preserving performance and safety. J Heart Valve Dis 2015;24(1):101—9.

[26] Flameng W, et al. A randomized assessment of an advanced tissue preservation technology in the Juvenile sheep model. J Thorac Cardiovasc Surg 2015;149(1):340—5.

[27] Lehmann S, et al. Mid-term results after Epic xenograft implantation for aortic, mitral, and double valve replacement. J Heart Valve Dis 2007;16(6):641—8. discussion 648.

[28] Bavaria JE, et al. The St Jude Medical Trifecta aortic pericardial valve: results from a global, multi-center, prospective clinical study. J Thorac Cardiovasc Surg 2014;147(2):590—7.

[29] Westaby S, et al. The carbomedics "oxford" photofix stentless valve (PSV). Semin thorac cardiovasc surg 1999;11(4 Suppl. 1):206—9.

[30] Zilla P, et al. Improved ultrastructural preservation of bioprosthetic tissue. J Heart Valve Dis 1997;6(5): 492—501.

[31] Nimni ME, et al. Chemically modified collagen: a natural biomaterial for tissue replacement. J Biomed Mater Res 1987;21(6):741—71.

[32] Human P, Zilla P. Characterization of the immune response to valve bioprostheses and its role in primary tissue failure. Ann Thorac Surg 2001;71(5 Suppl. l):S385—8.

[33] Grunkemeier GL, Bodnar E. Comparative assessment of bioprosthesis durability in the aortic position. J Heart Valve Dis 1995;4(1):49—55.

[34] Nistal F, et al. Comparative study of primary tissue valve failure between Ionescu-shiley pericardial and hancock porcine valves in the aortic position. Am J Cardiol 1986;57(1):161—4.

[35] Dahm M, et al. Relevance of immunologic reactions for tissue failure of bioprosthetic heart valves. Ann Thorac Surg 1995;60(2 Suppl. l):S348—52.

[36] Stein PD, et al. Leukocytes, platelets, and surface microstructure of spontaneously degenerated porcine bioprosthetic valves. J Card Surg 1988;3(3):253—61.

[37] Malone JM, et al. Detergent-extracted small-diameter vascular prostheses. J Vasc Surg 1984;1(1): 181—91.

[38] Wilson GJ, et al. Acellular matrix: a biomaterials approach for coronary artery bypass and heart valve replacement. Ann Thorac Surg 1995;60(2 Suppl. l):S353—8.

[39] Kasimir MT, et al. The decellularized porcine heart valve matrix in tissue engineering: platelet adhesion and activation. Thromb Haemostasis 2005;94(3):562—7.

[40] Meyer SR, et al. Decellularization reduces the immune response to aortic valve allografts in the rat. J Thorac Cardiovasc Surg 2005;130(2):469—76.

[41] Naso F, Gandaglia A. Different approaches to heart valve decellularization: a comprehensive overview of the past 30 years. Xenotransplantation 2018;(1):25.

[42] Goldstein S, et al. Transpecies heart valve transplant: advanced studies of a bioengineered xeno-autograft. Ann Thorac Surg 2000;70(6):1962—9.

[43] Schneider AW, et al. Twenty-year experience with the Ross-Konno procedure. Eur J Cardiothorac Surg 2016;49(6):1564—70.

[44] Vesely I. Heart valve tissue engineering. Circ Res 2005;97(8):743—55.

[45] Sayk F, et al. Histopathologic findings in a novel decellularized pulmonary homograft: an autopsy study. Ann Thorac Surg 2005;79(5):1755—8.

[46] Simon P, et al. Early failure of the tissue engineered porcine heart valve SYNERGRAFT in pediatric patients. Eur J Cardiothorac Surg 2003;23(6):1002—6. discussion 1006.

[47] Ruffer A, et al. Early failure of xenogenous de-cellularised pulmonary valve conduits—a word of caution! Eur J Cardiothorac Surg 2010;38(1):78—85.

[48] Rieben R, et al. Xenotransplantation: in vitro analysis of synthetic alpha-galactosyl inhibitors of human anti-Galalpha1—>3Gal IgM and IgG antibodies. Glycobiology 2000;10(2):141—8.

[49] Naso F, et al. Alpha-gal inactivated heart valve bioprostheses exhibit an anti-calcification propensity similar to knockout tissues. Tissue Eng 2017;23(19—20):1181—95.

[50] Nuno-Conceicrao A, et al. Homologous dura mater cardiac valves. Structural aspects of eight implanted valves. J Thorac Cardiovasc Surg 1975;70(3):499—508.

[51] Joseph S, et al. Aortic valve replacement with frame-mounted autologous fascia lata. Long-term results. Br Heart J 1974;36(8):760—7.

[52] Love CS, Love JW. The autogenous tissue heart valve: current status. J Card Surg 1991;6(4): 499—507.

[53] Weinhold C, et al. Experimental studies of the anatomical and functional characteristics of Kangaroo aortic valve bioprostheses. Life Support Syst 1984;2(2):121—5.

[54] Doss M, et al. Aortic leaflet replacement with the new 3F stentless aortic bioprosthesis. Ann Thorac Surg 2005;79(2):682—5. discussion 685.

[55] Garcia Paez JM, et al. Chemical treatment and tissue selection: factors that influence the mechanical behaviour of porcine pericardium. Biomaterials 2001;22(20):2759—67.

[56] Paez JM, et al. Ostrich pericardium, a biomaterial for the construction of valve leaflets for cardiac bioprostheses: mechanical behaviour, selection and interaction with suture materials. Biomaterials 2001; 22(20):2731—40.

[57] Vola M, et al. Sutureless Medtronic 3f Enable aortic valve replacement in a heavily calcified aortic root. J Heart Valve Dis 2013;22(3):436—8.

[58] Webb JG, Binder RK. Transcatheter aortic valve implantation: the evolution of prostheses, delivery systems and approaches. Arch Cardiovasc Dis 2012;105(3):153—9.

[59] Oury JH, Angell WW, Koziol JA. Comparison of Hancock I and Hancock II bioprostheses. J Card Surg 1988;3(3 Suppl. l):375—81.

[60] Flomenbaum MA, Schoen FJ. Effects of fixation back pressure and antimineralization treatment on the morphology of porcine aortic bioprosthetic valves. J Thorac Cardiovasc Surg 1993;105(1):154—64.

[61] Vesely I, Noseworthy R. Micromechanics of the fibrosa and the ventricularis in aortic valve leaflets. J Biomech 1992;25(1):101—13.

[62] Barratt-Boyes BG, Ko PH, Jaffe WM. The zero pressure fixed Medtronic intact porcine valve: clinical results over a 6-year period, including serial echocardiographic assessment. J Card Surg 1991; 6(4 Suppl. l):606—12.

[63] Hansen B, Menkis AH, Vesely I. Longitudinal and radial distensibility of the porcine aortic root. Ann Thorac Surg 1995;60(2 Suppl. l):S384—90.

[64] Walley VM, Keon WJ. Patterns of failure in Ionescu-Shiley bovine pericardial bioprosthetic valves. J Thorac Cardiovasc Surg 1987;93(6):925—33.

[65] Vesely I. Transcatheter valves: a brave new world. J Heart Valve Dis 2010;19(5):543—58.

[66] Forcillo J, et al. Carpentier-Edwards pericardial valve in the aortic position: 25-years experience. Ann Thorac Surg 2013;96(2):486—93.

[67] Vesely I. The evolution of bioprosthetic heart valve design and its impact on durability. Cardiovasc Pathol 2003;12(5):277—86.

[68] Vesely I. The influence of design on bioprosthetic valve durability. J Long Term Eff Med Implant 2001;11(3—4):137—49.

[69] Vesely I. The role of elastin in aortic valve mechanics. J Biomech 1998;31(2):115—23.

[70] Vesely I, Lozon A. Natural preload of aortic valve leaflet components during glutaraldehyde fixation: effects on tissue mechanics. J Biomech 1993;26(2):121—31.

[71] Vesely I, Boughner D, Song T. Tissue buckling as a mechanism of bioprosthetic valve failure. Ann Thorac Surg 1988;46(3):302—8.

[72] Hiester ED, Sacks MS. Optimal bovine pericardial tissue selection sites. II. Cartographic analysis. J Biomed Mater Res 1998;39(2):215—21.

[73] Hiester ED, Sacks MS. Optimal bovine pericardial tissue selection sites. I. Fiber architecture and tissue thickness measurements. J Biomed Mater Res 1998;39(2):207—14.

[74] Lee JM, Corrente R, Haberer SA. The bovine pericardial xenograft: II. Effect of tethering or pressurization during fixation on the tensile viscoelastic properties of bovine pericardium. J Biomed Mater Res 1989;23(5):477—89.

[75] Lee JM, Haberer SA, Boughner DR. The bovine pericardial xenograft: I. Effect of fixation in aldehydes without constraint on the tensile viscoelastic properties of bovine pericardium. J Biomed Mater Res 1989;23(5):457—75.

[76] Lee JM, Ku M, Haberer SA. The bovine pericardial xenograft: III. Effect of uniaxial and sequential biaxial stress during fixation on the tensile viscoelastic properties of bovine pericardium. J Biomed Mater Res 1989;23(5):491—506.

[77] Klautz RJM, et al. Safety, effectiveness and haemodynamic performance of a new stented aortic valve bioprosthesis. Eur J Cardiothorac Surg 2017;52(3):425—31.

[78] Moore MA, et al. Shrinkage temperature versus protein extraction as a measure of stabilization of photooxidized tissue. J Biomed Mater Res 1996;32(2):209—14.

[79] Gallegos RP, et al. The current state of in-vivo pre-clinical animal models for heart valve evaluation. J Heart Valve Dis 2005;14(3):423—32.

[80] Yoganathan AP, Eberhardt CE, Walker PG. Hydrodynamic performance of the Medtronic Freestyle aortic root bioprosthesis. J Heart Valve Dis 1994;3(5):571—80.

[81] Dossche K, et al. Hemodynamic performance of the PRIMA Edwards stentless aortic xenograft: early results of a multicenter clinical trial. Thorac Cardiovasc Surg 1996;44(1):11—4.

[82] David TE, Bos J, Rakowski H. Aortic valve replacement with the Toronto SPV bioprosthesis. J Heart Valve Dis 1992;1(2):244—8.

[83] Krause Jr AH. Technique for complete subcoronary implantation of the Medtronic Freestyle porcine bioprosthesis. Ann Thorac Surg 1997;64(5):1495—8.

[84] Butany J, et al. Inflammation and infection in nine surgically explanted Medtronic Freestyle stentless aortic valves. Cardiovasc Pathol 2007;16(5):258—67.

[85] Nair V, et al. Characterizing the inflammatory reaction in explanted Medtronic Freestyle stentless porcine aortic bioprosthesis over a 6-year period. Cardiovasc Pathol 2012;21(3):158—68.

[86] Doss M, et al. In-vivo evaluation of the BioPhysio valve prosthesis in the aortic position. J Heart Valve Dis 2008;17(1):105—9.

[87] Doss M, et al. Mid-term follow up of a novel bioprosthesis in aortic valve surgery. J Heart Valve Dis 2012;21(6):753—7.

[88] Lazar HL. Management of coronary artery obstruction following TAVR-The importance of the heart team approach. J Card Surg 2017;32(12):782.

[89] Showkathali R, et al. Multi-disciplinary clinic: next step in "Heart team" approach for TAVI. Int J Cardiol 2014;174(2):453—5.

[90] Kaempchen S, et al. Assessing the benefit of biological valve prostheses: cumulative incidence (actual) vs. Kaplan-Meier (actuarial) analysis. Eur J Cardiothorac Surg 2003;23(5):710—3. discussion 713-4.

[91] Caimmi PP, et al. Twelve-year follow up with the Sorin Pericarbon bioprosthesis in the mitral position. J Heart Valve Dis 1998;7(4):400—6.

[92] Yankah CA, et al. Aortic valve replacement with the Mitroflow pericardial bioprosthesis: durability results up to 21 years. J Thorac Cardiovasc Surg 2008;136(3):688—96.

[93] Pellikka PA, Thaden J. Midterm sapien transcatheter valve durability: ready for prime time or waiting to fail? JACC Cardiovasc Imaging 2016;10(1):26—8.

[94] Daubert MA, et al. Long-term valve performance of TAVR and SAVR: a report from the PARTNER I trial. JACC Cardiovasc Imaging 2017;10(1):15—25.

[95] Kapadia SR, et al. 5-year outcomes of transcatheter aortic valve replacement compared with standard treatment for patients with inoperable aortic stenosis (PARTNER 1): a randomised controlled trial. Lancet 2015;385(9986):2485—91.

[96] Zilla P, et al. Prosthetic heart valves: catering for the few. Biomaterials 2008;29(4):385—406.

CHAPTER 11

Heart valve calcification

Linda L. Demer[1], Yin Tintut[2]

[1]Departments of Medicine, Physiology & Bioengineering, University of California, Los Angeles, Los Angeles, CA, United States; [2]Departments of Medicine (Cardiology), Physiology & Orthopaedic Surgery, University of California, Los Angeles, CA, United States

Contents

11.1 Native valves

11.1.1 Aortic valve

Aortic stenosis is the most prevalent heart valve disorder in developed countries occurring 0.4% in the general population and 1.7% in those over 65 years old [1]. The cup-like leaflets or cusps, which are normally thin and flexible, become progressively thickened, retracted, and stiff over the years, causing severe life-threatening obstruction to cardiac outflow [1]. Advanced disease is usually accompanied by calcified nodules and hence

Principles of Heart Valve Engineering
ISBN 978-0-12-814661-3, https://doi.org/10.1016/B978-0-12-814661-3.00011-3

the term calcific aortic valve disease (CAVD). The severity of calcium deposition provides incremental prognostic value for mortality beyond clinical and noninvasive indicators and independently predicts excess mortality [2]. Clinically, progression of aortic stenosis is monitored by frequent echocardiography, which identifies key features of the disease, including the degree of calcification, severity of stenosis, adverse left ventricular remodeling, reduced left ventricular longitudinal strain, myocardial fibrosis, and pulmonary hypertension [3]. The only effective treatment has been valve replacement, either surgical or interventional. With increased understanding of genetic and molecular factors involved in the disease process, potential targets for medical therapy include lipoprotein(a), bone morphogenetic protein-2, Notch1, autotaxin, serotonin, Von Willebrand factor, cyclooxygenases, Wnt family proteins, and osteoprotegerin.

11.1.2 Mitral valve

Mitral annular calcification is a strong and independent predictor of cardiovascular disease mortality, especially in the more severe cases of mitral annular calcification [4]. About 24% of patients undergoing repair of mitral valve prolapse have mitral annular calcification [5]. In the calcified annulus, it is common to find cartilaginous metaplasia, suggesting that cells there undergo chondrogenic differentiation. Calcium deposits also occur in the leaflets, though less prominently, and they increase stiffness and reduce tensile strength [6]. Unlike the aortic valve, the leaflets of which resemble collapsing cups, the mitral valve consists of broad leaflets held under tension by tendinous chords that resemble sails held in tension by ropes. Thus, in diseased mitral valves, the greatest amount of calcium deposition is found almost solely within the fibrous valvular ring that surrounds the orifice. However, as with aortic valve disease, growing evidence suggests that mitral annular and valvular calcification arises from metabolic mechanisms related to atherosclerosis and the mineral disturbances of end-stage renal disease.

11.2 Bioprosthetic valves

11.2.1 Surgical valve replacement

The leaflets of bioprosthetic valves are made from xenografts, specifically porcine valves or bovine pericardium. The tissue is sewn onto fabric-covered metal alloy valve frames. A key step in producing these valves is the treatment of the animal tissue with glutaraldehyde to cross-link the proteins to enhance their stability and promote immune tolerance. Nevertheless, progressive deterioration and dysfunction of bioprosthetic valves occur, usually through oxidation modifications and calcification. Prosthetic failure requires reoperation if recognized early enough, and, if treatment is delayed, it may cause irreversible functional changes in the heart and even death. Abrupt bioprosthetic valve failure is a medical-surgical emergency. Factors that increase risk for bioprosthetic calcification include chronic inflammatory disease and youth. Children are at greater risk of calcific degeneration and failure of bioprosthetic valves [7]. Both diabetes and coronary artery disease (CAD) requiring coronary artery bypass graft surgery (CABG) are associated with significantly earlier degeneration and need for replacement [8]. Failure occurs by two modalities: matrix breakdown and calcification. Measures to reduce prosthetic valve degeneration are under active development. Ethanol extraction of cholesterol and phospholipids, which may contribute to inflammatory matrix degeneration, has been used experimentally to prevent calcification of xenografts [9], and this may increase their long-term clinical performance [10]. An aldehyde reduction protocol was found to reduce inflammatory responses and calcification in animal studies [11]. Using the same principles on which kidney stone ultrasonic lithotripsy is based, investigators showed that pulsed cavitational ultrasound can be used to improve valve leaflet flexibility in calcified bioprosthetic valves explanted from humans and tested ex vivo or tested in vivo in sheep [12].

11.2.2 Transcatheter intervention

Transcatheter aortic valve replacement (TAVR) also known as implantation (TAVI) is a rapidly emerging therapy, yet it faces challenges with valvular calcification at two levels. First, their xenograft leaflets are subject to calcification and degeneration similar to bioprosthetic valves used in surgical implantation. Secondly, the implantation procedure is challenging when the native valve is highly calcified and stiff, as the TAVR valve is implanted without removing the diseased native valve. Thus, the implantation technique must work around the problem of physically forcing the native valve out of the way, against the aortic wall. A major adverse outcome arises when the native valve calcification is extensive. The balloon, which deploys TAVR valves within an expandable stent-like frame, may not be able to fully compress the native cusps against the aorta, leaving a circular valve in a noncircular orifice and causing blood to leak through openings. As a result, paravalvular regurgitation occurs in some [13] but not all cases [14].

11.3 Structure and pathology of aortic valves

11.3.1 Aortic valve anatomy

The native aortic valve is composed of three layers: the fibrosa, spongiosa, and ventricularis. The fibrosa, on the surface facing the aorta, is rich in collagen, which provides mechanical strength and rigidity. The ventricularis, on the surface facing the left ventricle, is rich in elastin, which provides flexibility and recoil, whereas the middle layer spongiosa is rich in proteoglycans, which absorb mechanical impact during the abrupt closure of the valve leaflets at the start of diastole [15]. Structure follows function in that the fibrosa, with its rigid collagen fibers, acts to resist deformation and leakage, whereas the ventricularis, with its resilient elastin, allows significant elongation and deformation to accommodate the strain imposed when the cusp halts the downward momentum of the column of blood in the ascending aorta.

To elucidate molecular mechanisms, in vitro systems from both endothelial and interstitial cells have been developed. Valvular interstitial cells (VIC) from both human and animals have been isolated and cultured in several laboratories. VIC isolated from humans form three-dimensional calcified nodules [16], similar to those formed by a subpopulation of vascular smooth muscle cells [17]. VIC isolated from murine aortic valves also form calcified nodules in response to the procalcific inflammatory cytokine, tumor necrosis factor-alpha (TNF-α) [18]. When cultured in high phosphate "mineralization" medium, VIC readily generates a calcified matrix [19]. This high phosphate condition is believed to model the clinical hyperphosphatemia and extensive cardiovascular calcification observed in patients with chronic kidney disease.

11.3.2 Biomechanical environment

The distribution of mechanical stress and strain has been shown to relate closely to the distribution of calcium deposits in the valve leaflets [20]. Not only do mechanical forces affect the cells and matrix of the valves but also the cells and matrix affect the mechanical forces [20]. Calcium deposits significantly reduce the tensile strength of valve leaflets, increasing their risk of rupture [21]. Clinically, both valve and aortic calcification interfere with the mechanics of surgical procedures, especially percutaneous catheter-based interventional procedures.

One notable biomechanical feature that may explain differences between CAVD and atherosclerotic calcification is that most vasculature is exposed to laminar flow, even with early stages of atherosclerosis, whereas valves are exposed to disturbed flow. This is because of vascular remodeling that maintains lumen size until the most advanced stages of disease, a process known as the Glagov phenomenon [22]. In contrast, valve leaflets, even without disease, are exposed to vortices of disturbed flow, which becomes more disturbed and even turbulent with progression of CAVD. The importance of fibrosis and stiffening of the valve matrix is evidenced by the finding that VIC differentiates along a range of mesenchymal lineages and that the compliance of the underlying matrix substrate determines the choice of lineage [23]. As such, alteration of the collagen matrix has been shown to promote calcification [24].

11.3.3 CAVD pathology

With progression of CAVD, the extracellular matrix of the cusps or leaflets thickens and calcifies. During the early disease stage, known as aortic sclerosis, the cusps thicken and start to stiffen. Once the stiffness increases to the point that the cusps cannot fully collapse and the valve cannot fully open, it has advanced to the stage known as aortic stenosis. As stenosis progresses, the pressure gradient across the valve increases, and it eventually reaches critical aortic stenosis, defined in terms of orifice diameter, at which point flow is critically impaired. Initially, histopathological changes resemble those of atherosclerosis: basement membrane disruption, lipid deposition, inflammation, fibrosis, and elastin degradation. Gradually, the fibrosis, primarily collagen synthesis, leads to thickening. With thickening, the leaflet becomes less compliant. Changes in extracellular matrix stiffness have been shown by Simmons et al. [23], and also described below, to associate with differentiation of valve cells into chondrogenic and osteogenic differentiation. In about 10%—15% of surgically explanted valves, fully formed bone or cartilage tissue is apparent [25].

11.3.4 CAVD mechanism

In a major paradigm shift, similar to that of vascular calcification, it is now increasingly recognized that CAVD is not a result of passive degeneration or simple wear and tear, as previously believed, but rather a result of a regulated process similar to atherosclerosis involving inflammation and osteogenic differentiation of valvular interstitial cells (VIC). These processes are regulated at the cellular and molecular levels by hyperlipidemia, oxidant stress, inflammation, and/or mechanical factors [26—35].

11.3.5 Hyperlipidemia models

Mouse models for CAVD have been developed and refined over the past few decades. Many of these models are based on hyperlipidemic conditions [26,36—39], which induces CAVD in mice whether dietary, genetic, or both. One of the most effective models in showing functional valve disease is the LDL receptor-deficient apolipoprotein B-100 overexpressing ($Ldlr^{-/-}Apob^{100/100}$) mouse. These mice initially develop aortic regurgitation, which is followed by stenosis as the mice age [39,40]. Remarkably, genetic reversal of the hyperlipidemia after development of CAVD dramatically reduced osteogenic differentiation signals in the valves and significantly reduced valve calcification, but did not result in improvement in valve function [41]. Interestingly, when hypertension is superimposed on hyperlipidemia, a purely fibrotic form of CAVD occurs [37], which may account for the fibrotic form seen in some patients. Oxidative degradation products of lipids, such as those in LDL, have been found in the fibrosa layer of diseased human aortic valves, in close proximity to calcified nodules [42,43].

11.3.6 Lipoprotein(a)

Despite the many similarities between atherosclerosis and CAVD, there may be important differences [38]. For instance, the established treatment for atherosclerosis—lipid-lowering treatment with the class of drugs known as statins—has failed to reverse aortic stenosis or slow down progression of CAVD [44]. One possible reason for the failure of statin therapy is that the valve disease in the patients in the clinical trials may have been too far advanced. Another possibility is the direct proosteogenic effects of statins in bone cells and tissue [45—48]. As VIC undergo osteoblastic differentiation and form bone-like tissue in many cases, they may respond to statins in the same manner as bone tissue, with increased osteogenesis. Thus, the upstream anticalcific effect of lipid lowering may be countered by direct downstream proosteogenic effects. A third, important possible explanation is the known increase in levels of lipoprotein(a) [Lp(a)] caused by statin treatment [49] because Lp(a), an LDL-like particle, is now considered a likely causative factor in CAVD. Genome-wide association studies have shown that the LPA locus contains a single-nucleotide polymorphism (rs10455872) that is tightly associated with aortic valve calcification [50—52]. Lp(a) levels have a strong positive association with progression of CAVD [53], and lowering levels of Lp(a) reduce the risk of CAVD [54]. The mechanism is unclear, but Lp(a) has a greater propensity to bind oxidized phospholipids than does LDL [55], accumulates in calcified valve leaflets [56], and carries lipoprotein-associated phospholipase A2 (Lp-PLA2), which converts oxidized phospholipids into the inflammatory molecule, lysophosphatidylcholine [57].

11.3.7 Inflammation

Evidence suggests that inflammation is central not only to atherosclerotic calcification but also to CAVD. Oxidized lipids are known to stimulate release of inflammatory cytokines, such as TNF-α and interleukin-6 (IL-6), in monocyte/macrophages [35]. Inflammatory infiltrates and increased expression of the inflammatory cytokines TNF-α and galectin-3 are found within human stenotic valves [58—61]. High circulating levels of both TNF-α and IL-6 also correlate with calcification in rheumatic valve disease [62]. Furthermore, human immunodeficiency virus (HIV) positivity, a high inflammatory state, confers increased risk of both aortic valve and mitral annual calcification [63]. In mouse models, the aortic valves of hyperlipidemic mice show increased TNF-α immunopositivity [64]. Inhibition of TNF-α by infliximab in hyperlipidemic mouse models [36] and antiinflammatory cyclooxygenase 2 (COX2) inhibitors in klotho-knockout mice attenuated valve calcification [65]. In a three-dimensional hydrogel VIC culture, TNF-α induces features of CAVD including retraction, stiffening, and nodule formation [18]. Inhibitors of another proinflammatory molecule, galectin-3, also reduce VIC calcification [60].

11.3.8 Molecular regulators

The multiplicity of molecular and cellular factors that regulate CAVD has been reviewed [66]. These include bone morphogenetic protein-2 (BMP-2), Notch1, autotaxin, serotonin, Von Willebrand factor, cyclooxygenase, and the wnt family of proteins.

11.3.9 Bone morphogenetic protein-2

BMP-2 was previously established as a potent osteodifferentiation factor driving bone formation in the skeleton and in the artery wall. It is now known to also govern calcification of VIC. BMP signaling is an underlying mechanism by which TNF-α induces VIC valve calcification, with evidence for both an increase in BMP-2 transcripts and a decrease in expression of its inhibitor, Smad6 [64]. Inactivation of the BMP type IA receptor in klotho-deficient aortic VIC results in the inhibition of calcification [67].

11.3.10 Notch 1

As a mechanosensory receptor, Notch1 plays a role in embryonic valve development [68], and it may be affected by increasingly disturbed flow in the context of CAVD. Human mutations in the gene NOTCH1 are known to cause CAVD [69], and diseased valves have reduced expression of Notch [70].

11.3.11 Autotaxin

Autotaxin, a lysophospholipase D enzyme that produces lysophosphatidic acid from lysophosphatidylcholine, is encoded by the ENPP2 gene, transported to the blood stream by lipids and enriched in Lp(a) particles. Autotaxin activity is found in VIC and its expression is increased in CAVD [42].

11.3.12 Serotonin (5-hydroxytryptamine)

Patients with endogenously high serotonin levels, such as in carcinoid syndrome, have a high risk of structural valve disease, especially regurgitation [71,72]. Serotonergic drugs, such as benfluorex, fenfluramine, and pergolide [73], are consistently linked with a several fold greater risk of aortic and mitral valvulopathy, especially regurgitation, in a dose- and duration-dependent manner in humans [74—79], and the aortic regurgitation regresses in patients after the serotonergic drugs are discontinued [80]. The relationship is so compelling that the drugs have been banned. Serotonin receptors are expressed in aortic VIC [81]. In rats, serotonin treatment causes aortic and mitral regurgitation, and the effects are reversible after discontinuation [82]. In rabbits, serotonin treatment induces ectopic cartilage formation (chondroid metaplasia) in both mitral and aortic valves [83]. Altogether, both animal and human studies clearly link activation of serotonin receptors with valve disease.

11.3.13 Von Willebrand factor

Von Willebrand factor (vWF) is a well-known mediator of hemostasis and vascular inflammation that is stored and released from endothelial cells, depending on hemodynamic conditions. It has been used as a biomarker and a prognostic indicator in patients with valve disease and those undergoing valve replacement or repair [84], though its role in pathogenesis remains under investigation. There is an as yet unexplained correlation between valvular heart disease and gastrointestinal bleeding that may relate to angiogenesis.

11.3.14 Cyclooxygenase activity

An important role of enzymatic activity has been demonstrated as well. Based on microarray analysis of valves from a mouse model of CAVD and its role in bone differentiation, the gene encoding COX2 was identified as a potential contributor to valve disease. COX2 is overexpressed in calcific human valves, isolated valve cell cultures, and in calcified valves of Klotho mice, a model of premature aging [65]. Moreover, inhibition of COX2 cyclooxygenase activity, by pharmacologic or genetic means, reduces both in vitro osteodifferentiation and in vivo osteodifferentiation [65].

11.3.15 Wnt

Oxidized lipids induce VIC calcification by activating the canonical Wnt signaling pathway [85]. Wnt5a, a noncanonical Wnt ligand, is expressed in the foci around calcified valves, and in vitro culture studies show that Wnt5a and Wnt11 promote calcification of human VIC [86]. Circulating levels of Wnt modulators are detected in patients with symptomatic aortic stenosis and are present in calcified aortic valves [87].

11.3.16 Other factors

The osteoclast inhibitory factor, osteoprotegerin, which acts as a soluble decoy receptor for the osteoclast differentiation factor, RANKL, has a protective effect against valve calcification [88]. Elevated serum phosphate, as expected, also correlates with CAVD [89] as it does with vascular calcification.

References

[1] Lindman BR, Clavel MA, Mathieu P, Iung B, Lancellotti P, Otto CM, Pibarot P. Calcific aortic stenosis. Nat Rev Dis Primers 2016;2:16006.

[2] Clavel MA, Pibarot P, Messika-Zeitoun D, Capoulade R, Malouf J, Aggarval S, Araoz PA, Michelena HI, Cueff C, Larose E, Miller JD, Vahanian A, Enriquez-Sarano M. Impact of aortic valve calcification, as measured by MDCT, on survival in patients with aortic stenosis: results of an international registry study. J Am Coll Cardiol 2014;64:1202–13.

[3] Lindman BR, Bonow RO, Otto CM. Current management of calcific aortic stenosis. Circ Res 2013; 113:223–37.

[4] Willens HJ, Chirinos JA, Schob A, Veerani A, Perez AJ, Chakko S. The relation between mitral annular calcification and mortality in patients undergoing diagnostic coronary angiography. Echocardiography 2006;23:717–22.

[5] Fusini L, Ghulam Ali S, Tamborini G, Muratori M, Gripari P, Maffessanti F, Celeste F, Guglielmo M, Cefalu C, Alamanni F, Zanobini M, Pepi M. Prevalence of calcification of the mitral valve annulus in patients undergoing surgical repair of mitral valve prolapse. Am J Cardiol 2014;113:1867–73.

[6] Pham T, Sun W. Material properties of aged human mitral valve leaflets. J Biomed Mater Res A 2014; 102:2692–703.

[7] Saleeb SF, Newburger JW, Geva T, Baird CW, Gauvreau K, Padera RF, Del Nido PJ, Borisuk MJ, Sanders SP, Mayer JE. Accelerated degeneration of a bovine pericardial bioprosthetic aortic valve in children and young adults. Circulation 2014;130:51–60.

[8] Lee S, Levy RJ, Christian AJ, Hazen SL, Frick NE, Lai EK, Grau JB, Bavaria JE, Ferrari G. Calcification and oxidative modifications are associated with progressive bioprosthetic heart valve dysfunction. J Am Heart Assoc 2017;6.

[9] Vyavahare N, Hirsch D, Lerner E, Baskin JZ, Schoen FJ, Bianco R, Kruth HS, Zand R, Levy RJ. Prevention of bioprosthetic heart valve calcification by ethanol preincubation. Efficacy and mechanisms. Circulation 1997;95:479—88.

[10] Raghavan D, Shah SR, Vyavahare NR. Neomycin fixation followed by ethanol pretreatment leads to reduced buckling and inhibition of calcification in bioprosthetic valves. J Biomed Mater Res B Appl Biomater 2010;92:168—77.

[11] Shang H, Claessens SM, Tian B, Wright GA. Aldehyde reduction in a novel pericardial tissue reduces calcification using rabbit intramuscular model. J Mater Sci Mater Med 2017;28:16.

[12] Villemain O, Robin J, Bel A, Kwiecinski W, Bruneval P, Arnal B, Remond M, Tanter M, Messas E, Pernot M. Pulsed Cavitational Ultrasound Softening: a new non-invasive therapeutic approach of calcified bioprosthetic valve stenosis. JACC Basic Transl Sci 2017;2:372—83.

[13] Khalique OK, Hahn RT, Gada H, Nazif TM, Vahl TP, George I, Kalesan B, Forster M, Williams MB, Leon MB, Einstein AJ, Pulerwitz TC, Pearson GD, Kodali SK. Quantity and location of aortic valve complex calcification predicts severity and location of paravalvular regurgitation and frequency of post-dilation after balloon-expandable transcatheter aortic valve replacement. JACC Cardiovasc Interv 2014;7:885—94.

[14] Staubach S, Franke J, Gerckens U, Schuler G, Zahn R, Eggebrecht H, Hambrecht R, Sack S, Richardt G, Horack M, Senges J, Steinberg DH, Ledwoch J, Fichtlscherer S, Doss M, Wunderlich N, Sievert H, German Transcatheter Aortic Valve Implantation-Registry I. Impact of aortic valve calcification on the outcome of transcatheter aortic valve implantation: results from the prospective multicenter German TAVI registry. Cathet Cardiovasc Interv 2013;81:348—55.

[15] Hinton Jr RB, Lincoln J, Deutsch GH, Osinska H, Manning PB, Benson DW, Yutzey KE. Extracellular matrix remodeling and organization in developing and diseased aortic valves. Circ Res 2006;98: 1431—8.

[16] Mohler 3rd ER, Chawla MK, Chang AW, Vyavahare N, Levy RJ, Graham L, Gannon FH. Identification and characterization of calcifying valve cells from human and canine aortic valves. J Heart Valve Dis 1999;8:254—60.

[17] Demer LL, Tintut Y. Inflammatory, metabolic, and genetic mechanisms of vascular calcification. Arterioscler Thromb Vasc Biol 2014;34:715—23.

[18] Lim J, Ehsanipour A, Hsu JJ, Lu J, Pedego T, Wu A, Walthers CM, Demer LL, Seidlits SK, Tintut Y. Inflammation drives retraction, stiffening, and nodule formation via cytoskeletal machinery in a three- dimensional culture model of aortic stenosis. Am J Pathol 2016;186:2378—89.

[19] Rattazzi M, Iop L, Faggin E, Bertacco E, Zoppellaro G, Baesso I, Puato M, Torregrossa G, Fadini GP, Agostini C, Gerosa G, Sartore S, Pauletto P. Clones of interstitial cells from bovine aortic valve exhibit different calcifying potential when exposed to endotoxin and phosphate. Arterioscler Thromb Vasc Biol 2008;28:2165—72.

[20] Merryman WD, Schoen FJ. Mechanisms of calcification in aortic valve disease: role of mechanokinetics and mechanodynamics. Curr Cardiol Rep 2013;15:355.

[21] Gunning GM, Murphy BP. The effects of decellularization and cross-linking techniques on the fatigue life and calcification of mitral valve chordae tendineae. J Mech Behav Biomed Mater 2016;57:321—33.

[22] Glagov S, Weisenberg E, Zarins CK, Stankunavicius R, Kolettis GJ. Compensatory enlargement of human atherosclerotic coronary arteries. N Engl J Med 1987;316:1371—5.

[23] Chen JH, Simmons CA. Cell-matrix interactions in the pathobiology of calcific aortic valve disease: critical roles for matricellular, matricrine, and matrix mechanics cues. Circ Res 2011;108:1510—24.

[24] Rodriguez KJ, Piechura LM, Porras AM, Masters KS. Manipulation of valve composition to elucidate the role of collagen in aortic valve calcification. BMC Cardiovasc Disord 2014;14:29.

[25] Steiner I, Kasparova P, Kohout A, Dominik J. Bone formation in cardiac valves: a histopathological study of 128 cases. Virchows Arch 2007;450:653—7.

[26] Aikawa E, Nahrendorf M, Figueiredo JL, Swirski FK, Shtatland T, Kohler RH, Jaffer FA, Aikawa M, Weissleder R. Osteogenesis associates with inflammation in early-stage atherosclerosis evaluated by molecular imaging in vivo. Circulation 2007;116:2841–50.

[27] Alrasadi K, Alwaili K, Awan Z, Valenti D, Couture P, Genest J. Aortic calcifications in familial hypercholesterolemia: potential role of the low-density lipoprotein receptor gene. Am Heart J 2009;157:170–6.

[28] Awan Z, Alrasadi K, Francis GA, Hegele RA, McPherson R, Frohlich J, Valenti D, de Varennes B, Marcil M, Gagne C, Genest J, Couture P. Vascular calcifications in homozygote familial hypercholesterolemia. Arterioscler Thromb Vasc Biol 2008;28:777–85.

[29] Miller JD, Chu Y, Brooks RM, Richenbacher WE, Pena-Silva R, Heistad DD. Dysregulation of antioxidant mechanisms contributes to increased oxidative stress in calcific aortic valvular stenosis in humans. J Am Coll Cardiol 2008;52:843–50.

[30] Sage AP, Tintut Y, Demer LL. Regulatory mechanisms in vascular calcification. Nat Rev Cardiol 2010;7:528–36.

[31] Scatena M, Liaw L, Giachelli CM. Osteopontin: a multifunctional molecule regulating chronic inflammation and vascular disease. Arterioscler Thromb Vasc Biol 2007;27:2302–9.

[32] Shanahan CM. Inflammation ushers in calcification: a cycle of damage and protection? Circulation 2007;116:2782–5.

[33] Shao JS, Cheng SL, Sadhu J, Towler DA. Inflammation and the osteogenic regulation of vascular calcification: a review and perspective. Hypertension 2010;55:579–92.

[34] Tintut Y, Patel J, Parhami F, Demer LL. Tumor necrosis factor-alpha promotes in vitro calcification of vascular cells via the cAMP pathway. Circulation 2000;102:2636–42.

[35] Tintut Y, Patel J, Territo M, Saini T, Parhami F, Demer LL. Monocyte/macrophage regulation of vascular calcification in vitro. Circulation 2002;105:650–5.

[36] Al-Aly Z, Shao JS, Lai CF, Huang E, Cai J, Behrmann A, Cheng SL, Towler DA. Aortic Msx2-Wnt calcification cascade is regulated by TNF-alpha-dependent signals in diabetic Ldlr-/- mice. Arterioscler Thromb Vasc Biol 2007;27:2589–96.

[37] Chu Y, Lund DD, Doshi H, Keen HL, Knudtson KL, Funk ND, Shao JQ, Cheng J, Hajj GP, Zimmerman KA, Davis MK, Brooks RM, Chapleau MW, Sigmund CD, Weiss RM, Heistad DD. Fibrotic aortic valve stenosis in hypercholesterolemic/hypertensive mice. Arterioscler Thromb Vasc Biol 2016;36:466–74.

[38] Weiss RM, Miller JD, Heistad DD. Fibrocalcific aortic valve disease: opportunity to understand disease mechanisms using mouse models. Circ Res 2013;113:209–22.

[39] Weiss RM, Ohashi M, Miller JD, Young SG, Heistad DD. Calcific aortic valve stenosis in old hypercholesterolemic mice. Circulation 2006;114:2065–9.

[40] Berry CJ, Miller JD, McGroary K, Thedens DR, Young SG, Heistad DD, Weiss RM. Biventricular adaptation to volume overload in mice with aortic regurgitation. J Cardiovasc Magn Reson 2009;11: 27.

[41] Miller JD, Weiss RM, Serrano KM, Castaneda LE, Brooks RM, Zimmerman K, Heistad DD. Evidence for active regulation of pro-osteogenic signaling in advanced aortic valve disease. Arterioscler Thromb Vasc Biol 2010;30:2482–6.

[42] Bouchareb R, Mahmut A, Nsaibia MJ, Boulanger MC, Dahou A, Lepine JL, Laflamme MH, Hadji F, Couture C, Trahan S, Page S, Bosse Y, Pibarot P, Scipione CA, Romagnuolo R, Koschinsky ML, Arsenault BJ, Marette A, Mathieu P. Autotaxin derived from lipoprotein(a) and valve interstitial cells promotes inflammation and mineralization of the aortic valve. Circulation 2015;132:677–90.

[43] Mohty D, Pibarot P, Despres JP, Cote C, Arsenault B, Cartier A, Cosnay P, Couture C, Mathieu P. Association between plasma LDL particle size, valvular accumulation of oxidized LDL, and inflammation in patients with aortic stenosis. Arterioscler Thromb Vasc Biol 2008;28:187–93.

[44] Mohler 3rd ER, Wang H, Medenilla E, Scott C. Effect of statin treatment on aortic valve and coronary artery calcification. J Heart Valve Dis 2007;16:378–86.

[45] Ghosh-Choudhury N, Mandal CC, Choudhury GG. Statin-induced Ras activation integrates the phosphatidylinositol 3-kinase signal to Akt and MAPK for bone morphogenetic protein-2 expression in osteoblast differentiation. J Biol Chem 2007;282:4983–93.

[46] Monzack EL, Masters KS. A time course investigation of the statin paradox among valvular interstitial cell phenotypes. Am J Physiol Heart Circ Physiol 2012;303:H903—9.

[47] Mundy G, Garrett R, Harris S, Chan J, Chen D, Rossini G, Boyce B, Zhao M, Gutierrez G. Stimulation of bone formation in vitro and in rodents by statins. Science 1999;286:1946—9.

[48] Pagkalos J, Cha JM, Kang Y, Heliotis M, Tsiridis E, Mantalaris A. Simvastatin induces osteogenic differentiation of murine embryonic stem cells. J Bone Miner Res 2010;25:2470—8.

[49] Yeang C, Hung MY, Byun YS, Clopton P, Yang X, Witztum JL, Tsimikas S. Effect of therapeutic interventions on oxidized phospholipids on apolipoprotein B100 and lipoprotein(a). J Clin Lipidol 2016;10:594—603.

[50] Arsenault BJ, Boekholdt SM, Dube MP, Rheaume E, Wareham NJ, Khaw KT, Sandhu MS, Tardif JC. Lipoprotein(a) levels, genotype, and incident aortic valve stenosis: a prospective Mendelian randomization study and replication in a case-control cohort. Circ Cardiovasc Genet 2014;7:304—10.

[51] Kamstrup PR, Tybjaerg-Hansen A, Nordestgaard BG. Elevated lipoprotein(a) and risk of aortic valve stenosis in the general population. J Am Coll Cardiol 2014;63:470—7.

[52] Thanassoulis G, Campbell CY, Owens DS, Smith JG, Smith AV, Peloso GM, Kerr KF, Pechlivanis S, Budoff MJ, Harris TB, Malhotra R, O'Brien KD, Kamstrup PR, Nordestgaard BG, Tybjaerg-Hansen A, Allison MA, Aspelund T, Criqui MH, Heckbert SR, Hwang SJ, Liu Y, Sjogren M, van der Pals J, Kalsch H, Muhleisen TW, Nothen MM, Cupples LA, Caslake M, Di Angelantonio E, Danesh J, Rotter JI, Sigurdsson S, Wong Q, Erbel R, Kathiresan S, Melander O, Gudnason V, O'Donnell CJ, Post WS, Group CECW. Genetic associations with valvular calcification and aortic stenosis. N Engl J Med 2013;368:503—12.

[53] Capoulade R, Chan KL, Yeang C, Mathieu P, Bosse Y, Dumesnil JG, Tam JW, Teo KK, Mahmut A, Yang X, Witztum JL, Arsenault BJ, Despres JP, Pibarot P, Tsimikas S. Oxidized phospholipids, lipoprotein(a), and progression of calcific aortic valve stenosis. J Am Coll Cardiol 2015;66:1236—46.

[54] Emdin CA, Khera AV, Natarajan P, Klarin D, Won HH, Peloso GM, Stitziel NO, Nomura A, Zekavat SM, Bick AG, Gupta N, Asselta R, Duga S, Merlini PA, Correa A, Kessler T, Wilson JG, Bown MJ, Hall AS, Braund PS, Samani NJ, Schunkert H, Marrugat J, Elosua R, McPherson R, Farrall M, Watkins H, Willer C, Abecasis GR, Felix JF, Vasan RS, Lander E, Rader DJ, Danesh J, Ardissino D, Gabriel S, Saleheen D, Kathiresan S, Consortium CH-HF, Consortium CAE. Phenotypic characterization of genetically lowered human lipoprotein(a) levels. J Am Coll Cardiol 2016; 68:2761—72.

[55] Tsimikas S, Brilakis ES, Miller ER, McConnell JP, Lennon RJ, Kornman KS, Witztum JL, Berger PB. Oxidized phospholipids, Lp(a) lipoprotein, and coronary artery disease. N Engl J Med 2005;353: 46—57.

[56] Torzewski M, R A, Yeang C, Edel A, Bhindi R, Kath S, Twardowski L, Schmid J, Yang X, Franke UFW, Witztum JL, Tsimikas T. Lipoprotein(a)-Associated molecules are prominent components in plasma and valve leaflets in calcific aortic valve stenosis. JACC Basic Trans Sci 2017;2: 229—40.

[57] Bergmark C, Dewan A, Orsoni A, Merki E, Miller ER, Shin MJ, Binder CJ, Horkko S, Krauss RM, Chapman MJ, Witztum JL, Tsimikas S. A novel function of lipoprotein [a] as a preferential carrier of oxidized phospholipids in human plasma. J Lipid Res 2008;49:2230—9.

[58] Cote N, Mahmut A, Bosse Y, Couture C, Page S, Trahan S, Boulanger MC, Fournier D, Pibarot P, Mathieu P. Inflammation is associated with the remodeling of calcific aortic valve disease. Inflammation 2013;36:573—81.

[59] Jung JJ, Razavian M, Challa AA, Nie L, Golestani R, Zhang J, Ye Y, Russell KS, Robinson SP, Heistad DD, Sadeghi MM. Multimodality and molecular imaging of matrix metalloproteinase activation in calcific aortic valve disease. J Nucl Med 2015;56:933—8.

[60] Sadaba JR, Martinez-Martinez E, Arrieta V, Alvarez V, Fernandez-Celis A, Ibarrola J, Melero A, Rossignol P, Cachofeiro V, Lopez-Andres N. Role for galectin-3 in calcific aortic valve stenosis. J Am Heart Assoc 2016;5.

[61] Syvaranta S, Alanne-Kinnunen M, Oorni K, Oksjoki R, Kupari M, Kovanen PT, Helske-Suihko S. Potential pathological roles for oxidized low-density lipoprotein and scavenger receptors SR-AI, CD36, and LOX-1 in aortic valve stenosis. Atherosclerosis 2014;235:398—407.

[62] Davutoglu V, Celik A, Aksoy M. Contribution of selected serum inflammatory mediators to the progression of chronic rheumatic valve disease, subsequent valve calcification and NYHA functional class. J Heart Valve Dis 2005;14:251–6.

[63] Rezaeian P, Miller PE, Haberlen SA, Razipour A, Bahrami H, Castillo R, Witt MD, Kingsley L, Palella Jr FJ, Nakanishi R, Matsumoto S, Alani A, Jacobson LP, Post WS, Budoff MJ. Extra-coronary calcification (aortic valve calcification, mitral annular calcification, aortic valve ring calcification and thoracic aortic calcification) in HIV seropositive and seronegative men: multicenter AIDS Cohort Study. J Cardiovasc Comput Tomogr 2016;10:229–36.

[64] Li X, Lim J, Lu J, Pedego TM, Demer L, Tintut Y. Protective role of Smad6 in inflammation-induced valvular cell calcification. J Cell Biochem 2015;116:2354–64.

[65] Wirrig EE, Gomez MV, Hinton RB, Yutzey KE. COX2 inhibition reduces aortic valve calcification in vivo. Arterioscler Thromb Vasc Biol 2015;35:938–47.

[66] Towler DA. Molecular and cellular aspects of calcific aortic valve disease. Circ Res 2013;113:198–208.

[67] Gomez-Stallons MV, Wirrig-Schwendeman EE, Hassel KR, Conway SJ, Yutzey KE. Bone morphogenetic protein signaling is required for aortic valve calcification. Arterioscler Thromb Vasc Biol 2016;36:1398–405.

[68] MacGrogan D, D'Amato G, Travisano S, Martinez-Poveda B, Luxan G, Del Monte-Nieto G, Papoutsi T, Sbroggio M, Bou V, Gomez-Del Arco P, Gomez MJ, Zhou B, Redondo JM, Jimenez-Borreguero LJ, de la Pompa JL. Sequential ligand-dependent notch signaling activation regulates valve primordium formation and morphogenesis. Circ Res 2016;118:1480–97.

[69] Garg V, Muth AN, Ransom JF, Schluterman MK, Barnes R, King IN, Grossfeld PD, Srivastava D. Mutations in NOTCH1 cause aortic valve disease. Nature 2005;437:270–4.

[70] Acharya A, Hans CP, Koenig SN, Nichols HA, Galindo CL, Garner HR, Merrill WH, Hinton RB, Garg V. Inhibitory role of Notch1 in calcific aortic valve disease. PLoS One 2011;6:e27743.

[71] Robiolio PA, Rigolin VH, Wilson JS, Harrison JK, Sanders LL, Bashore TM, Feldman JM. Carcinoid heart disease. Correlation of high serotonin levels with valvular abnormalities detected by cardiac catheterization and echocardiography. Circulation 1995;92:790–5.

[72] Sandmann H, Pakkal M, Steeds R. Cardiovascular magnetic resonance imaging in the assessment of carcinoid heart disease. Clin Radiol 2009;64:761–6.

[73] Rothman RB, Baumann MH, Savage JE, Rauser L, McBride A, Hufeisen SJ, Roth BL. Evidence for possible involvement of 5-HT(2B) receptors in the cardiac valvulopathy associated with fenfluramine and other serotonergic medications. Circulation 2000;102:2836–41.

[74] Andrejak M, Szymanski C, Marechaux S, Arnalsteen E, Gras V, Remadi JP, Tribouilloy C. Valvular heart disease associated with long-term treatment by methysergide: a case report. Therapie 2014;69:255–7.

[75] Dahl CF, Allen MR, Urie PM, Hopkins PN. Valvular regurgitation and surgery associated with fenfluramine use: an analysis of 5743 individuals. BMC Med 2008;6:34.

[76] Frachon I, Etienne Y, Jobic Y, Le Gal G, Humbert M, Leroyer C. Benfluorex and unexplained valvular heart disease: a case-control study. PLoS One 2010;5:e10128.

[77] Tribouilloy C, Rusinaru D, Marechaux S, Jeu A, Ederhy S, Donal E, Reant P, Arnalsteen E, Boulanger J, Ennezat PV, Garban T, Jobic Y. Increased risk of left heart valve regurgitation associated with benfluorex use in patients with diabetes mellitus: a multicenter study. Circulation 2012;126:2852–8.

[78] Van Camp G, Flamez A, Cosyns B, Goldstein J, Perdaens C, Schoors D. Heart valvular disease in patients with Parkinson's disease treated with high-dose pergolide. Neurology 2003;61:859–61.

[79] Van Camp G, Flamez A, Cosyns B, Weytjens C, Muyldermans L, Van Zandijcke M, De Sutter J, Santens P, Decoodt P, Moerman C, Schoors D. Treatment of Parkinson's disease with pergolide and relation to restrictive valvular heart disease. Lancet 2004;363:1179–83.

[80] Weissman NJ, Panza JA, Tighe JF, Gwynne JT. Natural history of valvular regurgitation 1 year after discontinuation of dexfenfluramine therapy. A randomized, double-blind, placebo-controlled trial. Ann Intern Med 2001;134:267–73.

[81] Xu J, Jian B, Chu R, Lu Z, Li Q, Dunlop J, Rosenzweig-Lipson S, McGonigle P, Levy RJ, Liang B. Serotonin mechanisms in heart valve disease II: the 5-HT2 receptor and its signaling pathway in aortic valve interstitial cells. Am J Pathol 2002;161:2209—18.

[82] Droogmans S, Roosens B, Cosyns B, Degaillier C, Hernot S, Weytjens C, Garbar C, Caveliers V, Pipeleers-Marichal M, Franken PR, Bossuyt A, Schoors D, Lahoutte T, Van Camp G. Dose dependency and reversibility of serotonin-induced valvular heart disease in rats. Cardiovasc Toxicol 2009;9: 134—41.

[83] Lancellotti P, Nchimi A, Hego A, Dulgheru R, Delvenne P, Drion P, Oury C. High-dose oral intake of serotonin induces valvular heart disease in rabbits. Int J Cardiol 2015;197:72—5.

[84] Gragnano F, Crisci M, Bigazzi MC, Bianchi R, Sperlongano S, Natale F, Fimiani F, Concilio C, Cesaro A, Pariggiano I, Diana V, Limongelli G, Cirillo P, Russo M, Golia E, Calabro P. Von Willebrand factor as a novel player in valvular heart disease: from bench to valve replacement. Angiology 2018;69:103—12.

[85] Gao X, Zhang L, Gu G, Wu PH, Jin S, Hu W, Zhan C, Li J, Li Y. The effect of oxLDL on aortic valve calcification via the Wnt/beta-catenin signaling pathway: an important molecular mechanism. J Heart Valve Dis 2015;24:190—6.

[86] Albanese I, Yu B, Al-Kindi H, Barratt B, Ott L, Al-Refai M, de Varennes B, Shum-Tim D, Cerruti M, Gourgas O, Rheaume E, Tardif JC, Schwertani A. Role of noncanonical wnt signaling pathway in human aortic valve calcification. Arterioscler Thromb Vasc Biol 2017;37:543—52.

[87] Askevold ET, Gullestad L, Aakhus S, Ranheim T, Tonnessen T, Solberg OG, Aukrust P, Ueland T. Secreted Wnt modulators in symptomatic aortic stenosis. J Am Heart Assoc 2012;1:e002261.

[88] Weiss RM, Lund DD, Chu Y, Brooks RM, Zimmerman KA, El Accaoui R, Davis MK, Hajj GP, Zimmerman MB, Heistad DD. Osteoprotegerin inhibits aortic valve calcification and preserves valve function in hypercholesterolemic mice. PLoS One 2013;8:e65201.

[89] Linefsky JP, O'Brien KD, Sachs M, Katz R, Eng J, Michos ED, Budoff MJ, de Boer I, Kestenbaum B. Serum phosphate is associated with aortic valve calcification in the Multi-ethnic Study of Atherosclerosis (MESA). Atherosclerosis 2014;233:331—7.

CHAPTER 12

Immunological considerations for heart valve replacements

Hamza Atcha[1,3], Wendy F. Liu[1,2,3]

[1]Department of Biomedical Engineering, University of California Irvine, Irvine, CA, United States; [2]Department of Chemical Engineering and Materials Science, University of California Irvine, CA, United States; [3]The Edwards Lifesciences Center for Advanced Cardiovascular Technology, University of California Irvine, CA, United States

Contents

12.1 Introduction

Recent advances in biological prostheses have resulted in improved postsurgical outcomes and quality of life. These technologies often rely on the use of naturally occurring or synthetic biomaterials and have made significant contributions to heart valve replacement therapies to treat diseased valves. While the goal of these procedures is to restore the functional capabilities of severely damaged or defective heart valves, the treatment approaches often evoke a host immune response that can be detrimental to the longevity of the device. Mechanical heart valves, for example, involve the use of metal or polymeric materials in a design that recapitulates valve functions, but the materials themselves cause blood clotting and thus patients are administered lifelong anticoagulants. Transplanted and tissue-based heart valves are derived from natural tissues and pose a significantly smaller coagulation threat, but they still have many immunogenic, antigen-related challenges. In this chapter, we discuss the immunological considerations surrounding various heart valve replacement technologies and provide insight into current efforts to mitigate immune-mediated complications and improve postsurgical outcomes.

Principles of Heart Valve Engineering
ISBN 978-0-12-814661-3, https://doi.org/10.1016/B978-0-12-814661-3.00012-5

12.2 Heart valve transplants

Heart valve transplants involve the replacement of damaged or diseased valves with functional grafts. Transplanted valves are derived from natural tissues and therefore have several advantages over synthetic substitutes. For example, the mechanical properties of transplanted valves tend to better match the surrounding tissue and, as a result, these valves display favorable hemodynamic characteristics compared with mechanical or bioprosthetic valves, both of which are discussed in later sections [1,2]. Grafts can include autografts, which are obtained from the patient itself, allografts, grafts obtained from another individual of the same species, and xenografts, or grafts from another species. Each tissue source elicits distinct immune responses. This section will focus on the immunological considerations for allografts and autografts, and xenografts will be discussed in a later section.

Allografts are commonly obtained from cadaveric donors and are used to replace diseased heart valves. As it is derived from an individual other than the recipient, allografts contain foreign antigens that are capable of eliciting an immune response, which is the most common mechanism for allograft failure [3]. It has been shown that the majority of patients with allografts develop antibodies against human leukocyte antigen (HLA) [4], a molecule present on most nucleated cells which presents peptides to T cells to initiate immunity. This system also plays an important role in self-recognition by preventing the recognition of host molecules as foreign [5]. Following transplantation, HLA molecules from donor tissues are recognized by the host immune system through either intact foreign HLA molecules on donor antigen-presenting cells or by foreign molecules bound to host HLA molecules expressed on host antigen-presenting cells [5,6]. This results in T cell activation, production of cytokines and chemokines, recruitment of natural killer cells and macrophages, and a rejection response [5,7]. Tissue typing is therefore of particular importance when considering implantation of allografts.

Several strategies have been explored to create more immune compatible allografts. Recent efforts using tissue engineering—based techniques such as decellularization aim to reduce the host immune response and have shown improved survival of aortic valve allografts. Decellularization removes donor cells and debris usually through the use of detergents and thus results in an antigen-free extracellular matrix that partially maintains the structural integrity of native tissue [8]. This method limits the immune response, while also providing a material scaffold for heart valve integration. Early and midterm evaluation of implanted decellularized aortic valve allografts have revealed promising results, including preserved structural integrity and reduced calcification, as well as adequate hemodynamic properties [8]. In addition, Bibevski et al. conducted a multiinstitutional study assessing the performance of SynerGraft decellularized allografts compared with standard cryopreserved allografts. This study found that SynerGraft resulted in significantly reduced valve dysfunction and need for reintervention, although no difference in 5- and 10-year survival rates were reported [9]. Decellularized grafts have also demonstrated the potential for growth and remodeling similar to native tissues, allowing these valves to be functional throughout a patient's lifetime [10]. When decellularized porcine valves were implanted as pulmonary valve replacements in four juvenile sheep, an increase in valve diameter was observed, the size of which coincided with growth of the sheep. Moreover, histology revealed an endothelial layer, fibroblast infiltration, and collagen production. Additionally, von Kossa staining and atomic absorption spectrometry suggested the absence of calcification and low calcium levels in the valve wall and leaflets, respectively [11]. Decellularization remains an important strategy in the use of allogeneic tissues, and continued efforts aim to improve the extent of cell and debris removal while maintaining tissue mechanical integrity.

Pulmonary autografts have also been explored as potential replacements for diseased aortic valves and present different immunological considerations. In the Ross procedure, the diseased aortic valve is replaced with a pulmonary autograft, which is subsequently replaced with a pulmonary or aortic allograft [12]. This procedure can be performed with low operative mortality and potential for growth, making it an ideal procedure for children and young adults [13—16]. Autografts are not susceptible to rejection as the transplant originates from the host. In addition, using the patient's own valve generally provides good hemodynamic properties, with low risk of thromboembolism, and could potentially eliminate the need for lifelong anticoagulation therapy [17]. The benefits associated with this method, similar to allografts, are in large part attributed to the use of natural tissues.

Despite its advantages, transplanted autograft valves also pose certain drawbacks, and the need for reoperation after failure is one of the major detriments to this method. Autograft failure is thought to be associated with the change in mechanical environment experienced in the transplant location. Although the replacement tissue is a native heart valve, pulmonary valves experience significantly different stresses when compared with aortic valves [17]: the pulmonary root is under limited stress as it is exposed to low pressure changes. As a result, when it is placed in the aortic position, it is forced to stretch beyond its normal transitional point of high to low extensibility [17]. This increase in applied stress is thought to lead to a loss of compliance and increased stiffening [18]. Mechanistically, decreased and fragmented elastin and smooth muscle cell hypertrophy are thought to be a direct result of cellular adaptation to the mechanical environment, where tissue remodeling may thus cause the degeneration of pulmonary autografts [17]. While transplant rejection is not a concern when using pulmonary autografts, tissue remodeling caused by changes in mechanics is likely mediated by immune cells, and controlling this process will be required for the successful integration of autograft valves. In addition, the procedure itself requires replacing a healthy pulmonary valve with an allograft, which, in turn, may also result in pulmonary valve dysfunction over time [17]. Both autografts and allografts, unfortunately, are susceptible to reoperation risks over time [19]. To overcome these challenges, biomaterial and tissue-engineered approaches have been developed to address the limitations of transplanted heart valves.

12.3 Mechanical heart valves

Mechanical heart valves were conceived over 50 years ago and are composed entirely of synthetic materials to form mechanical devices that replicate the function of natural heart valves. They are designed to be durable and are thus commonly used to treat patients younger than 70 years of age [20]. From their conception, significant progress has been made to improve the functions of mechanical valves, mostly involving iterations in design and materials to improve durability and hemodynamic characteristics. Current variations of mechanical valves include caged ball, tilting disc, and bileaflet valves [21]. These are composed of materials such as metal alloys, graphite, and polymers, which were chosen for their mechanical strength and durability as well as their relative ability to suppress coagulation and the immune-mediated foreign body response.

Following implantation, biomaterials adsorb proteins, activating the coagulation and complement systems to initiate clotting and inflammation, which leads to recruitment and activation of immune cells and ultimately scar formation [22–24]. In injured tissues, thrombogenesis or clotting initiates the formation of a provisional matrix that provides structural and biochemical components necessary for the regulation of wound healing. However, for materials implanted in the cardiovascular system, clotting that results from direct blood/material interactions can cause thromboembolism or stroke. Therefore, anticoagulants are typically an essential component of therapy following heart valve replacement. Activation of the complement system, which is critical in regulating defense mechanisms against infection and foreign elements, and subsequent infiltration of platelets and immune cells, including neutrophils, and mast cells around the material characterize the acute inflammatory response. This response results in the release of chemokines and other chemoattractants, which guide the recruitment of monocytes and macrophages to the biomaterial implant [22,25,26]. The presence and overall composition of adsorbed proteins is essential to the adhesion of monocytes/macrophages, their subsequent fusion into foreign body giant cells, and the overall reaction of a tissue to an implant [25]. Giant cells are thought to release reactive oxygen species, enzymes, and acids that degrade the implanted material [25,27,28]. The foreign body response, therefore, leads to degradation of materials and device failure [25], and thus material selection is not only dependent on mechanical properties to ensure proper function and durability but also resistance to the foreign body reaction.

Mechanical heart valves are composed of several different types of materials, each with properties needed for specific device attributes. Metal alloys, such as titanium or stainless steel, are often used to construct the struts of leaflets and the cage in caged ball valves

[21,29] because of their structural integrity, durability, and corrosion resistance [29,30]. However, caged ball valves generally display higher rates of thrombogenicity from valve-mediated turbulence [31,32], and the metal components are highly thrombogenic [30]. As a result, the tilting disc and bileaflet designs as well as pyrolytic carbon as a biomaterial have been used to create mechanical valves that display superior hemodynamic properties and biocompatibility. Pyrolytic carbon, a graphene-based material, has comparable mechanical characteristics to metal alloys with reduced thrombogenicity and can be used as a surface coating or to create valve leaflets and housing [30,33,34]. In addition, polymeric materials, such as Dacron (polyethylene terephthalate) and Teflon (polytetrafluoroethylene), are utilized to create the suture ring. These polymers have reduced clotting and foreign body response, making them ideal for use in a heart valve. Despite careful material selection, mechanical heart valves are still associated with a risk of thrombogenicity.

While mechanical heart valves provide excellent durability, one of their main disadvantages is the persistent risk of thrombosis which can lead to valvular dysfunction [35,36]. The most common cause of valvular dysfunction results from obstruction of the valve discs by thrombus, and to a lesser extent, the formation of pannus [36]. Pannus formation is characterized by fibrous tissue ingrowth and presence of endothelial and immune cells (Fig. 12.1). In addition, turbulence, while much improved compared with previous design iterations, still has the ability to promote thrombogenicity around the implanted heart valve [37]. Thrombosis may also result from compliance mismatch between the native tissue and implanted materials, where differences in mechanical compliance can result in disturbed flow patterns, as has been observed in vascular grafts [38]. To prevent thrombus formation, patients are required to remain on lifelong anticoagulation therapy [20], which has its own risks of excessive bleeding and lack of blood clotting when needed. Anticoagulants pose complications for younger women as they are associated with birth defects and other pregnancy-related complications [39]. Poor compliance, especially in elderly patients, and cost of treatment are additional challenges [40]. Recent efforts have attempted to address the risk of thromboembolism by modifying surface characteristics of mechanical heart valves. Bark et al. demonstrate that rendering a clinically available mechanical heart valve superhydrophobic through surface coating prevents platelet and leukocyte adhesion while maintaining hemodynamic performance [41]. Continued efforts in biomaterial surface development will be needed to find a completely nonthrombogenic surface and to enable the elimination of anticoagulation therapy required for patients receiving mechanical heart valves.

12.4 Tissue valves

Heart valve replacements composed of biological tissues have been developed to reduce immune-mediated complications related to synthetic materials. Unlike mechanical valves that are composed entirely of nonbiological materials, bioprosthetic and

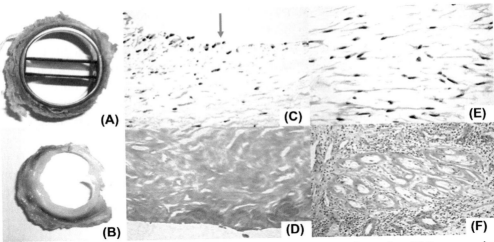

Figure 12.1 *Host response to mechanical heart valves showing pannus formation.* (A) Gross morphological image of Tri-technologies bileaflet mechanical valve following explantation. (B) Pannus formation of Tri-technologies bileaflet valve 4 years postimplantation. (C—F) Histological sections showing the host response to a St. Jude Medical bileaflet valve, with H&E showing pannus formation with presence of endothelial cells in the lumen layer indicated by the arrow (C), Van Gieson staining showing collagen and elastic fibrous tissues in the external media layer (D), H&E showing myofibroblasts in the internal media (E), and infiltration of inflammatory cells (neutrophils, lymphocytes, macrophages, foreign body giant cells, and mast cells) (F). *((A and B) were adapted from Cianciulli TF, Saccheri MC, Lax JA, Guidoin R, Zhang Z, Guerra JE, et al. Intermittent acute aortic regurgitation of a mechanical bileaflet aortic valve prosthesis: diagnosis and clinical implications. Eur J Echocardiogr J Work Group Echocardiogr Eur Soc Cardiol. May 2009;10(3):446—449. and (C—F) were adapted from Teshima H, Hayashida N, Yano H, Nishimi M, Tayama E, Fukunaga S, et al. Obstruction of St Jude medical valves in the aortic position: histology and immunohistochemistry of pannus. J Thorac Cardiovasc Surg August 2003; 126(2):401—407.)*

tissue-engineered heart valves (TEHVs) use allografts, xenografts, or a combination of biomaterial and native tissues or cells [42,43]. Bioprosthetic and TEHVs have mechanical and biochemical properties that more closely mimic native tissue and therefore display better hemodynamic properties and a reduced risk of thromboembolism when compared with mechanical heart valves [42]. Thus, patients do not require lifelong anticoagulation therapy. Nonetheless, immunological considerations are still significant and immune-mediated processes are largely responsible for the poor longevity of these heart valves. Because of a lack of available donor organs, xenogeneic tissues from porcine or bovine species are commonly used, and these tissues pose several challenges in their use in human patients.

One major barrier to consider for implantation of xenogeneic tissues is the so-called "Gal-barrier" [44]. The galactose-α1,3-galactose (α-Gal) antigen is abundantly expressed in most mammalian tissues, but not in humans. Therefore, when implanted into humans, the host immune system develops antibodies against α-Gal, leading to complement activation and hyperacute xenograft rejection [45]. This generally occurs with the activation of the classical pathway of the complement system, where antigen—antibody complexes bind the C1 complement protein complex, triggering activation of the complement cascade and generation of C5a and C3b [22,46,47]. C5a is a chemotactic protein that recruits inflammatory cells including macrophages and neutrophils, as well as platelets to the site of implantation [47]. To mitigate the α-Gal—mediated implant rejection, early efforts focused on antigen removal, for example, using glutaraldehyde to cross-link xenogeneic tissues. However, even with such treatments, porcine heart valves still result in an immune response that leads to degeneration and calcification of the graft, the mechanisms by which each occurrence is discussed in the following sections [48].

Gene therapy has been used to create transgenic pigs lacking the α-Gal gene. Lila et al. showed that α-Gal knockout porcine pericardium reduced calcification when compared with wild-type pericardium implanted in rats, when both xenografts were pretreated with human α-Gal antibodies [49]. This not only confirmed the role of α-Gal epitopes in eliciting an immune response to xenogeneic tissues but also provided a potential method to produce implantable tissues without harsh chemical treatments. Further work in understanding the role of α-Gal along with recognition of other foreign antigens will provide better insight into the mechanisms responsible for tissue rejection. In addition to challenges associated with xenogeneic tissue rejection, both bioprosthetic and TEHVs have other immunological barriers that affect their overall success in replacing native heart valve function.

12.4.1 Bioprosthetic heart valves

Bioprosthetic heart valves are typically composed of chemically treated leaflets derived from porcine or bovine pericardium attached to a stent mount, stentless mount, or an expanding stent. Relative to mechanical heart valves, bioprosthetic valves have a reduced risk of thromboembolism, thus making lifelong anticoagulation therapy unnecessary. In addition, bioprosthetic valves can be delivered through the minimally invasive transcatheter valve replacement technique, discussed in a later section, making them ideal for elderly patients [50,51]. Although bioprosthetic valves lack the need for lifelong anticoagulation therapy, they are not as robust and typically fail 10—15 years postimplantation. As a result, these valves carry the risk of reoperation [42,52,53]. The potential failure mechanisms associated with bioprosthetic valves include thrombosis, although the risk is lower when compared with mechanical valves, pannus formation, and calcification of leaflets, as exemplified in Fig. 12.2.

Thrombogenic complications resulting from bioprosthetic valves are infrequent, and thus there are differing opinions regarding anticoagulation management [54]. In a study conducted by Brown et al. only 8 of 4568 patients receiving a bioprosthetic

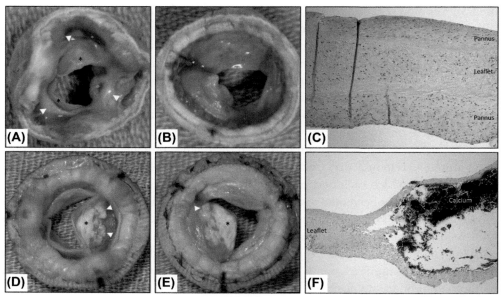

Figure 12.2 *Host response to bioprosthetic valves showing pannus formation and calcification.* (A and B) Gross morphological images of ventricular (A) and atrial (B) surfaces of porcine bioprosthetic mitral valve explanted 13 months postimplantation from a pediatric patient. * denotes leaflets and white arrows indicate pannus formation which is primarily seen on the ventricular surface. (C) H&E staining of valve leaflet displaying pannus formation. (D and E) Gross morphological images of ventricular (D) and atrial (E) surfaces of porcine bioprosthetic valve explanted 17 months postimplantation. * denotes one leaflet and white arrowheads indicate dense calcification. (F) H&E staining of calcified leaflet. *(All images were adapted from Gellis L, Baird CW, Emani S, Borisuk M, Gauvreau K, Padera RF, et al. Morphologic and histologic findings in bioprosthetic valves explanted from the mitral position in children younger than 5 years of age. J Thorac Cardiovasc Surg February 2018;155(2): 746–752.)*

aortic valve required reoperation due to thrombosis after a median postoperative time of approximately 1 year [55]. The risk of developing such complications has been associated with the incomplete endothelialization of the valve in the early stages following surgery [56]. Recellularization is thought to be impeded in glutaraldehyde-fixed valves because of cytotoxic effects of residual glutaraldehyde or the inability of the cross-linked matrix to allow cell migration and infiltration and thus appropriate tissue regeneration [42]. To prevent thromboembolic events before the initial endothelialization of the cloth sewing ring, anticoagulation therapy is often administered for the first 3 months postimplantation [57]. However, because of its rare occurrence, there is a lack of consensus about the duration and extent of the risk of thromboembolisms and the administration of anticoagulation therapy, particularly in patients without thrombogenic risk factors. Typically, anticoagulation therapy in the form of aspirin, an inhibitor of platelet aggregation but not coagulation, and/or warfarin, an anticoagulant, is administered following valve implantation. However, each drug has its own risks, including excessive bleeding [54,58,59], and therefore anticoagulation therapy should be administered on a case-by-case basis.

The endothelium lines the blood-facing walls of the heart and vasculature and is a key component of healthy functioning cardiovascular tissues, including heart valves. Endothelial cells provide an anticoagulating surface and play an important role not only in maintaining vascular tone but also in transporting molecules and immune cells out of circulation and into tissue. Normal, healthy endothelium does not express leukocyte adhesion molecules, but under inflammatory conditions expresses intracellular adhesion molecule-1 (ICAM-1, CD54) and E-selectin (CD62E) to facilitate attachment of peripheral blood mononuclear cells to the vascular wall and their extravasation [60–62]. Glutaraldehyde-treated xenografts or allografts contain residual donor endothelial cells or cell fragments that are thought to play a pivotal role in valve calcification. Existing phosphorous, primarily in the cell membranes that are preserved through glutaraldehyde treatment, and lack of mineralization inhibitors result in calcium phosphate crystal nucleation and growth. Over time, crystal buildup leads to valve failure with tearing or stenosis [42]. In vitro endothelialization of valves with autologous endothelial cells and preconditioning vascular grafts with shear stresses before implantation have been proposed to tame this response and prolong the lifetime of implanted bioprosthetic valves [60]. For example, valves treated with acetic acid–buffered urazole to remove glutaraldehyde in vitro displayed significant decreases in tissue inflammation and calcification when compared with glutaraldehyde-fixed valves without treatment [63]. Furthermore, endothelialization of valves significantly reduced thrombus formation when compared with valves without endothelialization [63]. Thus, proper endothelialization and removal of glutaraldehyde are both likely required for the long-term success of any bioprosthetic heart valve.

Lipid-mediated inflammation is also believed to be a central contributor to bioprosthetic valve failure. Previous studies have revealed an abundance of lipid droplets and macrophages in explanted bioprosthetic heart valves that exhibit calcification [64,65]. The majority of the lipids were oxidized low-density lipoproteins (ox-LDLs), which tend to co-localize with macrophages expressing scavenger receptor CD36 [66]. Scavenger receptors are essential to the uptake of lipids by macrophages and the subsequent formation of foam cells, a critical step in the development of cardiovascular disease [67]. Lipid accumulation is thought to further activate recruited macrophages, resulting in the production of inflammatory cytokines and matrix metalloproteases (MMPs) [66,68]. MMPs are involved in the degradation of the extracellular matrix weakening the overall integrity of the valve and, when combined with the mechanical stresses experienced by heart valves, may eventually result in leaflet tear [66]. Thus, the accumulation of ox-LDL within the leaflets of bioprosthetic valves leads to macrophage recruitment and their persistent inflammatory activation, foam cell formation, and production of MMPs, all of which contribute to eventual valve failure. To mitigate this, lipid lowering drugs including statins have been explored, but they have had mixed results. One study suggested that statins may slow down the progression of bioprosthetic degeneration [69], but a randomized control trial assessing the effects of statins on aortic valve calcification concluded no significant benefit [70–72]. The differing results could be attributed to the inability of statins to prevent valve fibrosis, an essential precursor to calcification, or the potential for statins to promote calcification through osteoblast differentiation [73]. Thus, while there has been significant progress in the development of bioprosthetic heart valves, continued efforts to improve endothelialization and reduce lipid-mediated inflammation and calcification are needed. Tissue engineering strategies have been employed to address these limitations and create more ideal heart valve replacements.

12.4.2 Tissue-engineered heart valves

TEHVs promise to provide an ideal long-term solution for heart valve replacement as they are designed to recapitulate the native heart valve and elicit minimal immune response, while also providing good hemodynamic characteristics and the ability for growth and regeneration. TEHVs include several essential components to ensure successful integration and functioning of the heart valve: a scaffold that is eventually replaced by newly regenerated native extracellular matrix, cells that are seeded in vitro or recruited in vivo, and biochemical and mechanical signals to encourage tissue growth [10]. Combinations of these components have been studied to better understand the parameters involved in creating an optimal artificial heart valve.

Biomaterials provide a three-dimensional scaffold on which cultured or recruited cells proliferate and differentiate to form functional tissue. The ideal scaffold promotes cell adhesion and differentiation, as well as deposition of extracellular matrix to improve remodeling postimplantation [10]. There are generally two types of scaffolds that have been used, including decellularized extracellular matrix and biodegradable synthetic materials. As described above, decellularized tissues are typically treated to remove cellular antigenic components and prevent an immune response to xenografts. However, the presence of residual cells, DNA, and the α-Gal epitope are still concerns for these materials [74]. In addition, the decellularization process can also degrade the matrix components and reduce its overall structural integrity [75]. Early studies using porcine-derived grafts resulted in severe complications with the death of three out of four pediatric recipients, one because of rupture of the valve 7 days postsurgery and the others because of valve degeneration 6 weeks and 1-year postimplantation. In all of these cases, the valves elicited an inflammatory response resulting in graft failure [76]. Complete recellularization and integration of these tissues can potentially improve the longevity and efficiency of decellularized valves; however, this still remains a major challenge.

To establish functional cell populations in decellularized scaffolds, in vitro and in situ recellularization strategies have been employed [77]. Early efforts used autologous bone marrow—derived endothelial and myofibroblast-like cells seeded on decellularized porcine pulmonary valves before implantation. These grafts survived following heart valve explant after 1 and 3 weeks, although incomplete endothelialization and regeneration were observed [78]. In addition, hybrid scaffolds composed of a metal mesh to provide structural support, enclosed by smooth muscle cell and fibroblast layers, and then covered by a layer of endothelial cells have been developed [79]. This tissue-engineered construct reduced the inflammatory response of monocyte-like cells in vitro and may thus improve implant compatibility [80]. In situ recellularization, or guided tissue regeneration, relies on host-mediated regeneration and repopulation of the scaffold. Scaffolds have been combined with biomolecules to promote cell adhesion to improve recellularization. For example, Flameng et al. coated decellularized ovine aortic homografts with stromal cell—derived factor-1α and fibronectin, its natural linker, and demonstrated that this combination prevented valve-mediated immune response as well as calcification and pannus formation while also stimulating re-endothelialization [81].

Polymer-based materials have also been explored as scaffolds to help promote integration and preserve valve function. Synthetic polymers provide several advantages including the ease of synthesis and reproducibility of materials, with minimal batch-to-batch variation, and a lack of foreign antigens. Several materials and cell—material combinations have been tested in ovine models with varied success. Implantation of polyhydroxyoctanoate-based grafts seeded with vascular cells resulted in minimal thrombus formation when compared with unseeded scaffolds. These valves also displayed the potential for progressive remodeling, as indicated by extracellular matrix formation, with mild valvular regurgitation [82,83]. PGA and poly(lactic-co glycolic acid) (PLGA) or PGA/poly-4 hydroxybutyric acid (P4HB) valves seeded with myofibroblasts demonstrated remodeling with limited stenosis, comparable extracellular matrix composition to native tissue, and degradation of polymeric scaffold within 6—8 weeks after implantation [84,85]. Degradation of polymer-based TEHVs must be controlled because rapid degradation of the graft could lead to early mechanical failure, whereas slow degradation may prevent timely remodeling and new tissue formation [86].

While several materials have been tested for their ability to create effective TEHVs, recent work suggests that differences in microstructure, rather than the material itself, play an important role in regulating the immune response to the implanted material [86,87]. For example, reducing the diameter of electrospun PLGA fibers to less than 1 μm in diameter resulted in less platelet activation and adhesion when compared with larger 2—3 μm diameter fibers of the same material [87]. Fiber alignment has also been implicated in platelet activation. Liu et al. showed that vascular grafts composed of aligned electrospun polyurethane fibers displayed a higher patency rate and reduced thrombus formation compared with a vascular graft with smooth topography, although no differences in cellular adhesion rates were reported [88]. Other studies evaluating the effect of surface topography on platelet adhesion have reported that smooth surfaces generally result in higher platelet activation compared with rough surfaces. This may be caused by a slower boundary layer velocity on smooth surfaces, which increases the number of platelet wall collisions [10,89]. Despite the several advantages offered by polymer-based TEHVs, there are currently no such systems that have advanced clinical trials.

12.5 Transcatheter valves

Transcatheter heart valves were first implanted in humans in 2002 and are similar to bioprosthetic valves in composition, as they consist of biological tissues mounted to an expandable stent [90]. However, transcatheter valves are deployed through a catheter in a minimally invasive procedure compared with open-heart surgery required for implantation of other valves. Examples of these valves include the Edwards SAPIEN 3, which is composed of bovine pericardial tissue mounted to a balloon expandable metal stent, and the Medtronic CoreValve, which uses porcine pericardial tissue mounted on a Nitinol self-expanding stent [91].

Transcatheter valves are an excellent option for patients who are at risk for surgical mortality and cannot withstand the major surgical procedures involved with other heart valve implants/transplants [92—95]. The relative noninvasive nature of this procedure has resulted in lower complication rates and faster recovery times. In addition, the development of valve—in—valve procedures to replace a failed prosthetic, where a catheter is used to expand a new functional valve in the opening of a degenerated valve, could potentially extend transcatheter valve application to both elderly and young patients [35]. However, despite these apparent advantages, valve—in—valve procedures are performed selectively in elderly populations as the complications arising from interactions of multiple bioprosthetic valves and the long-term efficacy of transcatheter valves is currently poorly understood, with the first comprehensive 10-year follow-up data expected by 2020 [35,96—98]. Transcatheter valves have similar immune-mediated failure mechanisms as bioprosthetic valves such as calcification (Fig. 12.3); however, the procedures used for valve deployment provide additional complications which affect its overall success.

Cardiac surgery, in general, has been shown to activate a systemic inflammatory response resulting in adverse postsurgical complications, referred to as systemic inflammatory response syndrome (SIRS) [99—101]. Transcatheter heart valve procedures often result in an increase in leukocyte counts within 72 h following valve replacement [102]. This increase in immune cell count might be associated with a systemic inflammatory response that negatively impacts patient health postprocedure. In a study conducted by Sinning et al. a systemic inflammatory response developed in 61 out of 152 (40.1%) patients 48 h following transcatheter aortic valve replacement. In general, these patients

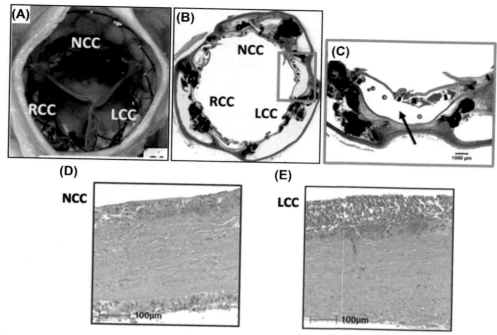

Figure 12.3 *Host response to leaflets of transcatheter heart valves showing calcification and inflammation.* (A) Gross morphological image of CoreValve showing aortic surface. (B and C) Von Kossa staining reveals heavy calcification, shown in dark purple, in the native aortic valve (B) and higher magnification section shows severe paravalvular gaps (C). (D and E) H&E staining shows inflammatory cell infiltration, which include macrophages and giant cells, on the aortic and ventricular side of the non-coronary cusp (NCC) (D) and aortic side of the left coronary cusp (E). Abbreviations: noncoronary cusp (NCC), left coronary cusp (LCC), right coronary cusp (RCC). *(Images were adapted from Yahagi K, Torii S, Ladich E, Kutys R, Romero ME, Mori H, et al. Pathology of self-expanding transcatheter aortic valves: findings from the corevalve US pivotal trials. Catheter Cardiovasc Interv Off J Soc Card Angiogr Interv September 12, 2017.)*

displayed elevated leukocyte counts, hyperventilation, tachycardia, and fever when compared with patients who did not develop a systemic inflammatory response [101]. Elevated inflammatory cytokines IL-6 and IL-8 were observed 4 h postsurgery with maximum differences occurring 24 h after valve replacement when compared with patients who did not develop SIRS. In addition, levels of procalcitonin, a biomarker indicating bacterial infections or tissue injury, was found to substantially increase 48 h after the procedure. Several predictive factors of SIRS were also identified, including the occurrence of vascular complications, major bleeding during surgery, the amount of contrast dye used during the procedure, repeated ventricular pacing runs, and postsurgical dilation of the implanted valve. The occurrence of SIRS is thought to result from decreased organ perfusion during balloon valvuloplasty and/or coadministration of cytokines, which accumulate in stored blood, during blood transfusions, and was also associated with a 7.4-fold increase in risk for 1-year mortality [99,101,103].

Transcatheter valves undergo a crimping process, typically immediately before implantation, to reduce the size of the valve which is critical for the catheter-based delivery method. This process exposes the valve leaflets to substantial mechanical stresses that are not experienced by bioprosthetic valves [104]. Recent studies have identified that mechanical stresses disrupt the collagen fibers in pericardial leaflets, which could lead to calcification and early valve failure [105,106]. In addition, Alavi et al. showed that crimping resulted in significant tissue damage both at the surface and through the depth of pericardial tissues when compared with uncrimped controls, and the damage was permanent and original tissue structure could not be recovered [107]. While the exact mechanisms are unknown, damage to leaflet collagen composition has currently been linked with an increase in calcification. Further work is needed to verify the relationship and mechanisms behind structural damage and calcification in pericardial leaflets.

Thrombosis may also cause complications after transcatheter procedures, although the occurrence is rare with only 26 cases reported out of 4266 (0.61%) patients between January 2008 and September 2013 following transcatheter aortic valve replacement [108]. Transcatheter heart valve thrombosis was typically observed within the first 2 years postprocedure, with exertional dyspnea being the most common clinical presentation. Echocardiography found increased mean aortic valve pressure gradient and thickening of the leaflets in patients who developed thrombosis. The occurrence of thrombosis and increased aortic valve pressure gradient was observed to significantly decrease when antithrombotic therapy was used following aortic valve implantation. Current clinical guidelines recommend that a minimum of aspirin should be used indefinitely following valve replacement, with coadministration of aspirin and clopidogrel, a more potent antiplatelet therapy, recommended for the first 3—6 months postprocedure [108]. The occurrence of thrombosis in transcatheter valves, while still unclear, is thought to be associated with prosthesis patient mismatch, where the diameter of the artificial valve is small relative to patient body size or underexpansion of the valve during deployment [108,109].

Tissue-engineered strategies are currently being explored to overcome the current limitations of transcatheter heart valves. For example, recent work conducted by several groups found that homologous cell—based transcatheter TEHVs, created using degradable synthetic scaffolds composed of PGA/P4HB seeded with vascular-derived cells for 4 weeks before decellularization, as shown in Fig. 12.4, displayed significant recellularization and extracellular matrix formation over time with similar collagen content observed 8 weeks postimplantation when compared with native leaflets [110]. In addition, after 6 months postimplantation, significant recellularization and elastin fiber formation was observed, whereas minimal thrombus and calcification was detected [110,111]. These studies demonstrate the ability of TEHVs to address

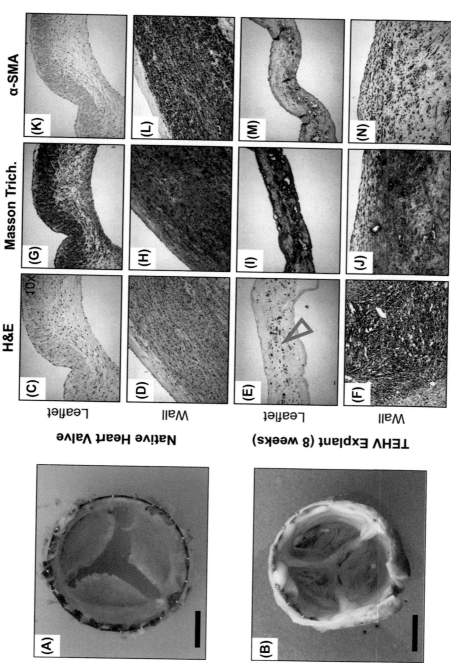

Figure 12.4 *Histological analysis of explanted tissue-engineered heart valve.* (A—B) Gross morphological image of explanted tissue-engineered valve composed of PGA/P4HB, seeded with vascular cells in vitro with mechanical conditioning for 4 weeks, decellularized, and then implantation into non-human primates (chacma baboons). (C—F) H&E staining reveals that comparable cellularity between explanted and native valves. (G—J) Masson Trichrome staining reveals similar collagen content in the walls and leaflet of the explanted valve. (K—N) Immunohistochemistry of smooth muscle actin shows no expression in the explanted valve leaflets, similar to the native valve leaflets, but significantly less expression in the wall of the explant compared to that of the native heart valve. *(Images were adapted from Weber B, Dijkman PE, Scherman J, Sanders B, Emmert MY, Grünenfelder J, et al. Off-the-shelf human decellularized tissue-engineered heart valves in a non-human primate model. Biomaterials October 2013;34(30):7269—7280.)*

the shortcomings of current transcatheter valves and can be used as a potential solution to prevent damage caused by crimping through utilizing the innate capacity of the host to remodel implanted tissues. While the results obtained are promising, further work is necessary to evaluate long-term functionality, durability, and clinical safety of tissue-engineered transcatheter heart valves [97].

12.6 Conclusions and future directions

Since the conception of the first artificial heart valve, significant progress has been made to create more durable and biomimetic valve substitutes. Several different strategies have been developed, each with its own unique advantages and risks. These strategies, however, all have immunological considerations which can affect heart valve function and longevity once implanted. For example, the constant risk of thrombosis or valvular dysfunctions resulting from structural degradation of the leaflets are detriments to the overall success of mechanical or bioprosthetic heart valves, respectively [35,66]. In addition, the rigors of open-heart surgery and the subsequent immune response have restricted the use of these and transplanted heart valves to relatively young patients. More recently, the development of transcatheter valves, where a bioprosthetic valve is deployed through a catheter in a minimally invasive procedure, and valve-in-valve procedures have provided a valve replacement strategy for use in both elderly and young patients. However, the durability and efficacy of these procedures remain unknown, with the first long-term follow-up data expected in the near future [97]. Despite this lack of experience with transcatheter valves, the procedure itself has been shown to result in heart valve damage before implantation and valve calcification. Applying tissue engineering strategies to heart valve production provides an exciting and novel opportunity to address the limitations of current heart valves. These TEHVs utilize both synthetic and biological materials which have selective mechanical properties and can harness the ability of the host to remodel tissues leading to better integration of the implanted heart valve. TEHVs can also modulate the immune response through selective recruitment of vascular cells, in vitro or in situ, and thus potentially prevent early valvular dysfunction or rejection. Despite much optimism surrounding TEHVs, substantial progress is still required to create functional, durable, and clinically relevant heart valves [10,77].

References

[1] Hasegawa J, Kitamura S, Taniguchi S, Kawata T, Niwaya K, Mizuguchi K, et al. Comparative rest and exercise hemodynamics of allograft and prosthetic valves in the aortic position. Ann Thorac Surg December 1, 1997;64(6):1753—6.

[2] Pibarot P, Dumesnil JG, Briand M, Laforest I, Cartier P. Hemodynamic performance during maximum exercise in adult patients with the Ross operation and comparison with normal controls and patients with aortic bioprostheses. Am J Cardiol November 1, 2000;86(9):982—8.

[3] Clarke DR, Campbell DN, Hayward AR, Bishop DA. Degeneration of aortic valve allografts in young recipients. J Thorac Cardiovasc Surg May 1993;105(5):934—41. discussion 941-2.

[4]   Lisy M, Kalender G, Schenke-Layland K, Brockbank KGM, Biermann A, Stock UA. Allograft heart valves: current aspects and future applications. Biopreserv Biobanking April 2017;15(2): 148–57.

[5]   Mahdi BM. A glow of HLA typing in organ transplantation. Clin Transl Med February 23, 2013;2:6.

[6]   Benichou G, Thomson AW. Direct versus indirect allorecognition pathways: on the right track. Am J Transplant Off J Am Soc Transplant Am Soc Transpl Surg April 2009;9(4):655–6.

[7]   Pratt JR, Basheer SA, Sacks SH. Local synthesis of complement component C3 regulates acute renal transplant rejection. Nat Med June 2002;8(6):582–7.

[8]   Costa FDA da, Costa ACBA, Prestes R, Domanski AC, Balbi EM, Ferreira ADA, et al. The early and midterm function of decellularized aortic valve allografts. Ann Thorac Surg December 1, 2010;90(6): 1854–60.

[9]   Bibevski S, Ruzmetov M, Fortuna RS, Turrentine MW, Brown JW, Ohye RG. Performance of Synergraft decellularized pulmonary allografts compared with standard cryopreserved allografts: results from multiinstitutional data. Ann Thorac Surg March 1, 2017;103(3):869–74.

[10] Blum K, Drews J, Breuer CK. Tissue engineered heart valves: a call for mechanistic studies [Internet] Tissue Eng B Rev January 12, 2018;91(5):947–55. Available from: http://online.liebertpub.com/doi/abs/10.1089/ten.TEB.2017.0425.

[11] Dohmen PM, da Costa F, Holinski S, Lopes SV, Yoshi S, Reichert LH, et al. Is there a possibility for a glutaraldehyde-free porcine heart valve to grow? Eur Surg Res Eur Chir Forsch Rech Chir Eur 2006; 38(1):54–61.

[12] Kouchoukos NT, Masetti P, Nickerson NJ, Castner CF, Shannon WD, Dávila-Román VG. The Ross procedure: long-term clinical and echocardiographic follow-up. Ann Thorac Surg September 1, 2004;78(3):773–81.

[13] Al-Halees Z, Pieters F, Qadoura F, Shahid M, Al-Amri M, Al-Fadley F. The Ross procedure is the procedure of choice for congenital aortic valve disease. J Thorac Cardiovasc Surg March 1, 2002; 123(3):437–42.

[14] Chambers JC, Somerville J, Stone S, Ross DN. Pulmonary autograft procedure for aortic valve disease: long-term results of the pioneer series. Circulation October 7, 1997;96(7):2206–14.

[15] Elkins RC. The Ross operation: a 12-year experience. Ann Thorac Surg September 1, 1999;68(3): S14–8.

[16] Kouchoukos NT, Davila-Roman VG, Spray TL, Murphy SF, Perrillo JB. Replacement of the aortic root with a pulmonary autograft in children and young adults with aortic- valve disease. N Engl J Med January 6, 1994;330(1):1–6.

[17] Takkenberg JJM, Klieverik LMA, Schoof PH, van Suylen R-J, Herwerden LA van, Zondervan PE, et al. The Ross procedure: a systematic review and meta-analysis. Circulation January 20, 2009; 119(2):222–8.

[18] Grotenhuis HB, Westenberg JJM, Doornbos J, Kroft LJM, Schoof PH, Hazekamp MG, et al. Aortic root dysfunctioning and its effect on left ventricular function in Ross procedure patients assessed with magnetic resonance imaging. Am Heart J November 1, 2006;152(5):975.e1–8.

[19] Ruzmetov M, Geiss DM, Shah JJ, Fortuna RS. Autograft or allograft aortic root replacement in children and young adults with aortic valve disease: a single-center comparison. Ann Thorac Surg November 2012;94(5):1604–11.

[20] Fioretta ES, Dijkman PE, Emmert MY, Hoerstrup SP. The future of heart valve replacement: recent developments and translational challenges for heart valve tissue engineering. J Tissue Eng Regenerat Med January 1, 2017 [n/a-n/a].

[21] Gott VL, Alejo DE, Cameron DE. Mechanical heart valves: 50 years of evolution. Ann Thorac Surg December 1, 2003;76(6):S2230–9.

[22] Gorbet MB, Sefton MV. Biomaterial-associated thrombosis: roles of coagulation factors, complement, platelets and leukocytes. Biomaterials November 1, 2004;25(26):5681–703.

[23] Nilsson B, Ekdahl KN, Mollnes TE, Lambris JD. The role of complement in biomaterialinduced inflammation. Mol Immunol January 1, 2007;44(1):82–94.

[24] Sinno H, Prakash S. Complements and the wound healing cascade: an updated review [Internet] Plast Surg Int 2013;2013. Available from: https://www.ncbi.nlm.nih.gov/pmc/articles/PMC3741993/.

[25] Anderson JM, Rodriguez A, Chang DT. Foreign body reaction to biomaterials. Semin Immunol April 1, 2008;20(2):86–100.

[26] Esche C, Stellato C, Beck LA. Chemokines: key players in innate and adaptive immunity. J Investig Dermatol October 1, 2005;125(4):615–28.

[27] Henson PM. The immunologic release of constituents from neutrophil leukocytes: II. Mechanisms of release during phagocytosis, and adherence to nonphagocytosable surfaces. J Immunol December 1, 1971;107(6):1547–57.

[28] Sheikh Z, Brooks PJ, Barzilay O, Fine N, Glogauer M. Macrophages, foreign body giant cells and their response to implantable biomaterials. Materials August 28, 2015;8(9):5671–701.

[29] Lam MT, Wu JC. Biomaterial applications in cardiovascular tissue repair and regeneration. Expert Rev Cardiovasc Ther August 2012;10(8):1039–49.

[30] Lancellotti P, éCile OC, Jerome C, Pierard LA. Graphene coating onto mechanical heart valve prosthesis and resistance to flow dynamics. Acta Cardiol June 1, 2016;71(3):235–55.

[31] Kroll MH, Hellums JD, McIntire LV, Schafer AI, Moake JL. Platelets and shear stress. Blood September 1, 1996;88(5):1525–41.

[32] Myhre E, Dale J, Rasmussen K. Erythrocyte destruction in different types of starr- Edwards aortic ball valves. Circulation September 1, 1970;42(3):515–20.

[33] Cao H. Mechanical performance of pyrolytic carbon in prosthetic heart valve applications. J Heart Valve Dis June 1996;5(Suppl. 1):S32–49.

[34] Goodman SL, Tweden KS, Albrecht RM. Platelet interaction with pyrolytic carbon heartvalve leaflets. J Biomed Mater Res October 1, 1996;32(2):249–58.

[35] Head SJ, Çelik M, Kappetein AP. Mechanical versus bioprosthetic aortic valve replacement. Eur Heart J July 21, 2017;38(28):2183–91.

[36] Ostrowski S, Marcinkiewicz A, Kośmider A, Walczak A, Zwoliński R, Jaszewski R. Artificial aortic valve dysfunction due to pannus and thrombus — different methods of cardiac surgical management. Kardiochirurgia Torakochirurgia Pol Pol J Cardio-Thorac Surg. September 2015; 12(3):199–203.

[37] Liu JS, Lu PC, Chu SH. Turbulence characteristics downstream of bileaflet aortic valve prostheses. J Biomech Eng April 2000;122(2):118–24.

[38] Spadaccio C, Nappi F, Al-Attar N, Sutherland FW, Acar C, Nenna A, et al. Old myths, new concerns: the long-term effects of ascending aorta replacement with Dacron grafts. Not all that glitters is gold. J Cardiovasc Transl Res 2016;9:334–42.

[39] Born D, Martinez EE, Almeida PAM, Santos DV, Carvalho ACC, Moron AF, et al. Pregnancy in patients with prosthetic heart valves: the effects of anticoagulation on mother, fetus, and neonate. Am Heart J August 1, 1992;124(2):413–7.

[40] Van Damme S, Van Deyk K, Budts W, Verhamme P, Moons P. Patient knowledge of and adherence to oral anticoagulation therapy after mechanical heart-valve replacement for congenital or acquired valve defects. Heart Lung J Crit Care April 2011;40(2):139–46.

[41] Bark DL, Vahabi H, Bui H, Movafaghi S, Moore B, Kota AK, et al. Hemodynamic performance and thrombogenic properties of a superhydrophobic bileaflet mechanical heart valve. Ann Biomed Eng February 1, 2017;45(2):452–63.

[42] Schoen FJ, Levy RJ. Calcification of tissue heart valve substitutes: progress toward understanding and prevention. Ann Thorac Surg March 1, 2005;79(3):1072–80.

[43] Vongpatanasin W, Hillis LD, Lange RA. Prosthetic heart valves. N Engl J Med August 8, 1996; 335(6):407–16.

[44] Cooper DKC, Ekser B, Tector AJ. Immunobiological barriers to xenotransplantation. Int J Surg Lond Engl November 2015;23:211–6. 0 0.

[45] Konakci K Z, Bohle B, Blumer R, Hoetzenecker W, Roth G, Moser B, et al. Alpha-Gal on bioprostheses: xenograft immune response in cardiac surgery. Eur J Clin Investig January 1, 2005; 35(1):17–23.

[46] Dalmasso AP, Vercellotti GM, Fischel RJ, Bolman RM, Bach FH, Platt JL. Mechanism of complement activation in the hyperacute rejection of porcine organs transplanted into primate recipients. Am J Pathol May 1992;140(5):1157–66.

[47] Huai G, Qi P, Yang H, Wang Y. Characteristics of α-Gal epitope, anti-Gal antibody, α1,3 galactosyltransferase and its clinical exploitation (review). Int J Mol Med January 2016;37(1):11−20.

[48] Bloch O, Golde P, Dohmen PM, Posner S, Konertz W, Erdbrügger W. Immune response in patients receiving a bioprosthetic heart valve: lack of response with decellularized valves. Tissue Eng May 10, 2011;17(19−20):2399−405.

[49] Lila N, McGregor CGA, Carpentier S, Rancic J, Byrne GW, Carpentier A. Gal knockout pig pericardium: new source of material for heart valve bioprostheses. J Heart Lung Transplant May 1, 2010;29(5):538−43.

[50] Barreto-Filho JA, Wang Y, Dodson JA, Desai MM, Sugeng L, Geirsson A, et al. Trends in aortic valve replacement for elderly patients in the United States, 1999−2011. JAMA, J Am Med Assoc November 20, 2013;310(19):2078−85.

[51] Dunning J, Gao H, Chambers J, Moat N, Murphy G, Pagano D, et al. Aortic valve surgery: marked increases in volume and significant decreases in mechanical valve use–an analysis of 41,227 patients over 5 years from the society for cardiothoracic surgery in Great Britain and Ireland national database. J Thorac Cardiovasc Surg October 2011;142(4):776−782.e3.

[52] Hammermeister K, Sethi GK, Henderson WG, Grover FL, Oprian C, Rahimtoola SH. Outcomes 15 years after valve replacement with a mechanical versus a bioprosthetic valve: final report of the veterans affairs randomized trial. J Am Coll Cardiol October 1, 2000;36(4):1152−8.

[53] Vesey JM, Otto CM. Complications of prosthetic heart valves. Curr Cardiol Rep March 1, 2004; 6(2):106−11.

[54] Cremer PC, Rodriguez LL, Griffin BP, Tan CD, Rodriguez ER, Johnston DR, et al. Early bioprosthetic valve failure: mechanistic insights via correlation between echocardiographic and operative findings. J Am Soc Echocardiogr October 1, 2015;28(10):1131−48.

[55] Brown ML, Park SJ, Sundt TM, Schaff HV. Early thrombosis risk in patients with biologic valves in the aortic position. J Thorac Cardiovasc Surg July 2012;144(1):108−11.

[56] Roudaut R, Serri K, Lafitte S. Thrombosis of prosthetic heart valves: diagnosis and therapeutic considerations. Heart January 2007;93(1):137−42.

[57] Brueck M, Kramer W, Vogt P, Steinert N, Roth P, Görlach G, et al. Antiplatelet therapy early after bioprosthetic aortic valve replacement is unnecessary in patients without thromboembolic risk factors. Eur J Cardiothorac Surg July 1, 2007;32(1):108−12.

[58] Brennan JM, Edwards FH, Zhao Y, O'Brien S, Booth ME, Dokholyan RS, et al. Early anticoagulation of bioprosthetic aortic valves in older patients: results from the society of thoracic surgeons adult cardiac surgery national database. J Am Coll Cardiol September 11, 2012;60(11):971−7.

[59] Whitlock RP, Sun JC, Fremes SE, Rubens FD, Teoh KH. Antithrombotic and thrombolytic therapy for valvular disease: antithrombotic therapy and prevention of thrombosis, 9th ed: American college of chest physicians evidence-based clinical practice guidelines. Chest February 2012;141(2 Suppl. l): e576S−600S.

[60] Simon A, Wilhelmi M, Steinhoff G, Harringer W, Brücke P, Haverich A. Cardiac valve endothelial cells: relevance in the long-term function of biologic valve prostheses. J Thorac Cardiovasc Surg October 1, 1998;116(4):609−16.

[61] Simon A, Zavazava N, Sievers HH, Müller-Ruchholtz W. In vitro cultivation and immunogenicity of human cardiac valve endothelium. J Card Surg December 1, 1993;8(6):656−65.

[62] Springer TA. Adhesion receptors of the immune system. Nature August 1990;346(6283):425−34.

[63] Trantina-Yates AE, Human P, Bracher M, Zilla P. Mitigation of bioprosthetic heart valve degeneration through biocompatibility: in vitro versus spontaneous endothelialization. Biomaterials July 2001;22(13):1837−46.

[64] Bottio T, Thiene G, Pettenazzo E, Ius P, Bortolotti U, Rizzoli G, et al. Hancock II bioprosthesis: a glance at the microscope in mid-long-term explants. J Thorac Cardiovasc Surg July 2003;126(1): 99−105.

[65] Grabenwöger M, Grimm M, Eybl E, Kadletz M, Havel M, Köstler P, et al. New aspects of the degeneration of bioprosthetic heart valves after long-term implantation. J Thorac Cardiovasc Surg July 1992;104(1):14−21.

[66] Shetty R, Pibarot P, Audet A, Janvier R, Dagenais F, Perron J, et al. sodi. Eur J Clin Investig June 1, 2009;39(6):471—80.

[67] Getz GS. Thematic review series: the immune system and atherogenesis. Immune function in atherogenesis. J Lipid Res January 2005;46(1):1—10.

[68] Ardans JA, Economou AP, Martinson JM, Zhou M, Wahl LM. Oxidized low-density and high-density lipoproteins regulate the production of matrix metalloproteinase-1 and -9 by activated monocytes. J Leukoc Biol June 1, 2002;71(6):1012—8.

[69] Antonini-Canterin F, Zuppiroli A, Popescu BA, Granata G, Cervesato E, Piazza R, et al. Effect of statins on the progression of bioprosthetic aortic valve degeneration. Am J Cardiol December 15, 2003;92(12):1479—82.

[70] Chan KL, Teo K, Dumesnil JG, Ni A, Tam J. Effect of lipid lowering with rosuvastatin on progression of aortic stenosis: results of the aortic stenosis progression observation: measuring effects of rosuvastatin (ASTRONOMER) trial. Circulation January 19, 2010;121(2):306—14.

[71] Cowell SJ, Newby DE, Prescott RJ, Bloomfield P, Reid J, Northridge DB, et al. A randomized trial of intensive lipid-lowering therapy in calcific aortic stenosis. N Engl J Med June 9, 2005;352(23): 2389—97.

[72] Rossebø AB, Pedersen TR, Boman K, Brudi P, Chambers JB, Egstrup K, et al. Intensive lipid lowering with simvastatin and ezetimibe in aortic stenosis. N Engl J Med September 25, 2008; 359(13):1343—56.

[73] Leopold JA. Cellular mechanisms of aortic valve calcification. Circ Cardiovasc Interv August 1, 2012; 5(4):605—14.

[74] Kasimir M-T, Rieder E, Seebacher G, Wolner E, Weigel G, Simon P. Presence and elimination of the xenoantigen gal (α1, 3) gal in tissue-engineered heart valves. Tissue Eng July 1, 2005;11(7—8): 1274—80.

[75] Gilbert TW, Sellaro TL, Badylak SF. Decellularization of tissues and organs. Biomaterials July 1, 2006;27(19):3675—83.

[76] Simon P, Kasimir MT, Seebacher G, Weigel G, Ullrich R, Salzer-Muhar U, et al. Early failure of the tissue engineered porcine heart valve SYNERGRAFT® in pediatric patients. Eur J Cardiothorac Surg June 1, 2003;23(6):1002—6.

[77] VeDepo MC, Detamore MS, Hopkins RA, Converse GL. Recellularization of decellularized heart valves: progress toward the tissue-engineered heart valve [Internet] J Tissue Eng August 25, 2017;8. Available from: https://www.ncbi.nlm.nih.gov/pmc/articles/PMC5574480/.

[78] Kim S-S, Lim S-H, Hong Y-S, Cho S-W, Ryu JH, Chang B-C, et al. Tissue engineering of heart valves in vivo using bone marrow derived cells. Artif Organs July 1, 2006;30(7):554—7.

[79] Alavi SH, Kheradvar A. Metal mesh scaffold for tissue engineering of membranes. Tissue Eng C Methods November 10, 2011;18(4):293—301.

[80] Alavi SH, Liu WF, Kheradvar A. Inflammatory response assessment of a hybrid tissue-engineered heart valve leaflet. Ann Biomed Eng February 1, 2013;41(2):316—26.

[81] Flameng W, Visscher GD, Mesure L, Hermans H, Jashari R, Meuris B. Coating with fibronectin and stromal cell—derived factor-1α of decellularized homografts used for right ventricular outflow tract reconstruction eliminates immune response—related degeneration. J Thorac Cardiovasc Surg April 1, 2014;147(4):1398—404. e2.

[82] Sodian R, Hoerstrup SP, Sperling JS, Daebritz S, Martin DP, Moran AM, et al. Early in vivo experience with tissue-engineered trileaflet heart valves. Circulation November 7, 2000;102(19 Suppl. 3): III22—29.

[83] Stock UA, Nagashima M, Khalil PN, Nollert GD, Herden T, Sperling JS, et al. Tissue-engineered valved conduits in the pulmonary circulation. J Thorac Cardiovasc Surg April 2000;119(4 Pt 1): 732—40.

[84] Breuer CK, Shin'oka T, Tanel RE, Zund G, Mooney DJ, Ma PX, et al. Tissue engineering lamb heart valve leaflets. Biotechnol Bioeng June 5, 1996;50(5):562—7.

[85] Hoerstrup SP, Sodian R, Daebritz S, Wang J, Bacha EA, Martin DP, et al. Functional living trileaflet heart valves grown in vitro. Circulation November 7, 2000;102(19 Suppl. 3):III44—49.

[86] Xue Y, Sant V, Phillippi J, Sant S. Biodegradable and biomimetic elastomeric scaffolds for tissue-engineered heart valves. Acta Biomater January 15, 2017;48:2−19.

[87] Milleret V, Hefti T, Hall H, Vogel V, Eberli D. Influence of the fiber diameter and surface roughness of electrospun vascular grafts on blood activation. Acta Biomater December 2012;8(12):4349−56.

[88] Liu R, Qin Y, Wang H, Zhao Y, Hu Z, Wang S. The in vivo blood compatibility of bioinspired small diameter vascular graft: effect of submicron longitudinally aligned topography. BMC Cardiovasc Disord October 1, 2013;13:79.

[89] Fan H, Chen P, Qi R, Zhai J, Wang J, Chen L, et al. Greatly improved blood compatibility by microscopic multiscale design of surface architectures. Small October 2, 2009;5(19):2144−8.

[90] Bourantas CV, Serruys PW. Evolution of transcatheter aortic valve replacement. Circ Res March 14, 2014;114(6):1037−51.

[91] Chambers J. Prosthetic heart valves. Int J Clin Pract October 1, 2014;68(10):1227−30.

[92] Leon MB, Smith CR, Mack MJ, Makkar RR, Svensson LG, Kodali SK, et al. Transcatheter or surgical aortic-valve replacement in intermediate-risk patients. N Engl J Med April 28, 2016;374(17):1609−20.

[93] Mack MJ, Leon MB, Smith CR, Miller DC, Moses JW, Tuzcu EM, et al. 5-year outcomes of transcatheter aortic valve replacement or surgical aortic valve replacement for high surgical risk patients with aortic stenosis (PARTNER 1): a randomised controlled trial. The Lancet June 20, 2015;385(9986):2477−84.

[94] Thourani VH, Kodali S, Makkar RR, Herrmann HC, Williams M, Babaliaros V, et al. Transcatheter aortic valve replacement versus surgical valve replacement in intermediaterisk patients: a propensity score analysis. The Lancet May 28, 2016;387(10034):2218−25.

[95] Thyregod HGH, Steinbrüchel DA, Ihlemann N, Nissen H, Kjeldsen BJ, Petursson P, et al. Transcatheter versus surgical aortic valve replacement in patients with severe aortic valve stenosis: 1-year results from the all-comers NOTION randomized clinical trial. J Am Coll Cardiol May 26, 2015;65(20):2184−94.

[96] Bapat V. Technical pitfalls and tips for the valve-in-valve procedure. Ann Cardiothorac Surg September 30, 2017;6(5):541−52.

[97] Dasi LP, Hatoum H, Kheradvar A, Zareian R, Alavi SH, Sun W, et al. On the mechanics of transcatheter aortic valve replacement. Ann Biomed Eng February 2017;45(2):310−31.

[98] Ye J, Cheung A, Yamashita M, Wood D, Peng D, Gao M, et al. Transcatheter aortic and mitral valve-in-valve implantation for failed surgical bioprosthetic valves: an 8-year single-center experience. JACC Cardiovasc Interv November 2015;8(13):1735−44.

[99] Cremer J, Martin M, Redl H, Bahrami S, Abraham C, Graeter T, et al. Systemic inflammatory response syndrome after cardiac operations. Ann Thorac Surg June 1996;61(6):1714−20.

[100] Sablotzki A, Friedrich I, Mühling J, Dehne MG, Spillner J, Silber RE, et al. The systemic inflammatory response syndrome following cardiac surgery: different expression of proinflammatory cytokines and procalcitonin in patients with and without multiorgan dysfunctions. Perfusion March 2002;17(2):103−9.

[101] Sinning J-M, Scheer A-C, Adenauer V, Ghanem A, Hammerstingl C, Schueler R, et al. Systemic inflammatory response syndrome predicts increased mortality in patients after transcatheter aortic valve implantation. Eur Heart J June 2012;33(12):1459−68.

[102] Nuis R-JM, Van Mieghem NM, Tzikas A, Piazza N, Otten AM, Cheng J, et al. Frequency, determinants, and prognostic effects of acute kidney injury and red blood cell transfusion in patients undergoing transcatheter aortic valve implantation. Cathet Cardiovasc Interv May 1, 2011;77(6):881−9.

[103] Izbicki G, Rudensky B, Na'amad M, Hershko C, Huerta M, Hersch M. Transfusion-related leukocytosis in critically ill patients. Crit Care Med February 2004;32(2):439−42.

[104] Ferrari E. Severe intraprosthetic regurgitation following trans-catheter aortic valve implantation-to crimp or not to crimp? this might be the problem. Eur J Cardio-Thorac Surg Off J Eur Assoc Cardio-Thorac Surg. April 2011;39(4):593−4.

[105] Khoffi F, Heim F, Chakfe N, Lee JT. Transcatheter fiber heart valve: effect of crimping on material performances. J Biomed Mater Res B Appl Biomater October 2015;103(7):1488−97.

[106]   Zegdi R, Bruneval P, Blanchard D, Fabiani J-N. Evidence of leaflet injury during percutaneous aortic valve deployment. Eur J Cardio-Thorac Surg Off J Eur Assoc Cardio- Thorac Surg. July 2011;40(1):257—9.

[107]   Alavi SH, Groves EM, Kheradvar A. The effects of transcatheter valve crimping on pericardial leaflets. Ann Thorac Surg April 2014;97(4):1260—6.

[108]   Latib A, Naganuma T, Abdel-Wahab M, Danenberg H, Cota L, Barbanti M, et al. Treatment and clinical outcomes of transcatheter heart valve thrombosis. Circ Cardiovasc Interv April 1, 2015;8(4): e001779.

[109]   Head SJ, Mokhles MM, Osnabrugge RLJ, Pibarot P, Mack MJ, Takkenberg JJM, et al. The impact of prosthesis—patient mismatch on long-term survival after aortic valve replacement: a systematic review and meta-analysis of 34 observational studies comprising 27 186 patients with 133 141 patient-years. Eur Heart J June 1, 2012;33(12):1518—29.

[110]   Weber B, Dijkman PE, Scherman J, Sanders B, Emmert MY, Grünenfelder J, et al. Off-the- shelf human decellularized tissue-engineered heart valves in a non-human primate model. Biomaterials October 2013;34(30):7269—80.

[111]   Driessen-Mol A, Emmert MY, Dijkman PE, Frese L, Sanders B, Weber B, et al. Transcatheter implantation of homologous "Off-the-Shelf" tissue-engineered heart valves with self-repair capacity: long-term functionality and rapid in vivo remodeling in sheep. J Am Coll Cardiol April 8, 2014; 63(13):1320—9.

[112]   Cianciulli TF, Saccheri MC, Lax JA, Guidoin R, Zhang Z, Guerra JE, et al. Intermittent acute aortic regurgitation of a mechanical bileaflet aortic valve prosthesis: diagnosis and clinical implications. Eur J Echocardiogr J Work Group Echocardiogr Eur Soc Cardiol May 2009;10(3):446—9.

[113]   Teshima H, Hayashida N, Yano H, Nishimi M, Tayama E, Fukunaga S, et al. Obstruction of St Jude medical valves in the aortic position: histology and immunohistochemistry of pannus. J Thorac Cardiovasc Surg August 2003;126(2):401—7.

[114]   Gellis L, Baird CW, Emani S, Borisuk M, Gauvreau K, Padera RF, et al. Morphologic and histologic findings in bioprosthetic valves explanted from the mitral position in children younger than 5 years of age. J Thorac Cardiovasc Surg February 2018;155(2):746—52.

[115]   Yahagi K, Torii S, Ladich E, Kutys R, Romero ME, Mori H, et al. Pathology of self-expanding transcatheter aortic valves: findings from the corevalve US pivotal trials. Catheter Cardiovasc Interv Off J Soc Card Angiogr Interv September 12, 2017.

CHAPTER 13

Polymeric heart valves

Megan Heitkemper, Lakshmi Prasad Dasi
Department of Biomedical Engineering, The Ohio State University, Columbus, OH, United States

Contents

13.1 Introduction

13.1.1 Scope

The main focus of this chapter will be to introduce the field of polymeric heart valve (PHV) engineering and to provide a brief overview of the current investigational technology and the challenges facing further development of PHVs. The discussion will be limited to flexible leaflet PHVs, intended for use in the aortic and mitral positions.

Principles of Heart Valve Engineering
ISBN 978-0-12-814661-3, https://doi.org/10.1016/B978-0-12-814661-3.00013-7

13.1.2 Need

Heart disease is the leading cause of death in the United States, killing more than 600,000 Americans each year [1]. With life expectancy on the rise and an aging population, the number of valve replacement surgeries can be expected to increase at unprecedented rates. Currently, over 290,000 heart valve replacements are performed annually worldwide and that number is estimated to triple to over 850,000 by 2050 [2,3].

Medical intervention for valvular heart disease is often a surgical valve replacement, where valve types used are either mechanical or bioprosthetic. Mechanical valves are often chosen for younger patients because of their superior durability, but they require lifelong anticoagulation therapy that can decrease patient quality of life and increase the overall risk of bleeding. Bioprosthetic valves, oppositely, do not require anticoagulation therapy, but have poor durability, and often need to be replaced within the lifetime of the patient. One alternative intervention for patients too weak or sick to undergo a surgical heart valve replacement is the less invasive transcatheter approach, where a prosthetic valve is crimped to a smaller size, inserted by way of patient vasculature, and expanded into the appropriate position. While transcatheter aortic valve replacement (TAVR) is becoming a routine procedure for high and intermediate-risk patients, the efficacy of transcatheter mitral valve replacement is still under investigation [4], although transcatheter mitral valve-in-valve and valve-in-ring have been performed successfully [5].

In addition to expanding valve replacement therapy to patients not able to undergo the traditional replacement surgery [6], TAVR has demonstrated faster patient recovery and reduced total hospital costs, with exception of the device cost, compared with patients implanted with surgical aortic valves (SAVs) [7]. While this technology provides desirable outcomes for many patients, the current transcatheter aortic valves (TAVs) are bioprosthetic and therefore are less durable [8–11] and have increased risk of thromboembolic events [12–14]. Furthermore, the manufacturing, quality control, and storage requirements for the fixed tissue components of these valves are not cost-effective, limiting TAV use to the high-risk and elderly populations, solely in industrialized nations. Similar concerns of durability, thromboembolism, and manufacturing costs should be expected as transcatheter mitral valve technologies progress.

Polymeric valves have emerged in the heart valve engineering field in the pursuit to provide a prosthetic valve that is as durable and as long-lasting as a mechanical valve, without the need for anticoagulation therapy. While polymeric prosthetic valves could change the standard of care for patients eligible for surgical valve replacements, their true potential lies within the transcatheter valve replacement market, where a more durable, cost-effective valve would allow for expansion of the technology to intermediate and low-risk patients, in both industrialized and developing nations.

13.2 History of polymeric valves

Flexible leaflet PHVs were first introduced in the late 1950s, with contributions from Akutsu [15], Berge [16], Braunwald [17], and Roe [18]. The first known mitral valve implantation occurred in 1960 by Braunwald [19] in which plaster casts of explanted human mitral valves were used to make molds for liquid polyurethane (PU) (Fig. 13.1A). In 1969, Roe [20] reported the first known polymeric aortic valve implantation, made from a silicone material (Fig. 13.1B). The progression of development of PHVs continued slowly following these trials, partially because of the success of the Starr–Edwards ball-and-cage valve [21] and partially due to the evolution of percutaneous mitral valve repair technologies [22]. From their start in the 1960s using silicone and PU, various designs of flexible leaflet PHVs have been developed from polymeric materials including polytetrafluoroethylene (PTFE), various PUs including polycarbonate urethane (PCU) and polyether urethane (PEU), polyvinyl alcohol (PVA), polydimethylsiloxane–polyhexamethylene oxide (PDMS-PHMO), polyhedral oligomeric silsesquioxane (POSS)–PCU, and poly(styrene-block-isobutylene-block-styrene) (SIBS). A summary of the outcomes of these materials for use in PHVs, including the advantages and shortcomings of each, can be found in Table 13.1. Each of these attempted to produce a valve with significant improvement over mechanical and bioprosthetic options. In 1968, a list of "Nine Commandments" for the development of a prosthetic heart valve was issued by Edwards laboratories [23], an adaption from Dwight Harken's famous "10 Commandments" in 1967 [24].

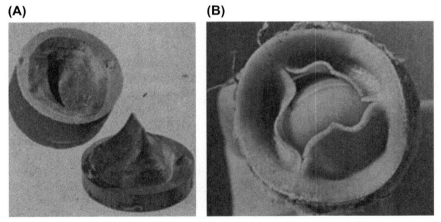

(A)   **(B)**

Figure 13.1 (A) Polyurethane bileaflet mitral valve by Braunwald et al. [19]; (B) aortic trileaflet valve made from silicone material by Roe et al. [20].

Table 13.1 Common polymers used in polymeric heart valve engineering.

Material	Advantages	Disadvantages	References
PUs	Viscoelasticity, resistance to tearing	Thrombosis, calcification, hydrolysis	35,19,44–51
Silicone	Biocompatability, elastic/flexural properties	Durability/tearing/stiffening, thrombus formation, fluid absorption	29,30,31,32,33,34
PTFEs	Hydrophobic, low coefficient of friction, low surface tension	Calcification, leaflet stiffening, instance of thromboembolism	26,30,35
PCU	Resistance to oxidation and hydrolysis	Calcification	51,52
PEU	Reduction in calcification and thromboembolic events, viscoelasticity	Low resistance to oxidation and hydrolysis	46,59
PVA	Nontoxic, biocompatible, biostable	Potential for foreign body response	56–59,62
HA-LLDPE	Biocompatible, high tensile and tear strength, reduced thromboembolic potential	None yet reported	63–66
XSIBS	Biostable, resistance to hydrolysis, oxidation, enzymatic activity	None yet reported	25,64,67

The "Nine Commandments," still of major importance in the engineering process today, are as follows:

1. Embolism Prevention
2. Durability
3. Ease and Security of Attachment
4. Preservation of Surrounding Tissue Function
5. Reduction of Turbulence
6. Reduction of Blood Trauma
7. Reduction of Noise
8. Use of Materials Compatible with Blood and Tissue
9. Development of Methods of Storage and Sterilization

While the progress in PHV engineering that meets these many requirements has been slow, most often because of limited in vivo durability as a result of material degradation, thrombosis, and calcification [25−28], the field is constantly expanding. Over the past two decades, there has been remarkable progress in polymer synthesis methods resulting in improved material properties [26], which have restored hope to the PHV engineering community.

13.3 Design considerations and challenges

Approval for use in humans is unlikely unless the next generation of PHVs meets or exceeds the functional durability and hemocompatibility of bioprosthetic heart valves currently on the market [25]. Therefore, the material, surface modifications, and geometric design of emerging technologies are extremely important. Design considerations for PHVs include the need for sufficient effective orifice area, jet velocities and pressure gradients within a normal physiological range, and minimal regurgitation, damage to blood cells, and thrombogenic potential [25,26]. Additional design considerations include how the PHV will anchor in the native environment, leaflet coaptation, commissure gap, leaflet thickness and geometry, biostability, and peak stresses on the valve components [25,26]. The following sections will introduce the most common materials, surface modifications, geometries, and manufacturing techniques that are considered for use in current experimental PHVs.

13.3.1 Material

13.3.1.1 Polysiloxanes

The earliest known implanted aortic valve was made of polysiloxane [20], a polymer with a backbone consisting of silicone and oxygen atoms [29,30]. The main advantages of this material are its elastic and flexural properties, as well as its good biocompatibility [31], while its greatest failure is through limited durability and tearing. In 1973, Mohri et al. reported good hemodynamic performance of silicone rubber, but was concerned about fluid absorption and thrombus formation [32,33]. Additional discussion around the importance of design and manufacturing for the consistency in durability, in conjunction with the material properties of a polymer, was addressed [32]. Chetta and Lloyd described a second mode of failure for silicone rubber, where the valve leaflets became stiff and thickened, eventually failing to open [33,34].

13.3.1.2 Polytetrafluoroethylene (PTFE)/expanded PTFE

PTFE and expanded polytetrafluoroethylene (eTPFE) commercially known at Teflon and Gore-Tex, respectively, are highly crystalline, hydrophobic, and highly stable polymers. Their main advantages for use in PHVs include good hemodynamic properties, mainly because of a low coefficient of friction, inertness, and low surface tension [26,30,35]. The use of PTFE and ePTFE for PHVs is limited by repetitive instances of calcification and leaflet stiffening, in addition to a low resistance to thromboembolism [26,30,35]. Quintessenza et al. have shown successful intermediate use of nonporous 0.1 mm PTFE for prosthetic bicuspid pulmonary valve implantation in patients with pulmonary insufficiency and/or pulmonary stenosis [36]. The use of nonporous 0.1 mm PTFE was shown to limit cellular ingrowth and thickening, improving the leaflet mobility and pliability [36].

13.3.1.3 Polyurethanes

PUs are among the oldest tested and most commonly used polymers chosen for use in investigational PHVs, as well as for all blood–contacting medical devices [35]. PUs are considered segmented block copolymers, containing soft and hard segments. A superior advantage of PUs is the capability to manipulate their mechanical and hemodynamic functionality by varying the type and/or molecular weight of the soft segment and coupling agents [35]. Since the initial use of PUs for investigational mitral valves in the early 1960s [19], there have been many varieties of PU materials, with soft segments containing polyester, polyether, polycarbonate, and polysiloxane [37,38]. The chemical differences between the soft and hard chains of the polymer are what give PUs the exceptional mechanical properties and biocompatibility that make the material so desirable, especially for biomedical applications [39,40]. Studies by Lyman et al. [41], Zia et al. [42], and Zu et al. [43] show the feasibility of enhancing the biocompatibility of PUs through minimal changes to the chemical structures. While earlier PUs have been primarily plagued by thrombosis and calcification [15,44−49], more recent trials have found that different grades of PEUs and polyether urethane ureas demonstrated no evidence of thrombogenic events or calcification [46,48]. Additional variations of PUs including POSS-PCUs [38,50], thermoplastics [30], and PDMS-PUs [51,52] have shown biocompatiblity and hemodynamic promise. The remaining challenge to realizing PUs as ideal polymers for PHVs is their long-term in vivo biostability [35]. There are multiple modes of biodegradation of PUs, including hydrolysis, oxidative degradation, enzymatic degradation, surface cracking, environment stress cracking, and calcification [53]. As more advances in the understanding of these modes of degradation are made, the bio-resistant properties of PUs can be tailored for use in PHVs [35].

13.3.1.4 Polyvinyl alcohol

PVA, a hydrophilic synthetic polymer, has been of interest in recent years as a potential material for both surgical and transcatheter PHV applications. The polymer is nontoxic, biocompatible, biostable [54−57], and has exhibited good mechanical properties [58]. Jiang et al. have shown that polyvinyl alcohol cryogels (PVA-C) have behavior similar to soft tissue [57,58], while Mohammadi has shown that polyvinyl alcohol-bacterial cellulose (PVA-BC)−based hydrogels have similar mechanical properties and anisotropic behavior as native porcine aortic valve leaflets [57,59]. One limitation to using PVA for PHV applications is that cell adhesion is not possible because of the hydrophilic nature of the polymer, which can cause a foreign body response [60]. Surface modifications to minimize this response by promoting endothelial cell attachment as described by Nuttleman et al. [56] will be discussed in Section 2.2.

13.3.1.5 Linear low-density polyethylene

Linear low-density polyethylene (LLDPE) is a hydrophillic polymer with a high tensile and tear strength and relatively low bending stiffness [61,62]. An interpenetrating network (IPN) between hyaluronan (HA) and LLDPE has been shown to increase the material strength and durability, while providing the added benefit of increased biocompatibility [62,63], making HA-LLDPE a highly attractive candidate for PHVs. The results of a study by Simon-Walker et al. demonstrated that IPNs of HA-LLDPE are nontoxic and reduce the thrombogenic potential of the material as compared to untreated LLDPE [64].

13.3.1.6 Poly(styrene-block-isobutylene-block-styrene)

SIBS by Boston Scientific [65], a thermoplastic elastomer, and the more recent polyolefin thermoset elastomer xSIBS by Innovia LLC [62] have been of recent interest for use in PHVs mainly because of their superior biostability and resistance to in vivo degradation [25]. These desirable properties are results of the polymer containing no reactive pendant groups, virtually eliminating any possibility of degradation because of hydrolysis, oxidation, or enzymatic pathways [25].

13.3.2 Surface modifications

As it has been described in the previous section, many polymers have been selected for use in an experimental PHV for one of their promising material properties, such as superior durability, and have failed because of another. While some of the failures have led researchers to discontinue pursuing a polymeric material, such was the case for the early polysiloxanes; it is becoming increasingly popular to engineer the surface of a biomaterial to enhance its biocompatibility. As biomaterial hemocompatibility relies heavily on the blood interaction with the surface, it is logical to move toward surface modifications that do not interfere with the bulk properties of a material [26,66].

Platelet adhesion and activation in response to injury in a blood vessel is a mechanism to minimize bleeding, essentially by covering the damaged portion and recruiting more platelets, initiating fibrin formation, and eventually developing a thrombus [67]. While this mechanism is important in a diseased vessel, risk of thromboembolism in response to an implanted biomaterial is of major concern especially for cardiac applications. As platelet activation is contact and flow-induced, the design and optimization of the mechanical function and the surface properties of PHVs is of extreme importance [25].

Surface characteristics including hydrophobicity/hydrophilicity, surface charge and surface free energy, and the topography of the surface all influence initial blood material interactions including protein absorption and denaturation [25,26]. Research on the effectiveness of surface modifications, including plasma immersion ion implantation [66,68], cholesterol modification [69], and peptide modification [70], have shown to enhance the affinity of surfaces to endothelial cells, which can protect the valve experiencing a foreign body response from the immune system. Additional surface modifications, including the creation of topographic features that attract specific cells as shown by de Mel et al. [71], have been shown to increase the hemocompatibility of a

Table 13.2 Surface modifications for polymeric heart valve engineering.

Modification	Advantages	Reference
PIII	Converts hydrophobic polymers to hydrophilic and improves biocompatibility	69,71
Cholesterol-modified PU	Increases surface energy and increases endothelial cell attachment and retention	72
Nanotopographic surface	Mimics natural extracellular matrix, stimulates cell adhesion and proliferization, and promotes endothelialization	26,74,75
HEPB-bound PUs	Reduces surface degradation, does not affect material properties, and decreases calcium permeation	51,76
RGD incorporation	Promotes endothelialization	63,73

PIII—plasma-immersion ion implantation, peptide mod 73; HEPB—hydroxyethane biophosphoric acid; RGD—R: arginine; G: glycine; D: aspartic acid.

material [26]. In addition to reducing thromboembolic risk, some surface modifications aim to reduce the amount of calcification deposited on the valve leaflets to improve the PHV durability and function. Among these, a study by Joshi et al. demonstrated that 2-hydroxyethane bisphosphonic acid (HEBP)—bound PEUs may serve as a valid calcification–resistant material for use in PHVs [72]. Table 13.2 summarizes popular surface modifications used for the improvement of biocompatibility of common biomaterials.

13.3.2.1 Geometry

Arguably one of the most challenging aspects of designing a viable PHV is optimizing the leaflet geometry. Added freedom in this phase of PHV design comes from the seemingly limitless potential geometric shapes and leaflet thicknesses that were not previously possible with bioprosthetic heart valves, which can be both an exciting and daunting challenge for engineers. As leaflet geometry is essential to the function of a PHV, including its durability, hemodynamic function, and biocompatibility, this is no small task. An idealized PHV geometry would allow for maximal effective orifice area throughout systole, sufficient flexibility to minimize resistance to forward flow [26], fully coapting leaflets that minimize backflow, and minimized and equal distribution of leaflet stresses to minimize blood trauma and maximize valve durability [25].

While the relative ease of manufacturing polymeric leaflets has provided the opportunity to design more complex valve designs, it is still extremely complex to mimic the anatomy of natural valves. For this reason, many designs have been proposed and tested

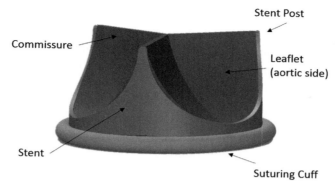

Figure 13.2 Detailed geometric features of a trileaflet prosthetic heart valve.

including single leaflet [73], bileaflet (including original bileaflet designs by Braunwald et al. [19] and other more successful trials [74,75]), and quadrileaflet designs [76]. However, the majority of modern PHV research is focused on trileaflet valves [26] because of studies that have shown superior mechanical efficiency, greater opening area [77], and improved stress distribution [78]. Detailed geometric features of a modeled trileaflet heart valve can be seen in Fig. 13.2.

There has been much debate over the shape of the native aortic leaflets, with descriptions of semilunar, sigmoid, paraboloids of revolution, and elliptical paraboloid shapes suggested, among others [30]. While some groups have put forth great effort into modeling and recreating realistic geometries, others have tossed aside the idea and have solely focused on recreating or even improving natural valve function, rather than form. In addition to leaflet shape, other important considerations include which position the valve will be manufactured in (closed, open, partially open), the leaflet thickness, and how the leaflets will be assembled to a stent frame (adhesive, stentless, or sutures). A significant amount of research has gone into optimizing leaflet thickness alone [45,47,79]. To achieve good hemodynamic performance including transvalvular pressure gradients, energy losses, and sufficient durability free from tearing, it has been shown that for PCU valves, the durability for a thickness between 100 and 300 microns is between 600 million and 1 billion cycles [80].

13.3.2.2 Manufacturing

Various methods to manufacture PHVs are commonly used, the choice of which most heavily depends on the material properties of the polymer. While the choice in manufacturing method is somewhat limited by whether the material is sufficiently soluble with heat or solvent processing, or if it is a thermosetting plastic [81], it should not be overlooked as a crucial factor to PHV performance [26,82].

13.3.3 Dip casting

Dip casting or dip coating is a manufacturing method commonly used for silicones and PUs. The technique requires a custom-shaped mandrel, in the shape of valve leaflets, to be dipped into a polymer solution. The desired thickness of the leaflets can be attained through subsequent dips, allowing the solution to air dry in between each. Dip casting allows for complex leaflet geometries and more precision with less cost than in conventional molding methods. Additionally, dip casting can provide a continuous integration of the leaflets and supporting frame as shown by Leat et al. [82]. One disadvantage of this manufacturing method is that the distribution of leaflet thickness is not easily controlled, leading to difficulty with reproducibility [26].

13.3.4 Film fabrication

In film fabrication, the leaflets and frame are manufactured separately. Leaflets are cut from polymer films and bonded to the frame [26]. In some cases, heat treatments can be applied to further shape the valve leaflets once they are attached to the frame [82]. While this method allows for increased control of leaflet geometry and ease of prototyping, its main drawback is the potential for decreased durability because of the boundary between the leaflets and frame.

13.3.5 Cavity and injection molding

Cavity molding is a process where a liquefied polymer is poured into a static mold and sealed, and then undergoes freeze thaw cycles in a water bath until a thin polymer is formed [58]. Similarly, an injection molding machine is used to fabricate valve leaflets through high-pressure injection of molten polymer, followed by hot and cold water baths [26,83]. Cavity and injection molding have increased precision and the ability to tailor the elastic properties of some polymers through the number and rate of freeze thaw cycles [84,85], but not without increased cost [83].

13.3.6 Three-dimensional printing

With the rapid advances in three-dimensional (3D) printing technology making printing some polymeric materials cheaper and faster than ever, it is no surprise that the technology has begun to take root in polymeric valve engineering. The ease of printing from computer-aided designs has the potential to move the manufacturing process towards leaflets enhanced with 3D surface structures and eventually fully 3D leaflets mimicking the native structures.

13.4 Investigational valves

The material, surface modifications, manufacturing methods, and geometries of three promising investigational valves will be discussed in detail below.

To begin, there have been promising material results from IPNs of HA and LLDPE by Prawel et al. [62]. The LLDPE films were blow molded by Flex–Pack Engineering Inc. (Union-town, OH) from Dowlex 2056 resin and had measured average thickness of 0.08 mm [62]. A swelling method was used to introduce HA into the LLDPE [62]. Characterization of the material showed that HA concentration affects clotting, significantly decreasing clotting as compared with plain LLDPE sheets. Additionally, the HA-LLDPE showed decreased platelet adhesion and activation as compared with traditional bioprosthetic and mechanical HV materials [62]. Yousefi et al. have manufactured trileaflet surgical aortic heart valves from HA-LLDPE sheets [86] and have described extensively the impact of arched leaflet geometries and stent profile on their hemodynamic performance and durability. Further work has been done to manufacture a trileaflet HA-LLDPE balloon-expandable transcatheter aortic valve (Fig. 13.3), and animal trials and accelerated durability testing are currently underway.

Claiborne et al. describe novel trileaflet surgical valves made from xSIBS [87], manufactured by custom compression molding. They describe an iterative design process, which lead them to choose hemispherical leaflet geometry, tapered leaflet thickness, and smooth stent edges, which significantly reduced the stresses in the valve leaflets from their previous models, which was a primary concern. In vitro evaluation of this valve showed promising hemodynamic performance, including low regurgitant fraction in comparison to commercial tissue valves, but a decreased effective orifice area and a higher peak velocity were observed [87]. The group has since designed a transcatheter aortic valve using the same polymer, xSIBS [25], as shown in Fig. 13.4.

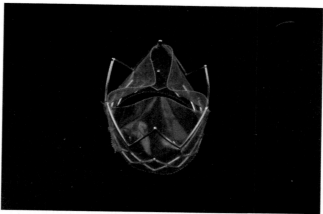

Figure 13.3 A balloon-expandable HA-LLDPE transcatheter valve developed by Dasi Cardiovascular Biofluid Mechanics Lab.

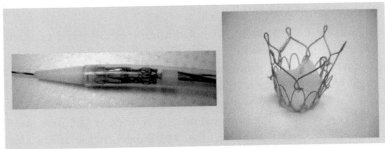

Figure 13.4 A self-expanding SIBS-Dacron-based transcatheter valve developed by Blustien Biofluids Research Group [25].

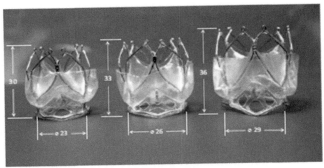

Figure 13.5 Self-expanding TRISKELE valve in sizes 23(left), 26 (center), 29 (right) manufactured from POSS-PCU leaflets [88].

In 2016, Rahmani et al. introduced the TRISKELE, an investigational trileaflet transcatheter aortic valve [88]. The valve, as shown in Fig. 13.5, was manufactured by automated dip coating of POSS-PCU, which had been previously validated in vitro for its hemocompatibility, antithrombogenicity, biostability, and resistance to calcification [88]. The frame design, made from self-expandable nitinol wire, was numerically optimized to incur minimal stress and improve anchoring at physiological pressure loads. Additionally, the group aimed to attain a single curvature in the open and closed positions to minimize energy dissipation. The TRISKELE valve demonstrated significant reduction in paravalvular leakage in comparison to two commercially available valves and has shown comparable hemodynamic performance. The valve is undergoing preclinical studies to investigate its durability and in vivo function.

13.5 Summary and conclusions

In this review, we have aimed to recognize the major contributions toward the goal of realizing a clinically successful PHV while also describing the challenges that have led many prototypes to fail in preclinical stages, namely degradation, calcification, and risk of thromboembolism. Despite their origins dating back to the late 1950s, there has not yet been any PHV that has proven to be better or even as good as the currently available bioprosthetic or mechanical valves. As the current valves improve, the quality and durability standards required to bring a PHV to market will contibue to rise. Even so, research aimed at developing PHVs is ongoing and many researchers are hopeful that their designs will outperform the existing technology.

The most promising avenue for the realization of PHV technology is within the TAV sector. Here, a prosthetic valve promising both superior durability and hemocompatibility in comparison to the fixed tissue components currently used would allow transcatheter aortic valve replacement to become a routine procedure for all patients suffering from aortic valve stenosis regardless of their eligibility for the traditional and more invasive SAV replacement. Additionally, polymeric transcatheter aortic valves have reduced costs associated with their manufacture and storage as compared to their bioprosthetic counterparts, which could reduce the cost of heart valve replacement therapy as a whole and bring this technology to developing and industrialized nations.

References

[1] Minino AM, Murphy SL, Xu J, Kochanek KD. Deaths: final data for 2008. Natl Vital Stat Rep 2011; 59:1−126.

[2] Yacoub N, Takkenberg J. Will heart valve tissue engineering change the world? Nat Clin Pract Cardiovasc Med 2005;2:60−1.

[3] Black MM, Drury PJ. Mechanical and other problems of artificial valves. Curr Top Pathol 1994;86. Springer-Verlag Berlin Heidelberg.

[4] Mack M, Smith R. TMVR with artificial cords in the treatment of mitral regurgitation: more questions than answers. J Am Coll Cardiol 2018;71:37−9.

[5] Evans AS, Weiner M, Patel PA, Baron EL, Gutsche JT, Jayaraman A, Renew JR, Martin AK, Fritz AV, Gordon EK, Riha H, Patel S, Ghadimi K, Guelaff E, Feinman JW, Dashell J, Munroe R, Lauter D, Weiss SJ, Silvay G, Augoustides JG, Ramakrishna H. The year in cardiothoracic and vascular anesthesia: selected highlights from 2017. J Cardiothorac Vasc Anesth 2018;32:1−13.

[6] Makkar RR, Fontana GP, Jilaihawi H, Kapadia S, Pichard AD, Douglas PS, Thourani VH, Babaliaros VC, Webb JG, Herrmann HC, Bavaria JE, Kodali S, Brown DL, Bowers B, Dewey TM, Svensson LG, Tuzcu M, Moses JW, Williams MR, Siegel RJ, Akin JJ, Anderson WN, Pocock S, Smith CR, Leon MB, Investigators PT. Transcatheter aortic-valve replacement for inoperable severe aortic stenosis. N Engl J Med 2012;366:1696−704.

[7] Reynolds MR, Magnuson EA, Lei Y, Wang K, Vilain K, Li H, Walczak J, Pinto DS, Thourani VH, Svensson LG, Mack MJ, Miller DC, Satler LE, Bavaria J, Smith CR, Leon MB, Cohen DJ. Cost-effectiveness of transcatheter aortic valve replacement compared with surgical aortic valve replacement in high-risk patients with severe aortic stenosis: results of the PARTNER (placement of aortic transcatheter valves) trial (cohort A). J Am Coll Cardiol 2012;60:2683−92.

[8] Dvir D. First look at long-term durability of transcatheter heart valves: assessment of valve function up to 10 years after implantation. In: Paper presented at: EuroPCR 2016; 2016 [Paris, France].

[9] Abbasi M, Azadani AN. The synergistic impact of eccentric and incomplete stent deployment on transcatheter aortic valve leaflet stress distribution. J Am Coll Cardiol 2015;66:B253−4.

[10] Sun W, Li KW, Sirois E. Simulated elliptical bioprosthetic valve deformation: implications for asymmetric transcatheter valve deployment. J Biomech 2010;43:3085−90.

[11] Dumesnil JG, Pibarot P. Prosthesis-patient mismatch: an update. Curr Cardiol Rep 2011;13:250—7.

[12] Makkar RR, Fontana G, Jilaihawi H, Chakravarty T, Kofoed KF, de Backer O, Asch FM, Ruiz CE, Olsen NT, Trento A, Friedman J, Berman D, Cheng W, Kashif M, Jelnin V, Kliger CA, Guo H, Pichard AD, Weissman NJ, Kapadia S, Manasse E, Bhatt DL, Leon MB, Sondergaard L. Possible subclinical leaflet thrombosis in bioprosthetic aortic valves. N Engl J Med 2015;373:2015—24.

[13] Latib A, Naganuma T, Abdel-Wahab M, Danenberg H, Cota L, Barbanti M, Baumgartner H, Finkelstein A, Legrand V, de Lezo JS, Kefer J, Messika-Zeitoun D, Richardt G, Stabile E, Kaleschke G, Vahanian A, Laborde JC, Leon MB, Webb JG, Panoulas VF, Maisano F, Alfieri O, Colombo A. Treatment and clinical outcomes of transcatheter heart valve thrombosis. Circu Cardiovasc Interv 2015;8:12.

[14] Mylotte D, Piazza N. Transcatheter aortic valve replacement failure deja vu ou jamais vu? Circ Cardiovasc Interv 2015;8:4.

[15] Akutsu T, D B, Kolff WJ. Polyurethane artificial heart valves in animals. J Appl Physiol 1959;14: 1045—8.

[16] Berge TL. A flexible cardiac valve prosthesis; preliminary report on the development of an experimental valvular prosthesis. Arch Chir Neerl 1958;10:26—33.

[17] Braunwald NS. It will work: the first successful mitral valve replacement. Ann Thorac Surg 1989;48: S1—3.

[18] Roe BB, Moore D. Design and fabrication of prosthetic valves. Exp Med Surg 1958;16:177—82.

[19] Braunwald NS, Cooper T, Morrow AG. Complete replacement of the mitral valve. Successful clinical application of a flexible polyurethane prosthesis. J Thorac Cardiovasc Surg 1960;40:1—11.

[20] Roe BB. Late follow-up studies on flexible leaflet prosthetic valves. J Thorac Cardiovasc Surg 1969; 58:59—61.

[21] Matthews AM. The development of the Starr-Edwards heart valve. Tex Heart Inst J 1998;25:282—93.

[22] Maisano F, La Canna G, Colombo A, Alfieri O. The evolution from surgery to percutaneous mitral valve interventions: the role of the edge-to-edge technique. J Am Coll Cardiol 2011;58:2174—82.

[23] Pierie WR, Hancock WD, Koorajian S, Starr A. Materials and heart valve prostheses. Ann N Y Acad Sci 1968;146:345—59.

[24] Harken DE, Curtis LE. Heart surgery—legend and a long look. Am J Cardiol 1967;19:393—400.

[25] Claiborne TE. Polymeric trileaflet prosthetic heart valves: evolution and path to clinical reality. Expert Rev Med Devices 577—594.

[26] Ghanbari H, Viatge H, Kidane AG, Burriesci G, Tavakoli M, Seifalian AM. Polymeric heart valves: new materials, emerging hopes. Trends Biotechnol 2009;27:359—67.

[27] Kütting M, Roggenkamp J, Urban U, Schmitz-Rode T, Steinseifer U. Polyurethane heart valves: past, present and future. Expert Rev Med Devices 2011;8:227—33.

[28] Kidane AG, Burriesci G, Cornejo P, Dooley A, Sarkar S, Bonhoeffer P, Edirisinghe M, Seifalian AM. Current developments and future prospects for heart valve replacement therapy. J Biomed Mater Res B Appl Biomater 2009;88B:290—303.

[29] Kheradvar A, Groves EM, Dasi LP, Alavi SH, Tranquillo R, Grande-Allen KJ, Simmons CA, Griffith B, Falahatpisheh A, Goergen CJ, Mofrad MR, Baaijens F, Little SH, Canic S. Emerging trends in heart valve engineering: Part I. Solutions for future. Ann Biomed Eng 2015;43:833—43.

[30] Bezuidenhout D, Zilla P. Flexible leaflet polymeric heart valves. 2014.

[31] Colas A, Curtis J. Silicone biomaterials: history and chemistry. 2004.

[32] Mori H, Hessel 2nd EA, Nelson RJ, Anderson HN, Dillard DH, Merendino KA. Design and durability test of Silastic trileaflet aortic valve prostheses. J Thorac Cardiovasc Surg 1973;65:576—82.

[33] Kiraly R, Yozu R, Hillegass D, Harasaki H, Murabayashi S, Snow J, Nose Y. Hexsyn trileaflet valve: application to temporary blood pumps. Artif Organs 1982;6:190—7.

[34] Chetta GE, Lloyd JR. The design, fabrication and evaluation of a trileaflet prosthetic heart valve. J Biomech Eng 1980;102:34—41.

[35] He W, Benson R. 8 - polymeric biomaterials A2 — kutz, myer applied plastics engineering handbook. 2nd ed. William Andrew Publishing; 2017. p. 145—64.

[36] Quintessenza JA, Jacobs JP, Chai PJ, Morell VO, Lindberg H. Polytetrafluoroethylene bicuspid pulmonary valve implantation: experience with 126 patients. World J Pediatr Congenit Heart Surg 2010; 1:20–7.

[37] Coury A, Slaikeu P, Cahalan P, Stokes K. Medical application of implantable polyurethanes: current issues. Prog Rubber Plast Technol 1987;3:24–37.

[38] Kidane AG, Burriesci G, Edirisinghe M, Ghanbari H, Bonhoeffer P, Seifalian AM. A novel nanocomposite polymer for development of synthetic heart valve leaflets. Acta Biomater 2009;5:2409–17.

[39] Blackwell J, Gardner KH. Structure of the hard segments in polyurethane elastomers. Polymer 1979; 20:13–7.

[40] Blackwell J, Lee CD. Hard-segment polymorphism in MDI/diol-based polyurethane elastomers. J Polym Sci Polym Phys Ed 1984;22:759–72.

[41] Lyman DJ, Brash JL, Chaikin SW, Klein KG, Carini M. The effect of chemical structure and surface properties of synthetic polymers on the coagulation of blood. II. Protein and platelet interaction with polymer surfaces. Trans Am Soc Artif Intern Organs 1968;14:250–5.

[42] Zia KM, Barikani M, Bhatti IA, Zuber M, Bhatti HN. Synthesis and characterization of novel, biodegradable, thermally stable chitin-based polyurethane elastomers. J Appl Polym Sci 2008;110:769–76.

[43] Xu D, Meng Z, Han M, Xi K, Jia X, Yu X, Chen Q. Novel blood-compatible waterborne polyurethane using chitosan as an extender. J Appl Polym Sci 2008;109:240–6.

[44] Bernacca GM, Mackay TG, Wilkinson R, Wheatley DJ. Calcification and fatigue failure in a polyurethane heart valve. Biomaterials 1995;16:279–85.

[45] Bernacca GM, O'Connor B, Williams DF, Wheatley DJ. Hydrodynamic function of polyurethane prosthetic heart valves: influences of Young's modulus and leaflet thickness. Biomaterials 2002;23: 45–50.

[46] Bernacca GM, Straub I, Wheatley DJ. Mechanical and morphological study of biostable polyurethane heart valve leaflets explanted from sheep. J Biomed Mater Res 2002;61:138–45.

[47] Mackay TG, Wheatley DJ, Bernacca GM, Fisher AC, Hindle CS. New polyurethane heart valve prosthesis: design, manufacture and evaluation. Biomaterials 1996;17:1857–63.

[48] Wheatley DJ, Bernacca GM, Tolland MM, O'Connor B, Fisher J, Williams DF. Hydrodynamic function of a biostable polyurethane flexible heart valve after six months in sheep. Int J Artif Organs 2001; 24:95–101.

[49] Wheatley DJ, Raco L, Bernacca GM, Sim I, Belcher PR, Boyd JS. Polyurethane: material for the next generation of heart valve prostheses? Eur J Cardiothorac Surg 2000;17:440–8.

[50] Kannan RY, Salacinski HJ, Butler PE, Seifalian AM. Polyhedral oligomeric silsesquioxane Nanocomposites: the next generation material for biomedical applications. Acc Chem Res 2005;38:879–84.

[51] Dabagh M, Abdekhodaie MJ, Khorasani MT. Effects of polydimethylsiloxane grafting on the calcification, physical properties, and biocompatibility of polyurethane in a heart valve. J Appl Polym Sci 2005;98:758–66.

[52] Simmons A, Hyvarinen J, Odell RA, Martin DJ, Gunatillake PA, Noble KR, Poole-Warren LA. Long-term in vivo biostability of poly(dimethylsiloxane)/poly(hexamethylene oxide) mixed macrodiol-based polyurethane elastomers. Biomaterials 2004;25:4887–900.

[53] Santerre JP, Woodhouse K, Laroche G, Labow RS. Understanding the biodegradation of polyurethanes: from classical implants to tissue engineering materials. Biomaterials 2005;26:7457–70.

[54] Nuttelman CR, Henry SM, Anseth KS. Synthesis and characterization of photocrosslinkable, degradable poly(vinyl alcohol)-based tissue engineering scaffolds. Biomaterials 2002;23:3617–26.

[55] Alves MH, Jensen BE, Smith AA, Zelikin AN. Poly(vinyl alcohol) physical hydrogels: new vista on a long serving biomaterial. Macromol Biosci 2011;11:1293–313.

[56] Nuttelman CR, Mortisen DJ, Henry SM, Anseth KS. Attachment of fibronectin to poly(vinyl alcohol) hydrogels promotes NIH3T3 cell adhesion, proliferation, and migration. J Biomed Mater Res 2001;57:217–23.

[57] Nachlas ALY, Li S, Davis ME. Developing a clinically relevant tissue engineered heart valve—a review of current approaches. Adv Healthc Mater 2017;6:1700918. n/a.

[58] Jiang H, Campbell G, Boughner D, Wan W-K, Quantz M. Design and manufacture of a polyvinyl alcohol (PVA) cryogel tri-leaflet heart valve prosthesis. Med Eng Phys 2004;26:269–77.

[59] Mohammadi H. Nanocomposite biomaterial mimicking aortic heart valve leaflet mechanical behaviour. Proc IME H J Eng Med 2011;225:718–22.

[60] Hersel U, Dahmen C, Kessler H. RGD modified polymers: biomaterials for stimulated cell adhesion and beyond. Biomaterials 2003;24:4385–415.

[61] Web M. Material property data.

[62] Prawel DA, Dean H, Forleo M, Lewis N, Gangwish J, Popat KC, Dasi LP, James SP. Hemocompatibility and hemodynamics of novel hyaluronan–polyethylene materials for flexible heart valve leaflets. Cardiovasc Eng Technol 2014;5:70–81.

[63] James SP, Dean IVH, Dasi LP, Forleo MH, Popat KC, Lewis NR, Prawel DA. Glycosaminoglycan and synthetic polymer material for blood-contacting applications. 2015.

[64] Simon-Walker R, Cavicchia J, Prawel DA, Dasi LP, James SP, Popat KC. Hemocompatibility of hyaluronan enhanced linear low density polyethylene for blood contacting applications. J Biomed Mater Res B Appl Biomater [n/a–n/a].

[65] Pinchuk L, Wilson GJ, Barry JJ, Schoephoerster RT, Parel JM, Kennedy JP. Medical applications of poly(styrene-block-isobutylene-block-styrene) ("SIBS"). Biomaterials 2008;29:448–60.

[66] Kondyurin A, Pecheva E, Pramatarova L. Calcium phosphate formation on plasma immersion ion implanted low density polyethylene and polytetrafluorethylene surfaces. J Mater Sci Mater Med 2008;19:1145–53.

[67] Andrews RK, Lopez JA, Berndt MC. Molecular mechanisms of platelet adhesion and activation. Int J Biochem Cell Biol 1997;29:91–105.

[68] Satriano C, Carnazza S, Guglielmino S, Marletta G. Surface free energy and cell attachment onto ion-beam irradiated polymer surfaces. Nucl Instrum Methods Phys Res Sect B Beam Interact Mater Atoms 2003;208:287–93.

[69] Stachelek SJ, Alferiev I, Choi H, Kronsteiner A, Uttayarat P, Gooch KJ, Composto RJ, Chen IW, Hebbel RP, Levy RJ. Cholesterol-derivatized polyurethane: characterization and endothelial cell adhesion. J Biomed Mater Res A 2005;72:200–12.

[70] Pierschbacher MD, Ruoslahti E. Cell attachment activity of fibronectin can be duplicated by small synthetic fragments of the molecule. Nature 1984;309:30.

[71] de Mel A, Bolvin C, Edirisinghe M, Hamilton G, Seifalian AM. Development of cardiovascular bypass grafts: endothelialization and applications of nanotechnology. Expert Rev Cardiovasc Ther 2008;6: 1259–77.

[72] Joshi RR, Frautschi JR, Phillips Jr RE, Levy RJ. Phosphonated polyurethanes that resist calcification. J Appl Biomater 1994;5:65–77.

[73] Hufnagel CA. Reflections on the development of valvular prostheses. Med Instrum 1977;11:74–6.

[74] Daebritz SH, Fausten B, Hermanns B, Franke A, Schroeder J, Groetzner J, Autschbach R, Messmer BJ, Sachweh JS. New flexible polymeric heart valve prostheses for the mitral and aortic positions. Heart Surg Forum 2004;7:E525–32.

[75] Daebritz SH, Fausten B, Hermanns B, Schroeder J, Groetzner J, Autschbach R, Messmer BJ, Sachweh JS. Introduction of a flexible polymeric heart valve prosthesis with special design for aortic position. Eur J Cardiothorac Surg 2004;25:946–52.

[76] Butany J, Fayet C, Ahluwalia MS, Blit P, Ahn C, Munroe C, Israel N, Cusimano RJ, Leask RL. Biological replacement heart valves. Identification and evaluation. Cardiovasc Pathol 2003;12:119–39.

[77] Ghista DN. Toward an optimum prosthetic trileaflet aortic-valve design. Med Biol Eng 1976;14: 122–9.

[78] Yee Han K, Lakshmi Prasad D, Ajit Y, Hwa Liang L. Recent advances in polymeric heart valves research. Int J Biomater Res Eng 2011;1:1–17.

[79] Bernacca GM, Mackay TG, Gulbransen MJ, Donn AW, Wheatley DJ. Polyurethane heart valve durability: effects of leaflet thickness and material. Int J Artif Organs 1997;20:327–31.

[80] Daebritz SH, Sachweh JS, Hermanns B, Fausten B, Franke A, Groetzner J, Klosterhalfen B, Messmer BJ. Introduction of a flexible polymeric heart valve prosthesis with special design for mitral position. Circulation 2003;108. II–134–II–139.

[81] Bezuidenhout D, Williams DF, Zilla P. Polymeric heart valves for surgical implantation, catheter-based technologies and heart assist devices. Biomaterials 2015;36:6–25.

[82] Leat ME, Fisher J. The influence of manufacturing methods on the function and performance of a synthetic leaflet heart valve. Proc IME H J Eng Med 1995;209:65—9.

[83] Rottenberg D, et al. Method for producing heart valves. 2000.

[84] Hui A, Duncan A, Wan W. Hydrogel based artificial heart valve stent. Asme-Publications-Bed 1997; 36:53—4.

[85] Wan W, Campbell G, Zhang Z, Hui A, Boughner D. Optimizing the tensile properties of polyvinyl alcohol hydrogel for the construction of a bioprosthetic heart valve stent. J Biomed Mater Res A 2002; 63:854—61.

[86] Yousefi A, Bark DL, Dasi LP. Effect of arched leaflets and stent profile on the hemodynamics of tri-leaflet flexible polymeric heart valves. Ann Biomed Eng 2017;45:464—75.

[87] Claiborne TE, Sheriff J, Kuetting M, Steinseifer U, Slepian MJ, Bluestein D. In vitro evaluation of a novel hemodynamically optimized trileaflet polymeric prosthetic heart valve. J Biomech Eng 2013; 135:021021. 021021-8.

[88] Rahmani B, Tzamtzis S, Sheridan R, Mullen MJ, Yap J, Seifalian AM, Burriesci G. In vitro hydro-dynamic assessment of a new transcatheter heart valve concept (the TRISKELE). J Cardiovasc Transl Res 2017;10:104—15.

CHAPTER 14

Regulatory considerations

Ivan Vesely[1,2]

[1]Class III Medical Device Consulting, Gaithersburg, MD, United States; [2]CroiValve Limited, Dublin, Ireland

Contents

The regulatory aspects of the heart valve business are profoundly polarizing. From the perspective of a medical device company, the regulatory requirements for bringing a product to market are almost always viewed as overly burdensome, time-consuming, costly, irritating and an impediment to bringing good products to market, saving lives, and making good money. If you are a medical device regulator, tasked with reviewing a design dossier that can be thousands of pages long, the regulatory requirements are principles that must be adhered to because they are the law. If you are a patient in need of a medical device, you really have no idea what studies were done on the device to demonstrate that it is safe, but at least there is some comfort in your mind that there is a process to "protect us" from "dangerous medical devices." In the court of public opinion, medical device manufacturers are only second lowest on the rank of mistrust and loathing, right behind the drug manufacturers.[1] Having spent 7 years bringing a heart valve product from concept to clinical testing and 2 years as a medical device and heart valve product reviewer at a regulatory agency, I have witnessed the regulatory process in all its elegance and ugliness from both sides. The purpose of this chapter, therefore, is not to review all the regulatory issues that a valve developer and manufacturer needs to consider, but rather to list a handful of aspects of the regulatory world that are interesting and instructive to the heart valve profession as a whole.

---

[1] https://www.nytimes.com/2015/09/21/business/a-huge-overnight-increase-in-a-drugs-price-raises-protests.html.

Principles of Heart Valve Engineering
ISBN 978-0-12-814661-3, https://doi.org/10.1016/B978-0-12-814661-3.00014-9

14.1 The sins of the father

For those interested in learning in part how and why the medical device industry became so heavily regulated, I would recommend a book entitled "Medical Device Accidents" by Leslie Geddes.[2] Many of the tragedies that befell patients, and still do today, are true accidents in that they are difficult if not impossible to prevent. But at the same time, the history of medical devices is littered with examples of outright fraud, negligence, and hubris. It is because of these examples that the outraged public demanded laws and a process by which new medical devices are vetted so that they can be considered sufficiently safe for widespread use in patients. One might think that in today's highly regulated environment, it is difficult to commit outright fraud. But as the PIP breast implant scandal[3] of 2010 clearly shows, there are still individuals out there who simply refuse to follow good manufacturing practice and commit crimes that may injure patients. When I first entered the world of heart valve manufacturing, I heard my colleagues complain in person and publicly in conferences as to how burdensome and "ridiculous" the regulatory requirements for getting a product approved are. What these good colleagues of mine somehow neglected to realize is that as an industry, we are collectively responsible for these regulations for the same reasons that there are regulations for food safety and speed limits on city streets—in the past, a few unsavory characters were responsible for a good number of very serious patient injuries. In the heart valve field, perhaps the most emblematic one is of the "convexo-concave" Bjork—Shiley tilting disc valve. Of the roughly 86,000 valves implanted, 619 experienced leaflet escapement because of fracture of the outflow strut and patient death in two-thirds of the cases [1]. Subsequent investigations into the cause of this tragedy identified falsification of manufacturing documents and the presence of a "phantom welder"[4] that was involved in the remanufacture of damaged valves that failed certain internal tests. Fraud had come to the heart valve field.

14.2 The need for documented procedures

Studies into medical device accidents found that in the absence of outright fraud, failures could be traced to the processes and procedures that manufacturers failed to follow during design and manufacture. The rules and requirements that are embodied in Good Manufacturing Practice, and in particular in ISO 13485, the main guidance document for a Quality System, is that whatever is done during the design, testing and manufacture of a medical device, needs to be in accordance to written procedures. Once procedures

---

[2] https://www.amazon.com/Medical-Device-Accidents-Illustrative-Cases/dp/1930056362.

[3] https://www.massdevice.com/pip-breast-implant-scandal-story-triggered-change/.

[4] http://www.ele.uri.edu/courses/bme462/handouts/Bjork-Shiley_valve.pdf.

are put down in writing, they cannot be changed unless they are formally reviewed and their change is duly justified. Having to document everything from the very beginning of the design process forces the individuals to think carefully about what might be the consequences of changing a specification for the sake of convenience, without doing a thorough analysis. It is not that written procedures cannot be changed but that if they are to be changed, there need to be a good reason for that change and an analysis of the consequences of the change done first. These requirements force medical device manufacturers to think rigorously about how they design and manufacture their product and how they source their materials, and thus minimize the chances of inadvertent changes to the specifications of the product, or the way that it is manufactured. Documentation simply forces everyone to slow down and think before acting.

14.3 Risk versus reward

As the main objective of laws and regulations is to protect the public from potentially defective or even dangerous medical devices, one of the main principles embodied in both the US regulatory system (FDA issued premarket approval) and the European system (Notified Body issued CE Mark) is the concept of safety and efficacy (US) and safety and performance (EU). The device needs to be safe in that it does no harm when used, but it also needs to have some purpose. In the US system, this purpose of action is stated as "efficacy." In the EU system, it is stated as a list of "performance specifications." Note that "benefit to the patient" is not stated outright in the EU system, but it is implied in the requirements to do some specific action.

Implied in the use of any new device is unknown risk. If a device is new and untested on patients, there is always the risk that some unknown problem may occur in the context of its use that was completely unexpected. With that in mind, the new device must have some additional beneficial function relative to what is already on the market. If there is no additional benefit to the patient, then the introduction of a new device offers only new risk. Such a device is thus unlikely to get approved. For example, if one were to make an exact copy of the Edwards Magna valve in a new, lower cost manufacturing facility and then try to bring it to the market at a lower cost and compete with Edwards, they would be faced with some regulatory challenges. Being an "exact" copy of the valve would offer no additional benefit to the patient. But having a new product manufactured under a new manufacturing system carries a finite risk that something might not be quite to the same level of quality and performance of the original Magna. The Risk/Benefit equation would therefore be weighed toward higher risk and no new benefit, and the regulators would push back. The argument that the valve is cheaper is typically not considered in the risk/benefit equation—only safety and performance. For these reasons, heart valve innovators always tout some additional benefit of a new product over the status quo and then argue that the new benefits of the device outweigh the new risks and that the overall risk/benefit equation is favorable.

14.4 Risk management

The regulatory environment is far more heavily weighted toward mitigating risk, rather than demonstrating that there is a new benefit to the patient. Risk and patient injury are always considered first and are of much greater importance. Perhaps this is where the European system with its CE Mark is fundamentally more "elegant" from a structural perspective than the US system. The US system for heart valve approval is more prescriptive, whereas the EU system is more flexible and structured around risk management. Generally speaking, to get a valve approved in the United States, one must follow the "draft guidance document" that is issued by the FDA. This document references the ISO 5840 standard, which is the main guidance document for heart valves, but includes additional requirements beyond what is written in the standard. In contrast, the EU process requires that all medical devices, including heart valves, meet the "essential requirements" (ERs) of the Medical Device Directive (MDD). However, because the ERs have been written so that they apply to all medical devices, ranging from bandages to X-ray machines to cardiac pacemakers, they can appear vague and confusing. This is where the standards come in. One of the ways that all the ERs of the MDD can be met is if the manufacturer follows all the requirements of all the applicable "harmonized standards," such as EN ISO 5840-2:2015. If a standard is harmonized with the MDD, it means that if one follows exactly what is stated in a harmonized standard then, by default, the device is in compliance to the ERs of the MDD. Harmonized standards thus take much of the guesswork out of being in compliance to the MDD. The other benefit of the EU system is that harmonized standards do not need to be followed exactly if compliance to the ERs can be demonstrated in other ways. This means that if you have a very simple device with low risk, you do not need to follow the same prescriptive approach that might be necessary for a high-risk device. Having a risk management—based approach is intended to provide greater flexibility for the medical device manufacturer.

Risk management is described in great detail in ISO 14971. Risk management is a process. It begins at the point of initial product conception, continues through design and development, into preclinical and clinical testing, and through manufacture, sales, and postmarket follow-up. Risk management never ends. It is a continuing process where ongoing clinical experience is fed back to the beginning, and the risk of the continued use of the product is compared with the initial and ongoing assessment of the safety and performance estimates.

Risk management begins with a listing of hazards that a particular valve might subject the patient to, such as risk of clotting and occlusion. A good way to list all these hazards is to first consider all the hazards that any generic heart valve can induce. Then you move on to hazards that are specific to a particular class of heart valves, such as tissue valves versus mechanical valves versus transcatheter valves, as each of these classes will have hazards that are unique to that particular class of device. Then you list hazards that are specific to your valve design, such as a valve that uses polymer leaflets instead of bovine pericardium, for example, or has multiple hooks that project into the tissue. Once all possible hazards are considered, one then begins the process of converting these hazards into a quantifiable risk and then coming up with ways to reduce that risk.

Risk is the product of harm and likeliness of occurrence, each of which can be estimated and assigned a numerical score. For example, the harm associated with flying on an airplane is a crash and death. So the harm level is thus very high—perhaps a 7. However, the probability of such a harm actually coming into fruition is very low, perhaps less than one in a million and is therefore assigned a probability rating of 1 (see Fig. 14.1). The resulting risk priority number (RPN) of 7 is therefore a relatively low number. In contrast, if one were to fabricate a transcatheter mitral valve with a particular frame design, one of the risks might be paravalvular leak. The harm of paravalvular leak might be estimated as being catastrophic—a 6. In the absence of any testing, whether this frame design does indeed lead to paravalvular leak, the probability of occurrence could be a very likely and hence a 7 (see table), and the risk would therefore be a 42—a considerably high value that would need to be mitigated. Some of the ways in which to mitigate this risk of paravalvular leak is to carry out bench testing and animal studies and measure the amount of regurgitation, potentially improve the design of the sealing cuff, and retest the design to assess regurgitation. After all these studies are done, one then updates the likelihood of regurgitation to perhaps a 3 and computes a new risk number, now 18. One then compares these final RPNs against preset acceptance criteria. If the risk has been reduced as much as possible and these final numbers are within the acceptance criteria, risk mitigation has been satisfied. If the numbers are not below the preset cutoff and no further risk mitigation activities can be done, then the final step is to do a risk/benefit analysis on each remaining unmitigated risk. In the case of the leaky mitral valve, one could argue that because there are no other transcatheter mitral valves on the market and the patient is inoperable, then the risk of persistent mitral regurgitation is better than the alternative, which is progressive congestive heart disease and eventual death.

Risk mitigation, in the example above, is best captured via a failure mode effects analysis (FMEA), which is a tabulated summary of the risk assessment and risk mitigation activities (see Fig. 14.2 for an example of a few entries). In heart valve development, there are typically three FMEAs—design, use, and manufacture. The design FMEA reviews all the failure modes that could occur as a result of inadequate design and how improvements in the design were used to mitigate the risk. For example, excessive pannus overgrowth can be mitigated via design by using industry-proven materials, and its mitigation verified via animal implants. What is thus initially very high risk can be mitigated down to a very acceptable risk.

The use of FMEA lists all the potential ways that the physician or assistants could make mistakes during the use of the device and how the manufacturer mitigated these risks via improvements in design, via alarms and training. For example, under expansion

Severity

Effect	Severity Rank	Description
Catastrophic	7	Probable patient death regardless of intervention
	6	Probable patient death without immediate intervention
Critical	5	Possible patient death or probable permanent disabling injury regardless of intervention
	4	Possible patient death or probable permanent disabling injury without immediate intervention
Serious	3	Possible permanent impairment of bodily function
Minor	2	Possible temporary impairment of bodily function
Negligible	1	Slight or no potential for patient injury

Probability

Probability Rating	Probability of occurrence during device life cycle	Effect	Criteria
7	> 0.1	Very High	Causes Happen Often
6	$10^{-2} - 0.1$	High	Causes Happen Sometimes
5	$10^{-3} - 10^{-2}$	Moderately High	
4	$10^{-4} - 10^{-3}$	Medium	Causes Happen Infrequently
3	$10^{-5} - 10^{-4}$	Low	
2	$10^{-6} - 10^{-5}$	Very Low	Causes Happen Rarely
1	$< 10^{-6}$	Remote	Causes Not Expected to Happen

RPN Range	Acceptability of Risk	Required Action
42 – 49	Intolerable	Risk must be reduced further
24 - 41	Undesirable	Risk should be reduced further. If not possible, carry out Risk / Benefit Analysis
10 - 23	Tolerable	Attempt to reduce risk further and then carry out Risk / Benefit Analysis
1 - 9	Acceptable	Attempt to reduce risk further if possible

Figure 14.1 Example tables of severity of an event, its probability, and a risk priority number table that guides subsequent risk reduction activities.

of a transcatheter valve can be mitigated by proper instructions provided to the user. However, per ISO 14971, the "instructions for use" is not a sufficient means of mitigating risk, as physicians may not read them. Accordingly, other engineered solutions must be provided, such as indicator bars on the injection syringes, etc. One way of mitigating the risk of releasing a transcatheter aortic valve replacement (TAVR) device before full expansion is to have an interlock in the catheter handle that prevents the release knob to operate before the expansion knob is fully actuated.

Similarly, manufacture FMEAs list all the possible ways that errors could be made in each and every step of the manufacturing process, all the way from selecting the wrong supplier of the metal frame to having the assembly technician put in fewer stitches to mixing up

Item	Failure Mode or Event	Potential Effect or Harm	Sev	Likely Cause of Event	PR	RPN	Action Taken to Mitigate Event	Confirmation of Mitigation	Supporting Documentation	Sev	PR	RPN
										Score after Mitigation		
Valve System	Valve Embolization	Sudden absence of a functional valve, leading to massive regurgitation and potential occlusion of aorta downstream of the valve	6	Inadequate design or deployment that leads to insufficient friction of frame against native aortic valve leaflets	5	30	Frame was designed to use a metal mesh with holes in it that provide a rough surface that grabs onto native valve leaflet tissue	Laboratory studies confirmed that frame does no dislodge at loads 3X greater than peak diastolic pressure	XXX-XXX Laboratory report on valve embolization simulation	6	3	18
	Valve Migration	Change in valve position or orientation, leading to loss of function and potential occlusion of downstream vessels	6	Inadequate design or deployment that leads to insufficient friction of frame against native aortic valve leaflets	5	30	Frame was designed to use a metal mesh with holes in it that provide a rough surface that grabs onto native valve leaflet tissue	Laboratory studies confirmed that frame does no dislodge at loads 3X greater than peak diastolic pressure	XXX-XXX Laboratory report on valve embolization simulation	6	3	18
Stent	Insufficient Radial Strength	Valve Migration	6	Inadequate design or deployment that leads to insufficient friction of frame against native aortic valve leaflets	5	30	Frame was designed to have radial strength after balloon dilation sufficient to avoid embolization. Target design stiffness after deployment is ___ Nm.	Mechanical testing of deployed stents shows stiffness of ___ Nm +/- ___%. Pull out testing confirms sufficient radial strength and no migration	XXX-XXX - Aortic Pericardial Bioprothesis Specifications; XXX-XXX - Design Input Document; XXX-XXX - Mechanical testing report; XXX-XXX - Pull out testing report; etc.	6	3	18
		Under-expansion of valve during deployment	6	Inadequate balloon expansion force	6	36	Baloon was designed to atain an expansion diameter of ___ mm when fully filled with saline.	Simulated deployment testing has verified that when a 23 mm balloon is filled with ___ ml of saline, it expands to a fully deploy the valve frame	XXX-XXX - balloon Specifications; XXX-XXX - balloon design Input Document; XXX-XXX - simulated deployment testing report; etc.	6	3	18
		Under-expansion of valve during deployment	6	Inadequate amount of saline delivered to baloon by physician	6	36	IFU written specifically to instruct physician on balloon deployment	Review of IFU by MAB members confirmed that instructions are sufficiently clear and appropriate	XXX-XXX - Report from MAB member on IFU	6	4	24
	Excessive pannus overgrowth	Leaflet immobilization leading to severe stenosis	7	Material used is not sufficiently biocompatible	6	42	According to modern industry practice, all non-biological valve components are covered by polyester cloth and tested for biocompatibility	5 month sheep implant studies show pannus overgrowth of device comparable to that for the Edwards Perimount, as evaluated by trained pathologist	XXX-XXX Laboratory report on sheep implantation studies	7	1	7
			7	Abnormal patient physiologic response	4	28	It is recognized that some patients may be predisposed to complications associated with biomaterial implants. Instructions for Use (IFU) have been developed taking into account necessary Warnings, Indications and contraindications	IFU documents have been reviewed for accuracy by Medical Director (Board Certified physician) and approved for use in CE Mark Clinical Trial	XXX-XXX - Report from MAB member on IFU	7	4	28
		Leaflet immobilization leading to regurgitation	4	Material used is not sufficiently biocompatible	4	16	According to modern industry practice, all non-bioprosthetic tissue components are covered by polyester cloth and tested for biocompatibility	5 month sheep implant studies show pannus overgrowth of device comparable to that for the Edwards Perimount, as evaluated by trained pathologist	XXX-XXX Laboratory report on sheep implantation studies	4	2	8

Figure 14.2 Example entries from a design failure mode effects analysis (FMEA) for a transcatheter valve. Note that failure modes are organized hierarchically, first for the valve as a whole, then for its components, and each identified failure mode can have many potential harms and thus different initial risk priority numbers.

sterile and nonsterile components. All of these potential failure modes can be mitigated via written procedures to which all aspects of the manufacturing processes must adhere.

Experience has taught the industry that only through such very detailed analyses of potential risks, and the strict adherence to written procedures, can lifesaving devices be produced to the highest levels of quality that we, as the patients, demand and deserve.

14.5 Objective performance criteria

Before the advent of transcatheter valves, surgical heart valves enjoyed a unique status in the medical device space in that their clinical testing did not require randomized studies with control devices. Single arm, nonrandomized studies were sufficient, as long as they were adequately powered and the new device met the minimum objective performance criteria (OPC). The driving force behind the concept of OPC has been, and continues to be, Dr. Gary Grunkemeier, a highly respected, well-published biostatistician. The concept of meeting minimum performance criteria was adopted for heart valves because of the extensive clinical experience with valves and a general understanding of what constitutes a valve with acceptable clinical performance. The OPC were based on the examination of thousands of data points from published clinical studies of valves on the market and from nonpublished reports submitted to the FDA during the course of prior clinical investigations. For example, the original OPC were derived from over 38,000 patient-years of clinical data. The updated OPC listed in table J.1 of Annex J in ISO 5840-2:2015 are derived from over 208,000 patient-years of clinical data obtained over the years 1999—2012 (see Fig. 14.3). The use of OPC as a benchmark means that a new valve does not need to demonstrate noninferiority against some control device, but rather be essentially equivalent to what constitutes the norms—the OPC. The standard is actually more generous than that. Any new valve does not need to have adverse event rates equivalent to the OPC, but rather event rates no worse than twice the OPC, as long as the data are supported by valid statistical tests. In prior versions of ISO 5840, examples of 400 or 800 patient-year clinical trials were provided as a means of how one could demonstrate with 80% statistical power that the true event rates of the valve were no worse than twice the OPC values. For example, if the measured event rate was two-third the OPC, then 400 patient-years was enough. If the event rate was greater, say equal to an OPC of 1.2%, then 800 patient-years would be required. These were difficult studies to do well because the lower the event rates, the greater the number of patients was needed to demonstrate with 80% power and $p < .05$ that the test was met. Also, one could not estimate the event rates in advance of the clinical trials and one could thus potentially enroll insufficient number of patients. For that reason, US-based clinical trials typically enrolled hundreds of patients and accumulated data of over 1000 patient-years.

In the current 2015 version of EN ISO 5840_2, the OPC guidelines have been updated and simplified. The updates involve a reduction of some of the OPC values because current generation valves have been found to be better and safer than

(A)

Table R.1—Objective performance criteria for heart valve substitutes

	Rigid	Flexible
Thromboembolism	3.0	2.5
Valve thrombosis	0.8	0.2
All hemorrhage	3.5	1.4
Major hemorrhage	1.5	0.9
All paravalvular leak	1.2	1.2
Major paravalvular leak	0.6	0.6
Endocarditis	1.2	1.2

NOTE—Values are in % per valve-year.

Data was produced from rigid and flexible valves. See Grunkemeier, et al.[8].

The formal statistical test is that the observed rates must be significantly less than x 2 the OPC.

The rates in Table R.1 were established by the United States Food and Drug Administration after a review of the literature. For events with an OPC of 1.2 %, if the rate equals the OPC, a sample size of 800 valve years will furnish approximately 80 % power for satisfying the formal statistical test (see Grunkemeier, et al.[8]). If in the same circumstance the rate is 2/3 of the OPC, a sample size of 400 valve years will also furnish approximately 80 % power. It is the responsibility of the manufacturer to propose a trial design that has adequate power, based on a risk analysis of the valve being tested.

(B)

Table J.1 — Objective performance criteria for surgical heart valve substitutes

Adverse Event (End point)	Bioprosthetic		Mechanical	
	Aortic	Mitral	Aortic	Mitral
Thromboembolism	1,5	1,3	1,6	2,2
Valve thrombosis	0,04	0,03	0,1	0,2
Major haemorrhage	0,6	0,7	1,6	1,4
Major paravalvular leak	0,3	0,2	0,3	0,5
Endocarditis	0,5	0,4	0,3	0,3

NOTE   Values are in % per valve-year.

The data in Table J.1 were derived using the same methodology as the original OPCs, an analysis of safety and effectiveness data submitted by manufacturers in pursuit of premarket approval of bioprosthetic and mechanical valves (yielding 38 359 follow-up years) combined with an analysis of recent literature from 1999 to 2012 (yielding 208 585 follow-up years). There was no significant heterogeneity between the two sources of data, either in methods of data collection or in complication rates. See Reference [19].

Figure 14.3 (A) Table of objective performance criteria (OPC) from the original 1994 FDA guidance and incorporated into ISO 5840-2005. (B) Table of OPC from the updated ISO 5840-2015.

prior devices. For example, for tissue valves, the OPC for thrombosis has been reduced from 0.2 to 0.04—existing surgically implanted tissue valves simply do not thrombose very much these days, and new tissue valves should not be expected to either. Similarly, the OPC for endocarditis has been reduced from 1.2 to 0.5, reflecting the better performance of sterilization protocols and operative sterility controls. The simplification to the requirements has been the elimination of "all hemorrhage" and "all paravalvular leak" with the need to report only "major" instances of hemorrhage and leaks. The statistical requirements have also been reduced. Rather than worrying about the sample size and its impact on statistical power for very low rates of occurrence, the requirement now states that if the mean value of the event rate is numerical no worse than twice the OPC, the valve has acceptable performance—no statistical power calculations are required. The caveat, however, is that the sample size of the test has now been prescribed to be 800 patient-years whether it is for a single valve position or multiple positions (i.e., aortic and mitral). The other important update to the standard is that it was clarified that these events start to be counted only 30 days after the procedure, rather than right away. This eliminates the early, perioperative issues from the equation, such as bleeding or stroke because of surgical complications.

OPC are therefore a unique feature of ISO 5840 and hence the surgical valve field. Because of the tremendous experience with the clinical performance of surgical valves, dating back several decades, we know what to expect. We do not need to do paired clinical studies and noninferiority tests. An inferior valve is simply one that does not match the historical norms encapsulated in the OPC table.

This is in stark contrast to what is required of transcatheter valves, as stated in EN ISO 5840-3:2013. ISO 5840 comes in three parts. 5840-1 lists general requirements, 5840-2 is specific to surgically implanted valves, and 5840-3 is the new standard that was developed in 2013 and pertains to transcatheter valves. Unlike ISO 5840-2, the clinical section of –3 is extensive and makes reference to testing the safety of the device over the course of the entire procedure, not simply recording events after 30 days like for surgical valves. This is because for transcatheter valves, much of the risks are associated with the implantation of the device, not just its performance afterward. From a statistical perspective, the standard recommends randomized control studies assessing noninferiority and superiority end points, clearly taking account of the unproven and far more risky nature of transcatheter valves. While there has been some talk about implementing OPC in the transcatheter valve version of ISO 5840, it is unlikely to be implemented any time soon because of the rapidly evolving nature of these technologies.

What is interesting about OPC is how long the original numbers have remained essentially unchanged. The concept of long patient follow-up began with Dr. Albert Starr who in 1960 implanted the first successful mechanical heart valve and introduced the concept of "lifetime follow-up." Interestingly, the longest surviving patient[5] with these early valves lived for 50 years [2], providing the longest ever clinical follow-up for a prosthetic valve. Grunkemeier, Starr, and Rahimtoola were the leaders with respect to long-term follow-up, eventually accumulating data on nearly 60,000 valves, 54,000 patients for over 220,000 valve years, and publishing a pioneering manuscript on this

---

[5] https://www.youtube.com/watch?v=BEEeYmQLiuY.

in 1992 [3]. In 1994, the US FDA implemented these data as a requirement for premarket approval for both mechanical (rigid) and bioprosthetic (flexible) heart valves (see Fig. 14.3A). These OPC were thus based on clinical data over the years 1985–90. Even as late as 2009, when ISO 5840 was updated, it used exactly the same OPC table that was created by Grunkemeier in 1994. Only in 2015 when ISO 5840 was updated with the new OPC values, aggregating clinical data from the years 1999–2012, we began to hold valves to a standard higher than what valves were in the late 1980s.

What happened in the late 1980s? That was when we finally understood why the first-generation Ionescu–Shiley valve failed from commissural tears [4]. The late 1980s was when some of the earliest data came out on the recently introduced Hancock II, low pressure–fixed, second-generation porcine xenograft valve [5]. The Edwards Perimount valve was introduced in the early 80s [6] and the Edwards Magna was not available until the turn of the century [7]. So it has been until as recently as 2015 that the clinical performance of new valves was held to a standard that was developed over 20 years ago, largely on a mix of first- and second-generation valves. Actually it is worse than that. Not long ago, some new surgical valves were still being judged by these outdated performance standards, simply because their long clinical studies have begun before the publication of ISO 5840-2:2015.

The update to the OPC is actually consistent with the concept of risk management. As noted in the previous section, risk management is a process that begins at the point of initial product conception and continues through postmarket follow-up. The intent of risk management is that clinical experience is fed back to the beginning where the initial assessment of risk was made. In the early days of heart valves, the risk of thromboembolism, for example, was estimated to be several percent, and the risk/benefit analysis concluded that valve thromboembolism rates no worse than twice the OPC value of 2.5 are acceptable. Over time, however, clinical data demonstrated that in practice, the rates of thromboembolism of tissue valves are actually lower, and the OPC value was adjusted down to 1.5. Clinical experience was thus fed back to risk management and the new expected rates of thromboembolism have been reduced. Any new valve that now comes on the market cannot have rates of thromboembolism no worse than 5% (twice the old OPC), and they must be no worse than 3% (twice the new OPC). As technology improves, safety improves, and with that the expectations of safety have improved also.

14.6 Making sausages

Otto von Bismarck is remembered by a quote that many in Congress like to repeat: *"Laws are like sausages, it is better not to see them being made."*[6] I expect that ISO standards

---

[6] https://www.brainyquote.com/quotes/otto_von_bismarck_161318.

are not much different. I had come across an interesting entry in ISO 5840 that I chose to investigate further.

Section 7.3.2.1 General requirements, pertaining to in vivo preclinical testing states in paragraph 4 the following:

> For long term studies, the duration of the observation period of the animals must be specified according to the parameter(s) under investigation. The observation period shall be appropriately justified in each study protocol, but shall be no less than 90 days.

Taken at face value, this implies that long-term chronic animal implants are recommended to be just over 90 days. In other words, if one wishes to do sheep implants to assess mineralization and other biologically relevant changes, a 3-month implant study will satisfy ISO 5840-2:2015.

This is quite misleading. In speaking with one of the participants who was involved in drafting ISO 5840, apparently the intent of that phrase was different. It was not intended to apply to new devices, but rather to devices that were modified and needed to be retested again in the sheep model. It should be perfectly clear to everyone that if one needs to determine the calcification potential of a device that is intended to be implanted into low risk patients, the appropriate model is as follows: (1) juvenile sheep model, (2) mitral implant preferred, and (3) duration of 20 weeks (5 months). This is the gold standard approach to preclinical testing [8]. Anything less than the gold standard approach is extremely risky, both in terms of patient safety and from a regulatory perspective. This is particularly important for early stage heart valve companies who might get the impression that all they need to do are 90-day sheep implants. This will not be sufficient. Based on my personal experience as a reviewer at a Notified Body, most other reviewers at Notified Bodies that review dossiers before CE Marking will not accept 90-day data, as it is not considered state of the art. Ninety days simply is not enough to show any propensity for valve tissues to calcify.

It is unclear how such a potentially misleading statement made it into the text of the standard. It is misleading because many regulatory bodies will not accept 90-day sheep implant data, no matter how much you argue that "it is in the standard." Who knows what sausage making went into this particular phrase when it was passed by the ISO committee. Risk management is still a matter of using the appropriate gold standard approaches for risk mitigation, regardless of what is written in the applicable standards.

14.7 Failure of preclinical models

With that in mind, let us review a handful of examples in the heart valve field where new valve designs apparently passed sheep studies, yet subsequently failed in humans. I can think of three illustrative examples: (1) Carbomedics Photofix-α, (2) St. Jude Silzone, and (3) Cryolife SynerGraft.

(1) Carbomedics Photofix-α: This valve apparently did well in animals and was approved for clinical usage. However, about 2 years after its introduction in the mid 1990s, it was removed from the market because of tears in the belly of the leaflet. A manuscript by noted pathologist Fred Schoen [9] noted that findings from 10 valves that failed after implantation for 8–23 months showed that each valve had one to several tears that extended from the belly toward the commissures. The microscopic findings and pattern of leaflet damage suggested that there was abrasion of the leaflets against the Dacron cloth. There was also evidence of leaflet sagging, which was consistent with the notion that leaflets abraded against the cloth-covered frame. Apparently, there was no significant inflammation, infection, thrombus, pannus overgrowth, or calcification. This type of failure was thus most likely entirely design-related.

So where did ISO 5840 fail? How did risk management fail? Were there absolutely no signs of this during preclinical bench and animal testing or were some of the signs overlooked? These questions can be answered only by those who worked on the project. But two things come to mind. Firstly, durability testing in an accelerated wear tester is done at opening and closing rates at least 10 times faster than physiologic. This means that even though pressures are set up and measured to have physiologic magnitude, they may not necessarily be "realistic." It is quite possible that at these accelerated rates of opening and closure, pressure and displacement are not in phase. It is thus possible that when the valve is fully closed and leaflets coapted, they may not experience full closing pressure—pressure and displacement could be somewhat out of phase. Or perhaps it is simply that full excursion of the leaflets could not be simulated during accelerated wear testing. Also, why did this not occur in the animal model? Two reasons are possible: (1) the animal tests lasted only 5 months, whereas the failure in humans manifested between 8 and 23 months, and (2) the sheep is a hyperfibrotic animal model, meaning that the cloth covering the valve frame grows over rapidly with a robust pannus sheath. Perhaps this pannus covering protected the valve leaflets from abrasion in the sheep, something that did not occur in human implants.

(2) St. Jude Silzone: In 1998, the FDA approved a modification to the St. Jude line of mechanical and bioprosthetic valves to incorporate a silver-coated sewing cuff. Approximately 36,000 of these prostheses were implanted before they were withdrawn from the market by the manufacturer in January 2000 because of a high incidence of explant because of paravalvular leak [10]. Silver has antibacterial properties and it was believed that impregnating the sewing cuff of prosthetic valves would eliminate the risk of sewing cuff-related infections. Unfortunately, the cytotoxic effects of the silver coating prevented the endothelialization of the cuff, making it remain thrombogenic long after the valve was implanted and perioperative anticoagulation was discontinued. Statistical comparisons of the Silzone-treated valves and the control, nontreated valves implanted into patients as part of a clinical investigation showed that 35% of the patients implanted with the Silzone valve suffered strokes within 3 months of implant, whereas non-Silzone treated valve patients had zero strokes. These strokes, as well as a number of cases of paravalvular leak, also due to the erosion of tissue adjacent to the cytotoxic cuff, led to the withdrawal of this valve from the market [11].

Why did this not show up during preclinical testing? Although it was found that the pannus overgrowth in the Silzone-covered cuffs was thinner than the controls, this finding was apparently interpreted as "a more mature state of tissue reaction" rather than a concern. Also, none of the valves showed "excessive thrombosis." Perhaps it should have been known that different animals have different clotting responses to thrombogenic surfaces. Sheep in particular are more tolerant to mechanical valves and typically do not require anticoagulation. A negative thrombogenic response in the sheep therefore does not translate to humans. Also, as noted in the prior example, young sheep show a far more robust healing response than human patients and likely endothelialized the toxic sewing cuff in a way that human patients did not. A similar phenomenon occurred with the old Braunwald-Cutter valve, where the cloth covered struts endothelialized well in sheep implants, but experienced cloth abrasion when implanted into humans [12]. Success in animals is thus not predictive of success in human patients.

(3) Cryolife SynerGraft: This was perhaps the first clinical use of a "tissue engineered" valve. In 2001, the SynerGraft was a decellularized, unfixed porcine aortic valve that apparently showed promise in sheep implants [13] and was implanted into four children following approval for a "compassionate use exemption" in children with congenital valvular abnormalities. Of the four patients (aged 2.5—9 years) implanted with these valves, one patient died on the seventh postoperative day because of graft rupture, two patients died 6 weeks later, and the final patient died 1 year later because of sudden cardiac death [14,15]. Pathological findings showed inflammation on day 2 of the implant, and all four grafts were encapsulated in a thick fibrous sheath. Histological exam demonstrated a severe foreign body reaction rich with neutrophils, granulocytes, and macrophages [16,17].

What went wrong? Perhaps this was the worst example of human experimentation without a clear understanding of how to mitigate risk. From an antigenic point of view, the pig is more close to the sheep than it is to humans. Nonhuman mammals express the α-gal epitope in their tissues [18]. Because the α-gal antigen is expressed by human intestinal bacteria, our immune system is continually challenged by this antigen [19]. Once the α-gal antigen on implanted xenogeneic tissue is recognized, the complement cascade is activated leading to thrombosis and a severe foreign body response. This is something that does not occur in nonhuman mammals and clearly did not occur when unfixed pig tissue was implanted in the sheep. Once again, lack of scientific rigor, a full investigation of the hazards, and means to mitigate risks led to the unfortunate death of patients.

Defenders of these incidents would argue that everything is clear "in hindsight." Perhaps so, but the lack of predictability of the sheep model has been known for many years via the prior examples above. It is also the best model that we have for heart valves, even with all its flaws and uncertainties. What is necessary here, however, is recognition of the model and its limitations, not its abuse via excuses that it is "prescribed" in ISO 5840. No procedure or test is prescribed. The only thing that is required is risk management. Only being honest with ourselves, and taking to heart the realization that heart valves can extend lives when they work well, and shorten them when they do not, can the principles embodied in the seemingly complex regulations be truly effective.

14.8 A case study

After I left Academia, I have had the pleasure of founding a heart valve company, guiding its technology through maturity, and witnessing my technology get implanted into patients. At the same time, I experienced the sorrow of watching the company die after its product did not get CE Mark approval, and the lenders foreclose on the company. The lesson to be learned is one about the regulatory process. Indeed, it is because of the death of ValveXchange due to regulatory failures that I decided to take a job with a Notified Body to witness the CE Marking process from the other side of the table.

The ValveXchange Vitality valve was a concept that most surgeons found very compelling—a two-part surgical valve wherein the base with the sewing cuff is installed first, and the leaflet set is snapped onto it afterward. The value proposition for the surgeon was a far more compact package for implantation via small incisions, great visibility into the subannular space to check the positioning of the pledgets, and half the stent post height of a normal valve which made getting fingers around the posts that much easier during knot tying. The configuration of the leaflets was identical to that of the Edwards Perimount Magna giving comfort that it would likely be durable, and sheep implant studies demonstrated resistance to mineralization at least as good, and potentially better, than the Edwards valve. In view of the good preclinical bench and animal data, and the similarity to the Edwards valve, we made the argument to our Notified Body that the risk to the patient is minimal. Of course the new and potentially risky features of the valve were its two-part design and thus the risk of separation inside the patient.

When we tested our clinical trial proposal with our Notified Body, we presented mechanical data demonstrating that no matter how much you try to pull the two pieces apart, you would end up destroying the valve during the test before the two pieces came apart. From a functional perspective, all external forces on the valve acted to keep the two pieces together, rather than pulling them apart. The risk of separation was therefore adequately mitigated via design and verified via extensive bench testing. We then needed to demonstrate safety and performance via the clinical investigation. Safety was to be demonstrated via OPC and performance via a hemodynamic comparison to the gold standard Edwards Perimount Magna valve. We proposed that the Vitality valve will be no worse than the Perimount with respect to mean transvalvular gradients. Literature values showed that the mean transvalvular gradient of the Edwards Perimount was 14.5 ± 3.2 mmHg, and statistical calculations demonstrated that only 34 patients will be necessary to demonstrate noninferiority with a 1.6 mmHg margin (half of the SD), with 80% power at a P value of 0.025 for a one-sided t-test. The mean gradients therefore needed to be lower than 16.1 mm Hg. For safety, we proposed to use the OPC published in Annex R of EN ISO 5840:2009, which at the time were unchanged from the original OPC published in 1994 (see Fig. 14.3A). Given the design similarity of the Vitality valve to the Edwards Perimount and the quality of our preclinical data, we convinced our Notified Body that we should not have to do the full 400 patient-years, but simply demonstrate that we are equivalent to the OPC in a numerical sense. This means that in our 34 patients, we will need to have 0% rate of all of the events listed in the OPC table. This was a very bold approach, but given the flawless nature of our first-in-man study (zero events), we felt confident that we could pull this off. Our Notified Body agreed with this approach and approved the trial. The various ministries of health in Germany, Austria, and Poland, where the trial took place, also approved the clinical investigational plan. In the end, we actually proposed to implant 45 patients so that we could get away with one instance of thromboembolism and still be numerically better than the OPC rate of 2.5%.

But this is not how it turned out. On the positive side, none of the valves came apart, as expected, and our mean transvalvular pressure gradient was 9.7 mmHg. Our valve was actually better than the Edwards valve, and we thus met our performance end point. We did not, however, meet our safety end point. In our study, we experienced the following complications:

- Two valves had to be removed intraoperatively because of sizing errors and replaced with different sized valves. The physician adjudication board concluded that these explants were nonvalve-related and instead were because of physician error.
- There was one stroke that was adjudicated to be due to a preexisting condition—an atrial septal shunt that predisposes the patient to thrombus. This condition was an exclusion criterion for enrollment into the trial, but it was apparently not caught by preoperative echo exam.
- There were two deaths. One was because of pancreatitis, which evolved into septicemia, multiorgan failure, and death. The second death was procedure-related. The patient was initially sized for a 19 mm valve but the physician believed that

root enlargement was necessary, and thus performed the procedure, and then implanted the Vitality valve. After the valve was implanted, the patient was unstable and it was found that the left main was occluded by the tight fit of the valve. The physician chose to do a saphenous vein graft to the left anterior descending coronary artery, but the patient remained unstable. The physician thus decided to explant the Vitality and did a complete root replacement with a Medtronic Freestyle valve, reconnecting the coronary arteries, etc. However, by that time, the patient had experienced several episodes of being weaned off cardiopulmonary bypass, began to suffer respiratory insufficiency and RV and LV failure, was put on extracorporeal membrane oxygenation and ultimately suffered a large cerebral event, and was taken off life support.

The second death was troubling because it was procedure-related. Our physician adjudication board concluded that this death was also not related to the Vitality valve, but rather due to the complex procedure to enlarge the root, and the potential residual tight fit that led to the occlusion of the left main coronary artery. The protocol also specified that concomitant procedures, such as root enlargement, were not permitted in the trial. But in the heat of the moment, it is impossible to tell the surgeon "Stop, you cannot do this!"

When the final clinical data were submitted to our Notified Body, they became very nervous: two deaths, two sizing errors, and a stroke, with so few patients done. Were they all really unrelated to the valve or was there something systematic going on? At the same time, the regulatory requirements were changing in the wake of the PIP breast implant scandal[7] and the concept of a 45-patient clinical trial for a Class III device was now thought to be ludicrous. We argued with our Notified Body that we had (1) met and exceeded the performance end points of the Edwards valve and (2) met the OPC in that we had no thrombosis, no infection, no paravalvular leak, and the single stroke was still acceptable in view of the 2.5% stroke rate specified in the OPC. But one cannot argue with a Notified Body. They were not convinced that the risk of procedural complications was sufficiently mitigated via a 45 patient trial. ValveXchange did not get its CE Mark, and the subsequent year was a futile effort to restart the company with new funding. Ultimately, our main creditor foreclosed on the company and that was the end.

In hindsight, had we done 100 patients or more, the two deaths, the stroke, and the occasional miss-sizing would not have played such a large role in the minds of the reviewers, and any lingering efforts regarding procedural issues could have been resolved via postmarket clinical follow-up. Indeed, during the postmortem negotiations on how to get the CE Mark, the Notified Body agreed that a few dozen more patients with no additional events would be acceptable for awarding of the CE Mark. The lesson here is

---

[7] https://www.massdevice.com/pip-breast-implant-scandal-story-triggered-change/.

simple—never skimp on a clinical investigation. There will always be complications, and there will always be departures from the enrollment criteria. There is no such thing as a perfectly executed clinical investigation, simply because we are dealing with patients who are sick and with physicians who want to save them. During surgery to save a patient's life, the objectives of the clinical investigation are secondary. A well-planned clinical trial will therefore always take into account unforeseen complications and will not allow these to affect the intended outcomes of the study.

14.9 Closing

Historically, the European CE Marking process has been the easier path to follow. This is why most new heart valves were approved for sale in the EU first. In some ways, this has been both good and bad. Germany has been the first to adopt TAVR because of the early clinical experience and their favorable device and procedure reimbursement policies. This has been a huge benefit for German patients, German physicians who have become prominent on the international stage as a result of their unique early experience, and for the country as a whole, as patients get access to beneficial technologies. The United States lags behind Europe in the dissemination of TAVR because of the more demanding regulatory environment. It could be argued that US patients thus suffer because of a lack of new product availability.

But let us consider the flip side. What if the conservative nature of the US FDA system turns out to be the smarter approach, particularly with respect to the dissemination of TAVR into the younger, operable patient. As described in greater detail in Chapter 11, TAVR devices may have serious durability problems. Early data based on 10 years of follow-up[8] suggest that TAVR devices might have a 50% failure rate at 8 years. It is well-known that once tissue valves begin to fail, they continue to fail faster. Up to now, TAVR devices have been implanted only in the elderly and those with limited survivability. But there is a movement to implant TAVR valves into younger, operable patients. In younger patients, tissue valves are known to fail faster [20]. The trend toward implanting valves into younger patients is already underway in Europe, particularly in Germany [21]. The suggestion that TAVR valves might have a failure rate as high as 50% at 8 years in the elderly implies that failure rates will only be worse in younger patients. Perhaps these early failures are a harbinger of what might be coming over the next decade and just the tip of the iceberg. Perhaps the more conservative approach for disseminating TAVR into the younger population being practiced in the United States is the smart way to manage this technology.

---

[8] https://www.mdedge.com/ecardiologynews/article/109297/interventional-cardiology-surgery/tavr-degeneration-estimated-50.

References

[1] Blackstone EH. Could it happen again? the Bjork-Shiley convexo-concave heart valve story. Circulation 2005;111(21):2717—9.

[2] Abad C, Hernandez-Ramirez JM, Caballero E. Patient lives almost 50 years after aortic valve replacement with a Starr-Edwards caged-ball valve. Tex Heart Inst J 2016;43(6):562.

[3] Grunkemeier GL, Starr A, Rahimtoola SH. Prosthetic heart valve performance: long-term follow-up. Curr Probl Cardiol 1992;17(6):329—406.

[4] Walley VM, Keon WJ. Patterns of failure in Ionescu-Shiley bovine pericardial bioprosthetic valves. J Thorac Cardiovasc Surg 1987;93(6):925—33.

[5] Oury JH, Angell WW, Koziol JA. Comparison of Hancock I and Hancock II bioprostheses. J Card Surg 1988;3(3 Suppl. l):375—81.

[6] Marchand M, et al. Twelve-year experience with Carpentier-Edwards PERIMOUNT pericardial valve in the mitral position: a multicenter study. J Heart Valve Dis 1998;7(3):292—8.

[7] Totaro P, et al. Carpentier-Edwards PERIMOUNT magna bioprosthesis: a stented valve with stentless performance? J Thorac Cardiovasc Surg 2005;130(6):1668—74.

[8] Gallegos RP, et al. The current state of in-vivo pre-clinical animal models for heart valve evaluation. J Heart Valve Dis 2005;14(3):423—32.

[9] Schoen FJ. Pathologic findings in explanted clinical bioprosthetic valves fabricated from photooxidized bovine pericardium. J Heart Valve Dis 1998;7(2):174—9.

[10] Ionescu A, et al. Incidence of embolism and paravalvar leak after St Jude Silzone valve implantation: experience from the cardiff embolic risk factor study. Heart 2003;89(9):1055—61.

[11] Jamieson WR, et al. Seven-year results with the St Jude medical Silzone mechanical prosthesis. J Thorac Cardiovasc Surg 2009;137(5):1109—1115 e2.

[12] Jonas RA, et al. Late follow-up of the Braunwald-Cutter valve. Ann Thorac Surg 1982;33(6):554—61.

[13] Goldstein S, et al. Transpecies heart valve transplant: advanced studies of a bioengineered xeno-autograft. Ann Thorac Surg 2000;70(6):1962—9.

[14] Kasimir MT, et al. The decellularized porcine heart valve matrix in tissue engineering: platelet adhesion and activation. Thromb Haemostasis 2005;94(3):562—7.

[15] Vesely I. Heart valve tissue engineering. Circ Res 2005;97(8):743—55.

[16] Sayk F, et al. Histopathologic findings in a novel decellularized pulmonary homograft: an autopsy study. Ann Thorac Surg 2005;79(5):1755—8.

[17] Simon P, et al. Early failure of the tissue engineered porcine heart valve SYNERGRAFT in pediatric patients. Eur J Cardiothorac Surg 2003;23(6):1002—6. discussion 1006.

[18] Naso F, Gandaglia A. Different approaches to heart valve decellularization: a comprehensive overview of the past 30 years. Xenotransplantation 2018;25(1).

[19] Rieben R, et al. Xenotransplantation: in vitro analysis of synthetic alpha-galactosyl inhibitors of human anti-Galalpha1—>3Gal IgM and IgG antibodies. Glycobiology 2000;10(2):141—8.

[20] Banbury MK, et al. Age and valve size effect on the long-term durability of the Carpentier-Edwards aortic pericardial bioprosthesis. Ann Thorac Surg 2001;72(3):753—7.

[21] Haussig S, Linke A. Transcatheter aortic valve replacement indications should be expanded to lower-risk and younger patients. Circulation 2014;130(25):2321—31.

APPENDIX

Bernoulli's equation, significance, and limitations

Niema M. Pahlevan
Department of Aerospace & Mechanical Engineering, University of Southern California, CA, United States

A.1 Introduction

The Bernoulli equation is among the most popular equations in elementary physics and is widely recognized as the most fundamental equation in fluid mechanics [1]. Use of this equation has become pervasive in medical diagnostics, particularly in cardiovascular medicine [2]. It can provide clinically relevant information regarding the function of the heart valves [2], severity of vessel stenosis [3,4], intraventricular pressure [5], and pulmonary hemodynamics [6,7]. Measuring pressure in cardiology presents numerous challenges as it requires an invasive procedure. In contrast, velocity can be easily and noninvasively measured using imaging modalities such as echoultrasound or magnetic resonance imaging. Despite its limitations, the Bernoulli equation is widely used in clinical practice because it employs a noninvasively measured velocity and computes an approximate value for pressure (e.g., pulmonary artery pressure [7]) and pressure gradient (e.g., transvalvular pressure gradient [2]).

Even in its most simplified form ($\Delta p = 4u^2$), the Bernoulli equation still provides considerable clinically significant information [2]. However, it is important to understand the underlying assumptions of different forms of Bernoulli equations and their limitations.

As an example, consider the following underlying assumptions of the so-called simplified Bernoulli equation, $\Delta p = 4u^2$, by examining the presence and/or absence of certain fluid dynamics parameters. First, there is no time or time derivative involved in the equation, which implies steady flow. Furthermore, this indicates that inertial or nonconvective forces are not considered in the equation. Second, the equation lacks viscosity or friction loss term, which presents inviscid flow assumption. Moreover, the equation does not include density, meaning incompressible flow and constant blood density assumption. Lastly, only one velocity term is present, which indicates that the terminal velocity is assumed to be an order of magnitude higher than the initial velocity.

Simplified models and equations play an important role in fluid mechanics and other branches of physics and engineering. However, the underlying assumptions of these simplified models can create significant errors if not appropriately addressed. Our goal in this chapter is to review the derivation of various forms of the Bernoulli equation and highlight their assumptions for a better understanding of their limitations.

A.2 Derivation of the generalized Bernoulli equation

Consider an infinitesimal cylindrical fluid element in a streamtube with its center along a streamline. The cross section of this element has an area A, length "dl," and side surface area "S". The tangential coordinate along the streamline is "l". The fluid element acceleration along the streamline will be

$$a_l = \frac{Du}{Dt} = \frac{\partial u}{\partial t} + u\frac{\partial u}{\partial l} \tag{A.1}$$

Here, $D(u)/Dt$ is the material derivative, and u is the component of the velocity vector along the streamline (Fig. A.1).

According to Newton's second law, the sum of all forces applied on this fluid element is equal to the mass of the fluid element times its acceleration. The external forces applied on this fluid element are (i) pressure force on the left cross section of the element (p), (ii) pressure force on the right cross section of the element ($p+\delta p$), (iii) shear forces related to the shear stresses (τ) applied on the side wall of the fluid element surface (S), and (iv) body forces (mf_b):

$$ma_l = Ap - A(p + \delta p) - \int_S \tau.dS - mf_b \tag{A.2}$$

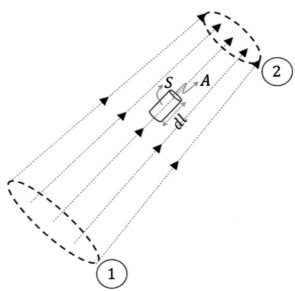

Figure A.1 Cylindrical fluid element along a streamline in a streamtube.

Here, m is the mass of the fluid element, and f_b is the body force per unit mass in the tangential direction (l) of the motion. By substituting for the mass of the fluid element ($m = \rho A dl$), acceleration of the fluid element (Eq. A.1), and variation of the pressure along the streamline $\left(\delta p = \frac{\partial p}{\partial l} dl \right)$ in Eq. (A.2), we will have

$$\rho A \left(\frac{\partial u}{\partial t} + u \frac{\partial u}{\partial l} \right) dl = Ap - A \left(p + \frac{\partial p}{\partial l} dl \right) - \int_S \tau . dS - \rho A f_b dl \qquad (A.3)$$

With some algebraic manipulation and dividing both sides by ρ, we will get

$$\frac{\partial u}{\partial t} dl + u \frac{\partial u}{\partial l} dl = -\frac{1}{\rho} \frac{\partial p}{\partial l} dl - \frac{1}{\rho A} \int_S \tau . dS - f_b dl \qquad (A.4)$$

By integrating along the streamline from l_1 to l_2, we will have

$$\int_{l_1}^{l_2} \frac{\partial u}{\partial t} dl + \int_{l_1}^{l_2} u \frac{\partial u}{\partial l} dl = -\int_{l_1}^{l_2} \frac{1}{\rho} \frac{\partial p}{\partial l} dl - \frac{1}{\rho A} \int_{l_1}^{l_2} \left(\int_S \tau . dS \right) dl - \int_{l_1}^{l_2} f_b dl \qquad (A.5)$$

The second equation of the right-hand side can be calculated without any assumptions regarding fluid flow. After rearranging the terms, we are left with the most general form of the Bernoulli equation:

$$\frac{1}{2} \left(u_2^2 - u_1^2 \right) + \int_{l_1}^{l_2} \frac{\partial u}{\partial t} dl + \int_{l_1}^{l_2} \frac{1}{\rho} \frac{\partial p}{\partial l} dl + \frac{1}{\rho A} \int_{l_1}^{l_2} \left(\int_S \tau . dS \right) dl + \int_{l_1}^{l_2} f_b dl = 0 \quad (A.6)$$

Here, u_1 is the velocity along the streamline at point l_1, and u_2 is the velocity along the streamline at point l_2 on the same streamline. Eq. (A.6) is the generalized Bernoulli equation for nonsteady, compressible, and viscose flow.

A.3 Generalized form of the Bernoulli equation for cardiovascular biofluid dynamics

Blood is an incompressible fluid. To represent incompressible fluids, the third term of Eq. (A.6) can be simplified to

$$\int_{l_1}^{l_2} \frac{1}{\rho} \frac{\partial p}{\partial l} dl = \frac{1}{\rho} (p_2 - p_1) \qquad (A.7)$$

Eq. (A.7) is valid, in general, provided that density remains constant along a streamline.

Assuming gravity is the only body force (reasonable assumption for cardiovascular applications) and considering that no body force is applied on the element if the fluid element moving direction is normal to the direction of the gravity, the body force integral (the last term in Eq. A.6) will be simplified to

$$\int_{l_1}^{l_2} f_b dl = gh_2 - gh_1 \tag{A.8}$$

Here, g is the earth gravitational acceleration and $h_i (i = 1, 2)$ is the normal height to the earth surface.

For simplicity, we will also replace the shear stress integral in Eq. (A.6) with $F_\mu \left(= \frac{1}{A} \int_S \tau . dS \right)$. By substituting shear force in Eqs. (A.6–A.8) and multiplying both sides by the density (ρ), we will get

$$\frac{1}{2} \rho u_1^2 + p_1 + \rho g h_1 = \frac{1}{2} \rho u_2^2 + p_2 + \rho g h_2 + \rho \int_{l_1}^{l_2} \frac{\partial u}{\partial t} dl + \int_{l_1}^{l_2} F_\mu dl \tag{A.9}$$

So far, we have not made any assumptions that are significant to fluid dynamics of blood flow in the cardiovascular system. Therefore, Eq. (A.9) can be considered as the generalized Bernoulli equation for cardiovascular biofluid dynamics. This equation is also applicable along a streamline to any nonsteady, incompressible, and viscous flow. It is noteworthy that F_μ is a function of blood velocity, blood rheological properties (e.g., hematocrit and red blood cell deformability), and characteristics cross section area. The choice of the characteristics cross section area depends on the application in which Eq. (A.9) is being used. As an example, for blood flow in a vessel, this area is just the vessel cross-sectional area; it is the valve orifice area for a transvalvular pressure gradient calculations.

A.4 Bernoulli equation and pressure drop calculation

We can derive an equation for pressure drop (Eq. A.10) along a streamline by simple rearrangement of the terms in Eq. (A.9)

$$p_1 - p_2 = \underbrace{\frac{1}{2} \rho \left(u_2^2 - u_1^2 \right)}_{I} + \underbrace{\int_{l_1}^{l_2} \frac{\partial u}{\partial t} dl}_{II} + \underbrace{\int_{l_1}^{l_2} F_\mu dl}_{III} + \underbrace{\rho g (h_2 - h_1)}_{IV} \tag{A.10}$$

The first term (I) on the right side of Eq. (A.10) is the so-called "convective acceleration" term, and it is related to the change of the kinetic energy of the fluid flow.

The second term (*II*) is related to the temporal variation of the blood flow velocity (unsteady term). This term has been also called inertial term and nonconvective term in literature [8]. Neglecting this term may introduce significant error especially in computation of transvalvular pressure drop or intracardiac pressure gradients [8−10].

The third term (*III*) of Eq. (A.10) is related to the viscose friction resistance. Neglecting this term means inviscid flow assumption. The contribution of the viscose forces to transvalvular pressure drop is negligible (around 0.14 mmHg [8]). However, neglecting this term can introduce considerable error in computation of the pressure drop in other hemodynamic applications such as pressure across a vessel stenosis [3] or computation of the fractional flow reserve [4].

The forth term (*IV*) is related to the potential energy stored in the fluid. This term is negligible as the height difference across the two points of interest is usually small in cardiovascular application. As an example, a height difference of 1 cm will only cause a pressure drop of around 0.8 mmHg.

A.5 Approximation of the Bernoulli convective term

The convective term (*I* in Eq. A.10) is linked to the cross-sectional area of the vessel or valve orifice through conservation of mass. Assuming the vessel wall is impermeable, the conservation of mass for the incompressible blood reduces to

$$\int_{A_1} u_1 dA = \int_{A_2} u_2 dA \tag{A.11}$$

This means that u_1 can be found as function of u_2, A_1, and A_2 ($u_1 = f(u_2, A_1, A_2)$) Hence, the convective term in the Bernoulli equation can be written as

$$\frac{1}{2}\rho(u_2^2 - u_1^2) = \frac{1}{2}\rho(u_2^2 - f^2(u_2, A_1, A_2)) \tag{A.12}$$

If we assume the flow is uniform over the cross section, u_1 will be simply equal to $u_2\frac{A_2}{A_1}$. Therefore, Eq. (A.10) will be reduced to

$$p_1 - p_2 = \frac{1}{2}\rho u_2^2\left(1 - \left(\frac{A_2}{A_1}\right)^2\right) + \rho g(h_2 - h_1) + \rho\int_{l_1}^{l_2}\frac{\partial u}{\partial t}dl + \int_{l_1}^{l_2} F_\mu dl \tag{A.13}$$

Eq. (A.13) is the Bernoulli equation for nonsteady, viscose, and uniform blood flow. This equation is handy, where upstream (or downstream) flow measurements are not possible. The error introduced by this equation (uniform flow assumption) can be reduced by using correction factors that take into account the effect of a nonuniform velocity profile.

A.6 Approximation of the Bernoulli viscose term

Assuming unidirectional, fully developed, steady flow in a tube, the Navier–Stokes equation will be simplified to [11]:

$$\mu\left(\frac{\partial^2 u}{\partial r^2} + \frac{1}{r}\frac{\partial u}{\partial r}\right) = \frac{dp}{dx} \tag{A.14}$$

Here, μ is blood viscosity, u is the velocity in the axial direction (x), and r is the coordinate in the radial direction. Note that the Newtonian fluid assumption is already included in the Navier–Stokes equation. Considering the assumption that pressure is invariant in r direction, we can integrate Eq. (A.14) across the cross r with boundary condition of no slip ($u = 0$) at the wall ($r = R$) and finite solution at the center ($r = 0$):

$$u = \frac{1}{4\mu}\frac{dp}{dx}\left(r^2 - R^2\right). \tag{A.15}$$

Furthermore, we can use the average velocity over the cross-sectional area to calculate an approximate expression for the viscose term:

$$\bar{u} = \frac{1}{A}\int_A u\,dA = \frac{1}{\pi r^2}\int_0^{2\pi}\int_0^R \frac{1}{4\mu}\frac{dp}{dx}\left(r^2 - R^2\right)r\,dr\,d\theta = \frac{1}{8\mu}R^2\frac{dp}{dx} \tag{A.16}$$

Here, $\bar{u}$ is average velocity over the cross section.

By integrating Eq. (A.16) along a streamline, we will get

$$p_1 - p_2 = \frac{8\mu\bar{u}l}{R^2} \tag{A.17}$$

Here, l is the distance between 1 and 2 along the streamline. Eq. (A.17) is like the Bernoulli equation without the inertia, convective, and the gravity term. Extensive assumptions are required to derive Eq. (A.17); however, this expression can be used as an approximation for the viscose term in the Bernoulli equation (Eqs. A.10 and A.13):

$$\int_{l_1}^{l_2} F_\mu\,dl = \Delta p_\mu = \frac{8\mu\bar{u}l}{R^2} \tag{A.18}$$

Sometimes, it is easier to measure the center velocity than the cross average velocity in clinical practice (e.g., continuous wave Doppler echo). So we can replace $u_{max}\left(= \frac{-1}{4\mu}\frac{dp}{dx}R^2\right)$ in Eq. (A.16) and derive the Poiseuille viscose pressure loss as

$$\Delta p_\mu = \frac{4\mu u_{max}l}{R^2} \tag{A.19}$$

Here, u_{max} is the center velocity assuming Poiseuille flow.

Eq. (A.18) demonstrates significant information regarding F_μ. The equation demonstrates that F_μ is a function of u, μ, and R. In general, the full detail knowledge about the velocity along the entire streamline is required to compute the viscose term. In engineering problems, this issue has been resolved by using friction factors (f) [1]. Initially, it was Reynolds [1] who observed that viscose losses are proportional to the square of the averaged velocity. Based on his observation, it is reasonable to assume that the shear term (F_μ) is proportional to $\bar{u}^2$. In cardiovascular problems, it is more convenient to replace $\bar{u}$ with the volume flow rate (Q):

$$F_\mu = f(\mu, R)\bar{u}^2 = (\mu, R)\frac{Q^2}{\pi^2 R^4} \Rightarrow \int_{l_1}^{l_2} F_\mu dl = \frac{Q^2}{\pi^2}\int_{l_1}^{l_2}\frac{f^2(\mu, R)}{R^4}dl \tag{A.20}$$

The friction factor function is usually determined through experiments, but it is possible to derive a closed-form formula analytically for certain cases as demonstrated above in the Poiseuille flow case. With the appropriate application of a friction factor (f), Eq. (A.20) can also be used for nonuniform flows and turbulent flows without significant errors.

A.7 Simplified versions of the Bernoulli equation

If we neglect the viscose term in Eq. (A.9), the Bernoulli equation will be simplified to

$$\frac{1}{2}\rho u_1^2 + p_1 + \rho g h_1 = \frac{1}{2}\rho u_2^2 + p_2 + \rho g h_2 + \rho \int_{l_1}^{l_2}\frac{\partial u}{\partial t}dl \tag{A.21}$$

This equation is valid along a streamline for any nonsteady, uniform, incompressible, and inviscid flow. Although blood flow through the orifice of a heart valve is not inviscid, this equation works well with good accuracy for most transvalvular pressure drops and intracardiac pressure gradients [8—10]. This is mainly due to negligible contributions of the viscose forces to a pressure drop, compared with convective and inertia (unsteady) components of the Bernoulli equation. This simplified version of the Bernoulli equation, however, is not valid for pressure drops across vessel stenosis as it may produce significant errors [3].

For steady, uniform, incompressible, and inviscid flow, the Bernoulli equation will be further simplified to

$$\frac{1}{2}\rho u_1^2 + p_1 + \rho g h_1 = \frac{1}{2}\rho u_2^2 + p_2 + \rho g h_2 \tag{A.22}$$

This equation lacks both viscose and inertia (nonsteady) terms. Therefore, its application to cardiovascular hemodynamics such as transvalvular flow, vessel stenosis, and intracardiac flow can contain significant errors.

Assuming that the blood density is the same for all patients and $u_1 \ll u_2$, Eq. (A.22) will be further simplified to

$$\Delta p = 4u_2^2 \qquad\qquad (A.23)$$

This equation is known as "simplified Bernoulli equation," and it is widely used for various clinical applications such as transvalvular aortic and pulmonary valve pressure drop [2]. In this equation, u is in m/sec and p in $mmHg$. The assumption of constant density for all patients is usually not a significant one, but it could introduce considerable error in certain circumstances such as hemorrhage [12].

A.8 Conclusion

The Bernoulli equation is a versatile and straightforward tool that enables clinicians to noninvasively assess pressure or pressure gradients in the cardiovascular system. However, it is important to understand the underlying assumption and limitation of each version of the Bernoulli equation. Careless usage of these equations can result in underestimations of the underlying disease or create false positive diagnostics.

References

[1] Synolakis CE, Badeer HS. On combining the Bernoulli and Poiseuille equation—a plea to authors of college physics texts. Am J Phys 1989;57(11):1013—9.
[2] Weyman AE. Principles and practice of echocardiography. 1994.
[3] Teirstein PS, Yock PG, POpp RL. The accuracy of Doppler ultrasound measurement of pressure gradients across irregular, dual, and tunnellike obstructions to blood flow. Circulation 1985;72(3): 577—84.
[4] Huo Y, et al. A validated predictive model of coronary fractional flow reserve. J R Soc Interface 2011. rsif20110605.
[5] Cortina C, et al. Noninvasive assessment of the right ventricular filling pressure gradient. Circulation 2007;116(9):1015—23.
[6] Fisher MR, et al. Accuracy of Doppler echocardiography in the hemodynamic assessment of pulmonary hypertension. Am J Respir Crit Care Med 2009;179(7):615—21.
[7] Parasuraman S, et al. Assessment of pulmonary artery pressure by echocardiography—a comprehensive review. IJC Heart & Vasculature 2016;12:45—51.
[8] Firstenberg MS, et al. Nonconvective forces: a critical and often ignored component in the echocardiographic assessment of transvalvular pressure gradients. Cardiol Res Pract 2012;2012.
[9] Yotti R, et al. Doppler-derived ejection intraventricular pressure gradients provide a reliable assessment of left ventricular systolic chamber function. Circulation 2005;112(12):1771—9.
[10] Falahatpisheh A, et al. Simplified Bernoulli's method significantly underestimates pulmonary transvalvular pressure drop. J Magn Reson Imaging 2016;43(6):1313—9.
[11] Zamir M. The physics of pulsatile flow. New York: Springer-Verlag; 2000.
[12] Nunez FA, et al. Variations in the density of blood during hemorrhage. FASEB J 2012;26(1_Supple ment):1132.9. 1132.9.

Index

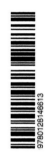